ROUTLEDGE LIBRARY EDITIONS:
GEOLOGY

Volume 8

ENVIRONMENTAL GEOMORPHOLOGY AND LANDSCAPE CONSERVATION

ENVIRONMENTAL GEOMORPHOLOGY AND LANDSCAPE CONSERVATION

Binghamton Geomorphology Symposium 1

Edited by
DONALD R. COATES

LONDON AND NEW YORK

First published in 1973 by Dowden, Hutchinson & Ross, Inc.

This edition first published in 2020
by Routledge
2 Park Square, Milton Park, Abingdon, Oxon OX14 4RN

and by Routledge
52 Vanderbilt Avenue, New York, NY 10017

Routledge is an imprint of the Taylor & Francis Group, an informa business

British Library Cataloguing in Publication Data
A catalogue record for this book is available from the British Library

ISBN: 978-0-367-18559-6 (Set)
ISBN: 978-0-429-19681-2 (Set) (ebk)
ISBN: 978-0-367-45861-4 (Volume 8) (hbk)
ISBN: 978-0-367-46035-8 (Volume 8) (pbk)
ISBN: 978-1-00-302656-3 (Volume 8) (ebk)

Publisher's Note
The publisher has gone to great lengths to ensure the quality of this reprint but points out that some imperfections in the original copies may be apparent.

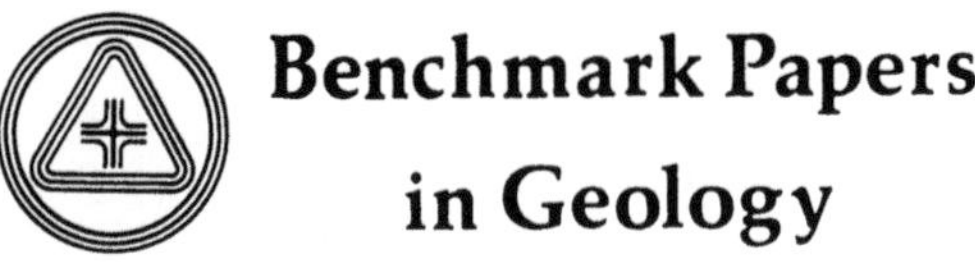

A *BENCHMARK*™ Books Series

ENVIRONMENTAL GEOMORPHOLOGY AND LANDSCAPE CONSERVATION

Edited by
DONALD R. COATES

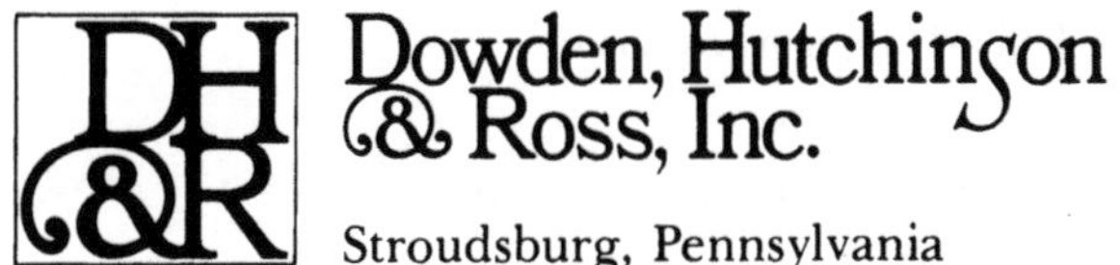

Stroudsburg, Pennsylvania

Benchmark Papers in Geology, Volume 8
ISBN: 0-87933-040-6

Library of Congress Cataloging in Publication Data

Coates, Donald Robert, 1922- comp.
Environmental geomorphology and landscape conservation.

CONTENTS: v. 1. Prior to 1900.--
v. 3. Non-urban.
1. Geomorphology--Addresses, essays, lectures.
2. Human ecology--Addresses, essays, lectures.
3. Conservation of natural resources--Addresses, essays, lectures.
GB406.C62 333.7'2'08 72-77882
ISBN 0-87933-005-8 (v. 1)

Manufactured in the United States of America.

Exclusive distributor outside the United States and Canada: John Wiley & Sons, Inc.

Acknowledgments and Permissions

ACKNOWLEDGMENTS

American Geophysical Union—*Transactions, American Geophysical Union*
"Subsidence of the Land Surface in the Tulare–Wasco (Delano) and Los Banos–Kettleman City Area, San Joaquin Valley, California"

American Water Resources Association—*Proceedings of the 2nd Annual American Water Resources Conference*
"Man's Role in Affecting the Sedimentation of Streams and Reservoirs"

Smithsonian Institution—*Smithsonian Institution Report*
"Permafrost"

United States Department of Agriculture
Soil Erosion a National Menace. Part 1: Some Aspects of the Wastage Caused by Soil Erosion, Circular 33
Silting of Reservoirs, Technical Bulletin 524
Prevention and Control of Gullies, Farmers' Bulletin 1813
Prevention of the Erosion of Farmlands by Terracing, Bulletin 512
Sand-Dune Reclamation in the Southern Great Plains, Farmers' Bulletin 1825
Restoring Surface-Mined Land, Miscellaneous Publication 1082

United States Department of Agriculture, Forest Service, Intermountain Region—*Landslide Hazards Related to Land Use Planning in Teton National Forest, Northwest Wyoming*
"Forest Land Use Implications"

United States Department of Agriculture, Soil Conservation Service—*Soil Conservation*
"Protecting Bottomlands from Erosional Debris: A Case Study"

United States Department of the Interior, Geological Survey—*Denudation and Erosion in the Southern Appalachian Region and the Monongahela Basin,* Professional Paper 72

PERMISSIONS

The following papers have been reprinted with the permission of the authors and the copyright owners.

American Association for the Advancement of Science—*Science*
"Ground Rupture in the Baldwin Hills"

American Society of Landscape Architects—*Landscape Architecture*
"The Inert Becomes 'Ert' "
"A New Technology for Taconite Badlands"

Association of Southeastern Biologists Bulletin—*Association of Southeastern Biologists Bulletin*
"Soil Destruction Associated with Forest Management and Prospects for Recovery in Geological Time"

Donald R. Coates—*Environmental Geomorphology*
"Impact of Highways on the Hydrogeologic Environment"

Melbourne University Press—*Water Resources Use and Management*
"Problems of Irrigated Areas"

Natural History—*Natural History*
"Landscape Esthetics"

Natural History Society—*Naturalist*
"River Systems: Recreational Classification, Inventory and Evaluation"

UNESCO—*Problems of the Arid Zone, Proceedings of the Paris Symposium*
"Basics Trends and Methods of Water Control in the Arid Zones of the Soviet Union"

The University of Chicago Press—*Man's Role in Changing the Face of the Earth*
"The Nature of Induced Erosion and Aggradation"

American Journal of Science (Yale University)—*American Journal of Science*
"Entrenchment of the Willow Drainage Ditch, Harrison County, Iowa"

To Cheryl, Eric, and Lark:
trust in today, and faith for tomorrow

Series Editor's Preface

The philosophy behind the "Benchmark Papers in Geology" is one of collection, sifting, and rediffusion. Scientific literature today is so vast, so dispersed, and, in the case of old papers, so inaccessible for readers not in the immediate neighborhood of major libraries that much valuable information has been ignored by default. It has become just so difficult, or so time consuming, to search out the key papers in any basic area of research that one can hardly blame a busy man for skimping on some of his "homework."

This series of volumes has been devised, therefore, to make a practical contribution to this critical problem. The geologist, perhaps even more than any other scientist, often suffers from twin difficulties—isolation from central library resources and immensely diffused sources of material. New colleges and industrial libraries simply cannot afford to purchase complete runs of all the world's earth science literature. Specialists simply cannot locate reprints or copies of all their principal reference materials. So it is that we are now making a concerted effort to gather into single volumes the critical material needed to reconstruct the background of any and every major topic of our discipline.

We are interpreting "Geology" in its broadest sense: the fundamental science of the Planet Earth, its materials, its history, and its dynamics. Because of training and experience in "earthy" materials, we also take in astrogeology, the corresponding aspect of the planetary sciences. Besides the classical core disciplines such as mineralogy, petrology, structure, geomorphology, paleontology, and stratigraphy, we embrace the newer fields of geophysics and geochemistry, applied also to oceanography, geochronology, and paleoecology. We recognize the work of the mining geologists, the petroleum geologists, the hydrologists, the engineering and environmental geologists. Each specialist needs his working library. We are endeavoring to make his task a little easier.

Each volume in the series contains an Introduction prepared by a specialist (the volume editor)—a "state of the art" opening or a summary of the objects and content of the volume. The articles, usually some thirty to fifty reproduced either in their entirety or in significant extracts, are selected in an attempt to cover the field, from the key papers of the last century to fairly recent work. Where the original works are in foreign languages, we have endeavored to locate or commission translations. Geologists, because of their global subject, are often acutely aware of the oneness of our world. The selections cannot, therefore, be restricted to any one country, and whenever possible an attempt is made to scan the world literature.

To each article, or group of kindred articles, some sort of "Highlight Commentary" is usually supplied by the volume editor. This should serve to bring that article into historical perspective and to emphasize its particular role in the growth of the field. References, or citations, wherever possible, will be reproduced in their entirety—for by this means the observant reader can assess the background material available to that particular author, or, if he wishes, he too can double check the earlier sources.

A "benchmark," in surveyor's terminology, is an established point on the ground, recorded on our maps. It is usually anything that is a vantage point, from a modest hill to a mountain peak. From the historical veiwpoint, these benchmarks are the bricks of our scientific edifice.

Rhodes W. Fairbridge

Contents

Contents by Author

Introduction

This is the last book in a three-volume series on the general theme of environmental geomorphology and landscape conservation. Owing to the breadth, depth, and scope of the literature encompassed by these topics, their coverage could not have been adequately handled in less than three volumes. Division of the materials, however, is difficult to organize and classify into three different volumes because of the interlocking nature of the subject. These problems are minimized by a first separation along temporal lines. Thus, Volume 1 deals with events and articles written during the period prior to 1900. Since, in environmental terms, urbanization (including population increase) has been the dominant trend and the greatest problem maker in the twentieth century, it was selected for separate treatment in Volume II. This should not be construed to mean there were no city-related problems before 1900. Emphasis has been placed on the accelerating deterioration of the urban environment, the effects that human concentration have on the land–water ecosystem, and some of the new problems that are just now reaching proportions of large magnitude in the latter part of the twentieth century.

The twentieth century has witnessed a series of events and changes unprecedented in previous human history. The range of problems, when contrasted to earlier periods of man's activity, differ in both magnitude and kind. Many of the same environmental problems that have always plagued man—flooding, erosion, and siltation, for example—continue to be vexing problems to the present day. In addition, new problems are occurring that had little or no reportable precedent before 1900, such as land collapse because of fluid withdrawal and encroachment of sea water in coastal freshwater aquifers. Because of a rapidly expanding population and the increase in machines of all varieties and descriptions, the quality of the environment is suffering great deterioration. Problems of pollution and waste disposal have emerged as some of the most troublesome facing man in the last half of the twentieth century. With minor exception, these problems are outside the realm of geomorphology, and since their elucidation would require separate treatment in volumes of their own, they cannot be adequately handled within these volumes.

The population explosion in the industrialized countries has caused rapid and widespread urbanization and has forced an entire series of revolutionary changes on the human scene. The dwindling of new land frontiers poses a space problem and also produces a psychological effect on man.

Energy requirements have grown even faster than population, and its creation often changes the landscape. The production of hydroelectric energy requires the construction of dams, and their impounding reservoirs introduce man-made changes in the topography. In some cases dams are multipurpose and create other desired benefits, such as water sources, recreational areas, and flood prevention. Coal continues to provide a significant source of energy, as in the United States, and its mining often produces vast changes in terrain. Prior to 1900, most mines were underground, but in the twentieth century strip mining has become possible with the development of an entire new family of monster machines that makes open-pit mining possible not only for coal, but also for other resources, such as copper and iron. The scars and desolation in the mined areas commonly present a scene of geomorphic catastrophy. Such man-made incisions introduce an entire sequence of deleterious events. Thus man himself is becoming a geomorphic force for landscape modification.

The development of extensive road networks is another major change ushered in during the twentieth century, and vastly proliferated the man-made scars of the ninteenth century railway networks. Their necessity was the result of the development of the internal combustion engine and a successful technology that was applied to the location and recovery of petroleum products. Thus the combination of cars and trucks using petroleum products has opened an entire new era in transportation as well as bringing about a high level of societal mobility. Wherever railways and roads were built, modification of terrain was necessary. For example, to maintain gentle grades, many roads either have to be filled in the low spots or lowered in the high spots. Natural drainage is blocked or diverted in the low areas, and hills are cut away in high areas. Stream regimes are changed and water supplies are upset. In addition, to provide road-building materials, sand, gravel, and rock quarries are developed, which create ugly land scars. Not only do regular highways create many geomorphic changes, but roads into mines and in lumbering areas also cause accelerated erosion. Snowmobiles pose a new threat to natural areas, as they can traverse many different terrains and may leave indelible trails.

Just as the need for energy has required the development of natural resources, man's concentration of population in congested urbanized areas has created a new set of problems. In big cities the population may become so great that it outstrips the local water resources and requires the importation of water from great distances. Also with the advent of air conditioning and other comforts that make living in such otherwise inhospitable regions as deserts possible for many people, man no longer settles near water as he used to. Where there is a paucity of water, he demands that water be provided him in these dry but often climatically attractive environments. Thus the twentieth century has seen in the United States development of vast reservoir systems to serve metropolitan areas, such as New York City, Los Angeles, and Phoenix, that are located hundreds of miles from the city. The historical "first" in this respect was in the infant state of Western Australia, where in the 1890s a 400-mile pipeline was constructed across the arid interior to permit the growth of the gold mining city of Kalgoorlie. Thus there are interlocking aspects of urban and nonurban problems, in which the urban area is causing changes to be made in nonurban regions.

Another problem that bridges both urban and nonurban areas is waste disposal. Since in the United States each person produces on the average 4.5 pounds of waste per day, its disposal poses a serious problem for cities, which generally haul it away to less populated areas. New York City is the greatest source of the sedimentation in the Hudson River. Cities with open areas in the environs may bury the debris in huge landfill areas (the question of "sanitary landfill" was explored in Volume II).

Thus there are many joint problems that have an urban–nonurban interface. In the organization of these volumes several rules had to be established to provide guidelines for the choice of subject matter. One rule is the requirement to avoid duplication as much as possible. So, with few exceptions, it was decided that certain topics would be largely covered in one of the volumes and not repeated in the others. Floods and coastal problems are examples of this decision and are discussed primarily in Volume II. It is realized that floods also damage nonurban areas, but flood-control structures are usually built on the basis of protection of urban areas, and when flooding does occur the damages are greatest in the cities. In somewhat similar fashion, damages in coastal areas are always most serious in areas of built-up communities. It is the community that requests preventative measures for beach stabilization, for structures cannot be justified on a benefit–cost ratio in nonurban areas, where there is only occasional habitation. A second coastal problem, that of intrusion of salt water into freshwater aquifers, is also generally only associated with highly urbanized areas, where there have been massive groundwater withdrawals and "groundwater mining." Therefore, it is only logical that such topics should be treated in Volume II, which deals with urbanization problems. Materials in Volume II are mainly concerned with the building activities of man and changes created by population concentration, whereas this third volume deals with soil and landscape considerations in the less densely inhabited regions.

Another rule for organization was to concentrate the legal aspects of environmental affairs into a single book, Volume II. There are laws that relate to nonurban areas and which affect it, but the urban areas require more management and stricter codes and enforcement procedures. A partial exception concerns landslides necessitating the inclusion of somewhat similar topics in Volumes II and III. The phenomenon of gravity movement on hillslopes is so ubiquitous that it merits some discussion in both volumes. It is such an important geomorphic event and hazard that management for minimizing landslide impact and destruction occurs in both urban and nonurban regions. The California grading ordinances illustrate the principle, for example, that in highly congested areas stronger controls have to be developed because what is done on one property may influence what happens to a contiguous property. Thus governmental action becomes more forceful. There has also been the evolution of the town–city–county planning boards or commissions, which are starting to have an acute sense of public trust.

A twentieth-century innovation has been man's growing awareness of his impact on the environment. With very few exceptions, such as the case of George Perkins Marsh (see Volume I), man before 1900 had a hostile view of the earth; his need was to conquer and subdue nature. Man placed himself on a pinnacle: he could do no wrong and held that the earth was created for his benefit and pleasure, to do with as he chose. The new attitude is expressed by Osborn (1948, p. 201):

> Man must recognize the necessity of cooperating with nature. He must temper his demands and use and conserve the natural living resources of this earth in a manner that alone can provide for the continuation of his civilization. The final answer is to be found only through comprehension of the enduring processes of nature. The time for defiance is at an end.

As Volume I illustrates, the number of nineteenth-century people who deplored degradation of the earth was minimal. The authors telling of pre-1900 waste were mostly twentieth-century scientists commenting in retrospect. Another entrenched idea prevalent prior to 1900 was that nature could always heal herself, so that a massive infusion of help from man in desecrated areas was unnecessary. However, the late twentieth century has witnessed a series of conservational awakenings unprecedented in human history. Along with this new awareness has come a greater realization on the part of the government of the fact that it has a responsibility to the people not only to do something (i.e., take action and remedial steps), but also to educate the general public. Although this pattern has been somewhat similar in other industrialized countries, the United States will be used as an example of governmental interest in environmental affairs.

The most heavily involved agencies empowered to undertake, construct, and manage projects are the Bureau of Reclamation (U.S. Department of Interior), Corps of Engineers (U.S. Army), and Soil Conservation Service (U.S. Department of Agriculture). In addition, such agencies as the U.S. Geological Survey gather, process, disseminate, and provide much of the basic data for environmental programs conducted by other parts of the federal and state governments. Involvement by academic institutions in environmental affairs is currently supported in grants by a broad range of governmental departments and agencies such as the Office of Water Resources (U.S. Department of Interior), National Oceanic and Atmospheric Administration (U.S. Department of Commerce), and the National Science Foundation (in the RANN program). Government action on the environmental scene also takes the form of laws and legislation that are aimed at the protection and preservation of the land and water, as well as the rehabilitation of damaged environments.

Literature of Environmental Geomorphology and Landscape Conservation

There has been an exponentially accelerating growth in all scientific literature during the twentieth century, but a publication explosion has occurred in the area of environmental affairs and ecology during the last several years. In spite of this, there is a scarcity of integrative information that links these ideas together.

> Only a relatively small part of this literature has stressed man's role as a geomorphic agent, exerting a noticeable effect on the earth's crust. On the whole, specific attention to this geomorphic rather than ecologic aspect has been neglected . . . The study of man's geomorphic activities, then, is one of the paradoxical "recognized yet unrecognized subjects" . . . (Golomb and Eder, 1964).

George Perkins Marsh's classic and pace-setting *Man and Nature* (1964) has gone through several editions in the twentieth century, but the subject matter is drawn from examples before 1900. Prior to the 1940s most of the literature relating to the land–water ecosystem was to be gleaned only from government bulletins and/or was written by government scientists. Many of the selections used in this volume indicate this trend. In the late 1930s and in the 1940s, books were published that provided basic principles of and gave status reports on man's stewardship of the land. Soil conservation textbooks were written by Ayes (1936) and Bennett (1939); geomorphology textbooks were authored by Worcester (1939) and Lobeck (1939). The U.S. Department of Agriculture devoted their yearbook series to outstanding topics of contemporary interest, for example, *Soils and Man* (1938). A book that provided a somewhat popularized account of man's relation to the soil is *Vanishing Lands* (Jacks and Whyte, 1939; the English edition was more strikingly titled *The Rape of the Earth*). Related works include *Our Plundered Planet* (Osborn, 1948) and *Deserts on the March* (Sears, 1947).

The decade of the 1950s was largely the calm before the publication storm. One notable and remarkable exception was the book *Man's Role in Changing the Face of the Earth* (Thomas, 1956). This was the first book of its type and provided the most comprehensive treatment of the topic yet written.

In the 1960s the paperback technology widened to embrace the environmental field, which has continued into the 1970s in an ever-increasing proliferation. Many of these are written in a reportorial style for popular appeal. They provide a very broad spectrum of topics: urban and nonurban problems, air–water–land pollution, economics, sociopolitical affairs, and so on. However, each touches upon elements of man's impact on the terrain and waters. Typical of this approach are such books as *The Frail Ocean* (Marx, 1967), *Moment in the Sun* (Rienow, 1967), *America the Raped* (Marine, 1969), *The Diligent Destroyers* (Laycock, 1970), *Eco-Catastrophe* (Ramparts, 1970), *The Environmental Crises* (Helfrich, 1970), and *Patient Earth* (Harte and Socolow, 1971). Another group of paperbacks stresses the do-it-yourself approach and provides information on how to be a conservationist/activist. Representative books include *The Environmental Handbook* (De Bell, 1970), *Ecotactics* (Mitchell, 1970), and *The Users Guide to the Protection of the Environment* (Swateck, 1970).

Although the usage of the term "environmental" did not really catch on until the late 1960s, Dasman was employing the term as early as 1959 in his book *Environmental Conservation.* The first U.S. books on the topic that were largely geologic–geomorphic did not start to appear until the 1970s. They include *Environmental Geology* (Flawn, 1970), *Man's Impact on Environment* (Detwyler, 1971), *Environmental Geomorphology* (Coates, ed., 1971), and *Man and His Physical Environment* (McKenzie and Utgard, 1972). Although many foreign books do not use this terminology, several of UNESCO's publications fall within this category, as does the book, *Water, Earth and Man* (Chorley, 1969).

The present volume contains facsimile copies of what are considered to be 26 of the most relevant and finest presentations on the theme of nonurban environmental geomorphology and landscape conservation in the twentieth century. One exception is the first publication of a manuscript by Jacob Aghassy, which is admirably suited to this theme.

A large number of factors were considered in making article selection to provide as comprehensive a coverage of the subject matter as possible and to reflect the vast panorama of man's activities in the nonurban environment. Although most articles deal with the United States, some articles concern situations in Australia, China, Israel, Japan, and Russia, in order to show the world-wide extent of environment affairs. Additional references and text materials enhance the global character and document the idea that no single country has a monopoly on either problems or solutions. It was also important to include articles by scientists such as H. H. Bennett, the Craigheads, R. F. Black, L. B. Leopold, W. C. Lowdermilk, A. N. Strahler, J. F. Poland, and others, who have been, or still are, acknowledged leaders in the field. To properly reflect the important role that U.S. governmental agencies have taken in matters of the landscape, articles were chosen from publications of the U.S. Geological Survey, Forest Service (U.S. Department of Agriculture), and Soil Conservation Service (U.S. Department of Agriculture). The publication dates of articles range from 1911 to 1971. Where possible, the first significant article on a particular topic was chosen, even though more recent and longer papers may have been written on the same subject. The reason for including several items from the 1960s and 1970s is that they either present new ideas or synthesize information in an outstandingly lucid fashion. Thus the articles provide as many dimensions as possible in representing influential ideas that have helped to focus thinking and action patterns on the environment.

Organization of This Volume

This volume recognizes three central themes which comprise the subject material for the three different parts.

Part I. Man-Induced Terrain Degradation. This section considers the various ways in which man has caused destruction on the earth's surface. Even in his usual activities, where no deliberate change is intended, as in many agricultural practices, soils are eroded, gullies created, and abnormal sedimentation rates occur. Additional significant terrain changes occur where man deliberately chooses to alter the landscape in such ways as road building or development of raw materials. These events create a different family of land scarification.

Part II. Soil Conservation. This topic deserves a separate section because it constitutes one of the most vital aspects of man's inheritance and survival on earth. The history of the conservation movement is traced, using examples primarily from the United States. A series of six articles provide information on various erosion-control methods in a variety of different terrains.

Part III. Landscape Management. This field includes the companion ideas of planning environmental safeguards and the rehabilitation of damaged terrain, as well as measurement and evaluation of environmental quality for use in determining procedures for management of the lands and waters as a recreational and aesthetic resource.

Man-Induced Terrain Degradation

I

This part of the volume deals with man as a destructive force in creating changes on the landscape. From that viewpoint man becomes a geomorphic process, just as running water, groundwater, oceans, glaciers, wind, and downhill gravity movement are: all working to transform the earth's surface. In this case, man is a biogenetic factor. All these processes have in common the erosion, transportation, and deposition of soil and rock materials. The landscape is constantly changing, and it is the interrelationships of these changes with the means by which the processes operate in producing landforms that constitute the field of geomorphology. Such forces of change are called "exogenic" to differentiate them from the "endogenic" forces, which operate from within the earth and involve such geological disciplines as structural geology, tectonics, magnetism, volcanism, and igneous petrology.

A variety of terms have been used in the literature to describe man-created changes as contrasted with changes resulting from other processes. For example, erosion that has been created by man has often been termed "anthropogenic," "accelerated," "man-made," or "man-induced." The literature also contains suggestions that the process might be called "anthropogenesis," or, tongue in cheek, "bulldozergenesis." "Erosion created by other, i.e., non-man-made geomorphic processes have been termed "geologic erosion," "natural erosion," and "normal erosion." Such terminology introduces the concept of man as being unnatural, abnormal, and outside the realm of geologic or geomorphic factors in landscape development. But man is a biologic agent. Nothing is more natural than man. To assume otherwise would introduce a semantic dilemma which could produce dangerous errors. Nevertheless, for convenience, it is usual to say "natural erosion" for non-man-induced effects.

The distinction between anthropogenic and non-man-made erosion is most apparent when placed in the dimensions of a time and rate-of-change context. Thus, man-induced terrain changes are commonly several orders of magnitude more rapidly achieved than are those in the usual geologic time reference. When a massive short-range geologic event occurs to transform the land, the term "hazard" or "natural hazard" is appropriately employed for such phenomena as flooding, hurricanes, and landslides (see "Floods," Vol. II). Other organisms besides man produce terrain modifications: beavers build dams, termites build hills, and bacteria rot rock, but man is so much more efficient!

Man affects the terrain in numerous ways. Since his usual abode is on the land, nearly everything he does produces some kind of change, either on the earth's surface or on the materials he uses—sand, gravel, soil, water, and the entire host of raw resources. Part I is divided into three sections, which indicate different areas of man-induced changes. The section on soil erosion and siltation deals with man's changing of soil and water regimes, largely through agricultural endeavors. Here he is not deliberately trying to produce any ill effects but rather he is exploiting the soil, and he may well enrich it; whereas the sections on construction activities and resource development deal with man's deliberate altering and modifying of the landscape and removing of constituents that comprise the earth's surface.

Editor's Comments on Papers 1, 2, and 3

1 **Strahler:** *The Nature of Induced Erosion and Aggradation*

2 **Glenn:** *Denudation and Erosion in the Southern Appalachian Region and the Monongahela Basin*

3 **Bennett and Chapline:** *Soil Erosion a National Menace. Part I: Some Aspects of the Wastage Caused by Soil Erosion*

Soil Erosion and Siltation

As pointed out on page 129 of Volume I of this trio of volumes, erosion of the land always produces the twin problem of siltation (or "sedimentation" or "aggradation"). This double jeopardy of landscape debasement constitutes the greatest environmental problem of all. In this section the Strahler, Glenn, and Bennett articles emphasize erosional aspects of terrain, and the Eakin and Stall articles stress the deposition of materials after their removal from hillslopes.

A Governor's Conference called by President Theodore Roosevelt in the early 1900s produced some of the earliest literature about soil erosion and was a beginning for the twentieth-century conservation movement (see Part II).

> When our soils are gone, we too must go, unless we shall find some way to feed on raw rock or its equivalent. The immense tonnage of soil-material carried out to sea annually by our rivers, even when allowance is made for laudable wash, and for material derived from the river channels, is an impressive warning of the danger of negligent practices. Nor is this all; the wash from one acre is often made the waste-cover for another acre, or for several. Sometimes one's loss is another's gain, but all too frequently one's loss is another's disaster; and the 1,000,000,000 or more tons of richest soil-matter annually carried into the sea by our rivers is the Nation's loss (Chamberlin, 1908, p. 78).
>
> Thousands of acres in the East and South have been made unfit for tillage. North Carolina was, a century ago, one of the greatest agricultural states of the country and one of the wealthiest. Today as you ride through the South you see everywhere land gullied by torrential rains; red and yellow

clay banks exposed where once were fertile fields; and agriculture reduced because its main support has been washed away. Millions of acres, in places to the extent of one-tenth of the entire arable area, have been so injured that no industry and no care can restore them (Hill, 1908, p. 67).

Numerous articles have been written in the United States and other countries decrying the great abuse of the lands by man. The following statements typify such remarks in the United States:

> Modern man still assumes that because for thousands of years nature seemed to have the ability to absorb an increasing number and variety of environmental insults it will continue to be able to do so. Man still does not clearly understand that he lives in a delicate equilibrium with the biosphere—upon the precious earth crust, using and re-using the waters, drawing breath from the shallow sea of air. Because he has not understood this, he is now faced with an urgent necessity for defining a hospitable environment (Linton, 1970, p. 337).

> Erosion, like many another curse of humanity, grows by what it feeds upon. It behaves like compound interest. Beginning first on the exposed surface of normal soil, it first removes the sponge-like, water-holding layer of dark humus, which normally is held in place by the roots of plants and protected by their tops (Sears, 1947, p. 98).

> Today the land of our forefathers is an unhappy land—scalped, mauled, defiled, and poisoned (Rienow and Rienow, 1967, p. 283).

Similar concern was world-wide, as indicated by the following:

> That soil erosion is extending rapidly over many parts of the Union . . . that besides slooting, there is a great deal of surface erosion, both by water and wind . . . that the soil of the Union, our most valuable asset, irreplaceable and part of this soil and valuable plant food is lost for ever . . . that great damage is done by the eroded material silting up reservoirs and that soil erosion caused greater irregularity in the flow of our rivers, thereby increasing the cost of irrigation works and the cost of producing feeding stuffs . . . that soil erosion is caused, mainly by deterioration of the vegetal cover brought about by incorrect veld management (Union of South Africa, 1923, p. 15).

Jacks and Whyte (1939) present an international appraisal of erosion in their pace-setting book:

> . . . in South Africa, according to General J. C. Smuts, erosion is the biggest problem confronting the country, bigger than any politics (p. 5).

> The results of land (in Africa) misuse are only now becoming apparent in a grave form, as much of the land in the settled areas has been cultivated for only fifteen to twenty-five years.

> Some areas in Kenya have already reached such a state of devastation that nothing short of the expenditure of enormous and quite impossible sums of money could restore the land for human use . . . (p. 63).

> The most urgent problem in New Zealand, however, is the control of floods and the prevention of the excessive washing of soil down the short river courses into the sea, a process which threatens to leave the country like an "emaciated skeleton." Deforestation by cutting, burning, or overgrazing of the undergrowth in the mountain areas by sheep, cattle, deer and other animals has greatly accelerated run-off and soil wash, and there is hardly a river in the country which is not affected by periodic flooding (p. 74).

> . . . erosion is the modern symptom of maladjustment between human society and its environment. It is a warning that Nature is in full revolt against the sudden incursion of an exotic civilization into her ordered domains (p. 11).

And more recently such scientists as Bouillenne (1962) have called attention to the continued progression of accelerated erosional problems caused by man:

> In the great forest massif of Central Africa, where the rate of regression is rapid, the present state of affairs is becoming disastrous, owing to the activity of the natives, who still start fires to clear the land for hunting and cultivation. This forest area, which is bounded on the north by the Sahara and on the south by the Kalahari, is receding with alarming speed. On all sides the deserts are advancing. It is estimated that in French Equatorial Africa alone the loss of fertilizing matter during the present generation has amounted to half a billion tons, and in a single cotton district in the Congo agronomists have shown that in 6 years 30,000 square kilometers of soil have been ruined (p. 704).

Earlier Sears (1947) referred to such man-induced changes as "deserts on the march."

In much of the literature any erosion of the land has generally been viewed as undesirable. Some pedologists, however, have urged the importance of separating the normal erosion, which may be necessary for the continued regeneration of soils, from the extraordinarily high erosion rates caused by man. For example, Smith and Stamey (1965) attempted to establish "the range of tolerable erosion." Although there are many variables that must be considered, they fix the amount between 0.5 and 6 ton acre annually. In the National Academy of Science's *Symposium on Dynamics of Land-Erosion,* Sharpe (1941) indicates some of the differences:

> Normal soil-development, normal erosion, normal balance of the several processes of denudation, and even normal sculpturing of the surface of humid lands, then, can take place only with the presence of a normal vegetal cover. Man, by removing the natural vegetation, has destroyed the balance of all of these relations. In clearing the forests and baring

> the land for cultivation, man has added his destructive powers to those of nature. Runoff has been greatly increased, both in rate and amount and as a result, streams now flow less regularly, their load is carried fitfully, and their floods reach higher crests. Changes wrought by man have increased sheet-erosion, rilling, gullying, and wind-erosion. These processes have accelerated (p. 236–237).

A problem arises concerning the determination of whether man has caused the environmental deterioration or whether it is due to natural processes, because even in nature there are great differences in hillslope erosion rates and sediment yields. In two undisturbed areas of Malaysia, Douglas (1967) found:

> . . . whereas a catchment with 94% of its area under natural vegetation had a sediment yield of 21.1 $m^3/km^2/yr$, a neighbouring catchment with only 64% of the area covered by natural forest had a yield of 103.1 $m^3/km^2/yr$ (p. 20).

As Leopold, Wolman, and Miller (1964) point out:

> . . . as a rule the geomorphic effects produced by man are the same as those produced without him. Usually man simply changes the magnitude of certain variables in the system. These in turn produce responses, perhaps only acceleration or deceleration, in the fundamental geomorphic processes. The appropriate principles are not abrogated (p. 434).

There is general agreement on the size of the increment in erosion caused by man. Brown (1970) has estimated world-wide natural erosion rates from 12 to 1500 $m^3/km^2/yr$ and man-induced erosion rates from 1500 to 85,000 $m^3/km^2/yr$. The range is explained in regional variations, both physical and human.

An interesting paradox can occur when man attempts to inhibit his own style of erosion. In studying the gullying problem in South Carolina, Ireland, Sharpe, and Eargle (1939) found that it ". . . usually has been brought about by roads and trenches, and by terraces that have been improperly constructed or maintained . . . ." Thus man has actually caused additional erosion by the very means with which he has attempted to control and prevent it.

The usual analytical method for differentiating natural from man-induced erosion involves the following: (1) significant change in quality, quantity, and texture of alluvial deposits; (2) change in plant cover, both in kind and amount; (3) development of steeper side slopes along stream channels; (4) stream channel or gully incisement into soils with previously undisturbed profiles; and (5) changes and upsets in drainage patterns and channel networks.

The first article, by A. N. Strahler, appraises many of these factors and provides a firm investigatory methodology for analyzing and understanding the erosional and aggradational (siltation) framework of fluvial systems. Strahler has been a leader in quantification of fluvial topography (Morisawa, 1972) and was the first to see the significance of R. E. Horton's quantitative work and to extend it into the realm of the dynamics and energy of erosional terrain. Such study provides a requisite mathematical base which is necessary for engineers and those who must predict fluvial behavior and construct devices to accommodate streamflow. For example, a thorough knowledge of such properties could have prevented most of the events described by Daniels and Aghassy in their papers. The significance of the Strahler article lies in its presentation of fundamental guidelines that provide analytical information for improved understanding of fluvially produced landscapes. The Ducktown, Tennessee, case (also discussed by Glenn) constitutes an unusual example of behavior of these processes in a man-disturbed terrain. Although this landscape scarification is a by-product of mining and fumes from the smelter, it is an inadvertent part of the operation. However, it clearly shows the delicate balance in nature and the quick response that can occur when the normal soil regime is upset. As Glenn points out: "The Ducktown region is . . . an emphatic warning of the extent and character of the disaster that may result in these southern mountains from the thorough destruction of the forests."

Vegetation kill by industrial fumes and the resulting destruction of soils and their erosion occurs also in other countries. Linzon (1971) has calculated the amount of forest destruction and loss in timber-cutting income because of sulfur dioxide fumes from the smelters in the Sudbury mining district of Canada. Here 2,000,000 tons of noxious gases annually affect a 720-square-mile area, and white pine now exists on only 7.6 percent of the productive area. Annual losses are $117,000.

This problem is not unique to the twentieth century, however, as demonstrated in studies of the Swansea Valley, England. Coal-burning smelters have been operating in this mining area for hundreds of years and into the twentieth century. For example, Bridges (1965) estimated that the fumes from the coal-burning industries in the lower Swansea Valley were instrumental in causing 1.5 million tons of soil materials to be eroded to depths of 30–36 cm over a 2-km^2 area.

> The continuous envelopment of the valley in fume for nearly a hundred years resulted in almost a complete destruction of its vegetation. The indigenous sessile oak and birch woodland of Kilvey Hill and all grass and heather in the area disappeared. The topsoil, no longer held by plant roots, was washed off the valley sides leaching the subsoil to be eroded into gullies. The area became a virtual desert and the effects are still visible today (Hilton, 1967, p. 26).

The paper by C. H. Glenn is one of the most comprehensive early twentieth-century analyses of erosional processes influenced by man. Owing to space limitations the article has had to be greatly abridged; the interested reader might refer to the many excellent photographs contained in the original. Glenn makes little distinction between the concepts of erosion and denudation, and generally uses them interchangeably. He discusses both natural and man-made erosion and shows how they have influenced groundwater levels, changes in stream regimen creating higher flood peaks, and gullying and stripping of hillslopes. Glenn also provides suggestions for remedial action and for controlling erosion. This article, when coupled with that of G. K. Gilbert (Vol. 1), provides some of the first appraisals by the U.S. Geological Survey of terrains that have been greatly altered by man. Survey publications have increasingly considered man as a geomorphic agent of change in the latter part of the twentieth century; their earlier publications generally dealt with "natural phenomena."

Any book on the role of man in erosion would be incomplete without reference to H. H. Bennett and the article "Soil Erosion a National Menace" (with W. R. Chapline). Bennett has been a major force in erosional studies and in the conservation movement (see Part II), and this article was probably the most influential ever written in the United States, determining the direction the federal government was to take in these matters. The article gained national prominence and set the stage for governmental action. Only the Bennett part of the article is reproduced here. Chapline, in his section, described soil erosion on western grazing lands. He showed how drought, fires, mining, and overgrazing affected the lands by producing flood damages, silting of reservoirs, and reduced productivity of range lands. Remedial steps that should be taken include reestablishment of the vegetative cover, regulation of grazing, and artificial erosion-control devices.

A second reason why this article is so important, was the inclusion in it of quantitative documentation of the extent of areas undergoing erosion. Thus, it contained some of the earliest statistical information on amount of damage caused by man's agricultural practices.

In another classic article that has served as a basis for quotations for the past 35 years in numerous papers and textbooks, Bennett and Lowdermilk (1938) provide figures for their estimates of the status of erosion in the United States.

	Acres
Total land area (exclusive of large urban territory)	1,903,000,000
Erosion conditions not defined (such as deserts, scablands, and large western mountain areas)	144,000,000
Total land area (exclusive of mountains, mesas, and badlands):	
Ruined or severely damaged	282,000,000
Moderately damaged	775,000,000
Cropland (cropland harvested, crop failure, and cropland idle or fallow):	
Ruined for cultivation	50,000,000
Severely damaged	50,000,000
One-half to all topsoil gone	100,000,000
Erosion process beginning	100,000,000

Such an assessment was based on a combination of reports and studies such as that by Winters (1930) who stated:

> The results indicate that, of the total area of 44,000,000 acres in the State (Oklahoma), we have 16,000,000 acres in cultivation; of the 16,000,000 acres now in cultivation, nearly 13,000,000 acres are washing . . . suffering from sheet erosion, or from gullying, slightly or badly. Nearly 6,000,000 acres are in gullies, and 374,000 acres are so badly gullied that good farm machinery cannot cross them . . . in the last three or four years . . . over 1,359,000 acres abandoned . . . due to soil erosion.

and by Hartman and Wooten (1935) in Georgia describing upland areas:

> Two-thirds of the upland of the region has lost from three to eight inches of top soil (A Horizon); 20 per cent has actually been stripped of all the surface and part of the subsoil. About 44 per cent has reached the gullying stage and is severely damaged, if not permanently ruined for agricultural use . . . Deep gullies affect . . . more than 10 per cent of all upland in the Lower Piedmont (p. 99–101).

There are many good textbooks on soil erosion and soil conservation, and the one by Stallings (1957) classifies and summarizes the various types of erosional damages that occur, largely by the hand of man.

1. Soil deterioration of croplands. In the 1942–1951 period, erosion caused the annual abandonment of 400,000 acres of cultivated lands. Using 1947 prices, replacement of nitrogen, phosphorus, and one-fourth of the potassium removed by erosion would cost $7.75 billion.
2. Losses of grazing and forest lands. These damages in the 10-year period are $180 million for grazing lands and $25 million for forest lands.
3. Watershed damage, flood water, and sediments. Damages for the same period amounted to $557 million. "These agricultural losses are of several kinds: damage to crops and pasture, land damage in the form of flood plain scour, steambank erosion, gullying and valley trenching, infertile overwash or deposition of sediment and swamping; damage to farm buildings, fences, roads, stored crops, livestock and drainage facilities; and indirect losses, such as delays in field work, disruption or delays in marketing of farm products, and others" (p. 41).
4. Effect of erosion on irrigation. The annual cost of maintaining the 28,000,000 irrigated acres in the United States in 1940 was more than $43 million. A considerable part of this cost was for cleaning silt from the irrigation canals.

Additional damages occur from soil erosion, and sedimentation, in siltation of dams, losses in fisheries, and public health costs. For example, in western Tennessee, where the south fork of Forked Deer River was clogged and sedimentation occurred over 14,000 acres of bottomlands, deaths from malaria in the area increased 50 percent (Stallings, 1957, p. 45).

Reprinted from *Man's Role in Changing the Face of the Earth,* 621–638 (1956)

The Nature of Induced Erosion and Aggradation

ARTHUR N. STRAHLER*

1

INTRODUCTION

To set forth adequately the nature of induced erosion and aggradation would require that we summarize a sizable science developed in the past quarter-century by a large body of competent engineers, hydrologists, soil scientists, and geologists. To offer a somewhat different treatment of erosion and aggradation than is usually seen, an attempt is made here to synthesize the empirical observations of the engineer on particular cases and in restricted physical limits with the more generalized rational theories of fluvial erosion and deposition formulated by the geomorphologist, who views landforms as parts of evolving systems adjusted to given sets of environmental factors.

DEFINITION OF EROSION AND AGGRADATION

It will be necessary first to define the terms "erosion" and "aggradation" and then to make a suitable distinction between normal geological processes and those accelerated processes induced by man.

In the broadest sense generally acceptable to the geomorphologist, *erosion* is the progressive removal of soil or rock particles from the parent-mass by a fluid agent. Entrainment of the particles into the fluid medium of transportation is thus implicit, but the form of transportation—whether by suspension or bed-load traction—and the distance of transportation are not specified. By introducing mention of the fluid agent, the definition excludes mass-gravity movements such as thick mudflows, earthflows, landslides, and slump, which are plastic flowage or slip movements not requiring a suspending fluid medium. Excluding these phenomena by no means implies that they lack importance in the scheme of landmass denudation; they are excluded in this paper because of lack of space. The fluid media acting upon landforms are water, air, and glacial ice. Here, again, limitations of topic and space require that only water erosion be treated. The omission of wind erosion is serious from the standpoint of the total problem of induced erosion but does not otherwise interfere with a treatment restricted to water erosion.

In this discussion water erosion will be recognized as taking two basically different forms: (1) *slope* erosion and (2) *channel* erosion. The first is the relatively uniform reduction of a fair-

* Dr. Strahler is Associate Professor of Geomorphology in the Department of Geology, Columbia University, New York. In addition to giving graduate instruction and research supervision in geomorphology, he represents physical geography on the university's committee in charge of advanced degrees and instruction in geography. His research and publications deal largely with the development of quantitative and dynamic aspects of geomorphology and the application of statistical methods to landform analysis. He is author of *Physical Geography,* 1951.

ly smooth ground surface under the eroding force of overland, or sheet, flow which, although by no means uniformly distributed in depth or velocity, is more or less continuously spread over the ground and is not engaged in carving distinct channels into the surface. The second form of erosion consists of the cutting-away of bed and banks of a clearly marked channel which contains the flow at all but the highest flood peaks. Channel erosion takes place in and produces both the gullies and deep shoestring rills incised into previously smooth slopes and the valley-bottom stream channels which have long been permanent features of the landscape. Too often, in the writer's opinion, the distinction between slope erosion and channel erosion is not clearly drawn in soil-erosion publications deploring the high sediment yield from a watershed.

In defining slope erosion, the question of inclusion of surficial soil creep requires consideration. By surficial creep is meant the slow, usually imperceptible, downslope movement of unconsolidated soil or weathered mantle caused by the disturbance or agitation of the particles and their subsequent rearrangement under the force of gravity (Gilbert, 1909, p. 345). The importance of this creep is extremely great in normal geological land-surface denudation and is responsible in part for the form of the slope profile. Nevertheless, it is excluded here because the rates at which detritus would be supplied to a stream by creep are assumed to be extremely slow and to produce a negligible increment to stream loads in comparison with quantities involved in accelerated slope erosion induced by radical changes in surface treatment. One mechanism of the creep process is not so easily surrendered, however. This is the process of downslope movement induced by rain-beat (splash erosion) on a poorly protected soil surface (Ellison, 1950). On a perfectly horizontal surface the impacts of vertically falling drops would produce no general soil movement in one direction, but on a slope a vector is added to the otherwise random movement, and a residual downslope movement of particles is inevitable. On broadly rounded divides and gentle slopes, splash erosion by rain-beat may be considered quantitatively to be of prime importance.

Aggradation, like water erosion, occurs in two basic forms: (1) sheet deposition, or slope wash, at the slope base and (2) channel deposition. In the first form, a slow accretion of soil particles occurs on the lower parts of the valley-side slopes where the gradient diminishes. It is brought down by sheet runoff from the slopes immediately above. The second form, channel deposition, includes the building of bars of well-sorted and stratified grains on the channel floor. This type of aggradation may be extended to include alluvial deposits spread broadly upon alluvial fans and flood plains by overbank flow of streams in flood. Debris flows and mudflows, commonly interbedded with sorted and stratified deposits of stream flows, may be included.

The term "sedimentation" is broader than the term "aggradation" and includes not only slope wash and channel aggradation but also deposition of sediment in reservoirs, lakes, or the oceans in the form of deltas, fine layers of sediment produced by settling, or turbidity current layers. This type of sedimentation, while of major concern to conservationists and engineers, is not treated in this paper.

NORMAL VERSUS ACCELERATED EROSION AND AGGRADATION

Much has been written about a distinction between a certain benign regimen of erosion, transportation, and deposition that characterized our country prior to the white man's spread and a sharply contrasting, usually devastating or impoverishing, white-man-induced regimen termed "accelerated erosion," which has since set in. The first condition has been termed the "geologic norm" by Lowdermilk (1935), Sharpe (1941, p. 236), and others. In many areas the second is clearly shown in the surface forms—particularly gullying—and is obvious to the most untutored layman. Less obvious to the eye are rapid accelerations in sheet erosion and aggradation which seem to have set in following deforestation and cultivation.

A survey of papers discussing the distinction between normal and accelerated erosion gives the definite impression that doubt exists as to the validity of the distinction mostly in reference to the semiarid and arid lands of the West. Apparently nearly everyone accepts the severe gullying of the Piedmont and of the Middle West loess regions as an example of accelerated erosion brought about by man. Where the question is with arroyo trenching in the Navaho country, sedimentation of the Rio Grande Valley, or occurrence of mudflows at the foot of the Wasatch, however, some seriously raise the question whether such radical changes are actually "normal" for the prevailing climatic, topographic, and vegetative environment or whether man's activities have been only coincidental with upsets brought about by natural factors. Such skepticism is not entirely unwarranted. The sedimentary contents of many alluvial fan deposits show histories of debris floods and mudflows throughout the construction of the fans (Blissenbach, 1954). Bailey (1935) found evidence of a history of repeated epicycles of erosion and filling in certain valleys of the Colorado Plateau, though not in all valleys or to such great depths as the recent trenching which he attributes to overgrazing. Such periods of accelerated erosion and deposition are entirely expectable in regions where steep slopes and a naturally poor vegetative cover place large amounts of soil and rock debris at the disposal of sporadic heavy storms.

Taking these things into consideration, the adjective "accelerated" as applied to erosion and aggradation will be intended to mean merely a very considerable increase in rate of these processes, quite irrespective of whether the acceleration is brought about by natural causes or man-made causes or by a combination of both. Where caused by man or his livestock, the qualifying expression "man-induced" will be used, and this may be shortened to "induced."

What are the physical criteria for distinguishing between a present-day period of accelerated erosion and deposition and a previous regimen constituting the geologic norm? One is suggested by Bailey (1941) on the basis of conspicuous changes in texture in sections of alluvial deposits found at the base of the Wasatch Range. A change to coarser texture is evidence of accelerated watershed erosion, because, as we shall see later, increases in transporting capacity and competence of the stream result from increases in runoff. As a second criterion, the existence of any steep-banked stream channel or axial gully in whose walls fine-textured soils (often with soil profiles) are exposed is ample proof of a recent upset in stream regimen.

DYNAMICS OF SLOPE EROSION

The geomorphologist and hydraulic engineer see erosion and aggradation processes from somewhat different points of view, each appropriate to the aims of his field of study. To the engineer concerned with accelerated erosion on a cultivated hillside, the problem relates to a sloping patch of ground. This particular area is often regarded as an independent plot and is in fact often physically isolated with artificial sides and a trap for runoff and sediment at the lower edge. To the geomorphologist the plot is merely an indistinguishable part of a continuous geometric surface formed into a drainage basin bounded by a natural drainage divide and centered upon an axial stream. It is the gross aspect of the whole system operating over a long period of time that interests the geomorphologist, but he realizes that he cannot fully understand the morphology without a knowledge of the principles of erosion operative upon any given small plot within the whole. We shall therefore consider first the dynamics of the slope erosion processes and then turn to the total basin morphology.

Following the cue given by Knapp (1941, p. 255), we recognize that "fundamentally, erosion is a mechanical process, whose vital components are the forces which cause erosion, those which resist it, and the resulting motion of the eroded material." We add to this the concept of a natural slope as an open dynamic system tending to a steady state (Strahler, 1952*a*, pp. 934–35). A slope plot of unit width and of any desired segment of the length between the limits of a drainage divide and the axial stream channel at the base is considered to form the open system. Water and rock waste pass through the system in one general direction only (downslope, or vertically downward). Added to by direct precipitation and by rock disintegration at all points in the system, the water and debris thus proceed cumulatively to the line of discharge at the slope base. When the system has achieved a steady state of operation, the rates at which materials enter, pass through, and leave the system become constant, or independent of time, and the form of the system is stabilized. The nature of this steady state is determined by the relative magnitude of the forces of resistance and the forces tending to produce downslope movement.

Forces tending to produce entrainment by sheet flow are given by Knapp (1941, p. 257) as uniform boundary shear, local intensified shear from eddies, fluid impacts, and particle impacts. All these are expressions of the downslope component of the gravitational force. To this is added the force of buoyancy, which makes entrainment easier. If force of raindrop impact is added, this, too, is a result of the action of gravitational force. Other forces which tend to disrupt or otherwise weaken the soil or rock near the surface are molecular rather than gravitational in nature (Strahler, 1952*a*, p. 932). Adsorption of water by colloids, hydrolysis of silicate minerals, growth of capillary films at grain contacts, thermally induced expansion and contraction, reactions between acids and mineral crystals, direct solution, and growth of ice or salt crystals are all processes tending to reduce the strength of the soil and rock. The forces responsible for them are not related to the gravitational field but are of the general group of intermolecular or interatomic forces.

Forces of resistance to entrainment include intergrain friction in the coarse sediments (proportional to the component of gravitational force normal to the surface), capillary film cohesion in silts and clays, and forces of intercrystal cohesion in rocks. Breaking, or shearing, resistance of plant roots and stems and of organic litter is a major force and differs from the inorganic forces in being continually restored through photosynthesis and conversion of solar energy. Equivalent to a resistive force is capacity of the soil to transmit precipitation through the surface to the ground-water system and thus to reduce or prevent the accumulation of surface runoff. This type of "resistance" is analogous to a fighter side-stepping a blow and letting it go harmlessly past. Volcanic cinders and permeable coarse sands may prove surprisingly resistant to sheet erosion simply because runoff cannot form.

We may say, then, that there exist two major groups of opposed forces: those which tend to produce movement or shear and those which tend to resist movement or shear. These may be formed into a dimensionless ratio with resistive forces in the denominator. Where this force ratio exceeds unity, entrainment will set in, and, in general, the higher the ratio, the more rapid will be the rate of erosion.

Horton (1945, p. 319) has cited the DuBoys formula as a rational expression relating the eroding force to slope and to depth of runoff:

$$F = wd \sin \alpha \ , \qquad (1)$$

where F is force per unit area (stress) exerted parallel with the soil surface; w is the specific weight of water (weight per unit volume); d is depth of overland flow; and α is angle of slope. Horton (*ibid.*, p. 320) sums the forces of resistance to erosion in the term "resistivity," R_i, expressed in pounds per square foot of surface. He assigns to R_i values ranging from .05 on newly cultivated bare soil to as high as 0.5 for well-developed grass sod. He observes that erosion will not occur on a slope unless the available eroding force exceeds the resistance of the soil to erosion (*ibid.*).

In addition to the resistivity, R_i, as a measure of susceptibility of a surface to erosion, Horton (*ibid.*, p. 324) introduces an "erosion proportionality factor," k_e, which we shall define as mass rate of removal per unit area divided by force per unit area:

$$k_e = \frac{e_r}{F} \ , \qquad (2)$$

where e_r is mass of soil removed per unit time per unit area and F is eroding force per unit area. Horton defines depth of soil erosion in inches per hour, whereas, for reasons which will be apparent later, it is here defined as mass removed per unit time per unit area. With terms as defined above, the proportionality factor has the dimensions of inverse of velocity. The proportionality factor is the ratio between the erosion rate and the eroding force being applied to produce that rate and will increase in value as susceptibility of the surface to erosion increases. Resistivity, by contrast, will decrease in value when susceptibility to erosion increases.

If eroding force is proportional to both depth and sine of slope angle, as the DuBoys formula requires, and if we assume for the moment that a particular surface plot has a fixed slope, then the next step in our analysis is to consider what factors control depth of runoff. Depth of overland flow may be related to runoff intensity and slope length, for a plot of unit width, by the following dimensionally correct equation (*ibid.*, p. 309):

$$d = \frac{aLQ_s}{V} \ , \qquad (3)$$

where d is depth of flow; a is a dimensionless numerical constant; L is length of slope; Q_s is runoff intensity, defined as volume rate of flow per unit area; and V is average velocity of flow.

From this equation we see that depth of overland flow will increase directly with both length of slope and runoff intensity but is inversely proportional to the velocity. We shall therefore wish to transfer our attention to the term Q_s, because this is most directly related to precipitation intensity.

The empirical Manning formula states that

$$V = \frac{b}{n} R^{2/3} S^{1/2} \ , \qquad (4)$$

where V is mean velocity; R is hydraulic radius or, in this case, equivalent to depth of flow, d, of equation (3); S is slope, measured as tangent of slope here (but approximately the same as sine of slope for angles up to 15°); n is Manning's roughness number; and b is a numerical constant. Horton (*ibid*., pp. 309–10) shows that by simple substitution of the value of V from the Manning equation (4) into equation (3) and by combining slope, the Manning number, slope length, and the numerical constants into one constant, K_s, runoff intensity may be expressed as a function of depth by

$$Q_s = K_s d^{5/3} \, . \qquad (5)$$

Although this relationship applies to turbulent flow only, it has been generalized by Horton to include laminar and mixed flow by a simple power function of depth:

$$Q_s = K_s d^m \, , \qquad (6)$$

where m is an exponent which has the value 5/3 only for turbulent flow. Horton (*ibid*., p. 311) found that the power function agrees remarkably well with values obtained from experimental plots: the values of m ranged between 1.0 and 2.0.

Continuing the chain of analysis, consider next what factors determine the magnitude of runoff intensity, Q_s. Expressed as discharge per unit area of ground surface, Q_s is dimensionally the same as precipitation intensity, I; both are independent of total surface area of the given plot or watershed. If no rainfall were lost through infiltration and evaporation, intensities of runoff and precipitation would be equal. Neglecting evaporation and other losses, the principal loss with which we are concerned is infiltration, expressed also in the velocity dimension, length (depth) per unit time. Horton's (*ibid*., pp. 306–9) infiltration theory of runoff states that runoff can occur only when precipitation rate exceeds infiltration rate and that the latter is itself a function of time such that

$$f = f_c + (f_0 - f_c)^{-Kt} \, , \qquad (7)$$

where f is infiltration rate at a given time, t; f_c is constant infiltration rate approached with time; f_0 is initial infiltration rate; and K is a numerical constant. The sharp drop-off in infiltration rate, described in equation (7) as exponential, is well known from field observations (Sherman and Musgrave, 1949, p. 247). The term f_c has been shown to vary considerably with different surface conditions of the same soil and is a major factor in determining whether or not rapid erosion will occur under a given precipitation regimen. Whereas rainfall intensity rarely exceeds infiltration capacity in undisturbed soils under dense forest cover and surface litter, it does so readily on bare soils whose infiltration capacity has been greatly reduced by rain-beat or livestock trampling. A prime cause of man-induced accelerated soil erosion is therefore the reduction in permanent infiltration capacity, because man has no control over the other variable, precipitation intensity, which also determines runoff intensity, or over slope (unless by terracing), which, along with runoff intensity, determines eroding force.

If, as the foregoing discussion has indicated, the rate at which erosion proceeds depends upon susceptibility of the surface to erosion, expressed as the erosion proportionality factor, k_e; upon the runoff intensity, Q_s, determined, in turn, by ratio of precipitation intensity to infiltration capacity; and upon the slope of the ground surface, S, we may conveniently combine these three terms into one dimensionless product:

$$N_H = Q_s k_e S \, . \qquad (8)$$

I propose that the number N_H be designated the *Horton number*, after the investigator who contributed so extensively to the research on erosion and surface runoff.

Although the Horton number has not been quantitatively investigated, this appears feasible and will lead to a series of numbers summarizing the erosional qualities of given regions. To obtain k_e, measurements of sediment yield will supply values of the term e_r in equation (2); the eroding force, F, may be estimated from the DuBoys formula, which requires data on runoff depth and slope. Runoff intensity can be measured directly from field installations, but a representative value must be stated in terms of storms of a particular intensity. Slope may be generalized for a given watershed by a mean slope or similar statistic derived from slope maps or random slope sampling by methods discussed elsewhere (Strahler, 1956*b*). We might anticipate that the observed range of the Horton number would show a critical value, analogous to the critical value of the Reynolds number, above which severe erosion would set in.

RELATION OF SLOPE EROSION TO TOTAL DRAINAGE BASIN MORPHOLOGY

In the preceding discussion slope erosion was considered as operating on a patch of ground, or slope plot, regarded as an open system within arbitrary boundaries. To satisfy the geomorphologist, this scope should be expanded to treat all the surface of a complete drainage basin lying within the limits of the watershed and terminated at the lower end by a stream channel through which all water and rock waste is discharged. The question now is: What does man-induced, accelerated sheet erosion and gullying mean in terms of the adjustment of the drainage-basin morphology to a steady state of operation?

A concept of the drainage basin as a natural open system whose geometrical form is delicately adjusted under normal, or prevailing, geological and climatic conditions to maintain a steady state of operation was adapted by the writer (Strahler, 1950, p. 676) from analogous biological systems described by Bertalanffy (1950). Attempts to determine the characteristic hypsometric curves and integrals of basins in the steady state have been carried out (Strahler, 1952*b*). Further studies of the nature of slope frequency distributions over an entire drainage basin (Strahler, 1956*b*) have added to existing knowledge on the characteristic morphology. All these studies are based upon Horton's (1945, pp. 281–82) theory that drainage systems can be analyzed according to a system of orders of magnitude of the channel branches and that certain exponential laws of dimensional increase relative to increase in order of magnitude are observed. Fundamental to such a treatment is the concept of the *first-order drainage basin*, or *unit basin*, which may be likened to the cell in living organisms. The first-order basin is defined as that basin contributing directly to a permanently situated fingertip stream channel to the downstream point where this channel joins with another permanent channel, whether of similar order or larger. The first-order basin is usually elliptical or ovate in outline.

The principal distinction between any two first-order basins taken from each of two regions differing in climate, vegetative cover, underlying lithology, and relief is in dimensions. Basin outlines and general surface forms are often surprisingly similar despite great inequalities in scale. While there are various ways in which scale of a basin can be stated, the most valuable general scale index has proved to be *drainage density*, D_d, defined by Horton (*ibid.*, p. 283) as total length of stream channels divided by total area, or

$$D_d = \frac{\Sigma L}{A}, \qquad (9)$$

where ΣL represents the summation of all channel lengths in miles and A is the area in square miles. Drainage density thus has the dimensions of inverse of length and increases in value as the channel network becomes finer in texture with channels closer together. Because of its dimensional quality, a linear scale must be specified; Horton used miles, and this unit has subsequently been adopted. Observed drainage density values range from as small as 2–3 miles per square mile in regions of massive sandstones to as high as 500–1,000 or more for badlands in weak clays.

Our problem now will be to tie together the principles of erosion with drainage density, which is an index of the scale of the basic landform units. From this we may be able to secure some insight into the fundamental meaning of accelerated erosion.

First of all, there are rational grounds, substantiated by field observation, for supposing that drainage density is a function of several variables already shown to be factors controlling the intensity of the slope-erosion process. The prevailingly high values of drainage density in regions underlain by impervious clays and marls, in contrast to density on more permeable rocks, is clearly seen where the differing rock types lie adjacent to one another, but at essentially the same level, under essentially the same vegetative cover, and subject to the same previous geologic history. This leads us to conclude that drainage density would be an increasing function of runoff intensity, Q_s.

The striking contrast in drainage densities where rocks of differing strength are exposed side by side is well known to all geomorphologists. Clays and marls invariably form fine-textured badlands in contrast to large, full-bodied slope forms of such dense hard rocks as sandstone and limestone. This we can attribute to differences in resistivity to erosion or, inversely, to the erosion proportionality factor, k_e.

Slope is also considered by geomorphologists to be an independent variable in controlling drainage density, although this is difficult to confirm by observation. The presumption is that, in regions of steep slope, channel gradients will also be steeper and hence that first-order streams can persist with smaller watershed areas than are required by a system of lower gradients. It is true that the highest drainage densities are recorded in badlands, and these are regions of excessively steep slope.

Relief, or average difference in elevation between divides and adjacent channels, is a scale factor which may vary independently of either drainage density or average slope. It is introduced here to give a characteristic vertical dimension to the drainage basin, analogous with channel depth or hydraulic radius in a stream channel. In general, it seems reasonable to suppose that regions of high relief are subjected to more intensive rates of erosion than are regions of low relief, because the potential energy of the former system will be higher and will tend to produce steeper ground slopes and channel gradients.

We are now prepared to introduce the following general equation showing the several variables of which drainage density is a function:

$$D_d = \phi\,(Q_s,\, k_e,\, S,\, H,\, \nu,\, g)\,. \qquad (10)$$

Definitions of the terms and dimensional analysis of each are given in Table 25. Kinematic viscosity of the runoff, ν, is added as a significant property of a system operated by hydraulic flow. Acceleration of gravity, g, is introduced as the force field within which the entire hydraulic system operates.

By means of the Buckingham Pi Theorem we may combine the variables in equation (10) into a set of dimensionless groups. Mathematical details of this procedure are presented elsewhere (Strahler, 1956*a*), and it must suffice here to give the final result only, which is a function containing four groups:

$$\phi\left(D_d H, Q_s k_e S, Q_s \nu H, \frac{Q_s^2}{Hg}\right) = 0. \quad (11)$$

TABLE 25

DIMENSIONS OF VARIABLES IN DRAINAGE DENSITY EQUATION

Symbol	Term	Dimensional quality	Dimensional symbol
D_d	Drainage density (Horton)	Length divided by area	$L/L^2 = L^{-1}$
Q	Runoff intensity (Horton)	Volume rate of flow per unit area of cross-section	$L^3T^{-1}/L^2 = LT^{-1}$
k_e	Erosion proportionality factor (Horton)	Mass rate of removal per unit area divided by force per unit area	$ML^{-2}T^{-1}/ML^{-1}T^{-2} = L^{-1}T$
S	Slope	Dimensionless	0
H	Average relief	Length	L
ν	Kinematic viscosity of fluid	Absolute viscosity divided by density	L^2T^{-1}
g	Acceleration of gravity	Distance per unit time per unit time	LT^{-2}

The first number, $D_d H$, is the product of drainage density and relief; it is here termed the ***ruggedness number.*** It gives a measure of over-all ruggedness, because it increases directly both with relief and with increasing detail of the drainage net. The second number, $Q_s k_e S$, is the Horton number, already derived in the discussion of dynamics of slope erosion, and is the measure of over-all intensity of the erosion process. The slope term, S, in this product is dimensionless and could be added to any one of the groups without disturbing their dimensionless quality. Nevertheless, it seems most desirable in the Horton number, where the mechanical influence of slope is required.

The third group, $Q_s \nu H$, is a form of the Reynolds number in which relief, H, is the representative length dimension of the system and Q_s is the velocity term. The fourth group, Q_s^2/Hg, is recognizable as a form of the Froude number. These last two numbers, while not the subjects of immediate concern in this discussion, are of prime importance in considerations of dynamic similarity of drainage systems of various scales (Strahler, 1956*a*).

Having decided that drainage density is a function of the various elements comprising the dimensionless groups, we may solve for D_d in equation (11):

$$D_d = \frac{1}{H} f\left(Q_s k_e S, Q_s \nu H, \frac{Q_s^2}{Hg}\right). \quad (12)$$

This tells us that drainage density is inversely proportional to relief times some function of the remaining three dimensionless groups. To determine this function, an extensive program of experimentation would be required. We can, however, now outline the principles relating accelerated slope erosion to the over-all basin morphology.

The essential point is that accelerated erosion in which a gully system is developed on previously smooth, unchanneled slopes is an adjustment of the drainage system toward a higher drainage density and a system of steeper slopes. When a relatively resistant vegetative cover is destroyed and the soil over a weak bedrock exposed, two changes occur in the Horton number, both tending to increase this number: the erosion proportionality factor, k_e, is greatly increased, which means simply that more material is eroded from the bare surface by a given runoff depth and velocity than in the previous condition of high resistance; and the runoff intensity, Q_s, is greatly increased for a given precipitation regimen because the stripping of the vegetative cover leaves the soil susceptible to direct rainbeat, which seals the surface openings and reduces infiltration rates. Additional breakdown of the soil structure by man or animals may further reduce infiltration capacity.

We must assume that the relations between increased sediment load entrained and increased overland flow are such that the flow at points where it is concentrated by favorable undulations of slope is capable of deep scour and begins to carve a gully, which is actually an extension of the stream-channel system and is the means by which drainage density is increased. As gullies are deepened and ramified, a system of steep slopes is formed, and these quickly replace the original surface. The rising grade of the newly formed channels lowers the relief, where this relief is measured from channel to immediately adjacent divide. Each gully end now becomes a new first-order stream segment, and the slopes draining into it are molded into new, small, first-order drainage basins. Presumably, the newly formed system would now achieve a new steady state of operation, which would persist as long as the ground surface continued to be barren and the physical quality of the material beneath the surface remained unchanged. Neither of these qualifications would be likely to be met for long in the humid regions of the eastern United States. Removal of the layer of weathered rock, even though thick, would in most places expose a more resistant bedrock. Vegetation would probably take hold long before the steady state could be achieved.

An interesting case illustrating the drainage-density adjustment theory of severe slope erosion is the Ducktown, Tennessee, locality (Fig. 123). Here the long-continued production of noxious fumes has prevented the recovery of vegetation and allowed the development of a nearly complete erosional topography in a stage that would be termed "mature" by the geomorphologist.

In general, it may be noted that severe erosion leading to extensive gullying can be expected in any region where the contrast in resistivity of a vegetated surface and a barren surface is great, because the low-resistivity surface will demand a high drainage density and because the erosion system will "take" such measures as it "needs" to adjust its morphology to a new steady state. Where difference in surface resistivity before and after surface denudation is not so marked and there is introduced only a small increase in runoff intensity, we may expect the uppermost ends of existing stream channels to be increased in length but with only limited occurrence of new gullies on side slopes.

DYNAMICS OF AGGRADATION

In the introductory definition aggradation was considered to take two basic forms: (1) slope-wash, or colluvial, deposition and (2) channel deposition and associated valley-bottom deposits. In both cases the dynamics are fundamentally similar. The problem is to explain why a sheet or stream of water engaged in bed-load transport should on the average drop more particles than it picks up from a given area of the bed and therefore raise the level of its bed. We assume that, if the transportation system is in a steady state of operation, it would neither aggrade nor degrade but would transport the load supplied to it through the system without changes in vertical position of the bed and without changes in transverse form of the bed or channel. This steady state is essentially the same as the state of operation of a graded stream with an equilibrium profile, to use the geomorphologists' conventional terminology.

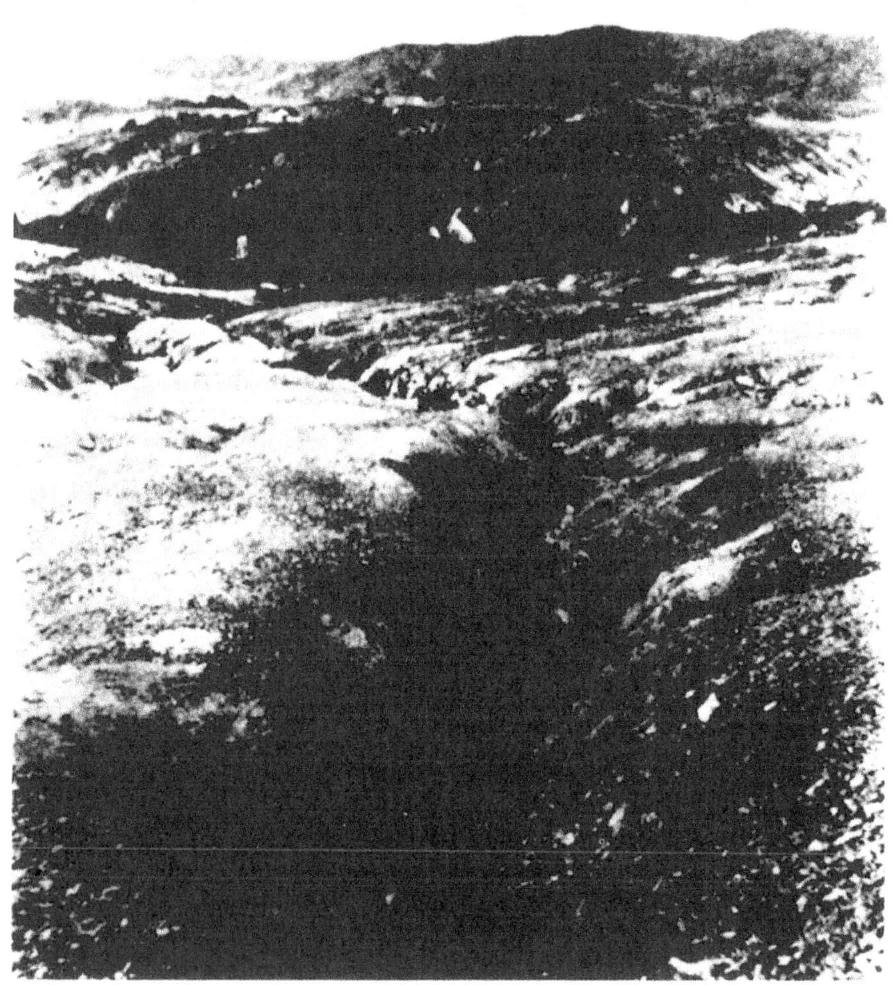

FIG. 123.—Severe erosion near Ducktown, Tennessee. Complete denudation of vegetation has been followed by transformation to a morphology characterized by high drainage density and steep slopes.

The occurrence of persistent aggradation is itself primary evidence of a change in the factors whose balance determines a steady state and is a manifestation of the self-adjustment of any open dynamic system to restore a steady state. If raising the height of the bed alone were involved—that is to say, merely lifting the entire profile by addition of a constant increment at all points—the operation of the system would not be appreciably affected in a short period of time. Instead, the aggradation must be viewed as increasing in depth upstream or downstream to produce a wedgelike deposit which increases or decreases the slope of the bed and therefore influences the velocity of flow and hence also the ability of the stream to transport. Only by producing changes in slope can aggradation restore a steady state of operation. Upstream increase in depth of aggradation is the characteristic change in fluvial systems, and we may say that aggradation is generally associated with increase in slope of the fluvial system. Reduction of slope may conceivably be accomplished for a short period by downstream increase in depth of aggradation but requires the special case of introduction of a raised or rising base level in the path of the flow (Mackin, 1948, pp. 496–97). As a general rule, in the absence of such base-level changes, reduction of slope is accomplished by erosional reduction of the bed, the depth of removal increasing upstream to yield a wedge of removal which thickens upstream.

Both the bed-load capacity for a given size grade and the competence (ability to move, stated in terms of particle size) are functions of bed velocity, which in a general way varies with mean velocity in a stream (Gilbert, 1914). Aggradation can thus be initiated by a reduced velocity of flow, and we shall need to inquire into those external changes which may on the average substantially reduce the velocity of flow, both overland and in channels. One such change is reduction of discharge, which, through a decrease in depth of flow, reduces velocity. Reduction in magnitude of flood discharges may be general over an entire watershed through climatic change in which runoff-producing rains of given high intensities become less frequent. Aggradation reflects not so much the inability of reduced discharges to keep the channel cleared of debris as it does a steepening of gradient undertaken by the system to restore its transporting ability and to restore a steady state of operation on a reduced budget of discharge. Loss of discharge by influent seepage through the stream bed and by evaporation is important in lower reaches of channels and on fans in the dry climates. Progressive loss of stream discharge by increasing diversion of flow to underground solution channels is important in limestone regions (Strahler, 1944) and is accompanied by aggradation in valley floors. An improvement in infiltration capacity on watershed surfaces would also tend to reduce peak discharges and hence might seem to be a cause of aggradation, but, because this is normally part of a sequence of events in which vegetation becomes more dense and consequently holds back sediment load, it is generally followed by channel erosion rather than aggradation.

Excessive decrease of slope of the stream bed or ground surface in the downstream direction is a second cause of velocity reduction. The concavity commonly present at the base of a valley-side slope provides a continually lessening declivity over which sediment-ladened sheet flow must pass. Deposition of sediment in this zone is easily attributed to the decrease in velocity forced by the decrease in slope, but in terms of open-system dynamics it would be more meaningful to say that the aggradation is an attempt to steepen the slope to the point where increased velocity will permit the sediment to be transported across the slope with no further aggradation.

In a channel the downstream decrease in slope, which is approximately of a negative exponential form, may be delicately adjusted for a steady state of operation under a given regimen of climate and watershed characteristics. If discharge is reduced, this slope will prove to be insufficient. Aggradation will provide the means of steepening the stream slope, thereby increasing its transporting power, and consequently will tend to restore a steady state in which the debris is carried through the entire system on a fixed grade.

Quite apart from changes in velocity of flow, the load itself is the fundamental independent variable to which the stream slope is appropriately adjusted when a steady state of transportation exists. Necessity for aggradation which will steepen the slope is brought about either by (1) an increased rate of supply of debris of a given size distribution (because the capacity for bed-load transport is limited) or by (2) an increase in coarseness of the debris (because the competence is also limited). Assuming that the overland flow and channel flow are adjusted to transport a given quantity of debris of a given coarseness per unit time, an increase in either the quantity or the coarseness will require that the flow be adjusted to supply greater transporting ability. This can be done only by an increase of slope through aggradation. Normally, the depletion of watershed vegetative cover accompanied by cultivation or other disturbance of surface will simultaneously increase not only the quantity of debris but also its mean grain size.

Discharge and load are not normally varied independently of each other. The same depletion of vegetative cover that causes greater quantity and larger caliber of load will also be expected to be associated with reduced infiltration capacity of the soil, and this in turn yields greater peak runoff and stream discharge. What is of prime importance is therefore the ratio between load and discharge. In general, an excessively high load-to-discharge ratio will be met with aggradation and steepening of slope; a low ratio, with scour of the bed and lowering of the slope. Whereas accelerated sheet erosion with severe gullying of slopes is most commonly associated with aggradation in valley floors, this would not necessarily have to be so in all cases. If the acquisition of debris by the runoff did not continue to excess in the downstream direction not only might there be no aggradation down the valley but erosional deepening of the entire drainage system (diminishing of course to zero at the mouth) might conceivably result.

Velocity may be reduced independently of changes in discharge and load by deterioration in the efficiency of the channel form and by increase in the irregularity of the bed (Mackin, 1948, pp. 487, 504). In the case of overland flow on slopes, or of subdivided flow over alluvial fans and alluvial valley flats, the development of a cover of grass or small bushy shrubs will be expected to retard the flow appreciably and to cause the velocity drop necessary to produce aggradation. While this effect may be deemed beneficial on hillside and valley-side slopes, where retardation of sediment movement is desired, it forms part of a vicious cycle in aggradation of valley bottoms, as described below.

In streams, deterioration of channel efficiency may be expected to reduce velocity and bring on aggradation at increased rates. Assuming first that a marked increase in bed load relative to discharge has caused rapid aggradation in the channel of a stream, the building of sand and gravel bars will be expected to decrease the depth of flow at the same time that the channel is broadened or actually subdivided. Leopold and Maddock (1953, p. 29) point out that an alluvial channel which is broad and shallow probably carries a relatively large bed load. A steep slope would, however, be required in compensation for reduced depth. The temporary loss of transporting power would be expected to intensify the rate of aggradation still further and to set off a vicious cycle wherein the flow spreads over the banks and is distributed over the entire valley floor. The rapid formation of channel-plug deposits with upstream sediment accumulation and overbank spreading has been described by Happ *et al.* (1940, pp. 71–73, 94–95). In the absence of a deep, strongly scoured channel in the valley axis vegetation can take hold, further impeding the flow and at the same time increasing the resistance of the valley floor to scour. Eventually, however, the valley-floor slope is built up by aggradation to the point where an unusual flood, deficient in bed load, can quickly incise the alluvium and initiate cutting of a relatively narrow, deep channel requiring a lower slope. That such cycles of aggradation and trenching are the normal deviations from a uniformly maintained steady state of operation in a semiarid environment has been under consideration by geomorphologists for many years, one of the earliest discussions being by McGee (1891), who introduced the term "varigradation" for normal development of alluvial accumulations in the course of a stream.

The role of man in such cycles of aggradation and intrenchment may prove, most rationally viewed, to be one of setting off changes that might otherwise be long delayed and of intensifying the extremes of the cycle. Deforestation, cultivation, or overgrazing might be expected to increase the load so greatly in relation to discharge as to force channel aggradation and initiate the aggradation cycle. In humid regions responsibility for such aggradations seems to rest almost solely upon man; in semiarid regions the responsibility is less clearly his. By the reasoning invoked above, trenching of alluvial fills to produce narrow arroyos is a reflection of either increased discharge or reduced bed load, or a reduction in ratio of load to discharge, and is thus the diametric opposite of the change invoked to explain aggradation. Under these circumstances it is difficult to accept the blanket explanation of man-induced depletion of vegetative cover to account for valley aggradation in one part of the country and for valley incision in another part.

Another aspect of aggradation dynamics deals with textural differences between parent-material and aggraded products. In general, the slope-wash and valley alluvium produced in the aggradational cycle will be the coarser fraction of the parent-material from which the fine clays and silts, readily carried in suspension to distant downstream points, have been sorted. The initial erosion velocity required to entrain the fine-grained materials is great, because of initial cohesion. Once in suspension, however, they remain in motion at relatively low velocities and will settle out only in reservoirs or lakes or when flocculated in contact with sea water. This accounts for the coarseness of aggradational deposits formed as aprons at the base of valley-side slopes and in the floors of the smaller stream valleys. Happ *et al.* (1940, pp. 86–88) state that sand or coarser sediment causes most of the sediment damage to valley agricultural land.

RELATION OF AGGRADATION TO TOTAL DRAINAGE-BASIN MORPHOLOGY

In a previous section of this paper dealing with relation of slope erosion to total drainage-basin morphology the principle was developed that, when severe erosion accompanied by gullying sets in, following breakdown of initial surface resistivity and reduction of infiltration capacity, the channel system is undergoing a change from low- to high-drainage density. Streams are lengthening and are increasing greatly in number by development of new branches. Slopes are being steepened and at the same time greatly shortened in length from divide to channel. We shall now attempt to integrate into this general theory of drainage-basin transformation the occurrence of aggradation of the channel and valley floor of the main stream. Such a development is illustrated in Figure 124, representing the case where severe slope erosion is concomitant with channel aggradation.

The drainage basin outlined represents a single unit, or first-order basin, prior to the induced modifications. It has one axial channel without branches. The slopes are long and smooth from

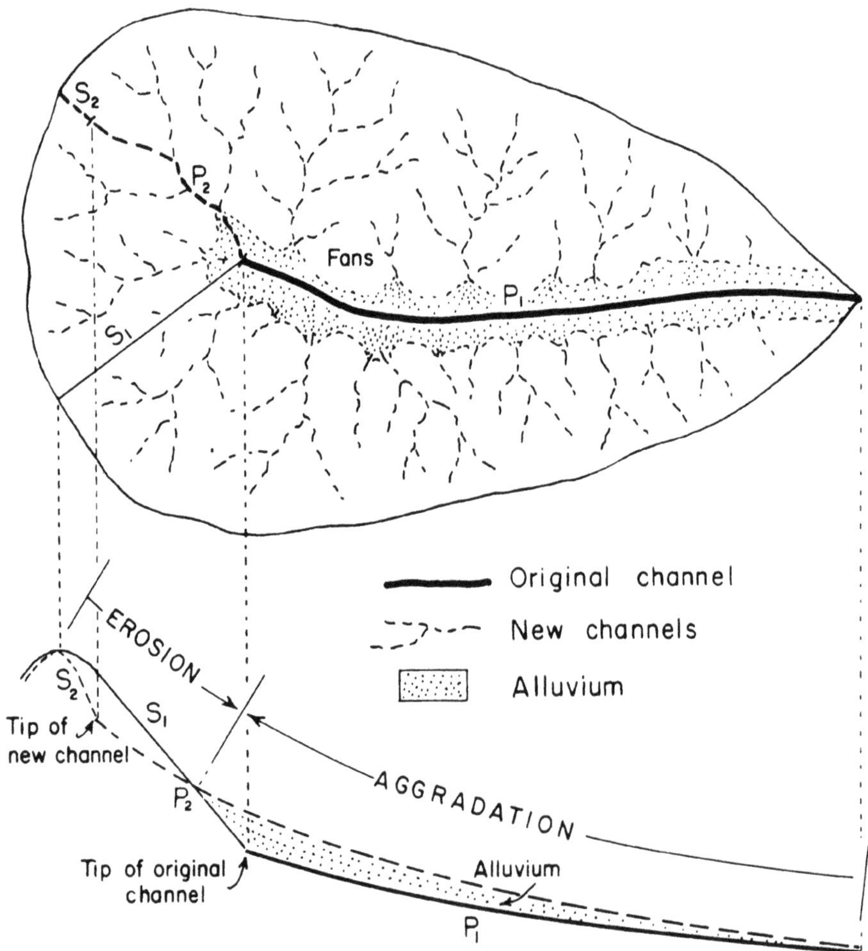

Fig. 124.—Drainage-basin transformation from low to high drainage density. Severe gullying and badland development on slopes accompanied by aggradation in main valley axis. Profiles show that channel gradient and slope gradient are both increased by the transformation.

divide to channel, although ephemeral shoestring rills may from time to time be carved on these slopes by unusual storms and subsequently rehealed. With the onset of severe erosion, a greatly ramified drainage network develops, with a large number of new first-order basins of small size replacing the single large basin. This is strikingly illustrated in the Ducktown, Tennessee, locality shown in Figure 125. The channels are extended headward into surfaces previously part of the smooth slopes. The new channel system transports greater quantities of debris, including coarser material than formerly, because the increased proportion of surface runoff greatly increases the transporting capacity of the system at the same time that the resistivity of the surface has been greatly reduced, and the load is more easily entrained by the runoff. As the gully incision progresses, more and more of the surface is transformed into steep badland slopes. Because the ratio of load to discharge has increased during this development, the slope of the channel of the main axial stream will have from the beginning required steepening, and this will take the form of coalescent fans built out from the base of each new gully into the floor of the main valley (*ibid.*, p. 92). This aggradation will be extended down the valley as a wedge until the entire gradient is steepened to allow discharge of the load through the system and out the mouth. At this point in time a stabilized, or equilibrium, profile is produced. If conditions of precipitation and surface resistivity remained constant thereafter, the only further change would be a slow lowering in elevation of the entire system as required by removal of the landmass. In time the aggraded material would be trenched and dissected, eventually to disappear from the entire basin.

Fig. 125.—Transformation of slopes by severe erosion near Ducktown, Tennessee, 1949. Smooth, long, vegetated slopes are being transformed by an extended channel network into regions of short, steep, bare slopes.

The transformation outlined above would represent an extreme case, comparable to change from landforms appropriate to humid climates with dense vegetative cover to badland forms in semiarid regions. We might suppose that moderate changes in resistivity of surface or in rainfall regimen would bring about correspondingly moderate changes in drainage density and that in some cases the growth of new channels and aggradation of the main valley floor would be observable only from careful observations of long duration.

CONCLUSION

The principal objective of this discussion has been to attempt to develop certain rational, qualitative principles of land erosion and aggradation in which the phenomena observed by engineers studying soil erosion and sedimentation on small plots of ground or on limited reaches of streams are related to a general geomorphological theory based upon consideration of an entire drainage basin as an open system tending to achieve a steady state of operation but responding by erosion and aggradation when the steady state is upset by man's treatment of the land or by natural changes in physical environment. Perhaps such a theory relating all parts of a fluvial system, if proved sound, will permit more accurate prediction of the downstream consequences which may be expected from changes in watershed conditions. Perhaps, also, such a theory will help to resolve some of the seemingly anomalous cause-and-effect relationships among erosion, aggradation, land use, and climatic change in the semiarid grazing lands of the West.

REFERENCES

Bailey, Reed W.

1935 "Epicycles of Erosion in the Valleys of the Colorado Plateau Province," *Journal of Geology*, XLIII, 337–55.

1941 "Land Erosion—Normal and Accelerated—in the Semiarid West," *Transactions of the American Geophysical Union, 1941*, pp. 240–50.

Bertalanffy, Ludwig von

1950 "The Theory of Open Systems in Physics and Biology," *Science*, CXI, 23–28.

Blissenbach, Erich

1954 "Geology of Alluvial Fans in Semiarid Regions," *Bulletin of the Geological Society of America*, LXV, 165–90.

Ellison, W. D.

1950 "Soil Erosion by Rainstorms," *Science*, CXI, No. 2880, 245–49.

Gilbert, Grove Karl

1909 "The Convexity of Hilltops," *Journal of Geology*, XVII, 344–50.

1914 *The Transportation of Debris by Running Water*. (U.S. Geological Survey Professional Paper No. 86.) Washington, D.C.: Government Printing Office. 263 pp.

Happ, Stafford C.; Rittenhouse, Gordon; and Dobson, G. C.

1940 *Some Principles of Accelerated Stream and Valley Sedimentation*. (U.S. Department of Agriculture Technical Bulletin No. 695.) Washington, D.C.: Government Printing Office. 134 pp.

Horton, Robert E.

1945 "Erosional Development of Streams and Their Drainage Basins: Hydrophysical Approach to Quantitative Morphology," *Bulletin of the Geological Society of America*, LVI, 275–370.

Knapp, Robert T.

1941 "A Concept of the Mechanics of the Erosion Cycle," *Transactions of the American Geophysical Union, 1941*, pp. 255–57.

Leopold, Luna B., and Maddock, Thomas, Jr.

1953 *The Hydraulic Geometry of Stream Channels and Some Physiographic Implications*. (U.S. Geological Survey Professional Paper No. 252.) Washington, D.C.: Government Printing Office. 57 pp.

LOWDERMILK, W. C.
1935 "Acceleration of Erosion above Geologic Norms," *Transactions of the American Geophysical Union,* 15th Annual Meeting, pp. 505–9.

MCGEE, W J
1891 "The Pleistocene History of Northeastern Iowa," pp. 187–577 in *U.S. Geological Survey: Eleventh Annual Report (1889–1890),* Part I. Washington, D.C.: Government Printing Office.

MACKIN, J. HOOVER
1948 "Concept of the Graded River," *Bulletin of the Geological Society of America,* LIX, 463–512.

SHARPE, C. F. STEWART
1941 "Geomorphic Aspects of Normal and Accelerated Erosion," *Transactions of the American Geophysical Union, 1941,* pp. 236–40.

SHERMAN, LEROY K., and MUSGRAVE, GEORGE W.
1949 "Infiltration," pp. 244–58 (chap. vii) in MEINZER, OSCAR E. (ed.), *Physics of the Earth,* Vol. IX: *Hydrology.* New York: Dover Publications, Inc. 712 pp.

STRAHLER, ARTHUR N.
1944 "Valleys and Parks of the Kaibab and Coconino Plateaus, Arizona," *Journal of Geology,* LII, No. 6, 361–87.
1950 "Equilibrium Theory of Erosional Slopes Approached by Frequency Distribution Analysis," *American Journal of Science,* CCXLVIII, 673–96, 800–814.
1952*a* "Dynamic Basis of Geomorphology," *Bulletin of the Geological Society of America,* LXIII, 923–38.
1952*b* "Hypsometric (Area-Altitude) Analysis of Erosional Topography," *ibid.,* pp. 1117–42.
1956*a* "Dimensional Analysis in Geomorphology," *ibid.,* LXVII, 571–96.
1956*b* "Quantitative Slope Analysis," *ibid.* (In press.)

Reprinted from *U.S. Geological Survey Profess. Paper 72* (1911)

DEPARTMENT OF THE INTERIOR
UNITED STATES GEOLOGICAL SURVEY
GEORGE OTIS SMITH, DIRECTOR

PROFESSIONAL PAPER 72

2

DENUDATION AND EROSION

IN THE

SOUTHERN APPALACHIAN REGION

AND THE

MONONGAHELA BASIN

BY

LEONIDAS CHALMERS GLENN

WASHINGTON
GOVERNMENT PRINTING OFFICE
1911

* * * * * * *

RELATION OF INDUSTRIES TO EROSION AND DENUDATION.

AGRICULTURE.

The population is largely agricultural. In the higher mountains grasses grow well and grazing is an important industry. Much of the cleared land there is kept in grass and so is prevented from washing badly, but some of the cleared slopes are so steep that even though kept in grass they soon erode to the bare rock and become useless wastes. During prolonged wet weather the turf and soil become thoroughly softened, and on steep slopes the hoofs of the cattle easily break the turf and start small landslides that quickly develop into gullies. Some of the so-called balds, once well sodded, have become bare, rocky wastes from erosion induced by the trampling of cattle. This conversion of sodded balds into rocky wastes is actively in progress to-day on the top of Roan Mountain.

When first cleared, the land is usually planted in corn for about two or three years, is then for two or three years put in small grain or grass to be mowed—not grazed—and then back into corn for several years. Unless it is well cared for the land has by this time become poor, for it has lost its original humus. The soil has become less porous and less able to absorb the rainfall and erosion begins. Means are rarely taken to prevent or check this erosion, so it increases rapidly and the field is soon abandoned and a new one cleared.

When the clearing is intended for grazing, the preliminary rotation practiced by the more progressive farmers is nearly the same. Corn is planted 2 years, then meadow for 2 years, then corn a year, then grass for permanent pasture. This first pasture sod usually lasts 8 or 10 years, when the land is again put in corn for a year, then oats and grasses are sown, and the sod is again as good as the first one. When properly cared for and not too steep, grazing land should be capable of indefinite use. The erosion of such lands is due chiefly to ignorance and neglect, and the remedy is not reforestation but education. Grazing lands, however, comprise but a small part of the entire area of cleared land, which has recently been estimated by several investigators at 24 per cent of the total mountain area. The writer believes that this estimate is approximately correct, but that it errs, if at all, in slightly overstating the cleared area.

In clearing land only the undergrowth and small trees are as a rule removed. The large trees are killed by girdling and left standing. Many fields are worn out and abandoned before the trees girdled in its clearing have all fallen. Then new grounds are usually cleared beside the abandoned field and the same destructive process is repeated.

The clearing of virgin forests for agriculture is going on steadily from year to year to replace worn-out, eroded, and abandoned lands. When the region was settled, the more level lands along and near the streams were first cleared and those that have been properly cared for and are out of the reach of the stream floods have remained in cultivation and are in good condition to-day. After these lands had been largely cleared the steeper slopes were next invaded by the axman and then still steeper slopes, so that very much of the land now being cleared is too steep for cultivation under present farm practice and should be kept in forests.

Numerous attempts have been made to estimate the percentage of the area of these mountains that might safely and profitably be cleared for cultivation. These estimates average about 15 per cent. It is difficult to give any definite estimate of such area, for the allowable limit of slope of lands that may be safely cleared—which is generally put at 10° and which alone has usually been considered—is not the only factor of the problem, for the nature of the soil, which is dependent on the geology of the underlying rock formations, and the intelligence and care of the cultivator should also be considered. On some soils 10° may be the maximum slope for safe cultivation; on other soils slopes of 20° do not wash. Slopes themselves may be changed by terracing, and education may so greatly increase the intelligence and care of the cultivator that estimates of cultivable area that consider these varying factors must of necessity vary, and the variation tends to increase the estimate of cultivable area as time passes. The increase, however, must be slow, and for present methods of cultivation 18 to 20 per cent is probably a liberal estimate for the area that may be cleared safely. The present cleared area, 24 per cent, is undoubtedly in excess of the limit of safety under the existing conditions of agriculture.

Tobacco growing has in the past been peculiarly injurious to the soil. The plant requires clean cultivation, and when planted year after year soon exhausts the fertility of the soil, depletes the humus, and leads to erosion and early abandonment. Some 15 or 20 years ago tobacco growing became quite profitable in some sections and lands were rapidly cleared, worked in tobacco a few years, worn out, and abandoned. The effect of this clearing and cultivation on the streams in general will be discussed later.

LUMBERING.

Lumbering has been so active in this region that practically all the timber that is easily accessible to the railway or to streams large enough to float it has been cut. As prices of lumber have advanced the mills have been moved farther back from the railroads until now a part of the product of the small mills is hauled 15 to 25 miles over rough mountain roads, and the owners of the larger mills have built their own logging railways or tram roads back into the heart of even the more distant mountains.

Most of the valuable timber has already passed into the hands of lumbermen, who are now actively buying lands in even the most remote and almost inaccessible localities. Prices of timber lands are rapidly advancing, and if cutting is continued at the present rate the primeval forests of these southern mountains will soon be a thing of the past.

A few large forest tracts are being preserved by private individuals or corporations or are being lumbered according to modern conservative methods, but except in these few areas lumbering is carried on so as to yield the largest possible immediate returns without thought or care for the future.

In recent years a number of large tanneries and several tanning-extract plants have been established in the region, and these insure the destruction of most of the chestnut, chestnut oak, and oak left by the lumbermen. Recently, also, one of the largest wood-pulp plants in the country has been built in the region and has begun to strip the forests and to pollute the stream on which it is located.

Forest fires may follow in the wake of the lumberman and complete the destruction of the young growth, and the blackened waste may be abandoned to the county as not worth paying taxes on. In many places destructive forest fires have not occurred and there the young growth retains possession of the land and begins the production of a second crop of timber, though only a few of the lumbermen in the region are looking ahead far enough to count this second crop as an asset of any value. Very many of them would part with the land for a small price when they have once thoroughly lumbered it.

MINING.

The southern Appalachians are noted for the great variety of their mineral deposits. Although many of these have never been developed, others, such as the deposits of gold, copper, marble, mica, corundum, talc, asbestos, slate, baryta, and kaolin, have become the basis of important mining operations, and annually yield mineral products to the value of several million dollars.

Mining is affecting erosion in several ways, one of which is locally important. The timbering required in underground mining creates a demand for mine props and other timbers. So far this has not been a serious menace to the forests, partly because underground mining operations are not generally extensive throughout the region and partly because the rocks are ancient crystallines that below the limit of surficial weathering are strong enough to stand without much timbering.

Much of the mining, especially of gold and monazite, is of the placer type. The gold mining is centered chiefly around Dahlonega, Ga. Stream gravels are washed to some extent, but mining is most active in the saprolite or surface rock that, though decomposed, is still in place. Entire hillsides are washed down into the streams by hydraulic giants. Careful examination and inquiry, however, failed to show that much damage is being done along the streams within the Dahlonega region itself as a result of this placer mining, partly because there are in most places little or no bottom lands susceptible of cultivation, but chiefly because most of the streams have much fall and sweep away the sand and clay as fast as the mines furnish it, or the frequent floods carry away any material that tends to lodge along the streams. Much of it lodges farther out along the Piedmont part of the streams, where their grade is less and their transporting power is correspondingly reduced, and there it may aid in doing damage.

In the Dahlonega region very little of the upland is cleared and the normal amount of waste furnished to the streams from agricultural lands is small. The streams are consequently able to transport the additional waste furnished by the hydraulic mines; but should the clearing of uplands be extended, as is to be expected, the increased agricultural waste, together with the mine waste, will soon overload the streams and cause the accumulation along their courses of sand and clay which they would then be unable to carry.

Dredging for gold is carried on in some of the streams of the Dahlonega region and large quantities of stream waste are moved in this way. The effect on the stream channels and on the adjacent bottom lands depends on the extent of the pay gravel and on the mode of disposing of the tailings. Where pay gravel underlies the bottoms they are worked and their future agricultural value is largely or entirely destroyed.

After the streams have left the Dahlonega region their slopes and velocity decrease and the waste carried out from the mining region is being deposited along the way, filling the stream channels, lodging on bars, and causing islands to form, thus further obstructing the channels and making overflows easier and the destruction of the bottom farming land more speedy, certain, and complete. With the further clearing of lands that may normally be expected this condition will become constantly worse.

The third and most injurious kind of mining studied in the region is that at Ducktown, Tenn., where sulphur fumes from the roasting and smelting of the copper ores have killed all vegetation for a number of miles around and the perfectly bare surface has eroded with wonderful rapidity (see pp. 24–25, 78–79).

POWER DEVELOPMENT.

Careful estimates made by competent engineers show that the streams of the southern Appalachians afford, in units of considerable size, about 3,000,000 undeveloped horsepower, and probably as many more in small units suitable for a great variety of minor industrial uses. Nearly all of the water power in the mountains has heretofore remained undeveloped because of its inaccessibility, but since the electrical transmission of water power has been made practicable, its utilization has become possible. As the opening of new railroads through the mountains renders the water powers more accessible, as population steadily increases and turns for a livelihood to new avenues of industrial activity, and as cotton mills and other manufacturing enterprises in the South utilize the available power sites in the Piedmont, the demand for this mountain stream power must yearly increase and will naturally lead to its utilization. At any reasonable valuation per horsepower the undeveloped power of these mountain streams is an important industrial asset. In Georgia and the Carolinas more than 100,000 horsepower has been developed and is being used by cotton mills alone, and public-service corporations in these three States are to-day developing 300,000 to 400,000 additional horsepower to turn the hundreds of mills and light the many towns and cities in the region. One of these power plants is shown in Plate I (opposite) and views of others are given in Plate XII, *A* and *B* (p. 36). In the operation of the power plants already constructed and in the financing and building of those yet to be developed the erosion problems of the region are becoming important factors, deserving serious consideration.

From the slopes along these streams a steadily increasing amount of waste is working its way down their channels, filling the dams and destroying their storage capacity; and this loss of storage means a decrease of efficiency that is calculated by the most experienced mill engineers to amount to 30 to 40 per cent in plants that have been built especially for storage and a somewhat less marked decrease in other plants, the exact amount depending on the topography of the basin and the regimen of the particular stream on which the plant is located. So universal is this silting of storage basins that a prominent mill engineer of wide experience in his reports on the construction of power plants no longer calculates on power or on anything except the flow of the stream, and he has increased his usual construction estimates by an allowance for increased storm waters that must be taken care of without endangering the dam or plant.

Experience has shown that storage basins constructed in this region in recent years are rapidly filled with sand and silt, through which the stream maintains a channel only large enough to carry the ordinary flow. On a few streams diversion dams without storage furnish the most economical development, but the dams at most of the cotton mills have been built to store water over night or over Sunday, and at one dam, at least, the original storage capacity was sufficient to hold a week's ordinary flow.

At one large plant storage basins that originally had capacity to hold the water accumulated by several days of ordinary stream flow have been so filled that they can not now hold even the flow of a single night. When the filling first began to cause serious trouble efforts were made to keep the material dredged out with centrifugal pumps, but the cost was soon found to be prohibitive. It has been necessary to maintain this dredging plant, however, in order to keep the channel to the wheel pits open.

At one dam where two years before, when the dam was first closed, there was a depth of 28 feet, an island had recently appeared. At another place, where a high dam had been built on a small stream, the pond has been so filled that its storage capacity has all been lost. At another place a pond in the Spartanburg region, cleared out by the bursting of a dam in the flood of June, 1903, and since rebuilt, was in 1905 again almost full of sand and silt. A pond about 4 miles long and 40 feet deep at the lower end was in four years entirely filled in its upper part and near the dam was three-fourths full. Another pond, seven-eighths of a mile long and 35 feet deep, filled within 300 feet of the intake gates in less than two years.

On one river eight dams were built within a few miles of one another. The upper pond soon filled with sand and clay; then the second; then the third; soon all the rest will be filled. The mills can then utilize only the unregulated stream flow. Auxiliary storage reservoirs built on the stream above the power plant have filled completely with sand and silt and are useless.

Wide ponds fill most rapidly because the current through them has less velocity. In other words, they act as settling basins. Long, narrow ponds fill less rapidly because of the greater velocity of current through them, and they scour out more during floods.

All the power plants recently built practically ignore storage and depend on the flow of the stream only. When sluicing out is attempted the stream cuts merely a vertical walled channel down through the material and removes only enough for a channel way. The expense of dredging has been found to be prohibitive. The dredge merely keeps open the way to the wheel pits.

Another result of erosion is the greater frequency, greater suddenness, and greater height of the floods, since the run-off from bare washed slopes is more rapid than from wooded slopes. These floods endanger dams and property of all kinds within their reach. During a single year recently the flood loss in these southern mountains reached some $18,000,000, and during the succeeding year the loss was about $9,000,000 more.

Still another result is a change in the normal stream flow. When so large a proportion of the rainfall rushes at once into the streams and is carried away, a correspondingly smaller proportion is left to soak into the ground and feed the springs that sustain the stream flow during periods of no rainfall. The flow of the springs is weakened, and low water in the streams follows more rapidly after floods than it formerly did. The period of high water is shortened and made abnormally high; the period of low water is correspondingly lengthened and made abnormally low.

NATURE, EFFECTS, AND REMEDIES OF EROSION.

PROCESSES.

Erosion may conveniently be regarded as of two kinds. One kind occurs along streams as a result of either normal or flood flow and may be called stream erosion; the other kind occurs over all other parts of drainage basins and may be called general surface erosion.

In the humid regions stream erosion results from the activities of the perennial streams, large and small, and hence is a continuous process, though its effects vary greatly in intensity, being at a minimum during low-water stages of the streams and at a maximum during flood periods. Its work is restricted to that portion of the basin within reach of the flood waters, or, in other words, to the flood plain. Stream erosion involves the transportation of the eroded material and the deposition of much of it along the reaches of lower slope and lessened velocity. During a single great flood a stream may first cut great runways across its flood plain or dig great holes in it, and then, in later stages, deposit upon it a thick covering of sand and bowlders.

General surface erosion results from the complex interaction of a number of forces, chief among which are the various atmospheric agencies, producing rock disintegration and decomposition and preparing the material for removal, and the more active effect of the rainfall striking the ground. The steepness of the slope, the nature of the soil, and the forested or cleared condition of the surface are also direct factors. In the region under consideration the actual erosion or removal of surface materials is due almost entirely to the work of the rain as it gathers into rills and rushes down the surface slopes to join the permanent streams. General surface erosion is therefore most active in periods of rainfall, and might be regarded as an intermittent rather than as a continuous process. This view, of course, disregards the erosive activity of the wind, since it is unimportant in a region like the southern Appalachians, and the slow but more important creep of the surface material down the slopes, due to freezing and thawing and to the action of plants and animals.

During heavy rainfall a grass sod will absorb less water than a good humus cover in a forested area and run-off from the grass land will therefore be greater than from the forest, and flood heights in areas left in grass must show a corresponding increase.

EROSION IN FORESTED AREAS.

In forested areas general surface erosion is at a minimum. The force of the blow struck by the raindrops is lessened by the forest cover, and, indeed, very many of the drops do not strike the ground at all, but hit the litter of fallen leaves and twigs or the mat of moss or other low vegetation that completely carpets and conceals the soil surface. The forest soil is

72091°—No. 72—11——2

thus effectively protected from the blows of the raindrops that would cause the loose soil on the bare surface of a freshly plowed field to melt away like snow before a summer sun.

As the fallen drops gather to flow down the slope they are checked in their movement by the leaves and litter or they are absorbed by the spongy mass of soft moss or turf, so that very much of the rainfall flows slowly and gently down beneath the actual surface and within or through the matting of leaves and moss. The movement is so gentle that scarcely any soil particles are removed, and so slow that for hours after a rain has passed the soil covering is filled with water and even days later is still moist.

This protective effect of the forest cover was observed by the writer many times while he was at work in the region. Forested surfaces showed no signs of erosion; bare gullies or other indications of general surface or sheet wash were notably absent. The bottoms of the gullies in many forested areas were covered with leaves, soft vegetable mold, or moss, and the vegetation growing in them showed that though some material was being removed along their immediate courses, this removal was proceeding at an exceedingly slow rate. A small gully under these conditions is evidently the result of long years of erosive activity. Indeed, under conditions of a stable forest cover the gullies or channels for the removal of surface rainfall doubtless slowly assume a slope and width that are in adjustment with the usual intensity of rainfall, and when this adjustment is once established the woodland rain-water channels become stable and would change exceedingly little, perhaps, even in hundreds of years, were it not for an occasional cloud burst or storm of extraordinary intensity which furnishes them a volume of surface water with which their cross section and slope are not commensurate. At such times erosion takes place along their courses and holes may be gouged out that during succeeding years once more become clad with vegetation.

In the course of the long geologic ages that these southern Appalachians have been subjected to atmospheric erosion there has come about, in all but a very small part of the area, such an adjustment between the steepness of the forested slope and the average rainfall and other erosive agencies that the wasting of the slope is reduced to a minimum. In places where the surface may have once had excessive slope erosion was rapid and the slope was gradually reduced; with lowered slope erosive power decreased until the two were brought into harmony. The material prepared by disintegration and decomposition for erosion then became just equal to the removing capacity of the erosive agents, and somewhat later, with further decrease of slope and of erosive power, more loose material was furnished than could be removed and it began to accumulate and form a soil covering. This soil covering has continued to increase in most of the area until now it is many feet deep over much of the region and mantles and conceals the underlying hard rock from which it has been derived.

This long-continued process of adjustment between surface slope and erosive action has given these mountains the rounded contours and softened outlines that characterize the southern Appalachian landscape and constitute one of its chief charms. Jagged ridge crests have long since become rounded domes and vertical cliff faces have given place to the gently flowing outlines of graded, forest-clad slopes. Cliffs are rare and are due to some rapid change in rock texture or structure or to a situation especially favorable for stream erosion.

On these graded forested slopes the removal of the soil cover of disintegrated material is perhaps due more to soil creep than to direct surface erosion, and it is probable also that under such conditions soil creep reaches its maximum of efficiency as an agent of general surface reduction. Movement of the soil layer due to such creep is necessarily slow, does not result in producing bare earth or rock surfaces and is not attended with danger to the surface, the forests, or the future industrial development of the region.

Streams that flow from such forest-clad graded slopes show certain well-defined characteristics. During rains they rise more slowly than similar streams in cleared areas; they also continue longer in flood and fall to normal stages again more slowly because of the retardation of the surface run-off by the leaves and litter of the forest.

Such streams, even when highest, are, as a rule, only slightly discolored, and this discoloration is due largely to macerated leaf fragments and decaying organic matter and only to a very slight extent to soil particles held in suspension. Some of the streams flowing from forested regions are perfectly clear, even when swollen by continued rains far beyond their normal size. Such, for example, are many of the streams in the Toxaway region, the headwaters of Pigeon River, the numerous small streams flowing from the Balsam and Great Smoky Mountains, the headwaters of Cane River flowing north from Mount Mitchell, and others that might be mentioned whose basins are unbroken forests.

Streams of this class carry almost no load of waste from the general surface of their basins and so are free to expend their energies in eroding materials from their bed and banks. Hence over all the territory examined such streams were generally found to have scoured their channels between steep and usually high banks, to be flowing over beds of cobbles and bowlders or across ledges of hard rock, and to have gouged out deep and quiet pools here and there in softer rock. Such stream channels change slowly. They are in stable equilibrium—an equilibrium or adjustment which would remain undisturbed for ages, during which the region would continue to waste away as slowly as in the past, were it not for the work of man in clearing the land and throwing out of equilibrium both the surface slopes and the stream gradients.

The removal of the forest brings the slopes that represent the adjustment of a forested surface to erosive agencies into unstable equilibrium, active erosion begins, the soil accumulated, it may be through centuries, is quickly swept down into the valleys, and the mountain side is left a scarred waste of bare rocks and bowlders. For such areas the only safety is the retention of the forest now there, for once removed neither the soil nor the forest cover can be replaced.

Bare rock surfaces are more numerous in granite or granite gneiss areas along the eastern face or scarp of the Blue Ridge than elsewhere in the southern Appalachians. Such, for example, are the surfaces of Table Rock, S. C., Mount Whiteside (Pl. II, p. 16), and numerous other bare, steeply domed or precipitous rock scarps in the Highland-Toxaway region, in the Hickory Nut Gap region and at other places northward. Rock surfaces of the same type are also found in the upper Davidson River country and at a few places elsewhere in the region west of the Blue Ridge. These granitic rocks weather spheroidally, and wherever erosion produces very steep slopes the surface soil and underlying rotten rock tend to shell off and leave a bare domed or precipitous granite surface. In these particular areas, where this tendency is so strong that nature has not been able wholly to counteract it, it would be especially regrettable should man cut away the forests which, even when unhindered, are scarcely able to hold a soil covering, and thus permit the few feet of soil now overlying the granite or gneiss to be swept away, as it would be swept away in a few years after extensive clearing, down to the bare rock.

The effectiveness of such a thin root-matted soil layer in protecting steep surfaces from erosion may be observed when its continuity is once broken by the clearing of a field, the cutting of a road, or even the upturning of a tree. On the lower side of the break the turfed soil layer—it may be a foot or even less in thickness—may be seen to slip off bodily in avalanche fashion or to roll back from the bare granite like the bark when stripped from a tree.

Fortunately for the region, this spheroidal type of weathering, with its peculiarly disastrous consequences when erosion becomes active, is restricted almost entirely to granitic and gneissic rocks and even in them is not found in all types and under all conditions, so that a large part of the mountain region, where such rocks either are not found or do not possess the requisite texture or mineralogical character, is free from that particular danger.

Another type of forested area in which the soil cover is thin occurs in the belt of steeply tilted and folded quartzite and slate rocks found chiefly along the North Carolina-Tennessee line. These rocks disintegrate so slowly that a soil mantle sufficiently thick to conceal the rock nowhere forms, but the meager residual material lodges in the chance crevices and irregularities of the surface, between which bare jagged ledges project. On such areas the growth of trees and other vegetation is too scant to aid greatly in soil development. The projecting ledges, as well as the poverty of the soil itself, render the land useless for agriculture, and it can be utilized only for the growth of such timber as it can support. The undergrowth is sparse, and during the summer the surface becomes very dry. These areas need guarding from forest fires and protection from deforestation.

Though forest tracts in the southern Appalachians are, under natural conditions, subjected to relatively little erosion, yet man's activities are causing destructive erosion. Much forest destruction may be caused by gullying attended by undercutting and caving (see pp. 19–20) starting in cleared areas and extending upward into higher forested areas. Where the slopes are too steep to be cleared, many of the more gently rounded crests are cleared as far down as possible, and the surface drainage from these ridge-crest clearings, gathering at the lower edges of the fields into a few channels, rushes down across the steep forested slopes below, excavating deep channels where otherwise the surface would remain unbroken.

A much more common and far more destructive cause of erosion in steep, forested slopes is the dragging of logs along the ground in lumbering. The logs quickly wear in the soft earth smooth grooves or trenches, which during rain are converted into chutes, down which the water rushes with great velocity, sweeping rocks and soil before it. The effects of this log dragging may be seen in the many runways opened through the forests and in the great amount of soil and loose rock swept down the slope and deposited as cones at the lower ends of the chutes on little flood plains or narrow stream channels. The principal injury is not usually confined to the lumbered area itself, though this becomes furrowed wherever logs have been dragged, but extends to the lands and property along the streams below, where the floods are increased in height and in destructive effect. These dragways exist wherever lumbering has been carried on, for few of the lumbermen have taken precaution to prevent this danger. Caring little for the land after the timber is removed, they feel neither moral nor legal responsibility for the condition of the lands and other property along the stream below.

EROSION ON CLEARED SLOPES.

The clearing of graded forest-clad surfaces destroys the slowly reached adjustment between slope and erosive power and gives great impetus to erosion. In this erosion of cleared slopes the influence of the geology on the rate and the results of erosion becomes plainly apparent.

Other things being equal, the extent and character of the general surface erosion observed varies with the geology. It may be said that in a large measure geology controls erosion, for the nature of the rock determines the character of the soil, and variations in the character of the soil determine corresponding variations in the liability, manner, and rate of erosion. Two areas that are equally steep may when cleared differ greatly in character and rate of erosion. One may not erode at all; the other may erode rapidly down to the barren subsoil or the bare rocks.

In some parts of the region the soil on even the steepest cleared slopes was observed to be scarcely at all affected by erosion, although tilled year after year, and inspection usually verified the truth of the general reply to questions concerning erosion asked in such regions—"Our mountain-side fields do not wash away." It was found that such soils were permanently more or less loose and porous, and that their resistance to surface erosion varied directly with their porosity or permeability to rain water. A large proportion of the rainfall immediately soaked into the ground and found its way downward through underground channels; very little flowed off over the surface. Generally such resistant soils were also characterized by many small undisintegrated rock fragments, which aided both in keeping the soil open-textured and in checking the velocity of the downward-moving surface portion of the rainfall by opposing themselves as obstacles to its flow and causing it to drop many of the finer soil particles it might have started to carry. Certain schists and gneisses, especially the more siliceous types with mica and hornblende and those of nonhomogeneous grain, were the rocks found to weather most commonly into erosion-resisting soils. Some quartzites were also found to produce open-textured stony soils that did not erode badly. In general it may be said that a porous or a stony soil, whatever the type of rock from which it has originated, will be apt to resist general surface erosion more or less effectively. Such soils may be cleared and cultivated with safety, even when they slope at an angle that would be quickly destructive to all other types of upland soil examined in the region.

The aggregate area of such erosion-resisting soils, however, is not great when compared with the total area of the mountain country. By far the larger part of the region is underlain by rocks that weather into soil that is easily eroded when exposed on deforested slopes.

In some places it was found that the entire surface wore away slowly, each heavy rain removing a thin layer or sheet of material, so that the fertile soil layer gradually wore thin and poor and the field was at last abandoned as worn out. Erosion of this type, which may be called sheet-wash, occurs characteristically on close-grained, compact clay soils whose particles cling together firmly and do not readily yield to corrasion by flowing water, so that channels are not readily cut beneath the surface and undercutting and caving are not possible. Soils of this kind usually result from the weathering of the finer grained and more basic varieties of igneous rocks, such as diorites, diabase, and gabbro, and also from the weathering of fine clay shales that are free from mica particles. Sheet-wash erosion is so slow and gradual that some farmers fail to recognize it and believe that their soils have deteriorated through exhaustion of the fertility, whereas they have slowly and almost imperceptibly worn away to the subsoil. Such farmers, ignorant of the process by which ruin has been wrought, clear other fields and start anew the same destructive process.

One of the most common types of general surface erosion is that which is characterized by the formation of parallel gullies, and which occurs on slopes covered with clay soils, homogeneous in texture and somewhat softer and more loamy than those described in the preceding paragraph. Such soils result from the weathering of sandy shales or, more commonly, of granite or other crystalline rocks which contain considerable quartz but no mica. Certain schists also produce soils of the same texture. (See Pls. III, *A*, and XVII, *A*, pp. 18, 78.) In such soils erosion begins by producing innumerable small parallel gullies that extend straight up and down the slope and divide the surface into a minute fluting of sharp grooves and ridges. As these gullies deepen they become wider, and the smaller ones are encroached upon and obliterated by their larger, more active neighbors until, instead of a dozen or more in a yard's width, their number is reduced to one in a yard or one in a number of yards. Their bottoms are sharp, their sides steep and convex, and the tops of the divides between adjacent gullies, at first jagged, afterwards become rounded and sharply or broadly convex, the differences depending on differences in the soil texture and in the rate of vertical downcutting along the gully ways. (See Pl. XVII, *B*, p. 78.) Such systems of gullies concentrate the surface rainfall so that it works very efficiently as an agent of corrasion, and the process once started is very difficult to stop. The fields attacked are soon abandoned and left as scarred wastes while new ones are cleared.

* * * * * * *

Of all forms of general surface erosion, gullying, characterized by rapid downcutting accompanied by undercutting and caving, is most rapid in its progress and most difficult to check, as well as most spectacular in its appearance and most destructive in its effects. Erosion of this type occurs in soils underlain by a deep, relatively soft micaceous subsoil. Deeply decomposed material of this kind results from the weathering of micaceous and feldspathic schists—rocks that cover considerable areas in the region and determine the type of surface erosion wherever they occur. The surface in such areas may have weathered to a fairly compact clay soil that offers moderate resistance to erosion, and if the slope is gentle careful attention may keep this clay surface intact; if it is once broken through, however, a gully quickly develops in the soft subsoil, deepens easily by corrasion, and the soft and arkosic material on the sides is rapidly undermined and slips or caves in, leaving vertical or overhanging walls. The head of such a gully is commonly more or less amphitheater-shaped and is rapidly extended up the slope by headwater erosion, in which caving plays a prominent part. Into these gullies many square yards of surface soil may cave during a single heavy rain, and as the decomposed micaceous and arkosic material in which such gullies grow is usually scores of feet in thickness they may become chasms of great depth and width. This type, which may be called chasm gullying, is illustrated in Plate III, *B*, page 18.

When begun in cleared land, chasm gullying may advance upward into a higher forested area and undermine and destroy even the largest trees. Once well started it is almost impossible to check.

Still another form of erosion in cleared areas originates in small landslides, which usually occur after the soil has been thoroughly saturated and softened during a period of prolonged rainfall.

On steep slopes the surface soil may begin moving of its own accord, either slipping bodily off in a mass or becoming so soft as to flow off, when once started, as a stream of soft mud. This change of soil into flowing mud is likely to take place where a small wet-weather spring on the steep hillside happens to occur in a certain kind of soil. More frequently, however, landslides are caused by the trampling of cattle on steep slopes after the soil has been softened by prolonged rain. One cow climbing up or down such a slope may cause a number of such landslides, each of which, when once started, usually grows both in width and in length until it becomes a great bare scar in the field.

Landslides are likely to occur where the soil is micaceous or where it has resulted from the weathering of slickensided granitic or gneissose rocks. The slickensided surfaces are preserved in the weathered soil and when wet become planes of easy movement. Such surfaces aid greatly in the extensive caving that occurs in some areas where chasm gullying is the characteristic form of erosion. Landslides are also likely to occur where weathering of the exfoliation type has produced a sharp line or division plane between the soil cover and the underlying hard rock. In such places the soil layer easily slips off. Landslides of this type may occur in the forest, but, like those of the other types, are more common in cleared areas. During prolonged wet seasons it would be well to keep cattle or other animals off steep slopes that show any tendency to slide.

Level or nearly level upland areas may in themselves be comparatively safe from erosion, but they may nevertheless be affected by intimate relation with erosion, for water from areas lying still higher may erode channels across them, and the waste from such higher areas may cover their surface soil and destroy its fertility; or their own drainage, after being gathered into a few channels, may be shed upon still lower lands with the same effect as that produced by water received by them from the higher areas.

EROSION ON FLOOD PLAINS.

The typical mountain stream of this region flows so swiftly in its upper course that it sweeps away all loose material in its path and flows over bare rock, which it is actively eroding. Somewhat farther downstream, as its slope and consequently its velocity decreases, the stream reaches a point where, though still swift, it can no longer sweep away all loose material, so that the coarsest material begins to accumulate, making what may be called a bowldery or a cobbly torrent plain. As the current varies in velocity at flood and at low-water stages, this material is continually being reworked, and in the course of the cut and fill characterizing this reworking the stream, largely because of the heterogeneous nature and different sizes of the material, constantly shifts its position as it tears to pieces and rebuilds its torrent plain. The surface of such a plain fashioned by an impetuous current of greatly varying power is characteristically irregular or hummocky.

Farther downstream, as the velocity of the current decreases, the coarseness of the material composing the torrent plain and the irregularity of its surface correspondingly decrease. The tendency to lateral shifting of the channel also diminishes, but, if from any cause the stream be overloaded with waste and forced to deposit much of the waste in its channel, this channel filling deflects the current and the tendency to undercut one bank and build a cobbly or sandy waste zone on the other becomes strong. An example of such filling and cutting may be seen in Plate IV, *B*, page 20.

Still farther downstream the velocity of the current diminishes until the stream carries and builds into its flood plain only fine material, and the flood-plain surface then becomes a level surface, which still lower may build in the familiar way a higher portion parallel and near to the stream.

If the stream cuts down as it swings laterally, it develops on the side away from which it swings a sloping flood plain, which will here be called a beveled or hanging flood plain, the exact slope of which depends on the ratio of down cutting to lateral swing. Only the lower edge of such a flood plain is ever covered and damaged by floods.

If the stream is not overloaded with waste it normally develops a channel whose width and depth are proportioned to its volume, velocity, and flood height, and this channel usually has a sufficient capacity not only to carry away the ordinary flow of the stream, but also to carry a considerable volume of flood waters. With increase of flood volume in excess of the capacity of the channel the waters spread beyond the banks and flood an additional area on which deposits are laid down and a flood plain is built.

The inspection of many flood plains shows that they may generally be divided into two classes, the forms of one class commonly differing from those of the other in width and invariably in position and height and in the frequency with which they are covered with floods. The stream channel is usually bordered on one or both sides by a more or less narrow shelf cut in the flood-plain material and rising only a few feet above ordinary stream level. The exact height and width of this flood-plain shelf vary with the size and character of the stream. This shelf, or notch, is covered by the ordinary floods of the stream and may be called the ordinary flood plain. It may be only a rod or a few rods in width, even on a large river, or it may be much wider.

On the side away from the stream this ordinary flood-plain shelf is bordered by a scarp that is usually regular and well defined and separates a higher part of the flood plain from the lower one. The height of the scarp or the difference in level between the two flood plains varies greatly on different streams and even on different portions of the same stream, but on all streams this higher flood plain is reached and covered only at comparatively long intervals—measured perhaps by years—by floods of extraordinary height. The floods that now and then cover this extraordinary flood plain deposit sediment on it in some places and so build it up somewhat; but more commonly they scour runways across it and thus start subaerial erosion lines along which it is ultimately cut to pieces. Because of these scourways its surface is usually uneven. Deposition during the rare flood periods does not equal erosion during the long intervals between floods, and the plain soon shows signs of wasting away.

On many streams there are to be found above the level of this extraordinary flood plain more or less perfectly preserved portions of other, still older flood plains or terraces, whose height places them above the reach of even the highest floods. These may be called fossil flood plains to distinguish them from the ordinary and extraordinary flood plains, which may be spoken of as living flood plains. Such topographic features are so common and in many places so prominent along the streams, large and small, of the southern Appalachians that it has been thought best to define the terms, as their use in this paper will give conciseness to the description of flood plains. A view of an extraordinary flood plain is shown in Plate IV, *A*; fossil flood plains are seen in Plate IV, *A*, and Plate V, page 20.

The ordinary flood plain may have runways cut across it by floods or may be cut or gouged into holes wherever the flood current becomes deflected downward, as by pouring over a rock wall or other obstruction, or by forcing its way under logs that have lodged on it.

In other places the stream may be overloaded with waste and forced to drop a part of it until it fills its channel. It then flows here and there in irregular shifting meanders across its flood plain, depositing great quantities of waste and building up or aggrading the flood plain (see Pl. VI, *A*, p. 22). Such flood plains rapidly become useless for agriculture.

Elsewhere the flood may strip off the alluvial soil down to the underlying bowlders or the bare rock (Pl. VI, *B*), or it may be checked so as to deposit a bed of sand on the former flood-plain surface and convert it into a barren sand waste (Pl. VII, *A*). Under other conditions the covering spread over the flood plain may be of cobbles or even bowlders.

The channel of a stream that is not overloaded retains its normal depth, and the great increase in its depth during floods gives a channel prism so large that the stream can remove a much greater volume of water; moreover, the speed of the current, which is greatly accelerated by the increase in depth, is still so much slower on the outer edges than in the middle of the stream that it deposits on the flood plain the finer material that was carried in suspension and thus benefits instead of injures the adjacent lands over which it flows. The Chattooga, for example, which has an unusually swift current, its average slope being 28 feet per mile, frequently floods its bottoms at Russells, but instead of cutting them to pieces deposits sediment that enriches them. The upper Yadkin is another good example. It is enlarging its channel by scouring away its banks in order to handle more efficiently the steadily increasing volume of flood water it is now receiving.

The cobble zones scoured out along the main channels of many of the streams examined are the results of the effort of the streams to adjust themselves to changed conditions by providing themselves with auxiliary channels to carry off flood water. Where man's selfishness or shortsightedness or an unusual combination of circumstances make these efforts of the streams inadequate, disaster results. Where man has increased erosion so as to overburden the streams with waste and to force them to fill their channels he has foredoomed his lands to be cut to pieces by floods. Wherever from any cause flood heights have increased, fences, bridges, dams, mills and all other property, as well as human and other lives within reach of the water, are endangered and liable to be destroyed, as they have so often been destroyed in the southern Appalachians in the last decade.

On the other hand, in regions such as the basins of the Holston, the New, and the Monongahela, where man's agency in clearing the soil has not greatly accelerated the erosion of upland slopes, and so has not so increased the supply of waste furnished to the streams from these slopes as to fill the stream channels, but has chiefly hastened the movement of storm waters into the streams and so augmented flood heights, the damage wrought by the streams will consist not so much in tearing lands to pieces as in sweeping away all classes of property within reach of the flood. Where stream channels have been filled by the progress of erosion, as in most of the southern Appalachians, the disaster includes lands as well as property and life.

REMEDIES FOR EROSION ON SLOPES.

The most obvious method of preventing erosion is to prevent the clearing of the steep slopes. Just how steep a slope may safely be cleared is a problem to which there is no one solution. It has often been said that no slope steeper than 15° should ever be cleared, and this is probably as good a general statement as can be made. In some places, however, under certain conditions of soil texture and rainfall, slopes less than 15° erode badly; in others slopes of 20° or even more may be cultivated without serious danger of erosion, the secret of the immunity being in the texture of the soil. In regions where slopes of more than 30° were cultivated the author was frequently told that the lands did not erode, and this statement was largely confirmed by ocular evidence.

The allowable limit of steepness for cleared lands varies, then, with local conditions, but it may be said that almost everywhere throughout the southern Appalachians this allowable limit is exceeded, and in many places so greatly exceeded that rapid erosion of the cleared slopes is inevitable.

All cleared slopes should be thoroughly terraced if they are to be cultivated. Terraces can be so constructed that little or no water will flow from the terraced fields during even the hardest rains. In fields cultivated under these conditions there is no removal of soil particles and no filling of stream channels and resultant destruction of flood-plain lands. Terracing can be far more generally practiced than it is all over the region examined and is necessary to insure a really safe system of agriculture. The best terraced region observed was in the Savannah River Basin, between Toccoa and Elberton, Ga. Good terracing was also noted about Greenville and elsewhere on the Piedmont in upper South Carolina. An example of crude but effective terracing is illustrated in Plate VII, *B* (p. 22).

Grass will effectually prevent erosion after the sod has once become well established. The farther south one goes, however, in the southern Appalachians, the more difficult it is to obtain and keep a good sod. In the Watauga and New River basins it is easy to protect the lands by grass, but on the Chattahoochee, in northern Georgia, it would be much more difficult, because the elevation there is much lower and the climate is therefore dryer and hotter and less favorable to the growth of grass.

If erosion has already occurred on steep slopes one of the best remedies, if conditions have not become too bad, is to cover the surface with straw, leaves, or forest litter, and over this cover to place brush to hold it in place. This mechanically protects the soil, much as does the humus cover in the forest itself, and holds it until plant growth has again taken possession and arrested rapid erosion. Except in a few places where the soil is exceedingly poor, weeds and briars will now spring up rapidly and will be followed in a short while by wind-sown tree seedlings that will start the new forest growth on the denuded slope. Natural reforestation in the southern Appalachians is usually rapid, but if for any reason it is desired to hasten it or to control more closely the kind of growth, artificial reforestation is easy, though of course this is more expensive than natural reforestation.

Where deep gullies have once started log or rock dams may be used to prevent their further growth and to catch material with which to fill them. The use of such dams is not always advisable, since the water tends to work beneath them and thus to concentrate and increase the vertical scour. Moreover, they become in themselves mechanical obstructions that may interfere with the subsequent use of the land for agriculture. If used they should be faced on the upper side with brush or straw to prevent the water from working through them. If thus made tight they soon become filled with the material that is being carried down the hillside, and the former gully or inclined trench is converted into a series of terraced steps. For a view of brush dams used to check erosion see Plate VIII, *A* (p. 24).

In places where the slope is not too great and the cutting has been shallow, gullies may be filled with straw so as to catch the sand and clay that is being removed. This device affords an effective remedy for gullies and has the added virtue of not leaving mechanical incumbrances on the reclaimed area.

Erosion may be checked by preventing the fires that usually follow in the wake of the lumberman and by guarding against the formation of erosion chutes by the dragging of logs down steep slopes. The mere removal of the mature forest itself need in no wise affect erosion or flood problems in a region like the southern Appalachians, where, if unhindered, the natural reproduction of the forest is rapid. In a few years the new growth makes an even denser cover of trunks, branches, and root mattings than the old, and protects the soil at least as effectually.

REMEDIES FOR EROSION ON FLOOD PLAINS.

On flood-plain areas the most important thing that can be done to the area itself to prevent erosion and flood damages is to maintain by all possible means a deep, straight channel for the rapid removal of flood waters and waste. The removal of obstructions and the straightening of the channels of small streams aid greatly in scouring out and improving the channels and hastening the run-off. Means of regulating and confining the shifting channel of a wild mountain torrent are widely used in Europe, but practically unknown in America. Retaining walls of stone or rock-filled cribs serve this purpose wherever the interstices may become filled with fine material and so closed to the free passage of water through the wall. Where such filling is impracticable a facing of sheet piling may answer the purpose. A rough rock-filled log crib used for restraining such a stream is illustrated in Plate VIII, *B*. In Europe the bottom and sides of some such channels are paved with stone, and dams or sills of rock are built at short intervals across the stream.

Another protection to banks that are being cut by streams that do not tend to shift so rapidly as torrents is to plant willow, aspen, balm of Gilead, or other easily propagated trees of rapid growth along the banks and over the entire denuded area. In starting such growth it is usually sufficient to plant small branches or sections of limbs at short intervals in the moist earth; they root readily and grow rapidly. The balm of Gilead throws up many shoots from its ramifying rootlets and soon makes an efficient bank protection.

In places protection may be had by building levees, but these are so expensive that their construction is warranted only when the lands or property to be protected are of considerable area and value. Lands that have been badly cut to pieces by floods, such as those illustrated in Plate IX, *A*, or deeply buried beneath sand (see Pl. VII, *A*, p. 22) or beneath gravel or cobbles, can usually be put to little or no immediate use. There is more ultimate hope of reclaiming lands that have been cut to pieces than those that have been buried beneath sand or stone, since later floods of less height may deposit material in the eroded places and in time may heal the scars and restore to use the once ruined land. A flood channel that is once established, however, is likely to be occupied by subsequent floods and to become a permanent flood runway. Planting trees and building walls across it may protect it and permit it to fill, but if the surface has been covered by stony or sandy waste there is little hope that it can soon be recovered.

In most places in the southern Appalachians both cutting and filling tend rather to grow worse than to improve by subsequent changes, since they have been started by floods whose tendency for a number of years has been to become gradually worse, because of the constant extension of clearing and of erosion on slopes that should have remained forested.

Now and then it is suggested that impounding reservoirs may be built with capacity sufficient to hold the flood waters, but this scheme is generally impracticable because of the great size and the cost of the necessary dams and the large area of land that would be submerged and rendered useless, as well as the danger that the dams might break and destroy life and property below. Furthermore, on many of our southern streams such reservoirs would become filled with sand and silt in a decade or two, as a number of private power dams in the Piedmont region of the South already have been filled.

Far the most efficient reservoir for impounding flood waters is a good forest humus over the steep slopes of the stream basins. The only thorough means to check and prevent the destruction now going on in the Appalachian region from erosion, floods, and droughts is to maintain such a humus cover where it already exists and to replace it where it has been destroyed.

EROSION AT DUCKTOWN, TENN.

* * * * * * *

Ducktown, Tenn., located in the southeastern part of the Hiwassee basin, is situated on an elevated and now deeply dissected plateau, presumably of the same age as the Asheville Plateau, and is walled in on almost all sides by mountain ridges. Copper is mined and smelted here, and in the immediate vicinity of the smelters all vegetation has been killed by sulphuric acid fumes. The region is peculiarly adapted for the study of erosion problems, since it affords an extreme example of the limits to which erosion may go in this climate when all protective vegetation has been removed. One smelter began operations about 12 years before the region was examined; another had started only 4 years previously; the complete destruction of vegetation had taken place within the latter period. During this short time the second-growth timber and the bushes and grass beneath it, which had covered most of the area, had been killed, and the dead trees had been removed for firewood; but though the ground is in many places still covered with small branches and twigs, the litter-covered surface is already cut to pieces by erosion.

The erosion starts near the bottom of a slope, and where the soil is porous rapidly cuts a steep-sided gully to a depth of 5 to 12 feet below the surface, where the underlying schist is as a rule still measurably firm. After a gully has reached its limit in depth it widens until its walls coalesce with the walls of adjacent gullies, by which time most of the soil has been removed. Where the soil is a more impervious clay, erosion begins, likewise, at the foot of a slope, and eats out amphitheater-like areas such as are shown on Plate XVI, opposite. Like the deep-grooved gullies, they rapidly grow headward until they reach the top of the hill and completely denude it.

* * * * * * *

The quantity of waste furnished by these bare slopes, being too large for the streams to remove, rapidly accumulates along the stream courses as flood or waste plains, which soon extend up even to the foot of the slopes at the head of the streamlets.

On Potato Creek this waste has been accumulating for a number of years at the rate of a foot or more each year, and has been built into a flood plain from 100 to 300 yards wide, in which telephone poles have been buried almost to their cross-arms, and highway bridges, road-beds, and trestles have either been buried by the débris or have been carried away by floods. At Isabella smelter recent floods have swept through the store and other houses, and the waters have risen to the level of the furnaces. This increase in the height of floods is due largely to the rapid building up of the flood plain, and extensive diking or other protective measure will soon be necessary to prevent serious damage to the smelter and other property.

Such great quantities of sand are carried into Ocoee River by each large flood as to prevent the running of the two ferries at the smelter until the river has had time to scour its channel clear again. Much sand has accumulated on the flood plain of the Ocoee in the few miles just above the river's entrance into the gorge below Ducktown.

This abnormal denudation and erosion has also affected the underground water level in the region. During the last few years wells have been going dry, and a number of springs, some of which supply water to the miners' families, flow less than formerly. This lowering of the ground-water level and decreased flow of springs can not be attributed to drainage effected by the deep mining shafts, for the mines are nearly dry. The normal flow of Potato Creek is said to be only about half as large as it used to be, and there can be no question that a much larger part of the rainfall now finds its way immediately into this stream and is carried off in floods, leaving a much smaller part to soak into the ground to supply the wells, springs, and streams during periods of dry weather.

Erosion about Ducktown to-day is limited only by the rainfall, the steepness of the slopes, and nature of the soil. The destruction of the soil on the slopes by erosion and of all of the alluvial land along the streams by flood-plain building is complete, and over an area comprising several square miles the entire country—flood plain and valley slope alike—has become a barren waste. The Ducktown region is, then, not only an impressive object lesson, but an emphatic warning of the extent and character of the disaster that may result in these southern mountains from the thorough destruction of the forests.

* * * * * * *

The result of erosion in the Ducktown region may be looked upon as the goal to which all erosion in the southern Appalachians is tending. If elsewhere in these southern mountains the cover of vegetation be removed as completely as at Ducktown, the rapid erosion that has wrought the destruction seen there may confidently be expected to follow, and with like results.

CHANGES IN STREAM REGIMEN.

Wherever mountain stream basins in the southern Appalachians have been extensively cleared the regimen or normal habit of flow of the streams has been changed. This change is believed to be due chiefly to increased erosion and consequent increased rapidity of run-off.

In the natural forested state comparatively little rock waste is furnished to the mountain streams, which therefore expend their energies largely in eroding their beds and cutting deep channels, such as may be seen in the Great Smokies, much of the Pisgah Range, or any other well-wooded part of the mountains. These deep, steep-sided stream channels are very effective in removing flood waters, and it takes considerable rain to fill them higher than their banks and produce a flood. They are amply able, even in flood, to carry away at once all the waste material furnished to them.

The removal of the forest on steep slopes generally increases the tendency to erosion. This increase may be very slight if the land is kept well sodded or if the soil is of a certain porous or stony type, but in a region like the southern Appalachians, with its deep soil and abundant, often torrential, rainfall, erosion is generally more rapid—it may be very much more rapid—on cleared than on forested slopes.

Erosion once begun, as a rule, soon develops gullies that furnish so much sand, clay, and cobble to the streams that they become overloaded and are unable to carry away all the waste that is brought to them. The excess waste is therefore deposited first in the channel, until that is practically filled, and then over the alluvial flood plain, which is thus converted into a barren waste of sand or loose stones. Such a sand-covered flood plain is illustrated in Plate VII, *A* (p. 22). The waste then begins working downstream, filling dams and pools as it goes, and soon gets down into the navigable parts of the great river systems, such as the Tennessee, making more difficult the problem of maintaining navigable channels. This condition may be readily understood by examining the Tennessee at low water and studying the growth of its sand, gravel, and bowlder bars and of its towheads and islands, and also by comparing the detailed reports made annually by the Army engineers concerning the improvements made by them in the open channel of the river. The knowledge gained by personal examination of the river adds much significance to the statements found in these reports. Some of the most notable changes on the Tennessee will be described briefly in connection with the general characteristics of this river on pages 79–83.

The waste filling a mountain stream causes an immediate change in the frequency and height of floods. When, under normal forested conditions, the channel is deep, a heavy rainfall is necessary to raise the stream to the bank-full stage; when the channel has become well-nigh filled with eroded material much less rainfall will put the stream out of its banks and cause a flood—in other words, floods will become more frequent.

The channel filling has another important effect. The same amount of rainfall will necessarily cause a higher flood when the channel is filled or partially filled with eroded material than when it was free from it, partly because the capacity of the channel to hold water is diminished, but more especially because the filled channel is not so efficient an agent for the rapid removal of flood waters. They pile up, as it were, and rise higher than before. Their height is further increased by the fact that the gullies on the eroded slopes and the bare surface of the cleared land deliver the storm waters to the streams almost as rapidly as they would flow from house tops or along city gutters. This rapid delivery of the storm water from the steep cleared slopes aids greatly in raising floods to abnormal heights. In other words, floods in steep, denuded, and eroded basins assume much the character of floods caused by cloudbursts, not only in their very rapid rise and great velocity, but also in their destructive violence and rapid decline to ordinary stages. Their height and velocity are increased and their length is diminished as compared with floods in the same stream when the stream basin was forested.

It might be thought that as rivers are but the sum or aggregate of their tributaries they will rise to the same height during floods and become correspondingly low during droughts. This, however, could be true only if stages of stream height varied synchronously on all the tributaries, and if all these tributaries were so adjusted as to position, length, and slope that they discharged into the main stream their abnormally high or low flows at the same time These conditions, however, are manifestly impossible in any large river system whose headwaters include areas that are diverse in geography and climate. Some tributaries may be in flood when others carry little water; some are long and of gentle slope, others are short and of abrupt descent. The river into which they flow represents, then, not the sum of their extremes, but more nearly their mean, and in its phenomena of flow will tend to preserve a mean.

The general variations in rainfall in all parts of any large river basin in the Appalachians are probably similar in kind though not exactly synchronous or of equal amount everywhere throughout the basin. The period of maximum rainfall, for instance, in this entire region usually occurs late in the winter or early in the spring, and the period of minimum rainfall late in the summer and autumn. The average stages of the small and large streams of the region vary in a general way in harmony with the rainfall, and just as the tributaries become, on the average, lowest late in the summer and fall, the main stream into which they flow becomes lowest then also, but this is a mean low, not an extreme low. In the same way the main stream is highest when the tributaries are highest.

It is probable that within recent years the mean low-water flow has been decreased on main streams whose tributary basins have been largely cleared. In the same way, the mean-high-water stages of these streams have probably been similarly increased by excessive clearing. By this increase in high-water stages a larger proportion of the total rainfall is immediately carried away as run-off, and consequently a smaller proportion is left to supply evaporation and maintain the flow of springs and other ground waters. Evaporation in cleared and cultivated areas is at least as great as in forested areas, and as the average annual rainfall has probably remained unaffected by the clearing of the forest, the increased run-off must constantly leave a decreased supply for the springs, which must therefore suffer partial or total failure during prolonged drought.

Occasionally it is maintained that in any large river system the flow during low-water stages is kept up by the run-off from frequent rain storms on small areas scattered here and there over the basin, and that the river is consequently not dependent on the flow of the springs or ground water for its maintenance. The failure of springs, it is argued, would consequently not materially affect the flow of the main stream of the basin, though it is admitted they would stop the flow of the streamlets to which they ordinarily give rise.

Local thunderstorms that involve considerable precipitation in small areas may, in times of general drought, contribute somewhat to the maintenance of the stream's flow, but these local storms are probably not so frequent in any particular large river basin as to maintain by themselves or even to nearly maintain the low-water flow of the main stream. In a basin like the upper Ohio, for instance, many days at a time will often elapse during prolonged dry seasons without the occurrence of enough local thunderstorms anywhere in the basin to affect materially the extreme low-water discharge of the river, and during some droughts in the Ohio basin there are so many days or even weeks together during which no rain falls anywhere within the entire basin that if the streams depended during such periods on the local thunder showers alone, the last water thus falling would have ample time to flow down the entire length of the river and out at its mouth and leave its bed entirely dry before another local shower would occur. The low-water flow of a stream during droughts depends very largely or almost entirely on springs and other ground water within its basin, and as they fail it will fail.

It has been suggested that, assuming that there has been no climatic change, and that rainfall is as heavy and thunderstorms are as frequent in forested as in deforested areas, it should be obvious that the discharge of springs plus the run-off in forested areas would exceed the run-off from thunderstorms alone in deforested areas and that of two streams the one from the forested area would show the larger flow.

* * * * * * *

The streams of the southern Appalachian Mountain region furnish many illustrations of the various phases of the change of regimen outlined in preceding pages and leave no doubt in the writer's mind that floods are now more frequent, rise higher, move downstream more rapidly, and are sooner succeeded by low water, and that this becomes lower and lasts longer than formerly. The writer contends that the belief that these changes have occurred rests on so large a body of observed facts that it can not be successfully controverted.

* * * * * * *

Increased flood heights and flood frequency are most evident on the headwaters and not down on the navigable middle and lower reaches of a large river system, and the chief destructive effect occurs both along the headwaters and at points where the streams first leave the mountains and run out upon the plains. It was in and above Pittsburg, for instance, and not down at Louisville or Cairo that the flood of 1907 on the Ohio was most destructive. It was at Elizabethton on the Watauga and at other places similarly situated on other Appalachian mountain tributaries of the Tennessee, and not down at Chattanooga, Florence, or Paducah, that the floods of 1901 and 1902 on that river were most destructive.

So far as the navigable portions of any of the larger river systems are concerned, the effects of erosion in filling stream channels with sand, gravel, and bowlders are of more importance than the increased height of floods. The increased frequency of floods on the navigable portions of streams is, however, of greater importance than increased height, since floods many feet lower than the highest floods known are able to destroy crops, and their increased frequency makes farming more hazardous on flood plains and decreases greatly the value of such lands.

There is abundant evidence on rivers such as the Tennessee of increased silting in the navigable portions of streams because of the increased erosion resulting from deforestation.

The various factors that regulate stream-flow and regimen may be classified under five heads—climate, topography, geology, vegetation, and artificial control. Since the effect of each of these factors can not as a rule be sharply distinguished, it is difficult to determine their relative magnitude, so that only their general tendency can be considered.

Many of those who have discussed the effect of forests on stream regimen have been disposed to consider one class of data or one group of conditions and to overlook others. Gage heights, for example, have been commonly used in presenting arguments for and against the effect of deforestation on stream regimen, without proper regard for the conditions that have caused the gage heights. Therefore, some of the deductions drawn from gage heights in regard to the effect of deforestation have been unwarranted and misleading.

The writer, though considering the various conditions affecting change of regimen, has based his conclusions chiefly on the record made during long ages in the flood plains, the slopes, and other features of the valley floors and sides.

Flood-plain deposits built up during long ages reveal the character of the floods by which they were formed. If the floods have been small or gentle the deposits will consist of fine alluvium; if they have been great or violent, the deposits in a region like the southern Appalachians will be coarser and will consist of sands, cobbles, or bowlders. If, then, the sands, cobbles, and bowlders that have been repeatedly strewn over their flood plains in the last decade by such rivers as the Watauga, the Doe, the Nolichucky, the French Broad, the Catawba, the Yadkin, and other southern rivers had been the kind of material those rivers had for ages been accustomed to deposit, their entire flood plains would be formed of such coarse material, instead of being composed, as they generally are, of fine sandy loam or clay. Had they at any time in the past been accustomed to carry material so coarse and built it into their flood plains, that material would be there to-day as a mute witness of the fact. Moreover, the normal change in the regimen of a river as the ages pass causes its flood-plain deposits to grow constantly finer. In these rivers, however, this process has been reversed; their deposits have recently grown coarser because there has been a recent increase in the height, velocity, and power of their floods.

This anomalous change in the regimen of these rivers is not due to any change of climate nor to any earth warping or other crustal movement. It is not due to the drainage of swamps and ponds, for these do not exist in the region. It is not due to road building, paving, or ditching, nor to the building of levees or dams or other engineering works, for such changes in the region are quantitatively insufficient to produce the results noted. It is, therefore, reasonably believed to be due to the denudation of the steep mountain slopes and their consequent erosion.

It is difficult to fix any definite period as the beginning of the change recorded, since it was more or less gradual and was not synchronous in all parts of the mountains. In general terms, however, the period extending from 1885 to 1890 may be taken as the start, for it witnessed the revival of industrial activity after the long period of exhaustion and slow recuperation that followed the Civil War. Railroad building then became more active; lumbering began to be an important industry; agriculture was stimulated, tobacco especially in some mountain sections becoming an important crop. Lumbering and clearing for agriculture have increased steadily since then, and their harmful effects began to be felt in certain areas within five to ten years. To-day reckless lumbering and careless and ignorant methods of agriculture are still potent causes of erosion.

PROBLEMS INVOLVED IN THE STUDY.

From the preceding discussion it is evident the subject presents two distinct problems, whose relative magnitude and importance differ in different parts of the region. One of these problems relates to agriculture and the other to forestry.

THE AGRICULTURAL PROBLEM.

The agricultural problem involves the selection of the areas best suited for agriculture because of fertility of soil and moderate slope of surface and the study of the ways in which such areas may best be handled to prevent their destruction through erosion and the destruction of other lands and property by the waste they yield and the floods they help to generate.

Much of the mountain area is properly agricultural land, and as the population increases more and more of this area must be brought under cultivation. This means that steeper and steeper slopes must be cleared, and that danger of erosion must increase unless improved methods of agriculture are introduced. Terracing, contour plowing and ditching, crop rotation, sodding to pasture or meadow, as well as the crops best adapted to the region, especially those most helpful in holding soil on steep slopes, should be studied, and to be of practical value this study must consider all these things as they are directly related to the specific and sometimes peculiar climatic, rainfall, soil, slope, labor, and other natural and economic conditions in the region. It can not profitably be a long-range or general study.

The study of the agricultural problem should also include a consideration of practicable methods of reclaiming eroded and abandoned lands, and of the effectiveness of brush, straw, or other filling for gullies, of brush, log, or rock dams across them, and of tree, vine, or other vegetative covering for bare areas. Such a study should also include a consideration of methods of regulating and restraining both the wild headwaters or torrent reaches, and the lower, but still rapid and easily changeable courses of the mountain streams along whose banks lie the most fertile agricultural lands of the region—lands that are now at the mercy of their uncurbed destructive activities in times of flood.

In studying these problems much could be learned of Europe, where for hundreds of years man has slowly won to agriculture area after area of steeper and steeper slope as population has pressed hard upon subsistence. Doubtless the methods employed in Europe should not be exactly followed, because of differences in climate, crop, soil, labor, and other factors, but, warned by their failures, and profiting by their achievements, we can adapt their successful methods to our own peculiar conditions. Examples of their methods of regulating mountain torrents and preventing erosion on steep slopes are shown in Plate X, opposite. The agricultural lands of the Appalachian Mountains are generally fertile, and if wisely handled will support safely and permanently a much greater population than now inhabits the region.

THE FOREST PROBLEM.

Much of the area is not properly agricultural land, and should not be cleared and forced into agricultural use, because that forcing means quick destruction both of the area itself and of the lower lying areas on the same stream ways. It means also slower, but none the less sure, interference with navigation on the more remote parts of the major stream systems.

The forester would protect steep slopes by keeping them clothed with timber, would coax back tree growth on denuded areas, keep down forest fires, protect and perpetuate the supply of hardwood, protect the game and fish, and enhance the beauty and charm of the region as a health and pleasure resort, as well as prevent the navigable streams that flow from these mountains from filling up with the sand and silt, whose removal is now costing annually large sums of money.

* * * * * * *

Reprinted from *U.S.D.A. Circular 33* (1928)

CIRCULAR No. 33 APRIL, 1928

UNITED STATES DEPARTMENT OF AGRICULTURE
WASHINGTON, D. C.

3

SOIL EROSION A NATIONAL MENACE

By H. H. Bennett, *Soil Scientist, Soil Investigations, Bureau of Chemistry and Soils,* and W. R. Chapline, *Inspector of Grazing, Branch of Research, Forest Service*

CONTENTS

PART 1. SOME ASPECTS OF THE WASTAGE CAUSED BY SOIL EROSION[1]

By H. H. Bennett

This circular is concerned chiefly with that part of erosion which exceeds the normal erosion taking place in varying degrees, usually at a slow rate, as the result of artificial disturbance of the vegetative cover and ground equilibrium chiefly through the instrumentality of man and his domestic animals. Removal of forest growth, grass and shrubs and breaking the ground surface by cultivation, the trampling of livestock, etc., accentuate erosion to a degree far beyond that taking place under average natural conditions, especially on those soils that are peculiarly susceptible to rainwash. This speeding up of the washing varies greatly from place to place, according to soil character, climatic conditions, vegetative cover, degree of slope, disturbance of the ground surface, and depletion of the absorptive organic matter in the soil under continuous clean cultivation. Under normal conditions rock decay keeps pace with soil removal in many places; under the artificial conditions referred to, soil removal by the rains exceeds the rate of natural soil formation over a vast area of cultivated lands and grazing lands, often working down to bedrock.

[1] This part discusses only the evils of erosion by rainwash. Much damage is also done by wind erosion, but this phase of the problem is not treated here. The details of checking and preventing erosion and restoring to use the recoverable areas are not included, since that important side of the problem deserves a full paper in itself.

88854°—28——1

GENERAL STATEMENT

Not less than 126,000,000,000 pounds of plant-food material is removed from the fields and pastures of the United States every year. Most of this loss is from cultivated and abandoned fields and overgrazed pastures and ranges. The value of the plant-food elements (considering only phosphorus, potash, and nitrogen) in this waste, as estimated on the basis of the chemical analyses of 389 samples of surface soil, collected throughout the United States, and the recent selling prices of the cheapest forms of fertilizer materials containing these plant nutrients, exceeds $2,000,000,000 annually. Of this amount there is evidence to indicate that at least $200,000,000 can be charged up as a tangible yearly loss to the farmers of the Nation. These calculations do not take into account the losses of lime, magnesia, and sulphur.

In this connection it must be considered that rainwash removes not only the plant-food elements but also the soil itself. The plant-food elements removed by crops (the crops do not take away the soil, but extract nutrients from it) can be restored in the form of fertilizers, manures, and soil-improving crops turned under; but the soil that is washed out of fields can not be restored, except by those exceedingly slow natural processes of soil building that require, in many instances, centuries to develop a comparatively thin layer. It would be entirely impracticable to replace even a small part of the eroded matter, which might be recoverable from stranded material not yet swept into the rivers.

A very considerable part of the wastage of erosion is obviously an immediate loss to the farmer, who in countless instances is in no economic position to stand the loss. Much of the wastage that perhaps might not be classed as an immediate farm loss is nevertheless a loss to posterity, and there are indications that our increasing population may feel acutely the evil effects of this scourge of the land, now largely unrestrained. A considerable part of the erosional débris goes to clog stream channels, to cover fertile alluvium with comparatively infertile sand and other coarse materials assorted from flood water, and to cause productive stream bottoms to become swampy and much less valuable. When the mellow topsoil is gone with its valuable humus and nitrogen, less productive, less permeable, less absorptive, and more intractable material is exposed in its place. As a rule this exposed material is the "raw" subsoil, which must be loosened, aerated, and supplied with the needed humus to put it into the condition best suited to plant growth. This rebuilding of the surface soil requires time, work, and money. In most places, this exposed material is heavier than the original soil, is stiffer, more difficult to plow, less penetrable to plant roots, less absorptive of rainfall and less retentive of that which is absorbed, and apparently its plant-food elements frequently have not been converted into available plant nutrients to anything like the degree that obtains in the displaced surface soil. This comparative inertness of the freshly exposed material is comparable to the lessened productivity brought about in some soils by suddenly plowing large quantities of the subsoil material to the surface. Such raw material must be given more intensive tillage in order to unlock its contained plant food, and on much of it lime and organic manures will be needed in order to reduce its stiffness sufficiently to make it amenable to efficient cultivation, to the establishment of a desirable seed-bed tilth. It bakes easier and, as a consequence, crops growing on it are less resistant to dry seasons, because of rapid evaporation from the hardened surface, and the many cracks that form deep into the subsoil to enlarge the area exposed to direct evaporation. Crops also suffer more in wet seasons because the material becomes more soggy or water-logged than did the original soil. On much of it both fertilizer and lime will be required for satisfactory yields.

Certain piedmont areas whose records are known have, within a period of 30 years, lost their topsoil entirely, 10 inches or more of loam and clay loam having been washed off down to the clay subsoil; and on this clay subsoil, substituted for the departed soil, from 400 to 600 pounds of fertilizer are required to produce as much cotton per acre as formerly was grown with 200 to 250 pounds of fertilizer of no better quality.

While these difficulties of tillage and the lowered productivity are being attended to by the farmer in those fields not yet abandoned, the unprotected fields continue to wash. Unfortunately the farmers in many localities are doing little or nothing to stop the wastage and much to accentuate it. (Pl. 1, A.) In many instances the farmer does not know just what to do to slow down erosion. In many other cases he does not even suspect that the waning productivity of his fields results from any cause other than a natural reduction of the plant-food supply by the crops removed. He does not recognize the fact that gradual erosion, working unceasingly and more or less equally at all points, is the principal thief of the fertility of his soil until spots of subsoil clay or rock begin to appear over the sloping areas.

SOME WASTING AREAS

The southern part of the great Appalachian Valley is an admirable place to see the evil effects of that gradual land washing known as sheet erosion. Here in thousands of areas of formerly rich limestone soil of loam, silt loam, and clay loam texture, the topsoil has been removed. The numerous galls or clay exposures that now splotch the slopes lose their moisture quickly in dry weather. The damaging effects of drought upon crops are felt much quicker than formerly, according to those who have witnessed these changes in the soil. A much lighter rain than formerly now turns the Tennessee River red with wash from the red lands of its drainage basin. Added to the severe impoverishment of a tremendous area of land throughout this great valley, and its extensions southward into Georgia and Alabama and northward into Virginia, are the gullied areas, which are severely impaired or completely ruined by erosional ravines that finger out through numerous hill slopes and even many undulating valley areas. Field after field has been abandoned to brush, and the destruction continues.

Much erosion of the same type has taken place over the smoother uplands of south-central Kentucky; that is, in the rolling parts of the highland rim country; over much of the Piedmont region, and through many parts of the Appalachian Plateau. Land destruction of even worse types is to be seen in the great region of loessial soils that cover the uplands bordering the Mississippi and Missouri Rivers and many of their tributaries, from Baton Rouge, La., northward. Numerous areas, small and large, have been severely impoverished and even ruined in the famous black lands of Texas. Even the drier lands of the West and the comparatively smooth prairies and plains of the North-Central States have not escaped damage. Erosion is wasting the fertility of the soil and even the whole body of the soil in many places where the slope is sufficient for rain water to run downhill. There are some exceptions to this, or rather some partial exceptions, such as the nearly level lands, the loose, deep sandy lands, the highly absorptive gravelly areas, the loose glacial till and morainic deposits in parts of the northern border of the country, the peculiar red lands of the northern Pacific coastal region, and a few others. Although the total area of these more or less erosion-resistant soils is large, the area of those lands which are susceptible to washing and which are being washed in a wasteful way, more disastrously in some places than in others, is very much larger. Save when the fields are frozen or are covered with a blanket of hardened snow, erosion goes on upon these vulnerable lands during every rain that is sufficiently heavy to cause water to run downhill. Even the gentle spring rains cause some erosion, and the surface water flows away from sloping fields muddied red, yellow, or dun, according to the color of the soils of the neighborhood. This color is caused by soil materials started en route to the sea. Most of this material comes from the surface layer, the richest part of the soil.

FIGURES ON SOIL WASTAGE

The estimate of the quantity of plant-food elements annually lost by erosion, as given above, is a minimum estimate based upon a yearly discharge of 500,000,000 tons of suspended material into the sea by rivers,[2] plus twice this amount stranded upon lower slopes and deposited over flood plains, in the channels of streams, and even in the basins of reservoirs, where it is not needed and not wanted. Often this overwash does much more damage than good to the lands affected. It gradually reduces reservoir storage capacity and makes water-power plants dependent more and more upon the flow of the stream rather than upon the impounded water.

It is obvious to all who are familiar with field conditions that the amount of erosional débris in transit to the sea, but temporarily stranded on the way, each year very greatly exceeds twice the amount that actually passes out the mouths of rivers into tidewater. Some soil scientists believe the amount thus annually washed out of the fields and pastures and lodged on the way to the oceans is more than a hundred times greater than that actually entering the sea. The figure used above has been used merely because no satisfactory data upon which to base conclusively accurate estimates are available.

The estimates given do not include the dissolved matter which is annually discharged to the sea, a very considerable part of which obviously comes from erosional products. Furthermore, it is not

[2] Dole and Stabler have estimated that 513,000,000 tons of suspended matter and 270,000,000 tons of dissolved matter are transported to tidewater every year by the streams of the United States (*6, p. 83*).[3] T. C. Chamberlin estimates that 1,000,000,000 or more tons of "richest soil matter" are washed into the oceans from the lands of this country every year (*5*).

[3] Italic figures in parentheses refer to "Literature cited," p. 35.

known how much erosional detritus enters the ocean as drag material swept along the bottoms of streams. This material is exceedingly difficult to measure. The débris thus swept along the bottoms of many streams travels rather after the manner of waves or of sand dunes drifting before the wind. This characteristic of many river beds was brought out before a commissioner appointed by the Supreme Court of the United States in the expert testimony relating to the recent Red River boundary dispute between Texas and Oklahoma. Gilbert (*9, p. 11*) makes the following interesting observations regarding the process:

Some particles of the bed load slide; many roll; the multitude make short skips or leaps, the process being called saltation. Saltation grades into suspension.

When the conditions are such that the bed load is small, the bed is molded into hills, called dunes, which travel downstream. Their mode of advance is like that of eolian dunes, the current eroding their upstream faces and depositing the eroded material on the downstream faces. With any progressive change of conditions tending to increase the load, the dunes eventually disappear and the débris surface becomes smooth. The smooth phase is in turn succeeded by a second rhythmic phase, in which a system of hills travel upstream. These are called antidunes, and their movement is accomplished by erosion on the downstream face and deposition on the upstream face. Both rhythms of débris movement are initiated by rhythms of water movement.

The amount of plant food in this minimum estimate of soil wastage by erosion (1,500,000,000 tons of solid matter annually) amounts to about 126,000,000,000 pounds, on the basis of the average compositions of the soils of the country as computed from chemical analyses of 389 samples of surface soil collected by the Bureau of Soils (1.55 per cent potash, 0.15 per cent phosphoric acid, 0.10 per cent nitrogen, 1.56 per cent lime, and 0.84 per cent magnesia). This is more than twenty-one times the annual net loss due to crops removed (5,900,000,000 pounds, according to the National Industrial Conference Board) (*16*). The amount of phosphoric acid, nitrogen, and potash alone in this annually removed soil material equals 54,000,000,000 pounds. Not all of this wasted plant food is immediately available, of course; but it comes principally from the soil layer, the main feeding reservoir of plants, and for this and for other reasons it is justifiable, doubtless, to consider the bulk of it as essentially representing lost plant food, without any quibbling about part of it having potential value only.

By catching and measuring the run-off and wash-off from a 3.68 per cent slope at the Missouri Agricultural Experiment Station, on the watershed of the Missouri River, it was found that for an average of six years 41.2 tons of soil material were annually washed from 1 acre of land plowed 4 inches deep, and that 68.73 per cent of the rainfall, the total precipitation amounting to 35.87 inches a year, was held back; that is, 24.65 inches of the 35.87 inches of precipitation were temporarily absorbed as an average for the six-year period. From a grass-covered area of the same slope and soil type less than 0.3 ton of solid matter was removed each year (or a total of 1.7 tons in six years), while 88.45 per cent of the rainfall was retained.

In 24 years this rate of erosion would result in the removal of a 7-inch layer of soil from the area tilled 4 inches deep; but for the removal of the same thickness of soil from the grassed area 3,547 years would be required.

At the Spur substation of the Texas Agricultural Experiment Station,[4] in the subhumid part of west Texas, 40.7 tons per acre of soil material were removed from a 2 per cent slope of fallow land by approximately 27 inches of rainfall. Of this precipitation only 55 per cent was retained (at least temporarily) on cultivated bare land of the same soil and slope without terracing, whereas 84 per cent was retained on an area covered with Buffalo grass.

The erosion station[5] in the piedmont region of North Carolina measured from an uncultivated plot a loss of 24.9 tons of solid matter to the acre each year, when the rainfall was only 35.6 inches, as against a normal of 43.9 inches. On the same slope and soil the erosion from grassland that year amounted to only 0.06 ton to the acre. In other words, the grass held back four hundred and fifteen times as much surface soil as was retained on untilled bare ground. It held back two hundred and fifteen times as much soil as was retained in the cotton plots on the same soil, having the same degree of slope. The uncultivated plot retained 64.5 per cent of the rainfall, the cotton plot 74.4 per cent, and grassland 98.5 per cent.

The agricultural scientists at the Missouri Agricultural Experiment Station have this to say of erosion (*7*):

> Most of the worn-out lands of the world are in their present condition because much of the surface soil has washed away, and not because they have been worn out by cropping. Productive soils can be maintained through centuries of farming if serious erosion is prevented. The soils of Missouri have become gradually less fertile during the last one hundred years due in large measure to the excessive cultivation of rolling lands. Many of the most fertile soils in the rolling prairies and timber lands of this state have been kept in corn until the "clay spots" are evident on nearly every hillside. So much soil has been lost from even the more gently rolling parts of the fields that the yields are far below those obtained by our grandfathers who brought the land into cultivation. The erosion of cultivated fields is taking place at such a rate that it is calling for a decided change in our system of soil management. If we are to maintain our acre-yields at a point where crops can be produced at a profit we must make every reasonable effort to reduce the amount of soil fertility that is carried away during heavy rains.
>
> Approximately three-fourths of the area of Missouri is subject to more or less serious erosion. The map . . . shows where these soils are to be found. It will be seen from this map that erosion is serious on many of the most fertile soils of the state. This is particularly true in the rich rolling prairie regions of central and northwest Missouri, where owing to the fertility of these soils much of the land is kept in corn a large part of the time. It must be remembered that not all the soils . . . erode at the same rate . . . in the Ozark region, they [the soils] are largely covered with timber so that erosion cannot be considered a serious problem.

A single county in the southern part of the piedmont region was found by actual survey (*4*) to contain 90,000 acres of land, largely cultivated at one time, which has been permanently ruined by erosion. The whole area has been dissected by gullies, and bedrock is exposed in thousands of places. Here and there islands and peninsulas of arable land have been left between hideous gullies, but most of these remnants are too small to cultivate. The land has been so devastated that it can not be reclaimed to cultivation until centuries of rock decay have restored the soil. It has some value, however, for

[4] Preliminary figures furnished by officials of the Texas Agricultural Experiment Station.

[5] BARTEL, F. O. SECOND PROGRESS REPORT, SOIL-EROSION EXPERIMENTS, EXPERIMENT STATION FARM, RALEIGH, N. C. (A project of the Div. Agr. Engin., Bur. Public Roads, in cooperation with the N. C. Dept. Agr.)

A.—Cotton field showing the early development of destructive gullies in the middles of rows extending up and down the slope
B.—Result of uncontrolled erosion on Greenville fine sandy loam
C.—Deep erosion of the lateral extension type in the loessial region of the lower Mississippi Valley

A.—Stack of wheat straw dumped in gully in field of Marshall silt loam, which during the fall rains of one year caught and held 430 tons of rich soil material washed from adjacent slopes

B.—An apple orchard in the region of loessial soils, in northeastern Kansas. The trunks of the trees and some of the branches have been buried by soil washed from the adjacent uplands, the surface having been raised about 5 feet growing shortleaf pine and for pasture. The extent of this devastated region unfortunately is yearly growing larger.

Another county in the Atlantic Coastal Plain (*12*) has 70,000 acres of former good farm soil, which, since clearing and cultivation, has been gullied beyond repair. In one place where a schoolhouse stood 40 years ago gullies having a depth of 100 feet or more are now found, and these finger through hundreds of acres of land, whose reclamation would baffle human ingenuity. (Pl. 1.)

The most severely eroded parts of this county are described as follows:

> The Rough gullied land includes areas which, as the result of erosion, are so steep and broken as to be unfit for agriculture. Much of the land classified under this head supports forest. Some areas are available for pasture, but a considerable total area is not even suitable for this use, as there are many deep gullies with steep or perpendicular sides on which no vegetation can find a footing. Providence and Trotman "Caves" . . . are examples of such areas. . . .
>
> In the southwestern part of the county in the Patterson Hills and in another large area . . . southwest of Spring Hill Church, a somewhat different condition is encountered. Here the Rough gullied land consists of narrow-topped ridges with precipitous slopes, covered with ferruginous sandstone fragments. No level land is found here and the slopes are generally too steep even to afford good pasture. . . . One of the largest [caves] within the county has developed in the memory of the present generation, having started with the formation of a small gully from the run-off of a barn. The caves, some of which are about 100 feet in depth and from 200 to 500 feet in width, ramify over large areas. There is little possibility of this gullied land being restored to a condition favorable to cultivation.

In the "brown loam" belt skirting the Mississippi bottoms on the east side, county after county includes 10,000, 20,000, or 30,000 acres of land which have been ruined by erosion. (Pl. 1.) Agriculture has been driven out of a very large part of the upland of several counties in northwestern Mississippi by the gullied condition of the upland. Hundreds of farms in these and many other counties of the region have been abandoned to timber and brush. Unfortunately, the kind of timber that has established itself over much of these dissected areas is largely worthless blackjack oak, simply because pine seed have not been distributed to start valuable pine forests or because black locusts have not been planted.[6] (Pl. 2, A.)

Not only have the uplands been widely and disastrously dissected, but large areas of former good alluvial land have been buried beneath infertile sands washed out of those upland gullies (pl. 2, B) which have cut down through the soil strata into Tertiary deposits beneath. Stream channels have been choked with erosional débris, and overflows have become so common that large tracts of highly

---

[6] In this connection W. R. Mattoon, of the United States Forest Service, says: "The State of Tennessee through its Division of Forestry has aided several hundred farmers and public organizations, particularly in west Tennessee, in checking gully erosion by the planting of black locust. This work has been done on a gradually increasing scale since its inception, about 1913. Practical methods have been developed of planting one-year-old locust seedlings, spaced about six feet apart each way, over the entire wash or gullied area. Preparatory to planting, the gully banks are plowed off and brush dams built across the channels at strategic points to catch the soil. The black locust produces a heavy surface root system adapted to holding the soil, it is a legume and enriches the soil, it is a vigorous grower and endures thin soils, and it ranks as the second most lasting fence-post timber in this country. Black walnut, yellow poplar, pines, and other trees have also been planted. In addition to checking erosion the land is put to profitable use by growing valuable fence posts and other timber crops and the blue grass that invariably comes in supports limited grazing. A large number of farmers by this method have realized excellent money returns from old gullied lands."

productive soil formerly tilled are now nothing more than swamp land.

The stream bottoms throughout the piedmont region from Virginia southward into east-central Alabama have been impaired by this process of overwash to an even greater extent. Here, probably, considerably more than 50 per cent of the bottom land has been converted into a nonarable swampy waste, entirely as the result of deposition of eroded material. In spite of the terracing that has long been practiced on many farms in the southern piedmont region, wastage of good agricultural uplands has gone on at a distressing rate, because many fields were not terraced and many terraces were not maintained. Thus, unleashed erosional waters, performing in the dual rôle of cutting away the topsoil of the uplands and depositing the less fertile assorted constituents of the eroded matter over the stream bottoms, have brought about an enormous amount of land impairment and destruction.

Some streams formerly navigable have been so choked with sand and mud, purely as a result of erosion, that they have not been plied by boats for a generation or more. E. N. Lowe, of the Mississippi Geological Survey, speaking of soil erosion and flood control in the Yazoo drainage basin, said five years ago (*13*):

> In many of our northern uplands [Mississippi] washing of the soil is progressing so rapidly without let or hindrance over large areas, that some necessary measures must be adopted soon to arrest the process, otherwise vast areas of formerly agricultural land will become hopeless wastes. Large areas in at least a dozen upland counties of north-central Mississippi have already reached such a condition of soil depletion that they are now hardly suitable for any kind of agriculture, and their taxable values are reduced accordingly.
>
> The erosion of these uplands has resulted not only in enormous losses of valuable agricultural soils, but also in concomitant stream-filling throughout those areas. Volumes of silt and sand after every heavy shower are poured into the streams from every furrow, gully and rill that trenches the hillsides, resulting in filling of their channels. The obliteration of their channels causes overflow of the streams after any considerable rain, with deposition of sand over valuable bottom lands, often doing irreparable damage.
>
> For years rapid and destructive filling has affected the Coldwater. Forty years ago boats of large size came up the river to Coldwater to load cotton. Now no kind of a boat can come up Coldwater River, so choked is it with sand bars.
>
> The Tallahatchie was formerly a navigable stream. Even as late as 1900 a small steamer drawing four feet of water plied on the Tallahatchie from Batesville downstream. Now the stream is choked with sand bars, and can be easily waded at almost any place.

In the great cotton-producing section of central Texas, known as the black waxy belt, white spots representing exposures of the basal chalk and marl beds that gave rise to the immensely productive black soil of this region, dot the landscape of the rolling areas. The same thing is to be seen in many parts of the Alabama-Mississippi prairie belt. (Pl. 2, C.) These exposures represent the products of erosion—nonarable land that has been substituted for some of the most productive cotton soil of the world. In one county of this region (*8*) 13.5 per cent of the total area was recently mapped as an eroded phase of the valuable Houston clay soil. It was found that much of this had been too severely washed to allow cultivation, whereas the remaining better parts become highly desiccated in dry seasons, giving lighter and lighter yields as the wearing off of the soil progresses.

SOIL EROSION NOT RESTRICTED TO THE SOUTH

The experts who recently completed the soil survey of Doniphan County, in northeastern Kansas, found that an average of at least 6 inches of soil had been removed from the rich uplands of the county. Nearly all tilled slopes have suffered, some much more severely than others because of variations in the surface relief and in the kind of soil. In one place examined an area of original timber had throughout its extent from 12 to 24 inches of rich soil overlying clay subsoil. This surface layer was so rich in humus, so moist and mellow, that it was possible to dig down through the dark-colored permeable soil with the bare hand, even to the depth of the subsoil. Cultivated soil of the same kind, having the same degree of slope, lying in immediate contact with this forested area had in most places no topsoil at all, as the result of erosion, and in some places even the exposed subsoil clay had been eroded off to a depth of 6 inches or more. Indeed, both soil and subsoil, it was found, had been washed off some areas down to the basal limestone that at one time was 4 feet beneath the surface.

These severe effects of erosion were found as a very common condition in a broad belt over the more rolling lands near the Missouri River; indeed, this condition, or a close approximation of it, was found to be the rule, not the exception, through this more rolling belt, where the virgin soil, the Knox and Marshall silt loams, were among the very richest upland soils in the United States. Apple trees were dying on the eroded hilltops where, seemingly, the soil moisture conditions had been unfavorably upset by the removal of the surface soil layer. In the depressions and on the gentler parts of the slopes and the bench positions some of the rich soil from above had lodged. In these places the apple trees were thriving. A farmer in this section said to the soil specialists:

> We have good apples on the deep soil of the flat places, but we have always had good apples in these places. These places did not need any more soil, they were already deep and rich. We want our soil to stay in the orchards and fields, but it is not staying there. In places 4 feet of soil has been washed off the land. The surface of the ground about our house has been gradually lowered more than a foot. I will show you washed places where not even weeds succeeded this year.

This terrific washing of the land has taken place in the memory of men living in the community. The wasted areas adjacent to the forested land referred to above were cleared about 40 years ago, according to the statements of men in the locality who said they had taken part in the clearing.

Wheat, alfalfa, and sweet-clover fields seeded in the fall of 1927 had been severely damaged by the fall rains. (Pl. 3, A.) In places each depression made by the seed drills had been converted into a small rill way or gully, and the wheel tracks of the seeder in some places had grown into ditches (pl. 3, B), which surely will expand rapidly into formidable gullies that will cause eventual abandonment of the areas affected. In small grain, alfalfa, and sweet-clover fields soil in excess of five tons to the acre was swept from the surface of numerous fields on these splendid soils, the Marshall and Knox silt loams, during a single period of rain last fall. In some fields of steeper slope the loss per acre as a result of this single rainy

88854°—28——2

spell was estimated as amounting to fully 40 tons. Much of this erosional débris passed down from the upland slopes into depressions and stream ways to be carried off in flood waters, although large quantities, as was readily determined, were deposited locally over depressional flats and the flood plains of small and large streams, where none of it was needed. In one place a few miles south of Troy, Kans., where newly planted grainfields had been severely dissected by the fall rains of last year a farmer had left a series of wheat-straw stacks in a depression at the foot of converging slopes. Against the upper side of these large quantities of rich silt had lodged, building up small flood plains, or alluvial fans, 4 feet deep in places. (Pl. 4, A.) One of these stacks had caught 430 tons of this rich soil matter during the single short rainy period referred to, even where vast quantities had been swept by to lower levels after the catchment basin formed by the straw bulwark had been filled. These straw stacks represented the sole attempt to check soil erosion that was observed through several Missouri and Kansas counties bordering on the Missouri River.

Along the outer edge of the Missouri River bottoms, in the northern part of Doniphan County, a farmer had constructed an 8-foot embankment some distance out in the bottoms, approximately parallel to the foot of the upland, in order to intercept soil material that was being brought out of the hills by small local streams. This erosional material was covering the farmer's rich Missouri River alluvium (Wabash and Sarpy soils), causing a reduction in the yield of corn, and was continually washing over the roadways, rendering them impassable. Within 10 years eroded material from the uplands had lodged here level with the dikes, from 5 to 7 feet deep. Thus had been formed a terrace averaging 6 feet deep over 40 acres; and the intercepted soil was not so productive as the land it had buried. The weight of this erosional placed material amounted to about 480,000 tons. It had accumulated at the approximate rate of 1,200 tons to the acre each year. It should be observed in this connection that not all of the erosional detritus brought out of the uplands had been held by the dike. At first, part of it had escaped downstream in the conveying flood waters. Finally drainage had been blocked in that direction, whereupon the material began to escape in the transporting water around the upstream end of the diked area.

This sort of thing is taking place in varying degrees up and down the Missouri River and its tributaries, and along many other streams of the central West. Recently, it was necessary for the soil surveyors working in this great region to recognize a new soil type in order to classify and map material derived from the regional uplands by erosion and freshly deposited over older stream alluvium.

In an apple orchard near Lookout Mountain in northeastern Kansas the trunks of the trees had been completely buried by overwash of silt from the adjacent uplands, and the level of the ground was among the branches of the trees. (Pl. 4, B.) The owner of this orchard stated that although the apple trees had not seemed to suffer by the filling in, the uplands had suffered very greatly from the gradual erosion that gave rise to the transported soil.

Near this orchard a gully is now advancing at a minimum rate of 150 feet a year, according to local information. This ravine is

60 or 75 feet deep, nearly 300 feet wide in places and almost three-fourths of a mile long. It is destined to destroy all the farm land in this fertile valley, including the apple orchard, and it may, with its deploying prongs, cut through the local hills.

Already in this new agricultural region fields (pl. 5, A) and even farms are beginning to be abandoned in the more rolling belts near the river, and land is being rapidly impoverished many miles back from the river. Indeed, all cultivated slopes are suffering to some extent. Nothing is being done to slow down the wastage, but considerable to accentuate it. In general, no effort is being made to cultivate along the slope contours; corn rows are run straight up and down hills as often as otherwise. It is a common practice in this region to plow furrows down the slopes in the spring, in order to allow water standing temporarily in corn "middles" to flow out. These furrows commonly develop into gullies that soon grow beyond control at anything like reasonable cost.

There are no terraces in this region; the farmers do not even know what they are. Erosion is gathering momentum. As the more absorptive topsoil is washed off down to the less absorptive subsoil, the rate of wastage increases. So, this region, which has already suffered seriously from rainwash, is really just upon the threshold of the most impoverishing kind of erosional wastage, and nothing is being done to conserve these splendid agricultural lands, the capital of the farmers living on them and a vital heritage to posterity.

It is not to be understood from the above that erosion in the north-central part of the United States is restricted to the Missouri River region. The wastage is taking place generally throughout this great region, most violently, of course, on the sloping areas. Soil displacement by this process is slow on the very extensive flat areas of the prairie regions that formerly were covered by a most efficient soil-conserving mat of native grass; but even here there is a much greater gradual removal of the rich surface material than is commonly recognized. (Pl. 5, B.) Since the clearing of the sloping and rolling areas and the destruction of the virgin sod, much costly washing has taken place in Missouri, Iowa, Nebraska, Illinois, Indiana, Ohio, Wisconsin, and other States. Recent soil surveys in southwestern Wisconsin have shown that the problem of erosion is a most serious one in many localities. It was found that slopes, especially on the Clinton and Boone soils, which were originally timbered or covered with brush, have been seriously gullied and damaged by sheet erosion from rain water and melting snow. Gullying was found even on bench lands of the valleys (Bertrand soils), and here as elsewhere the stream bottoms were being covered by overwash. These latter instances are mentioned to show that soil wash is a land menace even in parts of the northern border States.

EROSION IN THE DRIER REGIONS

Under the light rainfall of the western dry regions one might reasonably conclude, in the absence of the facts, that erosion is of negligible importance in comparison with that taking place in the humid regions. From the viewpoint of the extent of erosion, such a conclusion would be entirely contrary to the facts, at least for very large areas. The rivers entering the Mississippi from the west carry very large amounts of suspended matter. Some have ascribed this to the treeless condition of the western region. Doubtless this is a contributing factor; and certainly the vast extent of land used for clean-cultivated crops in the prairies and in the more humid eastern part of the Plains, is a most important factor. The peculiar structural behavior of the soils in the regions west of the prairies and eastern plain border, near the headwaters of the streams flowing eastward, coupled with the frequent dashing character of the rainfall, is also an important contributing factor to the heavily silted condition of these streams.

Much of the soil of the dry regions upon desiccation assumes a fluffy loosened condition or structure, to a depth of several inches. The structure is so loose, that the naturally pulverized surface material, even of heavy clays, can be scooped up freely with the hand. Heavy downpours cause this chafflike material to be swept ahead of the flowing water until the soil particles have become thoroughly saturated, disintegrated, and finally coalesced to form an emulsion which might appropriately be styled "liquefied" soil. The flow of this is at first slow, but it speeds up as the emulsion becomes thinner with the increasing proportion of water from rainfall.

In the cattle country of the southwestern dry region numerous places were observed and studied last year (1927), where the richest soil of the region, the areas of deeply accumulated valley-filling material (such as the Reeves silty clay loam), had been washed out entirely from valleys that formerly afforded excellent grazing. (Pl. 5, C.) This process was seen in all stages of development. The washes have their beginning in those places where the natural vegetative cover and normal ground equilibrium have been seriously disturbed. (Pl. 6, A.) Most of them have their start in cattle trails and the wheel ruts in roads. (Pl. 6, B.) In one place near Fort Davis, Tex., an area of approximately 1,000 acres had been so riddled by gullies, which had their beginning in a prairie dog town, that the ground, although once excellent grassland, was almost bare of vegetation. In another locality a gully 30 feet deep, 200 feet wide, and nearly a mile long was seen in the place of a former main highway. This gully was still growing, more rapidly than ever, acording to the ranchman who owned the land.

It was observed in this general region that many of the highways are protected from lateral erosion by retard and diversion dikes, and by diversion ditches dug along the slopes above the roadbed. This form of protection has been successful in some places, but not in others, the difference being due to soil variation and difference in adjustment of the ditch grade to the slope. On some soils of high vulnerability to rainwash, these ditches, built to protect the roadbed, have grown into erosional gullies, which have extended so far that the road has been undermined, necessitating its relocation.

The gullies that cut to pieces the valuable valley grazing lands of this dry country usually go down as incisions having perpendicular walls. When the gullies are cut to the underlying gravel or soft material, an undermining process begins that causes huge blocks of the upper strata to cave into the trenches. These blocks melt away rapidly with subsequent floods. In this region a peculiar soil property serves to accentuate erosion on some very extensive and valuable soil types. The clay soil, when it becomes dry, cracks and scales off from the sides of erosional trenches in such a way as to cause one gully to cut through to another. Thus, numerous natural bridges and caves are formed, and these help to speed up the invasion and destruction of the land.

A ranchman near Marfa, Tex., 15 years ago found that an important strip of his valley grazing land was in danger of being destroyed by an enlarging arroyo. To avoid disaster, he threw a small dam from one side of the valley out across the wash, and about halfway across the floor of the valley; from the end of this a wing dam was turned down the valley and carried for a long distance parallel to the arroyo. The result has been that the arroyo has nearly filled in, both above and below the dam, and the increased water carried over that part of the valley floor lying between the wing dam and the foot of the upland has caused greater subsoil storage of moisture, and thereby made the grazing value of the area affected 15 times as great as it was, according to the statement of the rancher. In addition, the alluvial soil, thus enriched in subsoil moisture has produced valuable crops of feed without irrigation, which in this region is very costly.

In parts of the western deserts railroad companies have found it necessary to construct numerous retard and diversion embankments along their road fills to prevent lateral erosion. The Southern Pacific Railroad in the desert between Niland, Calif., and the Colorado River, for example, has protected sections of its roadbed from erosion by a system of A-shaped embankments that catch the water on the upper side of the track and divert it or concentrate it to soundly constructed culverts beneath the track. Thus, a continuous line of earthen embankments connected like a rail fence, has been built to ward off the abrasive effects of silt and sand-laden desert flood water, and this line at no small cost, must be kept in repair against the erosion of the rainy seasons.

Recently, as the writer was informed, a brief heavy rainy spell (11 inches in three days) in the southwestern part of the United States caused deposition of a layer of infertile sandy material over a valuable orange grove. The trees quickly began to show signs of serious injury, and it was necessary to do something about it. It is said to have cost in the neighborhood of $100,000 to haul the deeper deposits of this inert material (that varying from 1 to 2 feet in depth) out of the grove, and to rake back the shallower deposits from the base of the trees.

In orchards observed by the writer last year (1927) in the valley east of Santa Paula, Calif., fruit trees had been planted on well-constructed terraces to prevent erosion, and, in addition, diversion ditches had been dug along the upper side to catch and divert injurious erosional débris coming out of the adjacent shale hills. To protect the diversion ditch itself, eucalyptus and tamarisk trees had been planted, not only along the ditch embankment, but along the hill slopes above the ditch.

RELATION OF SEDIMENTS TO FERTILITY

It is commonly believed that the products of erosion which do not actually go out to sea are not being wasted. It is believed that frequent deposition of flood alluvium enriches the land, and that floods, therefore, are beneficial, in respect to the productivity of the overflowed alluvial plains. There is some truth in this, of course; but in the main the conception is incorrect, and frequently the good accomplished is greatly overestimated. It has already been pointed out how disastrous overwash of inert sand has been to the alluvial lands of the piedmont region and the "brown loam" belt of the lower Mississippi Valley. This same condition, or an approximation of it, applies also to many other parts of the country. In the Ozark region, for example, the bottom lands of many farms, on which there was but little arable soil in the beginning, aside from the bottoms, have been seriously impaired or ruined by overwash of chert gravel washed down from the regional hillsides. The November flood of 1927 in the New England States laid down upon many of the productive bottom lands a blanket of relatively infertile loose sand and gravel, burying meadows and fields. In other parts of the bottoms the soil was ripped out and washed away by the swift, deep flood water.

The beneficial effects of the sediments deposited by the spring floods of the River Nile are often cited. It is not known precisely what the benefit amounts to in terms of money; but there is no doubt that some measure of soil enrichment does follow the floods of that river. It is obvious, also, that some enrichment of the soil is derived from the finer sediments laid down by the flood waters of the Mississippi. However, some damage is occasioned by the deposits of comparatively inert coarse sand scattered about in the "sand blows," or by patchy deposits that take form locally with every flood spreading over that great delta region. However, the damage and destruction to property and planted crops occasioned by the Mississippi floods quite obviously very greatly exceeds the net benefits accruing from sedimentation. When one thinks of possible benefits to alluvial land derived from deposition of flood-water silt, one should not lose sight of the damage done to upstream farm lands by the removal of the silt into the streams. Also the resultant increase in flood volume due to the additions of solid and dissolved products of erosion is dangerous, and one should not overlook the increased rapidity with which rain water flows off those areas denuded of their more absorptive topsoil. The alluvial soils of the flood plain of the Mississippi and most of its tributaries are naturally so rich that most of them could be cropped probably for many generations, without severe impoverishment of the soil. These alluvial soils are deep, many of them very deep, and exceptionally rich in plant food. The average chemical composition of "buckshot" soil samples taken from Coahoma and Issaquena Counties, Miss., is as follows (*3*): 0.28 per cent phosphoric acid; 0.80 per cent potash; 0.81 per cent lime; 1.31 per cent magnesia.

The phosphorus content of this soil, which is by far the most extensive soil of the lower Mississippi flood plain, is nearly twice that of the average surface soil of the country. It also exceeds the average soil considerably in content of organic matter and nitrogen. Material of this exceptionally good fertility extends to a depth of several feet with but slight change. The condition of fertility is so good that new sediments are not particularly needed, although,

A.—Land in northeastern Kansas formerly cultivated but now used for pasture because of gullying and sheet erosion. The gullies are constantly cutting deeper and fingering out over a wide area
B.—Sheet erosion on almost flat rich black Iowa soil where the first evidence of erosion is the exposure of patches of the clay subsoil
C.—Area in western Texas representative of the destructive effects of erosion on the dry lands of the West

A.—Absolute desert conditions brought about by erosion on a once-forested area in northern California. Every vestige of the topsoil and much of the subsoil are gone, and deep gullies have been cut into the soft basal rocks
B.—Intermediate stage of land destruction on Susquehanna soil having a stiff clay subsoil. The devastation has almost reached the stage at which any attempt at reclamation will be unprofitable
C.—Properly terraced field in the piedmont region. Unprotected soil of this type depreciates rapidly because of both sheet erosion and gullying

with their supply of lime and fresh organic matter, some temporary increase of productivity necessarily will follow the recession of the water. The degree of this increase can not be estimated with much accuracy with the small amount of available data; but probably its tangible money value does not greatly exceed 75 cents or a dollar to the acre of cultivated land each year for several years following a flood. It must be remembered that the Mississippi delta lands, particularly the predominant "buckshot" soil (Sharkey clay), as well as many of the other alluvial soils of the country, are among the richest soils in the world.

THE DANGER OF AVERAGES

The effect of erosion is extremely variable from place to place, on varying soil and varying slope, with varying vegetative cover and method of land usage. Hence, the average depth of surface denudation that has been commonly computed from river discharges alone, means very little. It implies that the surface everywhere, on steep hillsides and flat prairies, on sand dunes, loam and clay, is being planed down at an equal rate. This is far from the truth. The estimate so often read that erosion is lowering the Mississippi Basin at the insignificant rate of 0.0028 inch annually, is not only too small as an average, but since erosion does not operate over large areas of varying soils according to any plan of averages, such a statement is dangerous both for its inaccuracy and complacency.

A most important thing to know about soil erosion is the rate of cutting away the topsoil, and after that the subsoil of the individual soil types, in those regions of the more vulnerable lands, such as the region of loessial soils, the region of the Susquehanna soils, the Knox, Marshall and related soil regions, the Cincinnati soil region, the Houston clay soil region, and the regions where Orangeburg, Decatur, Cecil, Dekalb, Reeves, Vernon, Putnam, Fairmount and numerous other soils are important.

The studies already made in connection with soil-survey work show that there are many types of erosion, due to many variants, that have to do with the process, chief of which are, (1) soil type, (2) degree of slope, (3) climate, (4) vegetative cover, and (5) method of usage. Some soils can be cropped with a fair degree of safety on slopes having a gradient up to about 20 per cent, such as some of the very porous gravelly soils of the chert ridges in the southern part of the Appalachian Valley. On the other hand, some soils can not be cultivated without steady decline due to erosion, even where the slope does not exceed 1 or 2 per cent. The Knox silt loam, for example, is such a soil. On this soil erosion goes on in all tilled fields where there is any slope whatever.

On some soils greatest erosinal damage is done by gullying; on most soils, however, greater wastage results from that slow type of erosion called sheet erosion. On the Cecil soils of the piedmont region deep, broad V-shaped gullies form and finger out rapidly, whereas on the Orangeburg soils, the sides of the ravines are more nearly perpendicular, and they extend by a process of caving, when the loose sand of the substratum is washed out, so that rapid widening and head-on extension takes place. (Pl. 7, A.) On the Grenada soils of the "brown loam" belt, by reason of a compact subsoil layer peculiar to this group of soils, the washing extends more nearly equally in all directions, and rapidly invades broad areas of fine loessial land wherever the erosion has been neglected in its infancy. (Pl. 7, B.) On gravelly red land in northern California, where smelter fumes have annihilated the forests and destroyed almost every vestige of vegetation, extremely deep, narrow ravines have developed, which make travel over these areas difficult and even dangerous. (Pl. 8, A.) On soils like the Susquehanna, in which impervious heavy clay lies near the surface, the material of the cultivated soil is converted quickly into an approximate liquid condition during rains. This causes the surface substance to flow away rapidly. Following this skinning-off process, the exposed stiff clay is attacked by erosion and gradually cut to pieces by gullies that render the land absolutely unfit for further cultivation. (Pl. 8, B.) Numerous other variations of the manner by which soils erode could be given, but this will not be necessary for the purposes of this circular.

Although there is some erosion on most tilled and bare areas, and probably always will be, wherever water runs downhill, provided the soil is not frozen or protected by hard snow, the damage is greatest in the southern and central parts of the Temperate Zone and in the Tropics. So long as the ground is congealed freezing gives practically complete protection, save on those soils that "heave" badly. Slowly falling rains are everywhere much less destructive as an erosional agent than hard, beating rains. For example, no important effect of surface wash is observable on cultivated slopes of the Fairbanks silt loam, a wind-laid soil, in the Tanana Valley in northern Alaska, where the ground is frozen during eight or nine months and the light precipitation occurs almost entirely as drizzling rain and light showers.

By simple and well-known laws of mechanics the erosive power of flowing water increases enormously with increase of slope, but the destruction accomplished varies greatly with the soil type. Deep sandy soils, as a rule, do not wash severely, especially where the subsoil does not consist of impermeable clay or hardpan. However, some areas of sandy land, such as the Norfolk sand, do wash rather badly, and even gully on those slopes where there is impervious clay at a depth of 4 or 5 feet or less, as is true of areas in east Texas having a stiff subsoil like that of the Susquehanna clay.

RELATION TO FLOOD CONTROL

It is obvious that the erosional débris entering the streams adds to the volume of the water. It is equally obvious that those methods of soil conservation which have been found effective in slowing down or controlling soil erosion, chiefly terracing the land and the growing of trees, grass, and other soil-holding plants, are also methods which will cause more water to be retained in the surface soil and to be stored in the subsoil. Terracing of fields and the growing of trees, grasses, and shrubs on idle lands and areas too steep for cultivation, and upon soils that are highly susceptible to washing, as a combination of practices, will, it is believed, have considerable to do with flood reduction by decreasing the runoff and washoff from many land areas. Soil conservation is somewhat synonymous with moisture conservation. Nothing will hold back all the water, of course, but enormous quantities can be held temporarily or stored for summer-crop use, especially in the subhumid regions. At the same time the rich topsoil can be conserved by these proved implements of soil and water conservation. Soil conservation, therefore, should be an important adjunct of any long-continued system of flood control. To those who have seen the water from heavy rains rushing down unprotected cultivated slopes and bare areas, surcharged with soil matter, and carrying even gravel, cobbles, and bowlders, it is not necessary to argue about the effective contribution widespread use of these soil-conserving methods would make toward flood control as supplementary measures to protection with levees, spillways, and reservoirs.

Suspended material to the amount of 428,715,000 tons annually passes out of the mouth of the Mississippi River alone. This is but a part of the solid material that enters the river and its tributaries since much is left stranded somewhere along the pathway to the sea. In considering the relation of this water-transported erosional material to increased floods, it is necessary to take into account its full significance, along with that of a far greater amount stranded between the source of supply and the streams, in its relation to the increased amount of water flowing off the land areas which have contributed the material. So many tons of silt in the river stand, unmistakably, for so many denuded or partly denuded acres of sloping land somewhere upstream—land enabled by its denuded condition to contribute to the stream at a faster rate more of the rain that falls upon it.

In discussing the relation of forest and other forms of vegetative cover to run-off water and floods, it is frequently contended that, although the methods may have value, the time required for a forest to grow up is too great for this means of assistance to have any important relation to flood problems requiring immediate attention. In this connection the fact should not be lost sight of that the roots of trees and of other plants begin to function as effective agents for holding soil against erosion very shortly after the seedling begins to grow. Greatest efficiency in this respect will come, of course, when the forest or other vegetative cover, as grass, bushes, and chaparal, has made sufficient growth to develop an absorptive, spongy cover of vegetable litter. The immediate effectiveness of grass in holding both soil and water has been conclusively shown by results of the erosional test referred to above. It is said by those familiar with early conditions in the Prairie States that before the extensive cultivation of the land the matted turf of the prairies, in many places, hung like canopies over the banks of streams that carried clear water throughout the year. With the breaking of the land this situation was changed. The streams are more frequently dry in summer and are more heavily laden with silt when the rains come.

The following relates to the effects of rains on sloping areas in Orange County, Calif., following removal of a bush growth by fire:[7]

During the Orange County Farm Bureau Forestry Tour on November 19th, a remarkable demonstration of the effectiveness of chaparral cover in conserving water by preventing destructive erosion was seen at the Harding reservoir.

---

[7] Information furnished by C. F. Shaw, University of California, in a letter to the writer. Data obtained from Extension Service Report, December, 1927.

88854°—28——3

In October, 1926, the heavy chaparral cover on this watershed was almost entirely destroyed by fire, leaving the slopes unprotected. During November a heavy rain fell during a 24 hour period. Santiago Creek quickly became a turbulent mass of muddy water, containing over 60% solid matter washed from the burn. The Santa Ana River, which in the past had never had a peak flow of more than 8.000 second-feet during similar rains, showed approximately four times that at the height of the flood.

Harding reservoir was completely filled with rocks, silt and ashes from the burn, and a deposit of a half inch to an inch of this material was left over the entire bed of the Santa Ana River when the flood subsided. Other streams in the vicinity, where watersheds were untouched by the fire, showed scarcely any rise at all, and the water in all of them was clear throughout the storm. After many weeks of shoveling and washing, the capacity of Harding reservoir is less than one-fourth of its original volume.

LIMITED AMOUNT OF DATA AVAILABLE

In this country only a limited amount of information has been acquired concerning the rates of erosion on different soil types, the holding effect of terraces of different build or the possibility of reenforcing them with various stabilizers such, perhaps, as grass, shrubs, or vines, and the rate of alluvial deposition under varying conditions. Only three or four soil types of the many involved have had their susceptibility to erosion measured. It will be observed in reading this circular that little information other than estimates and observations have been given. This is because exceedingly little research work has been done on the subject. It is not known, for example, precisely what type of terrace or what degree of terrace slope is most applicable to the loessial soils of the Marshall, Memphis, and Knox series. It is known that some types of terrace have not given entirely satisfactory results on these peculiar friable soils of such exceedingly high silt content and such low content of clay to bind the silt. Possibly the Mangum terrace, if properly modified and given precisely the right slope, would effectively control erosion on these exceedingly vulnerable soils. Information greatly needed in connection with the problem of erosion should be made available through experimentation and research work as speedily as possible. If a particular type of terrace does not hold in one place and does hold elsewhere, the reason for the failure, as well as for its success, should be determined and the significance of the facts turned over to the farmers of the Nation in forms available for practical use.

As a Nation we are doing very little to abate the evil effects of erosion. Every one who knows anything about it admits the problem is a serious one, but few realize how very devastating is the wholesale operation of erosion. There is necessity for a tremendous national awakening to the need for action in bettering our agricultural practices in this connection, and the need is immediate. Terracing of sloping areas to prevent erosion has been carried on for a long time in the southeastern part of the United States. (Pl. 8, C.) Recently use of this method has extended across the Mississippi River and is being extensively and increasingly employed in Texas, Oklahoma, and Arkansas. The Federal land bank at Houston recently adopted the policy of requiring all vulnerable fields to be terraced before money is loaned on the land. The bank has employed an erosion expert, who, according to press dispatches, not only decides whether or not the property upon which a loan is asked needs terracing, but also goes out and instructs the farmer how to build a terrace if he is unacquainted with the engineering side of this method of conserving soil.

In the region north of Oklahoma and Tennessee the farmers, as a rule, do not know what a terrace is. Most of them have never seen one, and many of them have never heard of this valuable method of soil conservation. In other instances they are not used because the farmers have not been convinced that they are needed, having heard so much about plant food stolen by crops and so little about soil stolen bodily by erosion. Again, terraces are not used because the farmers have not known how to construct them.

Terracing is a very practical method of saving the land against rainwash. (Pl. 9, A and B.) In general the construction of these field embankments is not very costly, and when they are properly built they will pay the cost of construction and maintenance many times over. Although all the details relating to the best methods for terracing some of the peculiarly vulnerable soils are not fully known as yet, it is known that the broad-base, variable-graded ridge terrace, known as the Mangum terrace, properly laid out and built, is a highly efficient instrument for protecting vast areas of land now wasting through the effects of sheet erosion operating increasingly on unprotected slopes. Exceedingly steep slopes, of course, can not be saved by any method of terracing (pl. 9, C), save those expensive methods of building rock walls and huge retaining embankments, such as have been made abundant use of in parts of the Mediterranean Basin and other regions of the world where there is no excess of available farm land and where labor is abundant and cheap. In this country these steeper slopes should be used in accordance with their best adaptation from the economic viewpoint of America, i. e., for forestry and grazing. Probably in this country those erosive sloping lands which range in texture from silt loam to clay and, which have bedrock at a depth of a foot or less beneath the surface, should not be cultivated under any circumstances for generations to come, if ever. There are other soil conditions, also, where the land can not be economically saved by terracing, some even where the mere clearing off of the timber may be followed by wasteful washing. (Pl. 10.) All the details can not be given here.

In addition to terraces soil-saving dams, brush fillings and other obstacles to continuous washing have been successfully employed locally in combating erosion.

There are national associations for the preservation of wild flowers and for the preservation and propagation of wild life but none for the preservation of the soil. Conservation of this most fundamental and important of all resources is seldom seriously considered by any one not directly or indirectly associated with the ownership or management of a farm, and it is too infrequently considered even by the farmers themselves. Erosion is a very big problem. It is doubtful if the farmer can handle it alone.

SOIL-TYPE INFORMATION

The kind of information that is most needed about erosion is that which will apply to definite kinds of land—to soil types that vary from place to place, not only in their crop adaptations and requisite methods of cultivation, but in their resistance to erosion and in the means necessary for checking erosion. Any other method of procedure in studying the problem will be, in no small degree, wasted effort, as methods that may apply to one soil may injure a soil of different character.

As has already been pointed out, terraces must be adjusted carefully, not only to soil type but to slope. If the protection embankment is given too much or too little slope, there is danger of breaks and intensified washing that may exceed that prevailing before the terrace was constructed. In one instance terraces built on Granville sandy loam, in the southern piedmont region, broke with the first heavy rainfall, causing almost complete destruction of the area involved; whereas terraces of the same type made on the Wadesboro clay loam at about the same time and with the same slope withstood the rains that destroyed those on the other soil.

In this connection it is pertinent to refer to a statement of G. E. Martin, of Oklahoma (*14*):

> A half finished job of terracing is likely to result in wasted time, wasted effort, and wasted soil, and tends to bring into disrepute the most satisfactory means, so far determined, of preventing the enormous annual loss of soil fertility which now occurs. This loss constitutes a most serious drain upon the agricultural industry. It is very unlikely that any other industry could suffer such severe losses and survive.
>
> The importance of measuring the slope to determine the proper spacing of terraces can hardly be over emphasized. Too heavy a grade or too much fall along the terrace line, can defeat the moisture conservation objective and may result in hillside ditches instead of terraces.

As an illustration of the important rôle soil character plays in determining the rate of soil erosion, comparisons might be made between the results obtained at the erosion station in subhumid west Texas and those obtained in the humid piedmont of North Carolina. On a 2 per cent slope of the Abilene clay loam at Spur, Tex., 41 tons of soil matter were lost by erosion from 1 acre of land with 27 inches of rainfall; whereas at the North Carolina station only 25 tons of soil matter were removed from 1 acre on a 9 per cent slope with 35.6 inches of rainfall. In other words, although the slope in the latter instance was more than four times steeper than that in the former instance, the eroded material was very much less on the steeper slope.

It is not the purpose of this circular to go into the details of methods for preventing soil erosion, but rather to point to the evils of this process of land wastage and to the need for increased practical information and research work relating to the problem. Instructive bulletins have been published by the United States Department of Agriculture and by the States containing details relating to the best-known methods of checking soil erosion and of filling gullies. Bulletin No. 512, Prevention of the Erosion of Farm Lands by Terracing (*18*), published by the United States Department of Agriculture, is especially instructive in connection with the theory and the practical side of terrace construction; and Farmers' Bulletin No. 1386, Terracing Farm Lands (*19*), is another useful bulletin relating to the subject.

LOOKING FORWARD

It must be stated that this circular does not undertake to tell the full story of the appalling wastage being caused by soil erosion. It merely refers briefly to some of the working processes of this greatest enemy to the most valuable asset of mankind (the agricultural lands), to some minimum estimates relating to the damage wrought, and to the meagerness of fundamental data concerning the problem. To visualize the full enormity of land impairment and devastation brought about by this ruthless agent is beyond the possibility of the mind. An era of land wreckage destined to weigh heavily upon the welfare of the next generation is at hand. Indeed, what has happened already and what is going on at an ever-increasing rate of progress is pressing upon many thousands of farmers now struggling to win subsistence from erosion-enfeebled soil. That the evil process is gaining momentum is due to the wearing away of the topsoil, which was more productive and more resistant to rainwash than the subsoil that is taking its place. That some 15,000,000 acres or more of formerly tilled land has been utterly destroyed by erosion in this country is but an insignificant part of the story, for it is the less violent form of erosional wastage, sheet erosion, that is doing the bulk of the damage to the land. Land depreciation by this slow process of planing off the surface is of almost incalculable extent and seriousness, and since the denudation does not cease when the subsoil is reached, there must be in the near future, unless methods of land usage are very radically changed, an enormous increase in the abandonment of farm lands.

What would be the feeling of this Nation should a foreign nation suddenly enter the United States and destroy 90,000 acres of land, as erosion has been allowed to do in a single county? Any American of live imagination knows that the people of the United States would willingly spend $20,000,000,000 or as many billions as might be necessary, to redress the wrong. Because rain water was the evildoer in this instance, which is but one of many, is the act forgiveable and is there no occasion for concern about it?

It is not necessary to go to China or to some other part of the world for examples of what eventually happens to unprotected slopes of cultivated areas. There is an abundance at home, not yet so vast in area as in China, but just as bad, and by no means small. It is well to observe, however, that millions of human beings have been driven out of the wasted uplands of China into the valleys of the great rivers, where the population is so dense and the land so completely used that even the roots of grain crops are dug for fuel. China cut the forests from the uplands and made no provision for protecting the bared slopes. Erosional débris sweeping out of these wasting highlands has rapidly extended the river deltas and made floods ever more difficult to control. After 4,000 years of building dikes and digging great systems of canals, the Yellow River broke over its banks and brought death to a million human beings during a single great flood. During one flood that great river, known in China as the "scourge of the sons of Han," changed its channel to enter the sea 400 miles from its former mouth.

No one, of course, wants anything remotely like this to take place in this country, but "coming events cast their shadows before." That the greatest flood of which we have reliable records came down the Mississippi in 1927 was a prophetic event. G. E. Martin's statement (*14*) about erosion as an enemy to agriculture—"It is very unlikely that any other industry could suffer such severe losses and survive"—is prophetic. That bare land at the Missouri Agricultural Experiment Station was found to be wasting 137 times faster than land covered with bluegrass, on a slope of less than 4 per cent gradient, is prophetic. That many millions of acres of cut-over land lie bare and desolate and exposed to the ravages of fire and erosion, with but pitifully little done toward reforestation, is prophetic. That minimum estimates show that the rate of plant-food wastage by erosion is twenty-one times faster than the rate at which it is being lost in crops removed, is prophetic.

These shadows are portentous of evil conditions that will be acutely felt by posterity. Shall we not proceed immediately to help the present generation of farmers and to conserve the heritage of posterity?

The writer, after 24 years spent in studying the soils of the United States, is of the opinion that soil erosion is the biggest problem confronting the farmers of the Nation over a tremendous part of its agricultural lands. It seems scarcely necessary to state the perfectly obvious fact that a very large part of this impoverishment and wastage has taken place since the clearing of the forests, the breaking of the prairie sod, and the overgrazing of pasture lands. A little is being done here and there to check the loss—an infinitesimal part of what should be done.

* * * * * * *

LITERATURE CITED

(1) Barnes, W. C.
1925 *The Story of the Range*, U.S. Congress, 69th, 1st sess., Hearings Senate Res. 347, pt. 6, p. 1579–1640, illus.

(2) Bates, C. G.
1924 *The Erosion Problem*. Jour. Forestry 22: 498–505.

(3) Bennett, H. H.
1921 *The Soils and Agriculture of the Southern States*. 399 p., illus. New York, The Macmillan Co.

(4) Carr, M. E., Welsh, F. S., Crabb, G. A., Allen, R. T., and Byers, W. C.
1914 *Soil Survey of Fairfield County, South Carolina*. U.S. Dept. Agr., Bur. Soils, Field Oper. 1912, Rpt. 13: 479–511, illus.

(5) Chamberlin, T. C.
1909 *Soil Wastage*. *In* Proceedings of a Conference of Governors in the White House, Washington, D.C., 1908. U.S. Congress, 60th, 2d sess., House Doc. 1425, p. 75–83.

(6) Dole, R. B., and Stabler, H.
1909 *Denudation*. U.S. Geol. Survey Water-Supply Paper 234: 78–93.

(7) Duley, F. L.
1924 *Controlling Surface Erosion of Farm Lands*. Missouri Agr. Expt. Sta. Bul. 211, 23 p., illus.

(8) Geib, H. V.
1926 *Soil Survey of Rockwall County, Texas*. U.S. Dept. Agr., Bur. Soils, Field Oper. 1923, p. 123–152, illus. (Advance sheets.)

(9) Gilbert, G. K.
1914 *The Transportation of Débris by Running Water*. U.S. Geol. Survey Prof. Paper 86, 263 p., illus.

(10) Gray, L. C., Baker, O. E., Marschner, F. J., Weitz, B. O., Chapline, W. R., Shepard, W., and Zon, R.
1924 *The Utilization of Our Lands for Crops, Pasture and Forests*. U.S. Dept. Agr. Yearbook 1923: 415–506, illus.

(11) Jardine, J. T., and Forsling, C. L.
1922 *Range and Cattle Management During Drought*. U.S. Dept. Agr. Bul. 1031, 84 p., illus.

(12) Long, D. D., Beck, M. W., Hall, E. C., and Burdette, W. W.
1916 *Soil Survey of Stewart County, Georgia*. U.S. Dept. Agr., Bur. Soils, Field Oper. 1913, Rpt. 15: 545–606, illus.

(13) Lowe, E. N.
1922 *Reforestation, Soil Erosion and Flood Control in the Yazoo Drainage Basin*. South. Forestry Cong. Proc. 4: 10–11.

(14) Martin, G. E.
1927 *Terracing in Oklahoma*. Okla. Agr. Col. Ext. Co. Agt. Work Circ. 218, 19 p., illus. (Revised ed.)

(15) Munns, E. N.
1923 *Erosion and Flood Problems in California*. 165 p., illus. Sacramento. (Calif. State Bd. Forestry Rpt. to Legislature 1921 on Senate Concurrent Res. 27.)

(16) National Industrial Conference Board
1926 *The Agricultural Problem in the United States*. 157 p., illus. New York.

(17) Olmstead, F. H.
1919 *Gila River Flood Control*. U.S. Congress, 65th, 3d sess., Senate Doc. 436, 94 p., illus.

(18) Ramser, C. E.
1917 *Prevention of the Erosion of Farm Lands by Terracing*. U.S. Dept. Agr. Bul. 512, 40 p., illus.

(19) ———
1924 *Terracing Farm Lands*. U.S. Dept. Agr. Farmers' Bul. 1386, 22 p., illus.

(20) Reynolds, R. V. R.
1911 *Grazing and Floods: A Study of Conditions in the Manti National Forest, Utah*. U.S. Dept. Agr., Forest Serv. Bul. 91, 16 p., illus.

(21) Sampson, A. W., and Weyl, L. H.
1918 *Range Preservation and Its Relation to Erosion Control on Western Grazing Lands*. U.S. Dept. Agr. Bul. 675, 35 p., illus.

(22) Sheets, E. W., Baker, O. E., Gibbons, C. E., Stine, O. C., and Wilcox, R. H.
1922 *Our Beef Supply*. U.S. Dept. Agr. Yearbook 1921: 227–322, illus.

(23) Spencer, D. A., Hall, M. C., Marsh, C. D., Cotton, J. S., Gibbons, C. E., Stine, O. C., Baker, O. E., Valgren, V. N., Jennings, R. D., Holmes, G. K., Bell, W. B., and Barnes, W. C.
1924 *The Sheep Industry*. U.S. Dept. Agr. Yearbook 1923: 229–310, illus.

(24) Talbot, M. W.
1926 *Range Watering Places in the Southwest*. U.S. Dept. Agr. Bul. 1358, 44 p., illus.

Editor's Comments on Papers 4 and 5

4 **Eakin:** *Silting of Reservoirs*

5 **Stall:** *Man's Role in Affecting the Sedimentation of Streams and Reservoirs*

The next two articles show the other side of the erosion problem—siltation. Whenever soil or rock materials have been removed from the land, the resulting debris must ultimately be deposited elsewhere, causing sedimentation. Volume I provides some information on silt problems, but prior to the 1900s, few major dams were in existence anywhere in the world. In the twentieth century, building of dams and reservoirs went hand in hand with the development of large irrigation projects, with the response to the need for energy provided by hydroelectric power and for flood control, and with development of the multipurpose concept of use with recreation.

By the 1920s it was recognized that silt was becoming a problem of increasing severity in the American Southwest. One of the earliest articles linking siltation to human factors was by Fortier (1928). He provided some perspective on the amount of silt that could be anticipated and on costs for removal, which at that time amounted to $2.00 per acre in the Imperial Valley. The two articles selected for reproduction present two different approaches to the siltation problem. The paper by H. M. Eakin was the first comprehensive nationwide appraisal of the extent of silting in major reservoirs of the United States. The excerpt included here represents only a small part of the much longer published report. The article by J. B. Stall shows the magnitude of the problem 30 years later, and itemizes the costs involved. He also clearly recognized that most of the increased sedimentation rate was man-induced.

Happ (1937) shows the interrelationship of erosion and sedimentation, using the Wells district of northern Mississippi as a case study:

> The erosion problem was accentuated by the common practices of "turning out" land which has begun to decline in production, while clearing other new fields, and these became the prey of spreading and deepening gully systems (p. 197).

Sediments from these upland areas were deposited in the stream valley and drainage canals of the district.

Many estimates and measurements of the magnitude of the erosion–sedimentation problem have been made, and one that is often quoted showing man's influence in increasing the sediment load is that of Brune (1948, p. 16):

> . . . in the upper Mississippi River drainage basin where about 42 percent of the land is now cultivated or idle, the present rate of sediment production and erosion is approximately seventy-five times the geologic norm.

Brown (1950) groups the sediment problems into four categories:(1) suspended-sediment concentration in water; (2) in natural stream channels, improved river channels, and harbors, floodways, ditches, and canals; (3) on land, improvements, and habitats; and (4) in reservoirs. He estimated annual losses at $175 million. Sediment is also harmful to water supply, recreation, commercial fishing, flood control, and power generation.

Golze (1950) determined the cost for removing silt during the 1945–1948 period for several irrigation projects and showed that the costs per mile varied widely from $23 per mile in the Yuma project (Arizona–California) to $187 per mile in the North Platte project (Nebraska–Wyoming). The costs for silt removal generally amounted to 15 percent of all operations and maintenance costs.

Glymph and Storey (1967) summarize many aspects of sedimentation. For the artificial reservoirs built for all purposes to date in the United States, original capacity was about 5000 million acre-feet. Surveys for sediment production in 1069 reservoirs show an average rate of storage loss of 0.2 percent per year, or about 1 million acre-feet of siltation each year. In addition, sedimentation occurs in rivers and harbors, where each year it is necessary to dredge 450 million cubic years to maintain navigation. The Mississippi River deposits hundreds of millions of cubic yards of sediment (not necessarily all man-induced) and advances the 35-ft depth contour at the mouth of the river seaward 100 feet per year. During the 1930–1962 period, 30 million cubic yards were removed annually from the river between Cairo, Illinois, and Baton Rouge, Louisiana.

The U.S. Department of Agriculture maintains a surveillance service for sedimentation in reservoirs that contains data grouped according to the 79 drainage areas of the United States. In somewhat like manner, the U.S. Geological Survey maintains observation programs for wells (more than 10,000) and stream gaging stations (more than 8000) and various sediment-monitoring stations in rivers. However, no attempt is made to separate man-induced sediment rates and natural rates. These surveys indicate the wide variation of the sedimentation problem, ranging from those reservoirs where it is nil or almost nil to those where it is thousands of tons per square mile.

Siltation in reservoirs behind dams often has a deleterious effect on channels downstream. Here erosion occurs because the river is no longer burdened with its sediment load and thus has the capacity to do work. Leopold, Wolman, and Miller (1964) show that 35,000 acre-feet of channel erosion occurs in the Red River in a 100-mile reach downstream from the Denison Dam. Lowering of the stream bed was 5–7 feet in the first 10 miles.

> The sixteen years of record on the Red River suggests that erosion may continue for long distances and over long periods of time. The control of this process lies in the sediment budget of the reach; so long as the supply of appropriate sediments is deficient, the flow may continue to make up the deficiency.
>
> Approximately 386 million tons of sediment have been deposited in Texoma Reservoir behind Denison Dam. Of this, about 20% or 77 million tons, is sand. Sediment removal by erosion downstream from the dam amounts to about 67 million tons. These data and a similar example from the North Canadian River below Canton, Oklahoma, indicate that the amount of sand deposited in the reservoir is roughly equivalent to what is eroded from the channel bed and banks in the reach downstream from the dam (p. 457).

There is another side to the siltation picture, however—benefits may occasionally occur. One of these is the age-old story of the River Nile, and how the annual floods were necessary to replenish the soils and provide nutrients to the cultivated soils of the flood plain. In the Ganges–Brahmaputra delta region of the new country of Bangladesh, Ellis (1972) describes the flood–silt cycle:

> During monsoon rains, the rivers overreach their banks, embracing the land in lifegiving flood. Each year these inundations enrich and transform the terrain that in only a few areas rises more than 30 feet above sea level (p. 304).

In an analogous, but smaller example, Glymph and Storey (1967) report:

> Observations of Hooper Creek and Salt Creek in southeastern Nebraska, after a great flood in 1950, showed that corn yields were increased as much as 45% over a 3-year period on areas receiving from 4 to 6 inches of sediment (p. 208).

Reprinted from *U.S.D.A. Tech. Bull. 524* (1936)

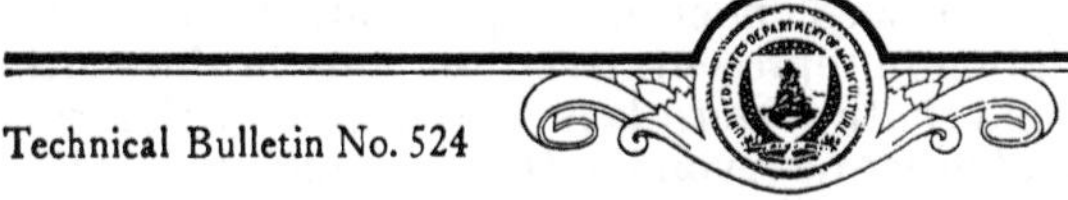

Technical Bulletin No. 524 July 1936

UNITED STATES DEPARTMENT OF AGRICULTURE
WASHINGTON, D. C.

4

SILTING OF RESERVOIRS[1]

By Henry M. Eakin

Head, Hydrodynamic Studies, Division of Research, Soil Conservation Service

CONTENTS

INTRODUCTION

THE PROJECT

The inherent relation of silting of reservoirs to problems of erosion control led the Soil Conservation Service to institute in July 1934 the first attempt at a general Nation-wide survey of the condition of American reservoirs with respect to reduction of storage by silting.

The ultimate objective of these studies is to accumulate information on the factors involved in the silting of reservoirs, including rates of silting, and to correlate the results with soil, slope, and climatic conditions and land use in watershed areas. It seems obvious that such a broad factual basis is necessary for sound determination of policy and practice of reservoir development and silt control in the various sections of the country.

[1] Progress report on reservoir surveys and investigations by the Soil Conservation Service in 1934-35 with notes on previous investigations of silting of reservoirs by other agencies.

63081—36

The work thus far accomplished under the project consists of detailed surveys of silt accumulation in representative reservoirs in the southeastern, south central, and southwestern type areas, along with reconnaissance examinations of other reservoirs in the same regions incident to selection of more broadly significant cases for special study.

The direct objective of each detailed survey has been to determine the volume and distribution of sediment deposits accumulated in the reservoir during a known period of time, either the entire period of the reservoir's existence or a shorter period between an earlier survey and the current resurvey. From these data and those on watershed areas the average annual rate of silting per unit of drainage area is derived as an important practical index to differences and changes in regional erosional conditions and expectancy of useful life of existing or contemplated reservoirs.

These studies are to be extended as rapidly as possible to the more northerly States in order to provide accurate records of the condition of silting in all important reservoirs of the country.

This initial report under the general project is designed to give a preliminary outline of the more important aspects of the problem, to summarize the results of studies made heretofore by other agencies, and to present the findings of the Soil Conservation Service during the fiscal year ended June 30, 1935.

ECONOMIC ASPECTS OF THE PROBLEM

The economic values involved in present and proposed reservoir developments are very large. They include not only investments in dams and basin lands of the reservoir properties themselves, but also the much larger values of appurtenant water supply and power facilities, industrial establishments, and irrigated-land resources and improvements.

A comprehensive inventory of the Nation's resources dependent upon water storage has never been made, but even rough calculations of Federal, State, municipal, and corporate expenditures for reservoir and auxiliary facilities and similar estimates of the aggregate value of dependent private holdings easily run to several billion dollars. All of these interests are bound together under the common menace of depletion of reservoir capacity by silting wherever accelerated erosion occurs.

It is true, of course, that some of these resources are less critically impaired by loss of reservoir storage than others. Power reservoirs, even when completely silted, still afford operating head for ordinary stream flow. Where stream flow is equable through the seasons, either naturally or by reason of regulation by other reservoirs in the same watershed, the injury from local reservoir silting may be almost negligible. In most cases, however, local storage is an important factor of power-plant operation in meeting daily and seasonal demands. The injury caused by silting increases with the dependence of the individual plant upon local storage, and this in many instances, particularly in more arid regions, has a critical relation to profitable power-plant operation.

In general, reservoir storage is vitally important to irrigation and municipal water-supply projects. In these, a depletion of storage by silting to bare equality with maximum seasonal or annual requirements puts the project in a precarious position, and any further depletion means actual deficiency, any degree of which is intolerable and must be covered by construction of additional storage facilities.

This leads to the question of values destroyed by reservoir silting. The view, all too frequently held, is that the destruction is to be measured by the cost of the original reservoir. This could be so only where additional storage could be developed indefinitely and at similar cost. There is, of course, an ultimate limit of feasible storage in every watershed. It is only natural that the initial reservoirs should be constructed at the most favorable and economic sites, and that substitute or supplementary storage facilities, to serve the same locality, are more costly. It is, therefore, probably correct to compute the more immediate harm of reservoir silting upon the basis of replacement rather than original storage costs and the ultimate harm in terms of the entire economic development dependent upon local water storage.

It is pertinent in this connection to consider the possibilities and relative costs of conservation of developed storage against depletion by silting.

In a few instances attempts have been made to vent silt from a reservoir through outlet gates and conduits in the dam. The typical results shown after complete draw-down have been that the normal river current had cut out a new channel of rather deep and narrow cross section from the outlet in the dam to the previous head of backwater. The amount of silt thus removed as shown by the volume of the channel has generally been only a small percentage of the total accumulation.

The proposal has been advanced, in view of the known tendencies toward underflow of heavily silt-charged waters to the dam, that systematic venting of these silt flows through properly arranged gates or conduits might be effective in removing a considerable proportion of silt load.

This proposal might be technically feasible, though its efficiency in removing a large percentage of the total silt load is yet to be proved. However, it is open to grave objections in the matter of silt disposal below the dam. The excess of silt thus released would seriously impair the utility of water below the dam, particularly for irrigation projects, adding to the cost of maintaining canals and field distribution systems.

Hydraulic dredging and mechanical removal of silt would generally cost from 5 to 50 times as much as the original storage. This, of course, is prohibitive except for special-purpose dams where there is no alternative.

In certain instances the rate of storage depletion in reservoirs has been reduced by silt detention in headwater and valley areas by engineering structures and vegetation screens. Such measures are admittedly temporary in nature and have the effect in general of building up alluvium deposits in valleys above the reservoir, thereby being of practical application only where such valley lands are comparatively worthless for other use.

In contrast with these inefficient or objectionable measures, the permanent reduction of silt content in the contributing streams by means of erosion-control practices in the watershed has been abundantly demonstrated. Erosion control not only has the effect of conserving lands in the watershed, but is outstanding as the one fundamental and permanently practicable means of reducing the rate of reservoir silting. It inhibits primary production of debris and thus involves no progressive and ultimately embarrassing accumulations above, or troublesome silt-laden discharge from the reservoir.

PHYSICAL ASPECTS OF THE PROBLEM

Practical understanding of the physical basis of the present-day silt problem requires, first of all, a definite recognition of the dependence of silting upon erosion of directly tributary watershed areas and of the critical fact that rates of erosion are neither uniform nor fixed in different sections of the country but are subject to material change under a civilized use and abuse of lands and various practical measures of erosion control. This is all too frequently a difficult advance in thought concerning erosion, particularly on the part of scientists and engineers who have studied erosion only from the viewpoint of primordial forces and processes of nature.

It is true, of course, that erosion has been active throughout geologic history in sculpturing and planating the lands of the earth and that where man has least interfered with soil and vegetative conditions, erosional processes are still in force at primeval or geologically normal rates. It is also true that, elsewhere, man has materially changed natural erosional conditions by deforestation, agriculture, grazing, and fire so that erosion has been variously accelerated from moderate to truly catastrophic degrees from place to place.

On account of this historic change in erosional conditions over much of the country it is essential to draw a clear distinction between natural and man-induced erosion. This has been well done by erosion specialists through adoption of the terms "geologic norm" and "accelerated erosion" for succinct reference respectively to natural and man-induced phases of erosion. Under geologic norms, erosion was generally far less intense and more regular over broad type regions of climate, soil, and vegetative cover. Acceleration of erosion by human activities has greatly increased average silt production over large sections of the continent and has introduced erratic distribution of erosional intensities quite different from the corresponding phenomena of nature.

The original geologic norms of erosion in each type region of the country reflected the natural balance between opposite factors of (1) erosional attack, determined for the most part by amount and intensity of rainfall, and (2) resistance of the terrain to erosional attack, determined for the most part by vegetative protection of the soil.

The potentials of erosion were naturally stronger in regions of humid climate but were generally countered by heavy forest growth and other vegetation, so that erosion tended toward minimum rates and streams generally ran clear even in flood. Under more arid conditions, potentials of erosion were relatively weak but were less effectively countered by natural vegetation, so that silt production ran relatively high and streams were generally more or less silt laden. Human acceleration of erosion has affected both humid and arid sections of the country, but has been proportionately somewhat more effective in humid regions where human activities have destroyed the balance between stronger natural factors.

Acceleration of erosion in the humid sections of the country has been brought about most extensively by land clearing and clean cultivation of sloping lands occupied by deep soils. It has been for the most part in such regions of greater rainfall that an aggregate of approximately 100,000,000 acres of formerly tilled land has been washed and gullied to the point of agricultural abandonment, or so severely affected by sheet washing as to support only a submarginal type of agriculture, and an even greater area has been seriously affected within the geologically brief period of civilized occupation. Acceleration of erosion in the more arid sections of the country, though proportionally less than in the humid regions, has nevertheless involved extensive wastage of topsoil and catastrophic development of arroyos and gullies over widespread areas.

To appreciate the full import of human acceleration of erosion in relation to the silt problem of the present and future, it is necessary to recognize the progressive and cyclical aspects of the phenomenon of erosion rejuvenation. It is a basic physiographic principle that rejuvenated erosion tends to spread progressively upstream and upslope throughout the affected watershed area. Each gully and arroyo lowers the elevation of controlling base level and carries the menace of potential degradation in like measure to its entire drainage area.

The entrenchment of rejuvenated drainage courses extends progressively upstream through headward erosion, and each entrenched section of trunk and tributary stream entails additional silt production through lateral planation of exposed banks and general slope readjustment. Acceleration of silt production thus advances at increasing rates through a long period leading up to maturity of a new erosion cycle. The new erosion cycles that have been recently developed so extensively throughout the country as a result of human activity are mostly still in incipient and immature stages. It thus appears that the present-day rates of silt production in many sections of the country, although now greater than those of the original geologic norms, may be destined to increase still further for a long time in the future unless countered by corrective interference with the natural progress of erosion through various measures already established or under development for erosion control. Natural revegetation will, of course, impede erosion on large areas of abandoned land, especially where fires are controlled.

PROCESSES OF RESERVOIR SILTING

The rate of sediment accumulation and forms of its distribution in a reservoir are determined by the volume and character of load carried below flow line by contributing streams. The transporting power of streams and the load they carry are, in turn, determined by the extent of the drainage area and erosional factors of climate, topographic relief, soils, vegetative cover, and land use over watershed areas. The total stream load derived from headwater sources is subject to variation, from time to time, by changes in factors controlling erosion, particularly through changes in extent and effectiveness of vegetative cover. The load at any given time is subject to change in character from place to place approaching the reservoir, through processes of wear of debris particles and exchange of material between load and alluvial environment in the course of travel, and to change in volume through processes of valley scour and deposition.

As a general rule, particularly where load includes a considerable proportion of coarse-grained material and the reservoir is frequently at full stage, the load of contributing streams is subject to radical reduction in both volume and average grain size through processes of valley aggradation above reservoir level. This process of load reduction tends to be still further augmented where such valley deposits come to support new vegetation that acts as a sediment screen by spreading and retarding stream flow. This process, like that of primary headwater erosion, is subject to change with time. Its characteristic trend is to become more and more effective with time and thus gradually to reduce the rate of silting of the reservoir proper.

This process is naturally a benefit to conservation of reservoir storage, but involves the possible ruin and sacrifice of riparian properties in the lower part of the affected valley. The deposits in the valley, due to this process that is attributable to the new base level created by the reservoir, tend to decrease in depth going up valley. Somewhat similar valley deposits may occur also as the result of accelerated headwater erosion, but the latter are generally characterized by decrease in depth of deposits going down valley. The possible use of this criterion to determine the origin of valley deposits, in many cases, may be of decisive economic importance.

The rate of reservoir silting is also subject to progressive reduction by escape of increasing amounts of debris over or through the discharge of dams as storage space is reduced by silting. This effect, however, is limited to the factor of finest fractions of suspended load in all but the final stages of reservoir filling and so cannot enter with any great importance into the problem of major storage conservation.

The processes and forms of sediment accumulation strictly within the storage space of reservoirs are dependent primarily upon differences in settling rates and mass effects of differently sized fractions of stream-borne load.

The relatively coarse-grained materials, that are carried on or settle rapidly toward the bottom, tend to accumulate in the form of deltas where velocity slackens at the mouths of tributary streams. Such deposits, typically developed without much change in reservoir stage, are composed of characteristic bottom-set, fore-set, and top-set beds of conventional delta form, limited in areal extent and confined in the vertical to the higher levels of original storage.

The aggregate volume of delta deposits reflects the presence and amount of relatively coarse-grained debris in total stream-borne load. If only coarse-grained materials are carried the deltas may truly represent the entire sedimentary accumulation. On the other hand, if no coarse-grained materials are carried by tributary streams, delta processes and features may be entirely lacking.

The processes and final forms of deltas are frequently complicated by changes in reservoir level. Such deposits formed at higher levels are subject to entrenchment and more or less complete removal at lower stages, the moved materials being redeposited in new delta features farther downstream and at lower elevation. The form of delta accumulations in reservoirs that vary in stage is thus subject to radical change from one time to another and the problem of comparative measurement of total delta volume by periodic resurveys is correspondingly complicated.

The building of deltas at the mouths of tributary streams is the most conspicuous and familiar phenomenon of reservoir silting and one that is all too frequently recognized to the exclusion of all others There are, however, just as definite and important phenomena of distribution of finer grained fractions of load within the storage space of reservoirs as those of delta building.

The finer grained materials, in contrast with the tendency of coarser debris to concentrate in limited higher areas of the reservoir, tend to spread broadly over the reservoir bottom and in many cases to accumulate selectively in the deeper portions of the basin. This tendency to concentrate beneath the deeper waters is due to the striking and widely prevalent phenomenon of underflow of heavier, silt-charged, inflowing waters beneath lighter, desilted waters already in storage. This is not a readily observable phenomenon and consequently it has not been broadly recognized in engineering literature. It is of signal importance to systematic silt studies in that the volume of finer grained bottom-set beds, more frequently than otherwise, has been found to exceed the total volume of delta deposits and that, contrary to customary thought, the depletion of deeper reservoir storage space in the vicinity of the dam may not await the gradual approach of growing deltas but may, and in most cases does, begin at selective rates from the very beginning of storage.

The velocity of such underflow would appear to be a natural function of the density of inflowing silt-charged waters compared with that of the overlying desilted waters already in storage. Toward the dam the velocity of the underflow currents probably tend to diminish as a result of progressive settling of the sediment involved. In extreme cases, such as that occasioned by flood flows from Puerco River, carrying up to 15 percent of solid matter into the Elephant Butte Reservoir, the rate of underflow has been computed at over 2 feet per second, or about 1½ miles per hour. Muddy underflows in this reservoir have repeatedly extended to the dam and have given muddy discharges through the outlet gates for longer or shorter periods while the surface waters of the lake were perfectly clear. The record includes a maximum period of 17 days of muddy discharge from a superficially clear blue lake with silt content of the discharge ranging up to 6 percent by weight.

The same phenomenon has occurred in connection with the operation of other southwestern reservoirs that receive very muddy flood flows from tributary rivers. The resultant deposits in the deeper parts of the reservoirs tend to be extremely flat in both transverse and longitudinal direction and so to increase in depth and in ratio to overlying water approaching the dam. Similar features of silt distribution have been noted in reservoirs of the more humid sections of Texas and even in the very humid section of the southern Piedmont, which would appear to indicate that the phenomenon of underflow is not strictly limited to cases of extremely muddy inflow but is a general phenomenon of stratification of different density liquids tending to be expressed quite generally in reservoir silting wherever an effective difference in silt content of inflowing and stored waters may occur.

Recognition of the importance and careful measurement of the widespread bottom-set beds in reservoirs is essential to accurate studies of general rates of reservoir silting. In the course of the present project, cases have been found where previous oversight of this form of deposit in the main reservoir area above the dam introduced very large errors in computations of total sediment.

PREVIOUS INVESTIGATIONS

Published results of actual capacity surveys to determine rates of silting appear to be limited to some 25 American reservoirs. Of this number, nine have been of basin type, with original capacities large enough to permit practically complete natural desilting of all inflowing water. The volumes of silt per unit of drainage area found in reservoirs of this type, therefore, afford a practical index to comparative average rates of erosion in the respective watersheds during periods of record. The rest of the list of previously surveyed reservoirs have been more strictly of channel type, with relative small storage in relation to inflow and, therefore, given to indeterminate wastage of silt past the dam in time of flood. In these cases the disclosed volumes of silt represent uncertain fractions of total silt delivered from the corresponding watershed areas.

Owing to this basic difference in practical significance of their data, these two classes of reservoirs are treated separately in the following discussion of previous investigations.

BASIN RESERVOIRS

All the reservoirs classed as basin reservoirs, with the single exception of Lake Michie, a small reservoir near Durham, N. C., for which a 3-year record was indicated by resurvey in 1930, are in the relatively dry sections of Texas, New Mexico, Arizona, and California, where flood flows of streams are generally conspicuously charged with sediment and the seriousness of the silt problem has been widely recognized for many years past.

Table 1 presents a summary of data on these reservoirs including computations of average annual rate of silt accumulation per 100 square miles of drainage area. This expression is essentially an index of comparative average erosional intensities in the directly tributary watershed area. Insofar as the local watershed is representative of average climatic and erosional conditions of a general region or erosion province, the determined rate of silting per unit watershed area would appear applicable to watershed areas of other reservoirs or prospective reservoir sites as perhaps the most practicable method of evaluating probabilities of storage depletion and useful life in the absence of individual reservoir silt surveys or adequate sediment records on the streams involved.

TABLE 1.—*Silting records of reservoirs of higher capacity-inflow ratio*

Reservoir	Stream	Location	Period	Amount of silt per 100 square miles of drainage per year	Percentage of original capacity per year depleted by silting	Original storage capacity per square mile of drainage area
				Acre-feet	*Percent*	*Acre-feet*
White Rock	White Rock Creek	Dallas, Tex	1923–28 1910–35	119.00 136.21	0.69 .86	159.28
Elephant Butte	Salt River	Hot Springs, N. Mex.	1915–25 1925–35	88.07 68.54	.88 .68	100.29
Roosevelt	do	Roosevelt, Ariz	1910–25	116.90	.41	284.25
Lake Michie	Flat River	Durham, N. C	1926–30 1926–35	17.60 26.55	.24 .36	74.54
Lake Worth	West Fork Trinity River.	Fort Worth, Tex	1915–28	57.07	2.26	25.00
Lake McMillan	Pecos River	Carlsbad, N. Mex	1894–1915 1915–25 1925–32	8.78 1.59 .95	2.14 .38 .31	4.09
Zuni	Zuni River	Black Rock, N. Mex	1907–17 1917–27	110.14 120.96	3.48 3.83	31.62
Sweetwater [1]	Sweetwater River	Sunnyside, Calif	1888–95 1895–1916 1916–27	89.74 74.03 95.83	.90 .65 .56	121.12 200.93
Lake Cabot	San Leandro River	Oakland, Calif	1875–1923	174.00	.54	404.76
Gibralter	Santa Inez River	Santa Barbara, Calif	1920–25 1925–31 1931–34	80.00 125.00 300.00	1.10 1.71 4.11	73.00

[1] Original capacity increased in 1895 by raising dam 5 feet, and in 1911 by raising dam an additional 15 feet.

Four of the reservoirs listed in table 1 were resurveyed in 1935, White Rock and Lake Michie by the Soil Conservation Service, Elephant Butte by the Soil Conservation Service in cooperation with the United States Bureau of Reclamation, and Roosevelt by the Salt River Valley Water Users' Association. In each of these cases, the 1935 as well as the earlier data are given.

* * * * * * *

CHANNEL AND OTHER RESERVOIRS OF SMALL CAPACITY-INFLOW RATIO

Table 7 lists the reservoirs of relatively low-capacity-inflow ratio that have been studied in the past by various agencies with respect to rates of silting. The computations of original storage capacity per square mile of drainage and average annual rates of silting expressed in acre-feet per 100 square miles of drainage and in percentage of original capacity, are based, for the most part, upon Stevens' (*14*) tabulation of records of American reservoirs for which amount of silting has been measured.

The arrangement of table 7 in the order of increasing original storage capacity per square mile of drainage brings out clearly the general tendency of silting to increase with the size of the original reservoir. The weight of evidence obviously indicates that the larger the res-

Table 7.—*Silting records of reservoirs of lower capacity-inflow ratio*

Reservoir	Stream	State	Period	Original storage capacity per square mile of drainage	Amount of silt per 100 square miles of drainage per year	Percentage of original capacity per year depleted by silting
				Acre-feet	*Acre-feet*	*Percent*
Coon Rapids Pond	Mississippi	Minnesota	1899–1931	0.42	0.44	1.08
New Lake Austin	Colorado	Texas	1913–26	.83	6.10	7.35
Old Lake Austin	do	do	1893–1900	1.29	9.16	7.10
Lake Penick	Clear Fork of Brazos	do	1920–27	1.37	6.13	4.47
La Grange	Tuolumne	California	1895–1931	1.55	3.60	2.31
Sterling Pool	Rock River	Illinois	1912–30	1.56	1.28	.82
Boysen	Bighorn	Wyoming	1911–24	2.07	12.92	6.25
Keokuk	Mississippi	Iowa	1891–1928	3.11	6.27	2.02
Guernsey	North Platte	Wyoming	1927–33	4.44	8.78	1.98
Furnish	Umatilla	Oregon	1909–31	4.58	17.08	3.73
Hales Bar	Tennessee	Tennessee	1913–30	7.16	12.34	1.72
Buckhorn	Buckhorn	Colorado	1907–25	9.15	24.15	2.64
Cheoah	Little Tennessee	Tennessee	1918–30	25.68	22.70	.88
Parkville	Ocoee	do	1912–30	161.67	185.00	1.14

ervoir the more complete the desilting of inflowing waters, and the greater will be the actual volume of silt deposited. At the same time it is shown that the average annual depletion of storage expressed as percentage of original capacity tends to decrease as original capacity relative to drainage area rises. This would appear theoretically to be a normal result of complete deposition of bed load and fractional deposition of suspended load, the latter only being the variable factor dependent upon capacity-inflow ratio.

The foregoing principle of dependence of rate of silting upon the capacity of the reservoir has been recognized by Taylor (*15, p. 37*) in his discussion of the silting of the reservoirs on the Colorado River at Austin, Tex. After stating that "the silting of reservoirs is generally erratic, and spasmodic, and is confined (mainly) to flood periods", he proceeds with the development of a mathematical expression of companionate waning of capacity and rate of silting in reservoirs receiving discharge of alluvial streams. Based upon the assumption, made to simplify analysis, that silting is a regularly progressive phenomenon, his formula indicates progressive reduction in rates of silting as capacity is depleted.

The general validity of this principle seems well supported by the histories of the two successive reservoirs at the same site on the Colorado River at Austin, Tex. The average total silt carried by the river has been computed by Taylor (*15*) as 18,000 acre-feet a year or 47.4 acre-feet per year per 100 square miles of drainage. The rates of silting in the old and new dams, respectively, have been computed as 8.84 and 6.18 acre-feet a year per 100 square miles of drainage or 18.7 and 13.1 percent, respectively, of total silt of the river. The higher rates of silting in the old reservoir relate to the original and final capacities of 49,300 and 25,777 acre-feet from 1893 to 1900 compared with corresponding capacities of 32,029 and 1,477 acre-feet for the new reservoir from 1913 to 1926. The original capacity of the new reservoir was 65 percent and its rate of silting 70 percent of corresponding features of the old dam.

The final limiting condition approached by such companionate waning of capacity and rates of silting apparently must be a moderate residual capacity and zero rate of silting. The final capacity would be that of the volume of an adjusted alluvial flood channel through the filled reservoir, subject, perhaps, to temporary reduction by sedimentation in low-water season and restoration to the same general dimension with each succeeding flood. Under these conditions the low-water silting would be temporary and result in no permanent increase in silt accumulation from flood to flood.

The rates of silting in the individual reservoirs (table 7) that appear to be more or less out of line with the general trend of variation of capacity-drainage area ratio are patently explicable by exceptional characteristics or vagaries of climate or unusual erosional conditions of the watershed or streams involved in each case. The discordantly low rate of the La Grange Reservoir is undoubtedly a reflection of the relatively dry climate and low annual rainfall of its general climatic province; that of Sterling Pool the result of a relatively flat watershed that is largely occupied by land of low erodibility. The relatively high rate of silting of the Boysen Reservoir on the other hand, undoubtedly relates, in the main, to catastrophic erosion of the Bad Water River drainage basin during the extraordinary floods of July and September 1923.

Inasmuch as the rates of silting of channel and other low-capacity-inflow ratio reservoirs relates to fractional desilting of inflowing waters, each reservoir of its general class is more or less a law of itself as to the relation of its life expectancy to past records. The data of periodic capacity surveys of these reservoirs are thus chiefly useful as a basis of individual forecast. However, in each case, they may be studied to advantage in relation to general land use and policy of erosion control, in the respective watershed areas. In final analysis, it appears that only through reduced production and delivery of erosional waste, particularly of coarser grained nonsuspensible character, by improved land use and erosion control in the watersheds of such reservoirs, can effective approach be made toward improving their condition and the prospects for longer periods of useful life.

* * * * * * *

Roosevelt Reservoir, Salt River, Ariz.: *A*, Mud cracks on drying silt surface exposed by lake draw-down; background shows the great width and depth of mud fill. *B*, Intermediate delta being reexcavated and moved into deeper parts of lake.

Roosevelt Reservoir, Salt River, Ariz.: *A*, Remnant of fill held between branches of tree and indicating former height of silt deposits. *B*, Silt deposits around large cottonwood trees; trunks entirely buried, branches only showing above surface.

SUMMARY

The foregoing data on significant reservoirs in the southeastern, southern Great Plains, and southwestern type areas of the United States show that silting of reservoirs is a practical problem of the first order of importance in all three regions, wherever accelerated erosion is in force. Observations made on the many other reservoirs, visited but not yet surveyed, and on erosional conditions along all routes of travel emphasize the dependence of undue rates of silting upon man-induced erosion and the general prevalence of these conditions over broad areas of the country.

In the Southeast, reservoir silting results chiefly from erosion of deep residual soils as influenced by human occupation. Lower rates obtain in mountainous and other sections wherever the natural forest cover is practically intact. Higher rates go with agricultural practices in the lower Piedmont country—practices that can be greatly improved upon from the standpoint of erosion. Organized cooperation of the agricultural population toward better terrace and crop practices and rededication of over-steep lands, to noncultivated crops or forest, is eminently in order. Also much benefit can come from stuctural and vegetational gully control.

In the southern Great Plains higher rates of silting relate to erosion of sedimentary soils under agricultural and grazing practices. Greater attention to terrace and contour cultivation, cover and strip cropping, and control of incipient gullies can contribute material improvement. More restricted grazing and range restoration is needed in many places. Silt detention above reservoir level in broad tributary valleys in most reservoir watersheds also can be effectively and profitably employed.

In the Southwest higher rates of silting are largely the result of overgrazing and its consequence of extraordinary sheet and gully erosion. Sheet erosion, perhaps, can benefit in time through further restriction of grazing beyond that already imposed by the sparseness of remaining grasses. Arroyo and gully production of sediment can be reduced or even prevented from increasing, only by direct engineering methods. Silt detention above reservoir level in broad tributary valleys by earth barriers and by growth of vegetative screens in delta areas is highly practicable in many cases.

It would also appear, in view of the indications of underflow and congregation against the dams in this region of muddy flood waters, that important reduction of rate of silting is probably feasible through selective wastage of heaviest charge waters through low-placed outlet gates. Such waters remain fluid after considerable settling and concentration of mud charge and capable of selective movement toward a lower outlet. Sludges carrying 20 to perhaps 30 percent of solid matter should be subject to such selective movement, under laws of stratification of liquids of different densities, for long distances above dams, provided that selective evacuation be practiced from the start to maintain bottom slope and deep sludge pockets approaching the dam. Silt deposits against the dam and in adjacent deeps of the reservoir reduce the forces of selective flow and the thickness of muddy water accumulation at the dam, and thus preclude the possibility of deferred development of optimum mud evacuation practice.

Without such practice it is not an unknown occurrence for clear waters, which have deposited their load in the reservoir to be wasted over the spillway, to make room at the bottom for a new body of very muddy water. This would appear to be the ultimate arrangement for shortening reservoir life. On the other hand, since the predominant volume of solid matter that encroaches upon capacity of these southwestern reservoirs is composed of finer grained silts and clays, long held in suspension before final settlement and semisolidification, it appears that a very notable lengthening of useful life of present and future reservoirs in this region should follow successful development of a practical system of mud evacuation. The question of disposal of mud thus passed through the dam would probably be the crux of the problem in most cases and positively preclude adoption of such a practice where clear water discharge is imperative.

In the California area higher rates of reservoir silting relate, for the most part, to watershed fires. Fire prevention, quick reseeding of burned-over areas and treatment of gullies in earliest incipient stage would appear to be generally in order to keep rates in nonagricultural areas at a minimum. When erosion of cultivated lands is involved, as at Lake Hodges and other reservoirs of the lower foothills, modern erosion-control methods are applicable.

In broad national view, it appears with unmistakable clearness that exorbitant rates of depletion of reservoir storage by silting are widely prevalent and that the problem of protection of reservoirs from this menace goes hand in hand with that of saving farm and range lands from impairment and destruction by uncontrolled erosion. While some supplementary practices of debris disposal may be employed to conserve storage from silt encroachment, they are generally subject to special difficulties and limitations.

Main reliance for material and permanent conservation of reservoir resources therefore resides in control of silt production at primary sources through more widespread and effective application of established methods of erosion control.

LITERATURE CITED

(1) ALLEN, R. T., THORNTON, E. W., and HILL, H.
1912. SOIL SURVEY OF CABARRUS COUNTY, NORTH CAROLINA. U. S. Dept. Agr., Bur. Soils Field Oper. 1910, Rept. 12: 297–339, illus.

(2) BALE, H. E.
1930. OIL AND GAS IN OKLAHOMA. LOGAN COUNTY. Okla. Geol. Survey Bull. 40, v. 2, pp. [225]–238; illus.

(3) DARTON, N. H.
1925. A RESUME OF ARIZONA GEOLOGY. Ariz. Univ., Ariz. Bur. Mines Bull. 119 (Geol. ser. no. 3), 298 pp., illus.

(4) FARIS, O. A.
1933. THE SILT LOAD OF TEXAS STREAMS. U. S. Dept. Agr. Tech. Bull. 382, 71 pp., illus.

(5) FIOCK, L. R.
1934. RECORDS OF SILT CARRIED BY THE RIO GRANDE AND ITS ACCUMULATION IN ELEPHANT BUTTE RESERVOIR. Amer. Geophys. Union Trans. Ann. Meeting 15 (pt. 2): 468–473.

(6) GOULD, C. N.
1925. INDEX TO THE STRATIGRAPHY OF OKLAHOMA. (With lists of characteristic fossils. By C. E. Decker). Okla. Geol. Survey Bull. 35, 115 pp.

(7) HUDSON, R. G., LIPKA, J., LUTHER, H. B., and PEABODY, D., JR.
1917. THE ENGINEERS' MANUAL. 315 pp., illus. New York.

(8) JURNEY, R. C., BACON, S. R., and MORGAN, J. J.
1931. SOIL SURVEY OF PERSON COUNTY, NORTH CAROLINA. U. S. Dept. Agr., Bur. Chem. and Soils, Ser. 1928, Rept. 14, 36 pp., illus.

(9) LAWSON, L. M.
1925. EFFECT OF RIO GRANDE STORAGE ON RIVER EROSION AND DEPOSITION. . . Eng. News Rec. 95: 372–374, illus.

(10) RANSOME, F. L.
1919. THE COPPER DEPOSITS OF RAY AND MIAMI, ARIZONA. U. S. Geol. Survey Prof. Paper 115, 192 pp., illus.

(11) SCHWENNESON, A. T.
1921. GEOLOGY AND WATER RESOURCES OF THE GILA AND SAN CARLOS VALLEYS IN THE SAN CARLOS INDIAN RESERVATION, ARIZONA. U. S. Geol. Survey Water-Supply Paper 450: 1–27, illus.

(12) SELLARDS, E. H., ADKINS, W. S., and PLUMMER, F. B.
1933. GEOLOGY OF TEXAS. Tex. Univ. Bull. 3232, v. 1, Stratigraphy, 1007 pp., illus.

(13) STABLER, H.
1925. DOES DESILTING AFFECT CUTTING POWER OF STREAMS? [Letter to editor.] Engin. News. Rec. 95: 968–969.

(14) STEVENS, J. C.
1934. THE SILT PROBLEM. Amer. Soc. Civ. Engin. Proc. 60: 1179–1218.

(15) TAYLOR, T. U.
1930. SILTING OF RESERVOIRS. Tex. Univ. Bull. 3025, 170 pp. illus.

(16) UNITED STATES CONGRESS.
1933. SANTEE RIVER, N. C. AND S. C. Letter from the Secretary of War transmitting report from Chief of Engineers. . . 73d Cong., 1st sess., House Doc. 96, 153 pp., illus.

(17) VAN ATTA, E. S., BRINKLEY, L. L., and DAVIDSON, S. F.
1924. SOIL SURVEY OF ORANGE COUNTY, NORTH CAROLINA. U. S. Dept. Agr., Bur. Soils Field Oper. 1918, Rept. 20: 221–264, illus.

(18) VEATCH, J. O., and STEPHENSON, L. W.
1911. PRELIMINARY REPORT ON THE GEOLOGY OF THE COASTAL PLAIN OF GEORGIA. Ga. Geol. Survey Bull. 26, 466 pp., illus.

(19) WILLIAMS, C. B., HEARN, W. C., PLUMMER, J. K., and PATE, W. F.
1917. COUNTY SOIL REPORT 4, REPORT ON CABARRUS COUNTY SOILS AND AGRICULTURE. N. C. Dept. Agr. Bull., v. 38, no. 8 (whole no. 235), 46 pp., illus.

Reprinted from *Proc. 2nd Ann. Amer. Water Resources Conf.*, 79–95 (1966)

MAN'S ROLE IN AFFECTING THE SEDIMENTATION OF STREAMS AND RESERVOIRS

5

by

John B. Stall

Engineer

Illinois State Water Survey

Urbana, Illinois

Abstract

The soil conservation job in America is now about one-third done but two-thirds remains to be done. Soil losses from land erosion contribute sediment to the streams of America. Sediment yields range from 3 to 10 tons per acre per year from large watersheds, for much of this country.

The value of the plant nutrients removed by soil erosion from the lands of America is $800 million per year. The damage caused merely by the presence of the sediment particles in the stream waters of America is estimated to be $260 million per year.

Strong new action programs are under way to control the pollution of streams. For most uses, the presence of sediment particles in water is injurious. Consequently sediment in streams is a pollutant, and it should be recognized as such. Pollution control programs should include efforts, regulations, and even laws to prevent stream pollution by sediment.

For 35 cities in Illinois hydrologic analyses showed that as a reservoir loses its capacity to sediment, the water supply yield of the reservoir is decreased proportionately.

Soil conservation programs can reduce sediment yields dramatically to allowable levels. In eight reservoirs throughout the country, sedimentation surveys have documented reductions in lake sedimentation; 28 percent reduction in one case and 73 percent in another.

The elements of a soil conservation program are well defined: contour farming, strip cropping, lengthened crop rotations to include more pasture or grass, and other specialized measures. These practices allow man to use the land effectively as a permanent resource. Further adoption of these practices will reduce sediment yields and will abate the pollution of our streams by sediment particles.

Sediment Damages

Erosion Damage

During the past 50 years man has scarred the face of America by using the land far beyond its permanent capability and by abusing the land in many cases. For the past 30 years man has been at work to carry out a national program of soil conservation which is aimed at the healing of erosion scars. This program has as its object the maximum productive use of each acre of land in accord with its capabilities. Such a program of soil conservation would reduce soil losses to allowable levels, and would create permanent harmony between man and the land.

Shown in Table I is a summary of the agricultural land resources of America today and a summary of the magnitude of the remaining soil conservation problem. These data are provided by the U.S. Department of Agriculture (11). In Table I it is shown that the total privately-owned agricultural land in the United States amounts to 1448 million acres. Of this, 447 million acres or about one-third is cropland. The lower portion of Table I shows that 38 percent of this cropland needs no treatment; of this, 17 percent of it has no conservation problem and 83 percent of it is already treated. Table I shows that 62 percent of the private cropland in the country today needs conservation treatment. This amounts to 272 million acres.

A simple way to summarize the scope of the soil conservation problem today as depicted in Table I is to say that about one-third of the cropland of the country has been treated and about two-thirds of the cropland of America still needs soil conservation treatment.

Sediment Yields

General information is available in America as to the variability of sediment yield from watersheds, sediment yield being the amount of sediment which leaves the watershed. Geiger (4) presented data on the sediment yields from 406 watersheds in the United States smaller than 100 square miles in size. These data have been generalized by the author and are shown in Figure 1. The map in Figure 1 shows the relative sediment yields for various parts of the United States. Sediment yields are shown in tons per acre per year. As can be seen from Figure 1 the highest apparent sediment yields in America are occurring in the midwest and in the south. The highest sediment yields shown in Figure 1 are about 10 tons per acre per year along the lower Mississippi River. A yield of 7 tons per acre per year predominates in most of the southeast. For most of the cornbelt the sediment yield is about 4 tons per acre per year. All of these yields are considered to be excessive; they could and should be reduced to an allowable rate of perhaps 1 ton per acre per year.

Magnitude of Sediment Damage

For the purpose of this paper an effort has been made to make an overall evaluation of the total annual damages caused by sediment in the United States. While the specification of these damages is nebulous, the results are adequate to furnish an appreciation of the magnitude of the sediment problem in America today. Table II is a summary of the annual sediment damages in the United States in 1966 dollars.

The first entry in Table II is an entry which is not counted as sediment damage. This item is the loss of agricultural productivity of farmland lost to erosion. This amount is estimated to be $800 million per year. This figure represents the present market value of the plant nutrients which are contained in the soil being lost to erosion. This available plant food is estimated to have a value of $1 per ton of soil. This $800 million annually is a monetary evaluation of the impoverishment of our land resources which we are suffering each year because of our

Table I. The Scope of the Conservation Job in America Today

LAND RESOURCES

Description	Amount in million acres
Total land area of USA mainland	1901
urban area	51
water area	7
federal land, not crops	396
total private agricultural land	1448
cropland	447
pasture or range	485
forest	450
other	66

CONSERVATION PROBLEMS

Description	Amount in million acres	Percent
Private cropland	447	
38% needs no treatment	164	
17% of this has no conservation problem	28	
83% of this is already treated	136	
62% needs conservation treatment	272	
Dominant conservation problems are:		
soil erosion	161	60
excess water	60	22
unfavorable soil	36	13
adverse climate	15	5
		100
Pasture land	485	
73% needs conservation treatment	364	
Forest land	450	
55% needs conservation treatment	241	
Other land	66	
17% needs conservation treatment	10	

Source: USDA National Inventory or Soil and Water Conservation Needs, Reference 1

present inadequate farming practices. It is to overcome this loss that the present national program of soil conservation is oriented.

The remainder of the items in Table II are separable items of sediment damage which are estimated to occur annually merely as the result of inert sediment particles being carried by the streams of America. These damages are caused by sediment and occur quite apart from the basic impoverishment of our soil by soil erosion.

Table II. Annual Sediment Damages in the United States (1966 dollars)

	Million Dollars
Loss of agricultural productivity of farm land lost to erosion $800 mil.	Not counted as sediment damage
Deposition of infertile sediments on river flood plains, from watershed studies covering 13% of the USA (Reference 2)	50
Sediment in reservoirs Cost of storage space destroyed (Reference 9)	50
Removal of sediment from drainage ditches, half the cost of maintenance, or 19¢ per acre or $93 per mile of ditch. (Reference 12)	18
Dredging sediment from inland navigation channels and harbors	83
Removal of sediment from irrigation ditches	16
Removal of excess turbidity from public water supplies	14
Flood damages, increment due to sediment in water, costs of cleaning out sediment, increased flood heights due to clogging of channel, 20 percent of flood damage (Reference 2)	20
Other damages not yet evaluated such as added maintenance cost of highways, damages to fish, damage to hydroelectric power turbines, impairment of public health (Reference 2)	11
Damage to recreation not evaluated	--
TOTAL	262

The results in Table II have been derived from a variety of sources and are mostly described within the table. Table II shows that the total annual damage in dollars caused by sediment in streams in the United States is about $262 million. Thus it can be seen from Table II that the damage caused by the mere presence of sediment particles in streams is about one-third the total cost of the basic loss of soil productivity which occurs each year in the United States.

One of the items in Table II shows that it costs about 16 million dollars each year just to remove sediment from irrigation ditches in the United States. Irrigation ditches represent a considerable portion of the investment in an irrigation project. Often it is economical to provide major measures for the removal or prevention of sediment in these ditches.

The outstanding example of large permanent works for the removal of

SEDIMENT YIELD

TONS PER ACRE PER YEAR

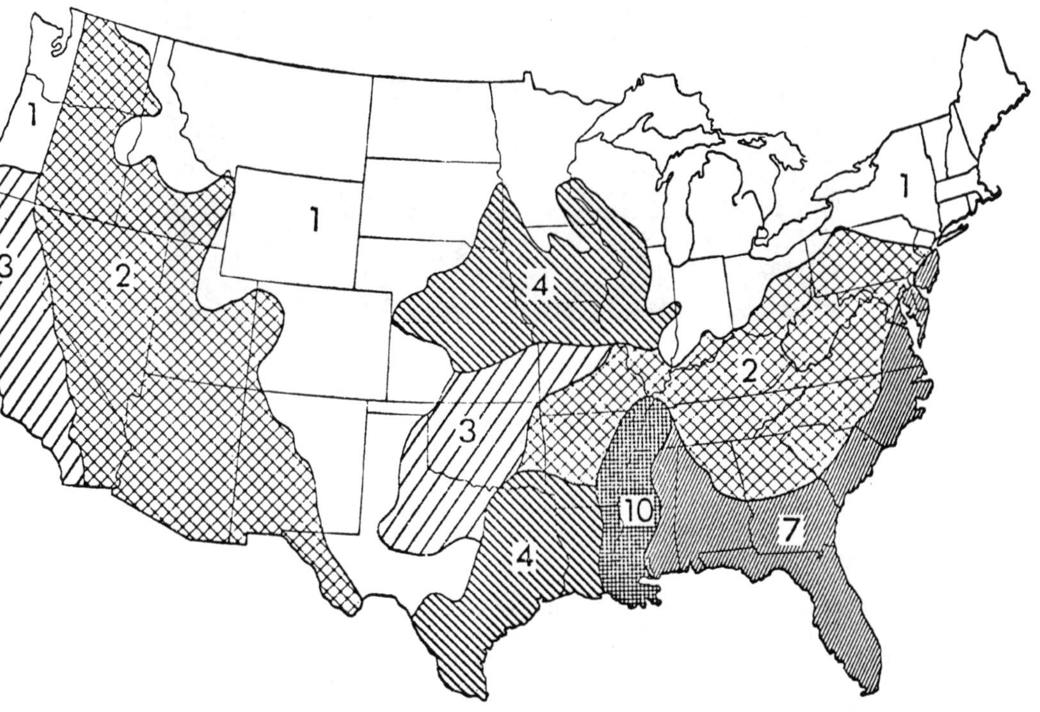

Figure 1. Generalized Sediment Yield Rates From Watersheds, in Tons per Acre per Year

sediment in this country is the installation made at the head of the All-American Canal at Imperial Dam on the Colorado River in California, shown in Figure 2. The construction of such permanent works for the artificial removal of sediment from streams represents the ultimate effort which can now be justified in removing sediment from streams. The series of six concrete basins provided reduces the sediment load from 60,000 tons per day to 20,000 in a water flow of 15,000 cubic feet per second. The basins are emptied artificially. Golze (5) reported that Imperial Dam cost $5.5 million and the desil ng works cost $4.5 million. Before the desilting works were built an average of $700,000 was spent each year for removal of sediment from the irrigation canal systems. The amortized annual cost of the desilting works is $200,000 so the net annual saving in sediment damage from the construction of these works is $500,000.

Figure 2. Man-made Desilting Works at the Head of All-American Canal, California

Sediment in Streams

The American Water Resource

Sediment problems are dominant in all water-resources development in the United States, as will be shown. In water resource planning the first required item is knowledge of the amount of the water resource. In Figure 3 is shown a map of the United States showing the average annual stream runoff in inches for various localities throughout the United States. This map is based on about 50 years of record from stream gaging stations in the United States. As such, this map furnishes a reasonable evaluation of the total quantity of surface water available from streams in this nation to be used by man. Inspection of Figure 3 shows that the Pacific northwest is the most water rich portion of the country. It enjoys an annual runoff of 40 inches per year. The Select Committee of the U.S. Senate (6) considered the total water supply of the United States to be the average sustained streamflow of 1081 billion gallons per day. This is equivalent to 7.5 inches of annual runoff, which appears to be a reasonable national average of the runoff as depicted in Figure 3.

Generally speaking the region east of the Mississippi River enjoys an annual runoff which exceeds 10 inches per year. Here in the eastern United States the annual runoff reaches as high as 30 inches per year in parts of New England, Eastern Tennessee and Southern Alabama.

About 1955 the Select Committee on Water Resources of the U.S. Senate

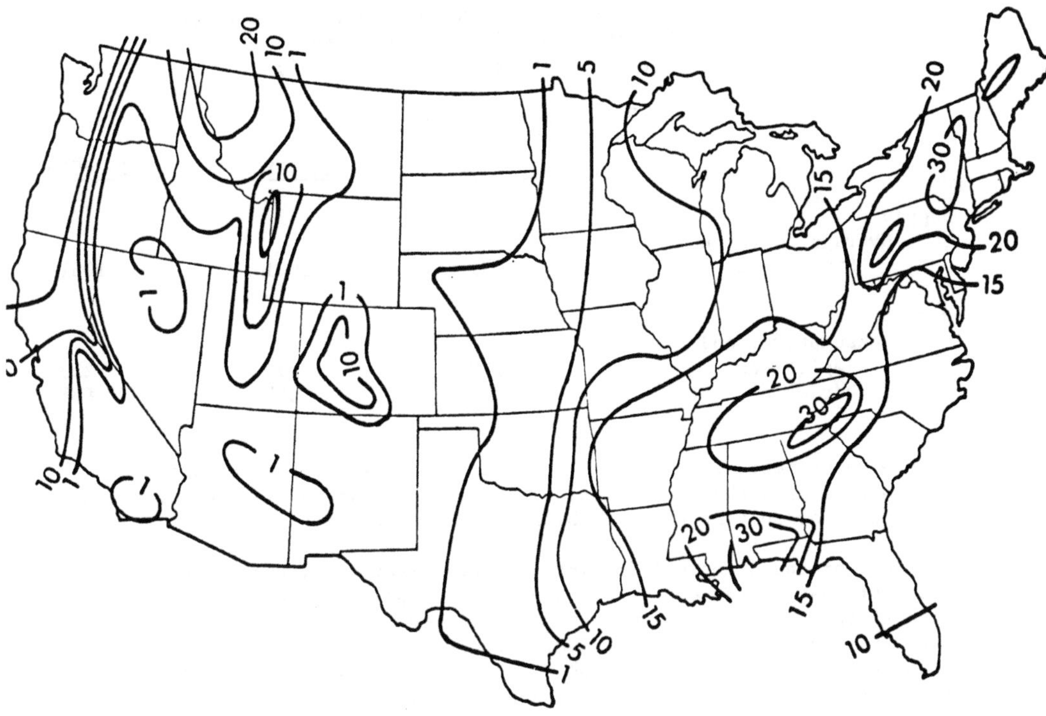

Figure 3. Average Annual Runoff, in Inches, of the Streams of America

completed the most comprehensive recent report (6) which balanced the water supply of the United States against present and future expected demands. Table III shows the general results of this study. By the year 2000 it is expected that the total consumptive use and depletion of water in the United States will total 253 billion gallons daily. The total supply considered to be available is equivalent to 1081 billion gallons per day. On a national basis this shows a supply of about four times the demand. Table III also shows that by the year 2000 the withdrawals for nonconsumptive use will total 888 billion gallons per day. This amount seems to reach dangerously close to the total supply available of 1081 billion gallons per day.[9]

The Water Quality Problem

The Select Committee also studied the water supply demand outlook for 22 water resource regions throughout the United States. These studies showed that one region in the South Pacific portion of the United States might be said to be out of water now. Present deficiencies are being made by importation of water from other regions. The Select Committee showed that for 5 of the 22 regions that full development of all water resources available will be required by the year 1980 or earlier, and by the year 2000, three additional regions would be utilizing their total water resources. The Select Committee reported that these results should not be construed as placing a ceiling on the growth of population and economic activity in these water-short regions. The problems involved in meeting

Table III. United States Total Water Requirements in Years 1954, 1980, and 2000 as Reported by Select Committee of the Senate (6)

in Billions of gallons daily

	in 1954	in 1980	in 2000
Supply			
Average Sustained Stream Flow	1081	1081	1081
Demand			
Consumptive, and depletions	110	190	253
Withdrawals	300	559	888

future water needs in these regions are considerable, but the national interest demands their solution. The Select Committee also indicated that the means for solving these problems are certainly available.

One of the principal findings of the Select Committee was the fact that for most of eastern United States the supply of water is adequate through the year 2000 to meet the anticipated needs in this part of the country. The amount of water available is adequate; the water resource problems in this region are problems of water quality. Shown in Figure 4 is a map of the United States showing the water supply outlook for the various portions of the country as reported by the Senate Select Committee. In the Pacific Northwest and in eastern United States, the supply is shown to be adequate to meet needs through the year 2000. In eastern United States, however, it was reported that the primary water resources problem will be maintaining adequate water quality. Because of this finding, strong new efforts have been undertaken in the past few years to take steps to overcome stream pollution in United States. Control of sediment damage must be adequately recognized in this pollution-control program.

Water Pollution

The Water Quality Act of 1965 represents strong new legislation to fight the pollution of streams in the United States. The law provides for the establishment of water quality criteria for all the streams of the nation by July 1, 1967. Strong new enforcement programs are being instituted to maintain these water quality standards and to prevent further deterioration of streams. One of the organizations in the United States which has shown the most effectiveness in the actual improvement of water quality in a river basin has been the Ohio River Sanitation Compact, better known as ORSANCO. As described by ORSANCO, pollution of water is an impairment of quality that is prejudicial or injurious to the suitability of water for the defined uses. It will be noted that this definition presupposes the determination of water use.

WATER SUPPLY OUTLOOK

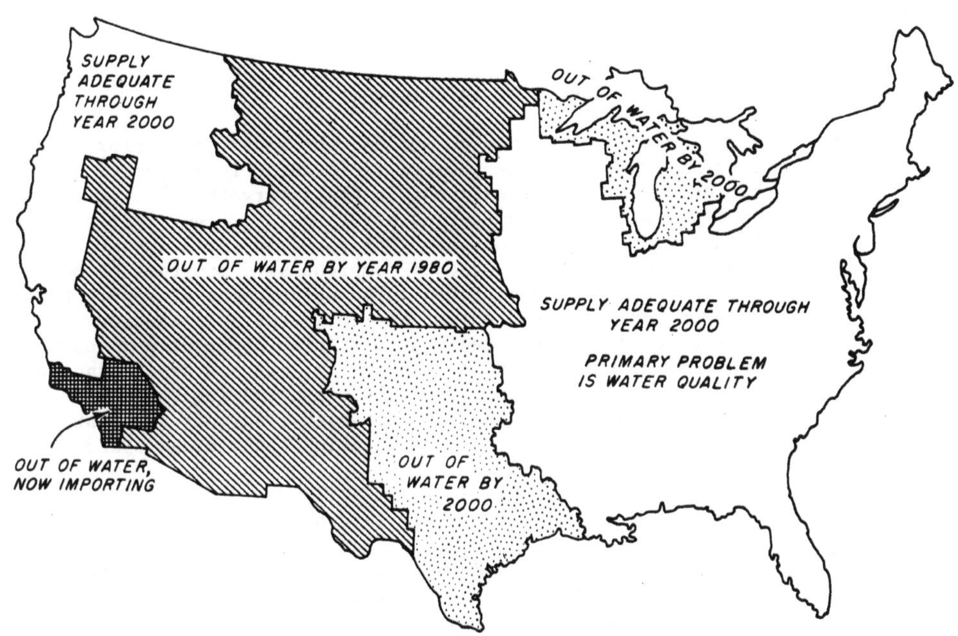

Figure 4. The Water Supply Outlook for Various Regions of America as Determined by the Senate Select Committee

Sediment as a Pollutant

Under the definition given for pollution it is obvious that the presence of inert sediment particles in stream water should be considered as a pollutant. This is because of the fact that the sediment particles are injurious to the water for several particular uses. As shown earlier in Table II sediment in water causes increased cost in treating water for public water supply. It is also shown in Table II that sediment particles deposit in reservoirs and occupy storage space. Sediment particles in flood waters cause clogging of channels and can cause the increasing of flood heights due to the bulking of flood flows. Sediment particles cause damage to hydroelectric power turbines, and interfere with fish propagation. It has been shown by Ellis (3) that the penetration of sunlight into sediment-laden water is halved or more. The presence of these sediment particles thus changes the temperature of the river water. The sediment blankets the stream bottom; it contains organic matter and other materials which create unfavorable bottom conditions for fish life.

Because of the magnitude of damage to water resources use created by sediment particles in water it is proposed that inert sediment particles be recognized as a pollutant of water. Additional studies are needed to evaluate more specifically than has been done in Table II the actual damages created by sediment particles in water. The sediment load of a stream is a subtle pollutant such as heat and other trace elements in water. The effects of these subtle pollutants need to be evaluated far more precisely.

Sediment in Reservoirs

Whenever a dam is built by man to impound the waters of a flowing stream the reservoir which is thus created is immediately subjected to destruction by the deposition of sediment in the reservoir. The Agricultural Research Service (1) has summarized the results of reservoir sediment deposition surveys in the United States. For about 30 years a number of federal and state agencies have conducted comprehensive programs of measurement of reservoir sediment deposition. The measurement of sediment in a reservoir furnishes concrete evidence of the fact that soil erosion is occurring on a watershed and that sediment particles are moving downstream from the farmer's land, through the stream system, and are depositing in a reservoir. By measuring the sediment in a lake and knowing the age of the lake it is possible to determine the rate at which reservoir sediment is depositing in the reservoir, and the rate at which the storage space is being destroyed.

The water supply yield of a reservoir-watershed complex is a function of the low flow hydrology of the stream, as has been shown by the author (7). Because the deposition of sediment reduces the storage capacity of a reservoir it also reduces the yield of the reservoir. It is not immediately obvious however that this is a direct relation. The low flow relations and the mass curve analysis to determine yield are complex. To determine the magnitudes of the actual reduction in yields caused by sedimentation, data were selected for 35 public water supply reservoirs in Illinois on which sedimentation surveys had been made. The total loss of capacity of each reservoir during its entire life up to the present was computed. This loss was expressed in percent of original capacity.

The net yield for each reservoir based on the original storage capacity was determined for the 40 year recurrence-interval drouth. A similar yield based on the present reservoir capacity, as reduced by sediment, also was determined. The comparison of these two yield values showed the percent reduction in reservoir yield caused by sedimentation. The graph in Figure 5 shows the percent reduction in yield as the reservoir capacity is reduced by sedimentation. The actual yield of the reservoir for water supply is reduced directly. Thus, in spite of the many variables involved in this yield function, there is a direct relation between the loss of reservoir capacity and the loss of yield.

Nature of Sediment

Experience in Illinois has shown that the sediment here is fine in texture and is carried into the reservoir as wash load. Thus the rate at which sediment enters the lake is more dependent upon the soil, farming conditions, and occurrence of erosion-producing storms on the watershed

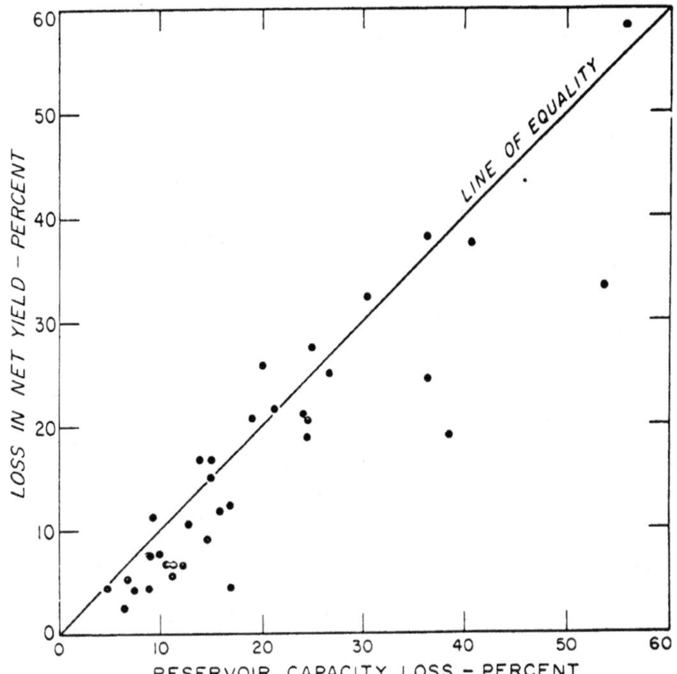

Figure 5. The Loss in Water Supply Yield Caused by Reservoir Capacity Loss Due to Sedimentation for 35 Illinois Cities

than it is upon the composition of the bed of the inflowing stream. The movement of sediment, and deposition patterns in Illinois reservoirs has been described by the author elsewhere (8). For 303 sediment samples collected from 32 reservoirs in Illinois analyses have been made of the distribution of sizes of the particles in the sediment. These results show that the sand content is usually only from 2 to 5 percent of the sample, that silt comprises about one-third of the sample, and clay comprises about two-thirds of the sample. Thus the sediment in Illinois reservoirs is predominantly silt and clay, fine-sized particles which move with the water into the lake and distribute themselves rather uniformly throughout the lake.

Sediment Reduction

Source of Sediment

The sediment which is carried by streams and interferes with many of man's uses of water, or which deposits in a reservoir, is comprised of soil particles which originate on the agricultural land of the watershed. Detailed studies of 101 reservoirs and watersheds in Illinois have shown repeatedly the relation between the sediment deposited in a reservoir and the soil erosion conditions on the watershed. An example of this relation is reported for Lake Springfield (10). The sediment which has deposited in Lake Springfield has been shown to be equivalent to about 1.0 ton per acre per year from the drainage area. By the mapping the 258-square-mile drainage basin of Lake Springfield and by use of the universial soil loss equation, it was possible to determine that the average annual soil losses from this watershed were equivalent to 3.3 tons per acre per year. This same soil conservation survey yielded information from which it was possible to specify the soil conservation program which was needed on this watershed to reduce soil erosion. This study showed that soil losses from the watershed could be reduced from 3.3 to only 0.7 ton per acre per year under a soil conservation program. In Figure 6 is a map of the watershed of Lake Springfield showing the five soil association groups on this watershed for which soil losses were computed.

One of the values of a complete watershed study such as reported for Lake Springfield is that the watershed sources of sediment are not only evaluated but are delineated. For the five soil groups shown in Figure 6, it was possible to determine the soil losses from each group. The relative importance of these five groups are shown in the bar charts in Figure 7. The bar chart on the left of Figure 7 shows the percent of the watershed area occupied by each of the five soil groups. For example, soil group one occupies 12 percent of the watershed, soil group two occupies 55 percent of the watershed, soil group three occupies 10 percent of the watershed, soil group four occupies 20 percent of the watershed, and soil group five occupies 3 percent of the watershed.

The bar chart at the right of Figure 7 shows the percent of total sediment produced from the various soil groups. It is seen that the group 1 soils contributed only 1 percent of the total sediment, group 2 soils contributed 33 percent of the sediment, group 3 soils contributed 4 percent of the sediment, group 4 soils contributed 62 percent of the sediment, and group 5 soils contributed an insignificant portion of the sediment.

The results of this analysis show that the group 4 soils which occupy only 20 percent of the watershed are contributing 62 percent of the sediment. This means that efforts to reduce sediment in Lake Springfield could reasonably be concentrated on the group 4 soils which are the large contributors of sediment. Reference to the map in Figure 6 shows the exact location of these soils; this will allow a concentrated effort at erosion control in these regions. It might be further noted that the group 4 soils are located largely directly surrounding Lake Springfield and along major streams. In this region around the lake much of the area is urbanized and soil losses are low in these lawns, yards and street areas. This means that the soil losses in the upstream portions of the group 4 soils must be the heavy contributors of sediment. This knowledge allows the concentration of soil conservation efforts in these regions.

Documented Cases of Sediment Reduction

At a number of places in the United States surveys have been made to measure reservoir sedimentation, and the watershed conservation needs have been determined. An effort has then been made to introduce the necessary soil conservation measures onto the lands of the watershed in order that soil erosion could be reduced. After these conservation measures, or a large part of them, have been introduced onto the watershed and have been

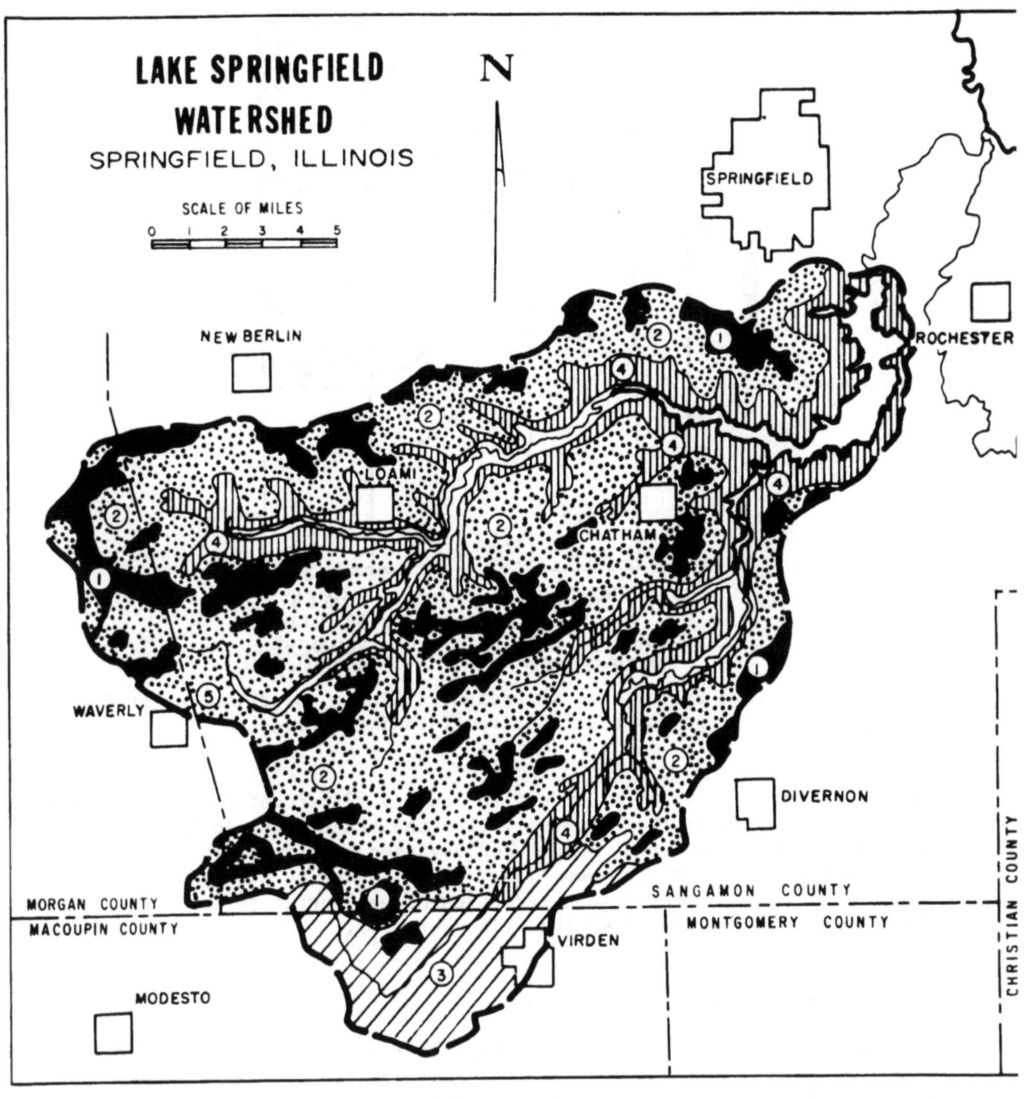

Figure 6. Watershed Map of Lake Springfield, Illinois, Showing Five Soil Association Groups

in effect for a number of years a second survey of the reservoir has been made to measure the reduction in reservoir sedimentation. In eight specific cases in the United States reductions have been shown. These eight cases are cited in Table IV. For example Table IV shows that the city reservoir number 3 at Fairfield, Iowa, was surveyed and found to be losing capacity to sediment. After a complete soil conservation program was instituted, another survey was made and the rate of sedimentation was found to be reduced by 73 percent. In this case the erosion losses from the

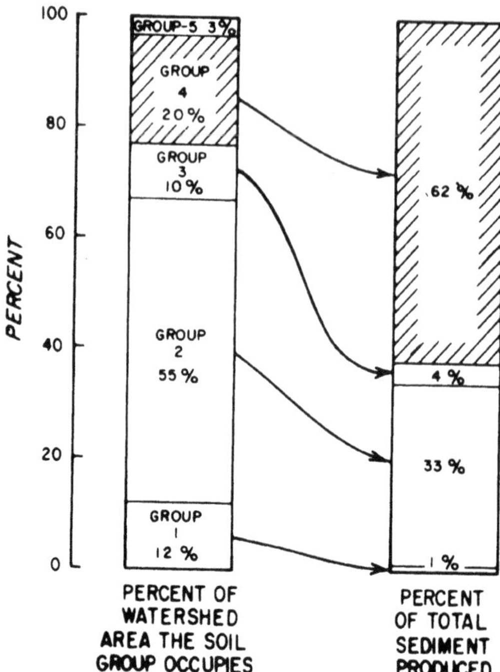

Figure 7. Lake Springfield Watershed, Illinois, Distribution of Soil Groups and Distribution of Sediment Produced

watershed were reduced by an amount to cause a 73 percent reduction in the rate of deposition in the reservoir. For the other lakes in Table IV similar reductions are shown by actual conservation programs. These cases tend to prove the fact that the reduction of soil losses on a reservoir watershed will actually accomplish a reduction of sedimentation in the reservoir. The author has also described this elsewhere (9).

Elements of a Conservation Program

It is an inescapable fact that the sediment which deposits in reservoirs and which causes damage in streams originates from soil erosion on watershed lands. A soil conservation survey of watershed land can determine specifically the land capability and the needs for various items of conservation treatment to reduce soil erosion and to protect the permanent land base. The elements of conservation programs are well known. For the eight cases described in Table IV the elements of the soil conservation programs which actually attended the reductions in sedimentation, are very similar in nature.

The elements of a soil conservation program are so well-known and well-documented as to be redunant. Measures which are often needed are: farming the land on the contour, strip cropping, terracing, providing grass

Table IV. Eight Documented Cases of Soil Conservation Effectiveness

Reservoir	Amount Sediment Reduction Percent
City Reservoir No. 3, Fairfield, Iowa	73
City Reservoir, Newnan, Georgia	70
Lake Issaqueena, Pickens County, South Carolina	50
White Rock Lake, Dallas, Texas	39
Lake Waco, Waco, Texas	38
City Reservoir, High Point, North Carolina	35
City Reservoir, Roxboro, North Carolina	35
Lake Crook, Paris, Texas	28

waterways, lengthening crop rotations to include more years of pastures and grassland, pasture improvement, drainage, shaping of waterways, stabilization of channels, reforestation, and the control of gullies and waterways by structures.

Man's role in reducing the sedimentation problems we face today is to accept the fact that the above soil conservation practices and measures are required in order that the permanent productivity of the land be preserved while it is being utilized.

Conclusions

In this paper an attempt has been made to sketch the role of man in affecting the sediment carried by the streams of America and depositing in reservoirs. In considering this general outlook the following can be concluded:

1) The present annual damage caused by sediment particles in the stream waters of America is about $260 million per year. This is about one-third of the estimated $800 million per year which is the value of soil eroded from the farms of America.

2) As concluded by the Senate Select Committee, the total quantity of water available in United States will meet future needs. However, in eastern United States, a major problem will be the maintenance of a suitable water quality. Attempts to evaluate and control stream pollution should provide more precise means of evaluation of the damage of <u>sediment as a water pollutant</u>. The value of this damage should be compared to the damage created by other organic and chemical pollutants.

3) The deposition of sediment in reservoirs in America today is destroying an extremelv valuable resource, the storage space of these reservoirs. As a reservoir fills up with sediment its usefulness for furnishing a water supply is diminished proportionately.

4) A detailed soil conservation survey of a reservoir watershed will provide the information needed to determine the soil losses from this watershed and the needed conservation treatment measures to reduce soil losses and to reduce reservoir sedimentation.

5) Soil conservation really works in reducing reservoir sedimentation. In eight cases cited, soil conservation measures reduced the rate of sedimentation as much as 73 percent.

6) The elements of a soil conservation program are well known. They include contouring, strip cropping, and other specialized measures which allow the land to be used in accord with its capability. Such measures will result in soil losses far less than those normally now occurring.

References

1. Agricultural Research Service, U.S. Dept. of Agriculture, Summary of reservoir sediment deposition surveys made in the United States through 1960, Misc. Publ. No. 964, U.S. Gov't. Printing Office, Washington, D.C., May 1964.

2. Brown, Carl B., Perspective on sedimentation, Proceedings, Federal Inter-Agency Sedimentation Conference, U.S. Bureau of Reclamation, Washington, D.C., 1948.

3. Ellis, M. M., Erosion silt as a factor in aguatic environments, Ecology, vol. 17, no. 1, January 1936.

4. Geiger, A. F., Sediment yields from small watersheds in the United States, International Association of Scientific Hydrology, General Assembly of Toronto, Vol. I, p. 269, Gentbrugge, Belgium, 1958.

5. Golze, Alfred R., Problems of irrigation canals, Chapter 21 of Applied Sedimentation, Editor, Parker D. Trask, Wiley and Sons, New York, 1950.

6. Select Committee on National Water Resources, Final Report, U.S. Senate, 87th Congress, 1st Session, Report No. 29, U.S. Gov't. Printing Office, Jan. 30, 1961.

7. Stall, John B., Low flows of Illinois streams for impounding reservoir design, Bulletin 51, Illinois State Water Survey, Urbana, Illinois, 1964.

8. Stall, John B., Sediment movement and deposition patterns in Illinois impounding reservoirs, Journal, American Water Works Ass'n., vol. 56, no. 6, June 1964.

9. Stall, John B., Soil conservation can reduce reservoir sedimentation, Public Works, September, 1962, p. 125.

10. Stall, John B., L. C. Gottschalk and H. M. Smith, The silting of Lake Springfield, Report of Invest. 16, Illinois State Water Survey, 1952.

11. U.S. Department of Agriculture, Basic statistics of the national inventory of soil and water conservation needs, Statistical Bulletin 317, Washington, D.C., 1962.

12. U.S. Department of Commerce, Bureau of Census, Drainage of agricultural lands, Census of Agriculture, 1959, vol. 4, p. 3, Washington, D.C., 1961.

Editor's Comments on Papers 6 Through 9

Land–Water Changes by Construction Activities

This section describes the topographic and hydrologic deterioration of the landscape that results from man's deliberate design to modify terrain by construction of roads, railroads, buildings, and new channels (referred to as geomorphic or physiographic engineering). The first paper, by J. Aghassy, is an original manuscript, printed here for the first time. It forms a bridge to the previous section on soil erosion because the first case history he cites involves terrain changes caused by improper cultivation methods. The second illustration concerns results stemming from railroad construction, and both examples provide dramatic proof of the enormous alteration of the earth's surface by man in creation of badland topography. Although these massive changes occur in the delicately balanced semiarid climate, rapid and significant changes can also occur in other climates. Probably the regions of climatic extremes are the most subject to extraordinarily widespread and rapid destruction—regions such as the arctic and semiarid tropical belts [see the article by Young (1968) on the importance of land surveys prior to development in tropical areas]—but this sort of damage can also occur in temperate and humid areas, as shown by the Daniels article (the last in this section) and by Sternberg (1956):

> That outright desert landscapes may be created by man, even in warm temperate rainy climates, is a well-known fact, which may be illustrated with a recent example from the pioneer coffee lands of southern Brazil. Incorrect agricultural practices that aggravate floods and droughts are certainly not restricted to the northwestern section of Brazil, but are the rule throughout the country (p. 203–204).

The R. R. Parizek article, somewhat shortened from the original, is the most definitive study to date detailing the broad range of environmental distortions caused by highway construction, especially when improperly planned. His examples are especially applicable in temperate climates. R. F. Black's classic paper on permafrost appears here in a somewhat abridged facsimile. (Separate volumes in this "Benchmark" series are being allocated to permafrost and arctic terrain problems.) This is the earliest and most comprehensive analysis in English of the range of man-induced terrain distortions in the unusually fragile environment of the arctic. Thus these papers provide details for a wide range of man-made topographic upsets in different climatic settings.

Since the arctic environment is receiving special attention, because of oil discoveries on the northern slopes of Alaska, northern Canada, and Siberia, it is becoming increasingly important to assemble data that will lead to better understanding of such regions. Also, it seems likely that increasing population pressures will require man to migrate to these regions in the future. The question of possible construction of a trans-Alaska pipeline has been a hotly contested issue between conservationists and commercial interests. A consortium of seven oil companies wished to transport oil from the fields at Prudhoe Bay southward for a distance of 789 miles to a terminal port at Valdez on Prince Williams Sound. Earth scientists studied environmental effects not only from the frozen ground viewpoint, but also because the planned pipeline would have to cross several earthquake-prone areas, and many rivers lay in its path. Such studies as that by Brice (1971) represent the type of geomorphic surveys necessary for appropriate planning to minimize stream channel distortions by man and nature. Haugen and Brown (1971) summarize some of the unusual characteristics of the terrain in terms of man:

> The extreme sensitivity of tundra–taiga vegetation and the underlying permafrost to stresses and disturbances is particularly evident in the ice-rich permafrost regions of Alaska. Many problems occur due to the unusually delicate equilibrium that exists between the surface cover, namely vegetation and associated peat and also snow cover, and the underlying wet soils, and permafrost. Natural disturbances such as lake drainage, fire and frost-heaving upset this balance. Man-induced stresses, which include disruption of the vegetation by tracked vehicular traffic, bulldozing of the vegetated mat, placement of buildings directly on the ground surface or on inadequate foundations, invariably cause accelerated degradation or melting of the underlying ice-rich permafrost. This generally leads to several destructive and often irreversible processes: the formation of water-filled channels; headward erosion in vehicular tracks and along stream channels and subsequent drainage of lake basins, and; slumping and cracking in foundations and other structures. Disturbances of even a minor nature can have a noticeable effect on permafrost terrane. For example, the comparatively slight disturbances associated with the summer staking-area of a dog and a foot trail were shown to cause significant thermokarst subsidence on the tundra (p. 139–140).

The last paper in this section, by R. B. Daniels, presents one of the first and clearest documentations of what can occur when man attempts to change river courses. The concept of "channelization," the artificial alteration of a stream bed, is taking on increased importance and has been drawing much attention in the last half of the twentieth century. Although the U.S. government has been rechanneling rivers since the 1870s, when the Corps of Engineers began working along the Mississippi River valley, it was not until the mid-1950s that large-scale alteration of waterways for agricultural purposes got seriously underway. The impetus was Congressional passage of the Watershed Protection and Flood Prevention Act of 1954 (U.S. Public Law 566). Through this program the Soil Conservation Service has helped farmers widen, deepen, and straighten more than 8000 miles of streams in every state. During the same time, the Corps has straightened another 1500 miles of waterways. SCS currently has congressional approval for ditching 13,524 more miles of rivers.

Many conservationists are strongly opposed to the principle of channelization and cite a variety of reasons, which include increased erosion in tributaries (Daniels), increased drought risk and lowered water tables, increased soil runoff, destruction of wetlands, and loss of resources (including fish, wildlife, and timber; downstream silting; loss of nutrients; destruction of wilderness and aesthetic amenities, and their loss to business). For example, Emerson (1971) cites the case history of the Blackwater River in Missouri. In 1910, a new channel was dredged for 29 km with a gradient of 3.1 m/km, shortening the river 24 km. The old river had been 53.6 km long with a 1.67 m/km gradient. He shows how meanders subsequently became entrenched in tributaries, serious erosion problems developed along the banks of the channelized stream, and headward erosion of gullies took place. Bridges have had to be widened because of increase in valley width. He concludes:

> Channelization has enabled more floodplain land to be utilized in the upper reaches of the Blackwater River. This benefit must be weighed against erosional loss of farmland, cost of bridge repair, and the downstream flood damage resulting from termination of the dredging project (p. 326).

The case against channelization has reached into popular periodicals with such articles as the "Rape on the Oklawaha" (Miller, 1970) and "Crises on Our Rivers" (Miller and Simmons, 1970). Indeed, feelings and activism against the Oklawaha project—a combined canal and river diversion in Florida—became so strong that even though $50 million in federal funds had been committed, construction on the project was halted by a presidential order. Thus this process of "geomorphic" or "physiographic engineering" draws mixed public reactions.

The Daniels article can be considered in the same vein as the Aghassy article because the construction activity of straightening and shortening the main channel of the river caused increased gradients throughout the tributary systems and ushered in entrenchment of their channels and a new cycle of erosion. It would be unwise to state categorically or even in broad generalization that all channelization projects lead to topographic catastrophe. Ruhe (1971) provides a case history of certain man-made changes on the Missouri River that seem to have improved the area from the human viewpoint:

> This case study of Otoe Bend shows that all man's manipulations of stream regimen are not bad. He put the river where he wanted it. He improved

the area by creating land for agricultural and other purposes, and he alleviated flooding. Facility of navigation of the river has been improved. But, man is also unpredictable. An unforeseen problem has arisen and involves law and ownership of land (p. 21).

6

Man-Induced Badlands Topography[1]

JACOB AGHASSY[2]

Abstract. This paper describes the man-induced evolution of badland topography in a semiarid environment. Two situations are considered: one resulting from Bedouins' ploughing techniques in the southern Coastal Plain of Israel during the first half of this century, and the second following the construction of a narrow-gauge railroad within the same area. In both instances, rill erosion activated the development of fine-textured drainage systems which dissected belts of land adjacent to major wadi-course in loess and loess-like soils leading to typical badlands landscape.

Introduction

The man-induced badland features (Fig. 1) discussed within this paper are located in the southern Coastal Plain of Israel, along the lower course of Nahal Besor (Fig. 2), and along some nearby tributaries. The conditions within this area under which evolution of badlands topography becomes possible seem to be very specific; therefore, a brief outline of these conditions is appropriate.

Structure, Lithology, and Topography

The Negev Coastal Plain is relatively flat on the east, whereas on the west it is dominated by recent and lithified dunes.

---

[1]Acknowledgment and thanks are given to Professor I. Schattner, Hebrew University, for advice and to Dr. Avitzur for permission to use illustrations from his book (footnote 9).
[2]Department of Geography, University of Pittsburgh, Pittsburgh.

Fig. 1. Partial view of badlands around Nahal Besor.

The aeolianite[3] ridges rising to 60–80 m above msl, intervene with the subsequent valleys 15–25 m lower. The ridges tend to become higher and wider away from the coastline and their relative relief decreases. The easternmost ridge merges with the flat plain at about 100 above msl and the dune-dominated relief gives way to fertile aeolian and fluvially redeposited loess cover. It is here that from the turn of the century until 1948 the Bedouins conducted intermittent land cultivation.

This region, higher in elevation than the adjoining plains to the north and south, was uplifted no earlier than the Upper Pliocene,[4] as indicated by continental clastic sediments.[5] Considerable field evidence indicates that uplift has continued through historical times.[6]

Nahal Besor, which is a well-established drainage element in the western Negev since the Cretaceous,[7] traverses the area from south to north and has cut water gaps through the sandstone ridges of this coastal plain. Nahal Besor and its major tributaries have adjusted to recent upheavals, graded their profiles, and seem to have reached a reasonable degree of slope stability, to the point that heavy grass sods became well established on all valley sides.

---

[3]S. Ravikovitch, The aeolian soils of the northern Negev, *Desert Res. Intern. Symp. Proc.*, 1953, pp. 404–431.

[4]L. and Solomonica Picard. On the geology of the Gaza-Beersheva district. *Bull. Geol. Dept.*, Hebrew University, Jerusalem, 1934.

[5]L. Picard, Structure and evolution of Palestine, *Bull. Geol. Dept.* Hebrew University, Jerusalem, 1934.

[6]Wells dug and built in historical times in wadi beds to collect water stand with their structure a few meters above today's bed level, etc.

[7]Bentor and Vroman, A structural contour map of Israel (1:250,000) with remarks on its dynamical interpretation, *Bull. Res. Council Israel*, Sept. 1954, and Bentor and Vroman, *The Geological Map of the Negev (1:100,000)*.

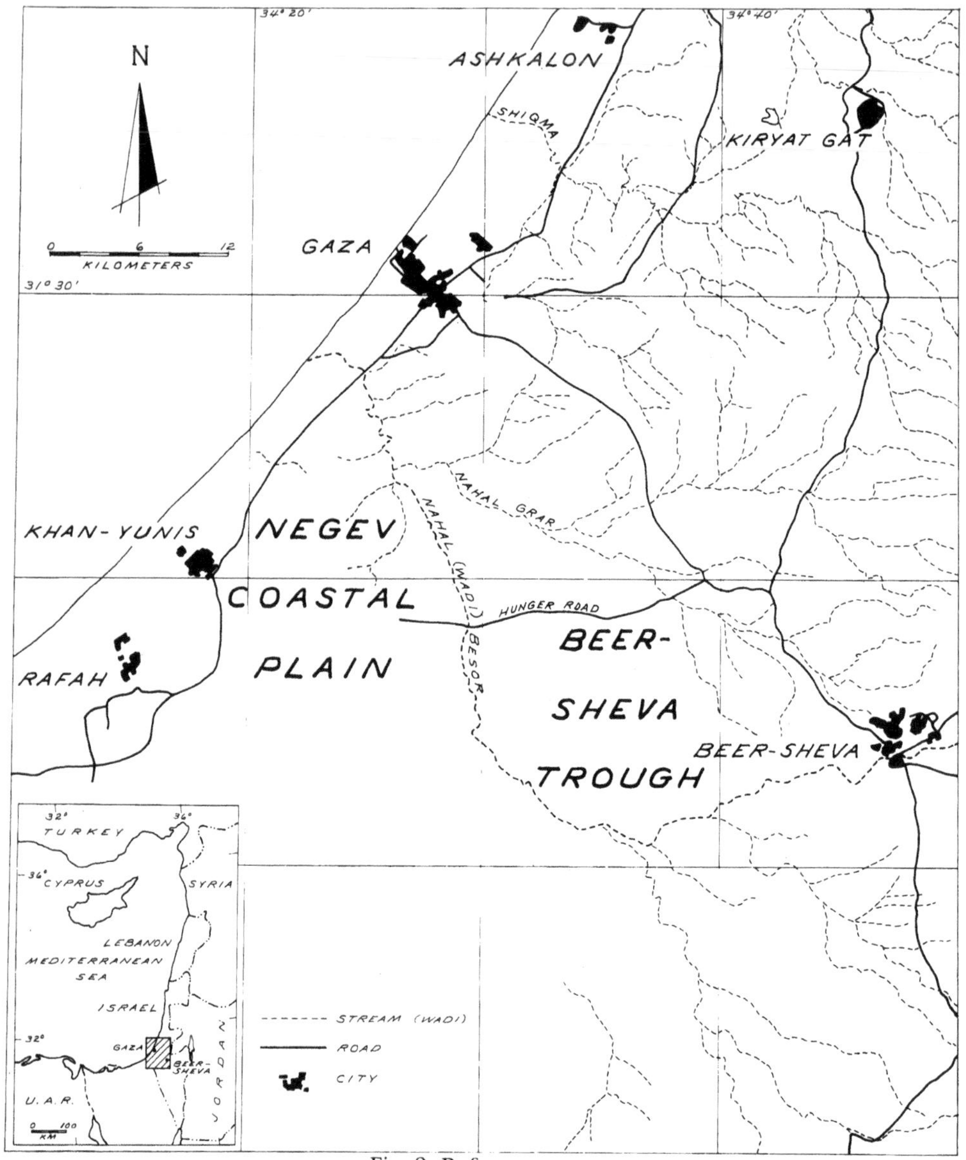

Fig. 2. Reference map.

Climatic Conditions

The Negev Coastal Plain is semiarid. Rainfall averages 200 mm/year, but very high fluctuations occur from year to year. Distribution of precipitation during the rainy season is uneven, and descends mostly in a number of heavy downpours, up to 55 mm in magnitude.

Drought drove Bedouins out of the area at times and in one instance, British authorities, between the two World Wars, hired the local Bedouins to construct a road leading nowhere,[8] solely for the purpose of preventing mass starvation.

The Bedouin Plough

In wet years, Bedouins harvested good yields of barley, used at first only for making bread until its discovery by the British beer industry. Barley then became an important cash crop, competing with the subsistence crop and creating new demands on the land. Bedouin plough[9] cultivation expanded from the flat plains to the slopes at the wadi margins, and there the disaster started. To the Bedouin

Fig. 3a. Bedouin camel-drawn plough, typical to the Negev. (Courtesy S. Avitzur.)

using a camel-drawn plough (Fig. 3), the concept of contour ploughing was virtually unknown. Once the grass sods on the slopes were transected by plough trenches perpendicular to contour lines, heavy rainstorms took over.

---

[8]Known locally as Hunger Road, branching from the Gaza–Beersheva highway.
[9]Information and illustrations taken from the book by S. Avitzur, *The Native Ard of Eretz-Israel, Its History and Development,* Hadassah Publishing House, Sifriat, Israel, 1965.

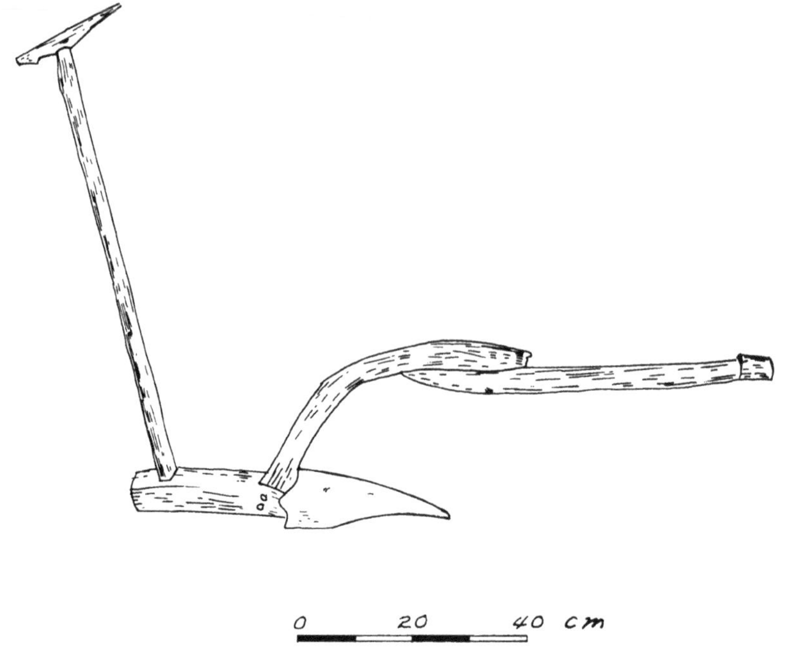

Fig. 3b. Bedouin plough typical to the Negev.

Slope Conditions for Evolution of Rills

The Besor channel, which is the local base level for the entire area, is at an elevation ranging from 45 to 80 m. From there, the slopes rise to the upper plain surface, which ranges in altitude from 75 to 110 m.

The horizontal distance between channel and upper surface ranges from 500 to 1000 m. Sections left untouched by either Bedouin ploughs or the rill system resulting from their use show that the 30-m drop per each 500- to 1000-m unit of horizontal distance seems to have been below the critical slope. These intact surfaces maintain reasonably rich grass sods. Figure 4 shows some selected slope profiles.

Stages in Evolution of Topography

At first, parallel channels seem to develop in the plough furrows (Fig. 5). For a time, these channels run independently until they widen enough to undermine the narrow ridges separating them and eventually eliminate them. Usually four to eight channels join downslope to make the next order channel, but when the original mini-interfluve surfaces are eliminated, only two or three channels will remain, adapting their initial courses to the new situation and establishing themselves as permanent first-order channels (Fig. 6). The newly generated first-order channels are very closely

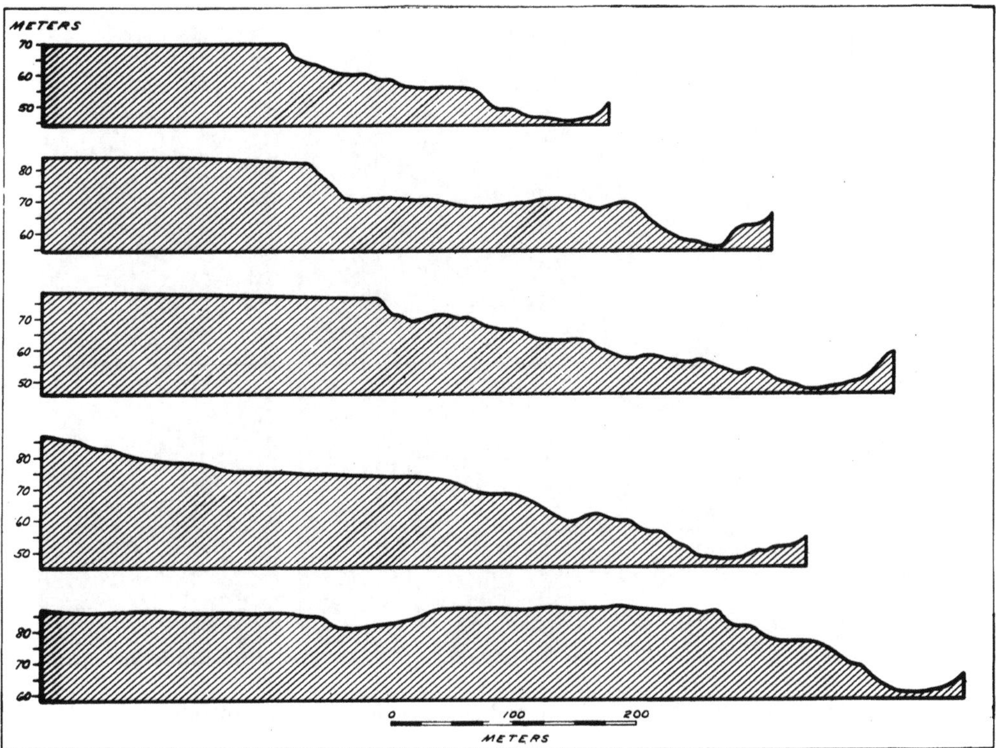

Fig. 4. Selected slope conditions under which badland channels develop (upper plain to wash bed).

spaced and leave very narrow interfluves (5–10 m wide). Their deepening is usually fast, as a result of a piping process which can increase their depth by as much as 2–3 m in one cycle, when a series of vertical pipes is developed along the channel leading to a horizontal underground tunnel (Fig. 7). The growth of vertical pipes leads to the collapse of the mini-bridges (Fig. 8), separating them with the result that deepening is accomplished within two or three rainy seasons. Several generations of piping incision are typical of the process (Fig. 9). Piping prevails also in second- and third-order streams of the newly generated system and accelerates the deepening process, but its effect becomes less important as the channel order becomes higher. Under favorable conditions, where up to fifth-order systems have developed as a result of the process, headward erosion of first-order tributaries penetrating into the upper plain surface can advance as far as 500–750 m from the original average position. The original location of the plain edge is detectable and retraceable by groups of "outlier" buttes on top of which sods and plough marks are still evident (Figs. 10 and 11).

The headward movement of first-order channels caused growth in the size of the fourth- and fifth-order drainage systems flowing to the Besor (Fig. 11). The result was widening of the "badlands belt" on both sides of the wadi at the expense of the upper plain. During this process, higher dissection depths and densities developed in the upper sections of the systems, whereas in the lower parts, adjacent

Fig. 5. Parallel channels developed along Bedouin plough furrows.

Fig. 6. Parallel channels resulting from Bedouin plough lines (foreground left) develop first-order channels once thin interfluves eliminated (right side).

Fig. 7. Development of piping in first cycle.

Fig. 8. Collapse of minibridges separating vertical pipes.

Fig. 9. Three stages in channel deepening resulting from three piping generations.

Fig. 10. "Outlier" residual buttes close to Besor Channel on top of which grass sod and plough marks are still evident.

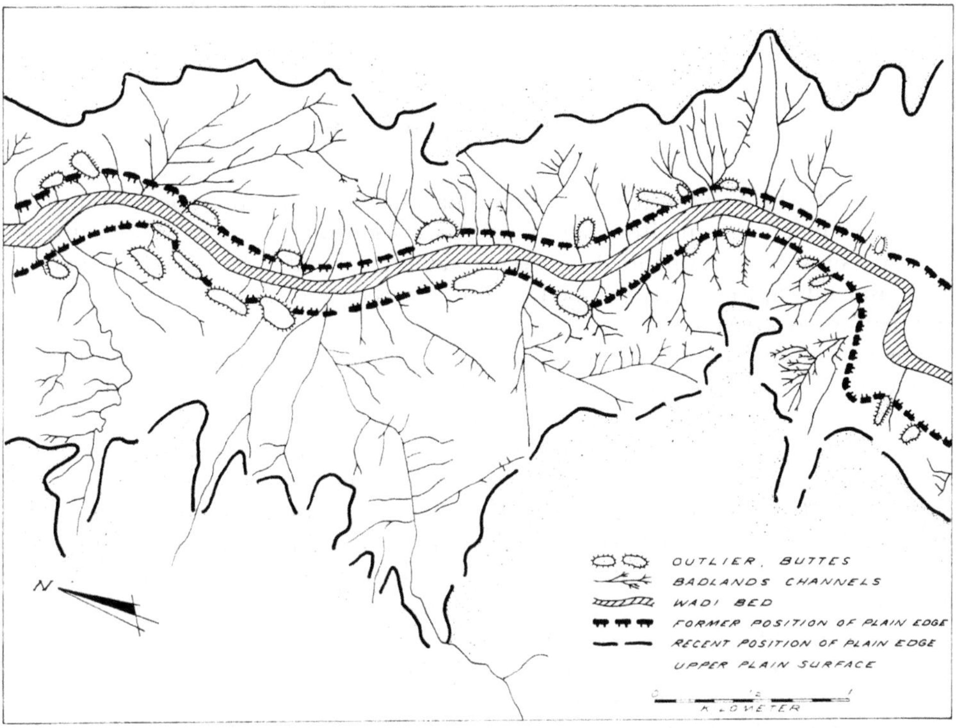

N

OUTLIER, BUTTES
BADLANDS CHANNELS
WADI BED
FORMER POSITION OF PLAIN EDGE
RECENT POSITION OF PLAIN EDGE
UPPER PLAIN SURFACE

0 ½ 1
KILOMETER

Fig. 11. Migration of the plain edge and expansion of the badlands belt.

Fig. 12. Lower section of channel systems adjacent to base level, showing signs of stabilization and residual buttes indicative of former position of the plain's edge.

to the Besor channel, the growth of relief was slowed and stability was attained (Fig. 12).

The Ottoman Railroad Track

Prior to World War I, when the area was part of the Ottoman Empire, a railroad system was built connecting Beer-Sheva with Gaza and other locations. This railroad system has since been dismantled, but remarkable erosional scars are left across its path. Most landscape damage occurred in approaches to bridges, over wadi crossings, and in cuts across hillsides.

The development of badlands following the railroad construction is particularly prominent near the major wadi crossings, since the wadi channel provides a "local base level" to the fine-textured drainage systems developing within the loess and associated materials. In most cases, the contact line between the horizontal compacted ramp leading to a bridge and the adjoining sloping loess becomes the most natural path for runoff in both the presence and the absence of drainage ditches. In such cases, a sizable trunk channel evolves, which attracts tributaries, usually on the side opposite the ramp, thus developing an asymmetric network that grows into a badlands system on one side of the railroad track. Eventually, deepening of the trunk channel adjacent to the ramp leads to serious slumps, and even dissection of the ramp, often requiring immediate improvised repairs when the railroad was operating. In turn, these repairs led to channel readjustments, which further increased the size of the dissected area. In most cases, a drainage system generated by this process would

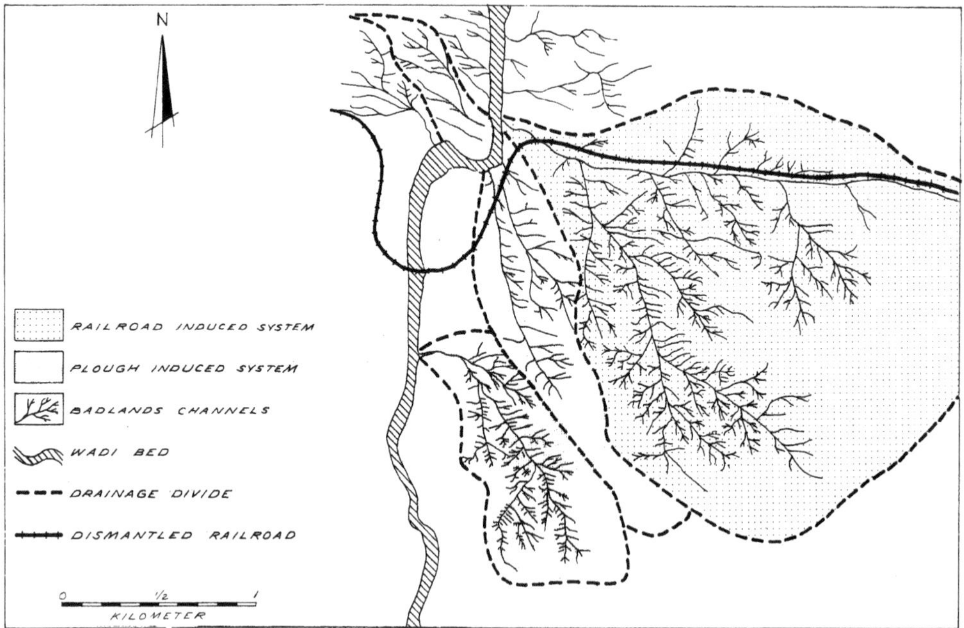

Fig. 13. Drainage-basin sizes for comparison between railroad and plough-induced badlands systems.

Fig. 14. Rill deepening by piping on a hillside railroad cut.

grow to occupy two to five times the area of drainage systems induced by the Bedouin plough (Fig. 13), thereby converting much more of the upper plain surface to badlands. With the abandonment and dismantling of the tracks, new channels became free to grow on the ramp side also, and to compete with channels of the adjacent systems on new "watershed" area. The process resulted in several "captures," and isolated sections of interfluve areas became residual hills, many of which retain traces of the original flat surface (Figs. 1, 10, and 12).

A somewhat different situation developed along hillside cuts, leading to limited "badland strips" along the tracks. Here the damage (i.e., extent of the badlands channel system and dissection intensity) was less than that caused where the railroad crossed wadis; however, the total area affected along the entire railroad system (which followed a very weird and winding path across the Beer-Sheva trough and the Negev Coastal Plain), was larger than that affected by wadi crossings.

The process by which rills and channels developed as a result of hillside cuts consists of two steps: (1) evolution of rills and pipes on the scarp of the fresh cut which joins the drainage ditch on either side of the compacted track. Above this new base level, rapid deepening of the rills occurs as a result of piping (Fig. 14), converting some of the rills to channels which bifurcate as they expand headward; (2) development of a truck channel along the drainage ditch which grows by collecting water, both from the newly evolved channels on the hillside cut and the previous drainage elements. Rejuvenation occurs as a result of the fast expansion and deepening of such trunk channels. From this point on, rapid destruction takes place and the area of the new fine-textured drainage system shows dissection of intensity similar to that of the other badlands of the Negev Coastal Plain. The basic difference is that there they form almost continuous belts along wadi sides, whereas in hillside cut areas, they appear as isolated local systems.

The man-induced features described here have all evolved prior to 1948. Destruction by railroad construction reached its peak prior to that period, and the Bedouin plough ceased operation in the area at about that time. Since then, man's impact has become quite different; conservation and modern agricultural techniques have kept the slopes around the badlands from experiencing destructive activities. And while more than two decades cannot reverse processes which long ago established their marks on the landscape, they have at least succeeded in preventing their expansion to larger areas.

Reprinted from *Environmental Geomorphology*, Donald R. Coates, ed., 151–199 (1971)

7

CHAPTER 9

IMPACT OF HIGHWAYS ON THE HYDROGEOLOGIC ENVIRONMENT

Richard R. Parizek

ABSTRACT

Highways may have favorable or unfavorable influences on our environment. These may be social, economical, or physical, chemical, and biological.

The presence or absence of deep cuts and extensive fills can produce a large variety of transformations on the terrain and in the hydrogeologic environment. Such possible changes include: (1) the beheading of aquifers when present in soils and shallow bedrock (2) the development of extensive ground-water drains where cuts extend below the water table (3) damage and pollution of water supplies (4) changes in ground- and surface-water divides and basin areas (5) reduction of induced streambed infiltration rates due to sedimentation (6) siltation of channels causing flooding, erosion, and reduction in recharge areas on flood plains (7) obstruction of ground-water flow by abutments, retaining walls, and sheet pilings, and (8) changes in runoff and recharge characteristics.

Unstable slopes may result by piping failures where cuts serve as ground-water drains, thus changing the seepage forces which may act outward or upward and either increase driving moments or reduce resisting moments.

Water quality changes may result from: (1) sediment damage to surface and ground-water supplies (2) pollution, resulting from use of highways and related accumulations of oils, chemicals, and hazardous substances through accidental spills, and (3) road side litter and debris. Changes occur by economic activity stimulated by the highway. Sanitary, industrial and solid wastes may be disposed of in an unsatisfactory manner.

Pollution may result from maintenance procedures used to control weeds, insects, snow and ice, and from exposed stock piles of chemicals used for these purposes.

Acid and other chemically polluted waters may be produced where road cuts and fills expose pyrite-bearing strata which normally cause serious mine drainage problems.

* * * * * * *

Previous Studies

Papers may be divided into two general types: (1) those dealing with the geological-physical-chemical impacts of highways; and (2) those dealing with the social-economical-political aspects.

The most relevant paper on the former subject (Scheidt, 1967) concerns the potential and actual pollution effects of highways including problems of erosion and sedimentation. Damage and possible problems resulting from chemicals used to control snow and ice have been outlined by personnel of the Salt Institute of Chicago (1962, 1966) by Weiner (1966), Sauer (1967) and others.

Studies of the physical effects of highways have largely been concerned with construction and maintenance problems resulting from ground-water flow, landslides, rock falls, etc. See for example Smith (1963). Other papers are concerned with causes and controls for frost heaves due to ice lenses. These concerns are relevant in design and construction of highways and related to the hydrogeologic and geomorphic setting.

Williams (1967) showed that highway cuts may become ground-water discharge areas, hence serve to aggravate ice-lens development.

Changes in the hydraulic regimen of rivers near bridge abutments were investigated by Neill (1964) as abutments influence channel scour and erosion, and local flooding.

In 1967 Cron investigated how snow drifts might be controlled through highway design. Where topographic conditions were favorable, the sweep of wind could be relied upon to control accumulation of snow drifts on highways.

Lierboe (1967) investigated the relationship between highway and reservoir locations and Wolman (1964), the relationship between urbanization and sediment production in Maryland.

Social-economical-political studies have been classified by McGough (1968) into three types: (1) those devoted to collection of primary data from a particular location to determine the economic effect of a particular highway development (2) those devoted to a general analysis of the impact of highway development on industry and other segments of the economy, and (3) those theoretical studies which investigated the consequences of highway improvements.

The economic impact of highways to farmers, problems caused, and adjustments required, have been reported by Rockey and Frey (1964). The impact of highways include changes in population, land use, property value, taxation, business volume, type of business, safety considerations, access problems, aesthetics, and potential change in the social composition of the population among other things (Dansereau, 1964). Matter of land-use control (Mariolo and Dansereau, 1965) and even protection of the highway itself must be considered (Frey, *et al.*, 1960). Highways have prompted changes in political and social organizations (Dansereau, 1964). Also rapid and unregulated growth often generated chaotic land use patterns and high governmental costs (Frey, *et al.*, 1960). The impact of highways on these and other factors have been under study by researchers of the Institute for Research on Land and Water Resources at The Pennsylvania State University and elsewhere.

Severance damage and impact studies have been conducted by Grace (1966), McGough (1968), and others.

Problem Types

Highways may influence either surface-water reservoirs or ground-water reservoirs both during and after construction. Influences may be either physical or chemical, favorable or unfavorable. Only the unfavorable influences will be stressed here. These influences may be related directly to the highways or to satellite land-use activities attracted to the region by the highway. Further, changes in the hydrologic regimen may be either transient or permanent. Examples of each are included but emphasis is placed on permanent changes in the hydrogeologic regimen.

PHYSICAL CHANGES

Beheaded Aquifiers

Shallow aquifer systems may be developed along the slopes of mountain flanks and dissected uplands. These may supply ground water to shallow wells and springs that serve private dwellings, villages and small towns in adjacent valleys. Mountain slope aquifers may provide soft water in comparison to water obtained from aquifers in adjacent valleys. These aquifers frequently contain water recharged within a forest environment rather than an agricultural or urban environment, hence have better quality water. Their upland setting permits less expensive sites for water storage tanks and gravity flow systems may be used.

Aquifers may occur in highly jointed top of bedrock that commonly underlies soils, in permeable bedrock units overlain by soils that serve as confining beds, in the more permeable deposits contained within the soils (Parizek, 1970), or in individual bedrock aquifers. Although these near surface aquifers are frequently thin and have limited total storage capacities, they may be reliable sources of water where they are widespread on the flanks of major uplands in humid regions. They may receive intermittent recharge from direct precipitation and influent surface runoff and persistent recharge from upwelling ground water derived from underlying bedrock along the lower reaches of slopes.

Colluvium and alluvium comprising some transported soils in Pennsylvania, may be highly permeable where either is composed of pebble- to boulder-sized fragments of rock derived from mechanically weathered bedrock. High on mountain slopes underlain by resistant sandstones, colluvial deposits may be composed largely of cobbles and boulders, have extremely high infiltration capacities and permeabilities that allow for complete infiltration of all available surface water. These permeable units overlie more intensely jointed bedrock which has a higher permeability compared to the same bedrock at depth. Normally colluvial deposits contain increased amounts of clay in the downslope direction which interfinger with these relatively clean boulder deposits. The interfingering relationship of these deposits and the presence of clay-rich colluvium above highly jointed bedrock combine to produce shallow but extensive artesian aquifers (Fig. 1). These aquifers together with their confining beds may range from 5 to 300 or more feet in thickness and blanket a few acres to many square miles of mountain slopes. Similar hydrogeologic settings may be afforded by glacial deposits.

Major highways located along the flanks of these slopes with immediately adjacent lanes or offset terrace-like lanes typically require extensive cuts and fills to maintain suitable grades. Cuts may extend near, into, or through these confined aquifers and intercept all or a significant percentage of ground-water flow (Fig. 2a). These road cuts and the drainage they provide represent permanent alterations in the hydrogeologic framework which may serve to behead aquifers and intercept their ground water, hence permanently change the ground-water regimen in the aquifers above and below the road cuts. Seepage pressures related to ground-water flow contribute to instability of slopes in this setting as will be discussed later.

Deep cuts below the water table in horizontal and dipping bedrock and glacial aquifers may do the same by diverting ground water that previously was discharged elsewhere to drainage facilities along the road bed (Fig. 2b).

The potential damage to water-supply facilities that results from beheaded aquifers can be appraised prior to construction provided that the source of ground water supplying downslope wells and springs is correctly determined. This requires information on aquifer distribution, hydraulic properties of the aquifer, direction of ground-water flow, definition of the ground-water basin

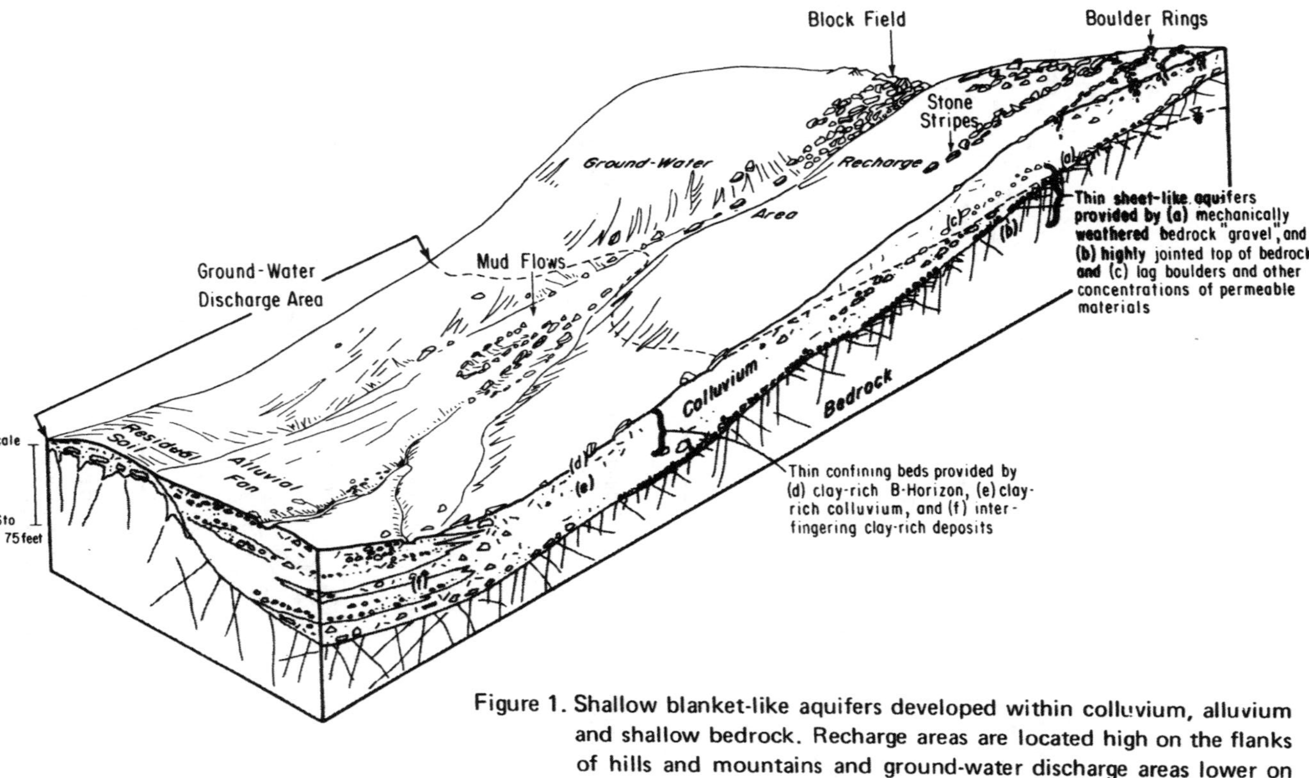

Figure 1. Shallow blanket-like aquifers developed within colluvium, alluvium and shallow bedrock. Recharge areas are located high on the flanks of hills and mountains and ground-water discharge areas lower on these slopes.

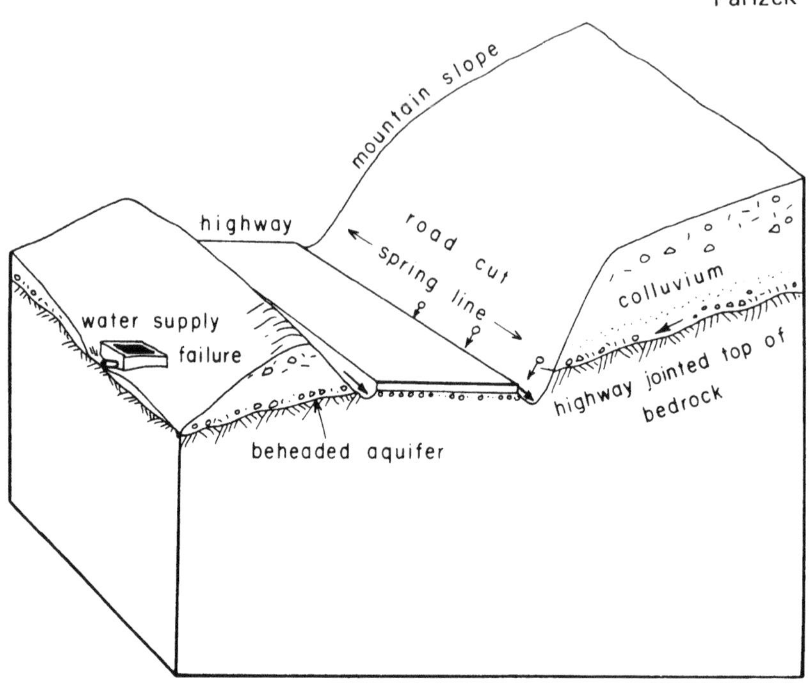

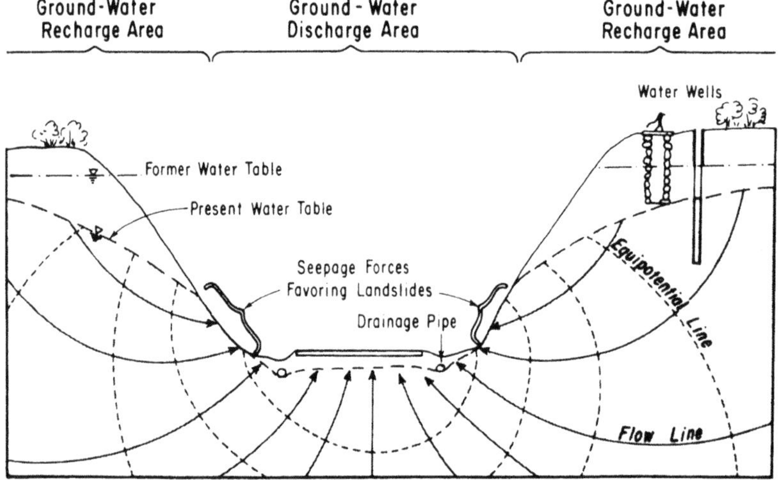

Figure 2. Shallow aquifer developed in colluvium and jointed bedrock beheaded by a highway cut (a) and a ground-water discharge area created by a road cut and drain facilities (b). The entire region may have been a recharge area with a shallow water table previously.

area before construction and change in area that will result after construction, definition of areas of recharge, and data of the proposed location and depth of cuts. The cost of obtaining these data should be weighed against that of the cost of probable damage to water supplies, or the cost of providing alternate water-supply facilities where damage is known to be inevitable.

The area of the ground-water basin previously serving wells and springs may be reduced. Hence, the total aquifer storage capacity and its potential recharge may be reduced after the hydraulic continuity of the ground-water system is disrupted. This may result for both confined and unconfined aquifers and for perched water table conditions.

Road Cuts as Ground-Water Sinks

Other problems may result from deep road cuts below the water table. As the excavation is extended, ground water is free to flow into the excavation and a new ground-water sink or drain is established. This causes the water table and/or piezometric surface to be depressed in the adjacent deposits. The hydraulic gradient in these strata increases as the water table or piezometric surface declines, and the volume of ground water discharged into the drain increases. Water level declines continue to increase with time and distance from the cut until a new equilibrium profile is established on the water table or piezometric surface. Ground water is diverted to the excavation and new ground-water sub-basins are established by this change in boundary conditions.

The end result is to reduce the volume of stored ground water, to change the configuration of the water table or piezometric surface and the directions of ground-water flow, and to cause a permanent decline in the water table and/or piezometric surface. Permanent declines in well yields may result. Problems develop where excavations are lower in elevation than water-yielding openings penetrated by the well or lower than the pump intake opening or bottom of the well. Dry holes, reduction in well yields, dewatered pumps, and increased turbidity are common changes that have been observed. The problem is analogous to well interferences that develop when two or more production wells are closely spaced, however, in the case of road cuts, drawdown is permanent.

To estimate the amount of drawdown that may result adjacent to a road cut, aquifer and confining bed distribution must be known along with their hydraulic properties, boundary conditions, thickness and related information. An equation defined by Ferris (1950) that describes the decline in artesian head at any distance from drains discharging water at a uniform rate may apply to this problem. It assumes that: (a) the aquifer is homogeneous, isotropic, and of semi-infinite areal extent, that is bounded on one side by a line sink (b) the discharge drain completely penetrated the aquifer (c) the aquifer is bounded by impermeable strata above and below, i.e., there is no vertical leakage (d) the flow is laminar and unidimensional (e) the release of water from storage is instantaneous and in proportion to the decline in head, and (f) the drain discharges water at a constant rate (Ferris, *et al.* 1962).

To use the Ferris equation in advance of construction, the discharge to the cut must be estimated using other hydrologic analytical procedures and a knowledge of the values of the coefficients of transmissivity and storage for the aquifer. The values for these coefficients may be determined in advance using pumping tests and various aquifer analysis procedures.

Not uncommonly, an aquifer exposed in a road cut may be overlain by confining beds and other aquifers. Vertical leakage through these confining beds from source beds to lower aquifers is common. If vertical leakage is excessive, use of the Ferris equations to predict drawdown within the aquifer as a function of time and distance may produce erroneous results because a basic assumption upon which the derivation of these equations are bases has been violated.

The possible solution to the problem of vertical leakage may be obtained using drain formulas that allow for removal of ground water and recharge derived from precipitation or irrigation.

The problem of unsteady seepage flow, including vertical leakage, to a confined aquifer, which might be likened to an unconfined aquifer receiving infiltration from either rainfall or irrigation, was considered by Karadi, Krizek, and Elnaggar (1968). They consider the time rate of water level change in the domain between two fully penetrating trenches or adjacent to one fully penetrating trench under varying boundary and recharge conditions.The method of solution is based on matrix mathematics and handles water level changes under complex boundary conditions that result when a fixed and known change in water level takes place, as is common for road cuts below the water table.

A still more recent solution to this problem of predicting water level declines within soil and rock masses with complex boundary conditions and variable permeability distribution is presented by Neuman and Witherspoon (1970). Their iterative approach to steady seepage of ground water with a free surface uses the finite element method which is applicable to heterogeneous porous media with complex geometric boundaries and an arbitrary degree of anisotropy.

The most frequent solution for "well" interference problems adjacent to highways has been to set pumps deeper, or drill deeper wells provided that underlying deposits contain water of favorable quality and yield sufficient volumes of water. This is not always the case.

Where cuts extend below the water table in uplands that served as recharge areas, line sinks may be produced that serve to direct ground water to new, local discharge areas, changing both the configuration of the ground-water flow system and reducing the total amount of ground-water recharge.

Changes in Drainage Divides

Drainage areas of small surface-water and ground-water basins may be altered by road cuts. This is common where valley bottom elevations vary on either side of a topographic divide transected by a highway and where deep and extensive cuts are required for favorable grades. This may have a greater impact on changing the location of ground-water compared to surface-water divides, depending upon the unique hydrogeologic setting. Ground water and surface water may be only slightly rerouted from their previous courses in the watershed without significant regional changes. However, for small drainage basins these diversions may represent a significant percentage of the total discharge from the basin and be accompanied by important changes in basin area. This causes changes in the flow and related characteristics of streams.

Reduction in Induced Streambed Infiltration

Ground-water supplies using induced infiltration of surface water as a source of recharge have been extensively developed especially in north central United States where valley train sand and gravel deposits are well developed. These deposits normally are confined to modern or partly buried valleys. Where the deposits are adjacent to major rivers and are undergoing extensive ground-water development, they are subject to recharge from streams by induced streambed infiltration.

* * * * * * *

Permanent or semi-permanent reductions in streambed infiltration rate, I_t, may result from highway construction in several ways. River channels may be diverted from their former course to provide access for highways. Channels may be located in less permeable deposits than previously. Streams and rivers frequently are rerouted along channels lined with concrete, rip rap, or closely-set boulders, and confined within large culverts or similar control facilities. The streambed infiltration rate in these settings may be eliminated entirely or significantly reduced.

Channels may be silted by sediments derived from cuts during construction, from borrow pits, newly placed embankments, sand and gravel washing operations and related sources. Upon completion of construction, rapid runoff derived from pavements, berms and cuts, etc., is normally routed along drainage facilities and discharged to adjacent waterways and land.

Increased and flashy flows in unprotected areas have upset the dynamic equilibrium of the land surface in many areas, and rapid erosion and sediment production have resulted. These materials may further reduce the streambed infiltration capacities after being redeposited in stream channels.

Quantitative values for these changes in infiltration capacities are not available to the author, but are probably significant. An analogy is provided where sewage effluent containing suspended solids is discharged to rivers or where the infiltration rate of slow and rapid sand filters is compared. Data presented by Walton (1963), and Fair and Geyer (1954) are included in Tables 1 and 2. In Table 2 it may be noted that the infiltration rate of slow and rapid sand filters was reduced significantly with time of use before flushing. These reductions are caused by the filtration of suspended solids which serve to plug the filters during use. Turbidity resulting from construction would have an analogous effect.

* * * * * * *

Prior to the construction of an industrial plant near University Park, Pennsylvania, computations indicated that streambed infiltration for an influent stream would accomodate 100% of plant effluent and cooling water discharged as overland flow in a karst drainage way. During plant construction, erosion around the plant site resulted in siltation of these drainage channels. The channel bed infiltration rate declined by more than 60% from that originally calculated, largely due to sediment damage.

The extent and significance of alterations in infiltration capacity of streambeds, hence potential damage to well fields, has gone largely unrecognized probably because detailed hydrologic analyses were not performed on well fields and adjacent streambeds before and after construction to allow comparison. A 5 to 80 percent reduction may be possible under some field settings. Scour during high flows might be expected to improve conditions with time, provided that siltation is not extensive or an ongoing process.

Damage to well fields resulting from streambed siltation may include decline in the water tables, and increased pumping lift costs, shutdown time to deepen pump settings, curtailment of pumping to minimize well interferences, abondonment of some wells in advance of their designed life, and significant reductions in the sustained yield of aquifers located in river bottom environments that rely on streambed infiltration.

At first glance the extent of this type of damage may appear insignificant until one realizes that valley bottom aquifers represent some of the most heavily used aquifers in some regions of the United States. Admittedly, it may be impossible to separate siltation damage resulting from highway construction from that of sediments derived from other land uses, particularly in urbanized areas.

Another direct damage to infiltration and ground-water flow may result where permanent sheet piling is required during construction or where pilings are not pulled after construction. Interlocking sheet piles are commonly used to provide support for excavation walls and embankments and to control ground water when excavating below the water table. These may be driven 5 to 60 feet through clay, silt, and sand deposits and provide a nearly impermeable steel curtain. They have been used along tunnel approaches, along channel walls where roads are placed directly adjacent to channels, beneath bulkheads, around excavations used for bridge and overpass abutments, along underpasses, etc. They may be left in place to minimize seepage to low areas after construction.

* * * * * * *

Highways may reduce recharge to upland aquifers by promoting flashy runoff in areas formerly receiving recharge. This contributes to the overall urban sprawl syndrome where ground-water recharge is concerned. Where sand and gravel, carbonates, basalts, and related highly permeable aquifers are present at the land surface the reverse may be true. The quick and nearly complete runoff of nearly all precipitation may be concentrated at fewer discharge points along the roadway that favor high rates of infiltration. In this manner, evapotranspiration and interception losses may be greatly reduced and local ground-water recharge facilitated. Storm water from roads, parking lots, etc., in the Miami area of Florida, for example, is routinely diverted to basins which serve to recharge the local carbonate aquifer. Similarly, snow fences, plowed snow, and drifting snow trapped by deep cuts further concentrate surface water which is available either for quick runoff or ground-water recharge depending upon the setting.

Highways in the flood plain environment may reduce the area of flooding during high flows, further reducing the recharge potential to underlying aquifers. The area available for infiltration from flood flows may further be reduced where highway embankments serve as levees for flood waters. It may be seen from equation 3 that the potential recharge from induced infiltration is directly related to the area through which infiltration occurs. Flood plains vary considerably in width, hence, the significance in reducing flood plain widths by highways varies. Between 10 to 35 acres of land per mile of divided highway may be lost to infiltration. For narrow flood plains, all but the river channel may be lost to infiltration.

In addition to restricting induced-infiltration, sediments may aggrade in channels downstream from construction and reduce channel transmission capacity. This not only reduces the efficiency of channels and water routing facilities but also promotes more severe flooding during peak flows, and alteration of the ecology of the stream. The latter two influences of highway construction are more obvious and are the ones surface water hydrologists and highway engineers tend to take into consideration.

Slope Stability Problems

Slope stability problems are commonplace along highways in many geologic settings throughout the nation. Rock falls, rock slides, slumps and related failures may be cited. The probability of slope failures increases when cuts are established below the water table. These failures are common within residual and transported soils–colluvium, alluvium, glacial drift, etc.–and within weathered bedrock. Three of the most common causes of slope failures in saturated soils in Pennsylvania are discussed by Parizek (1970).

Loss of stability may be brought about by piping failures (Fig. 3), where fine-grained soil constituents are selectively eroded from a stratified soil. Support for overlying deposits is removed as loss of fines continues. Grain-by-grain removal of matrix materials is accomplished by discharging ground water aided by seepage pressures acting outward from the unsupported slope. Where the cause of this type of failure is improperly diagnosed, annual maintenance inevitably results.

A second cause of slope failure is related to seepage pressures directed toward a road cut in the presence of flowing ground water. Seepage pressures are added to driving forces already present due to the weight of water, soil, and rock acting on a potential surface of sliding (Fig. 4a). Terzaghi and Peck (1948) indicate that percolating water exerts a seepage pressure on soil particles through which it flows due to its viscosity. This pressure acts in the direction of flow tangential to ground-water flow lines and its intensity increases in simple proportion to the seepage velocity and the hydraulic head to be dissipated. Where slopes are cut below the water table they establish free ground-water drains. Seepage velocities and seepage pressures are greater at the foot of slopes than higher up.

Terzaghi (1950) concluded as a consequence, that the state of failure is reached much earlier at the toe than higher up, and once the lower part of the slope has failed, the upper part follows in the absence of support.

* * * * * * *

Three possible corrective measures are presented by Parizek (1970) to combat these potential slope failures. They are in more common use today than in the past in highway construction.

Water-table level changes are seasonal in these settings hence the possible future role of ground water in promoting slope failures may be ignored or overlooked during highway construction. Corrective action is usually taken

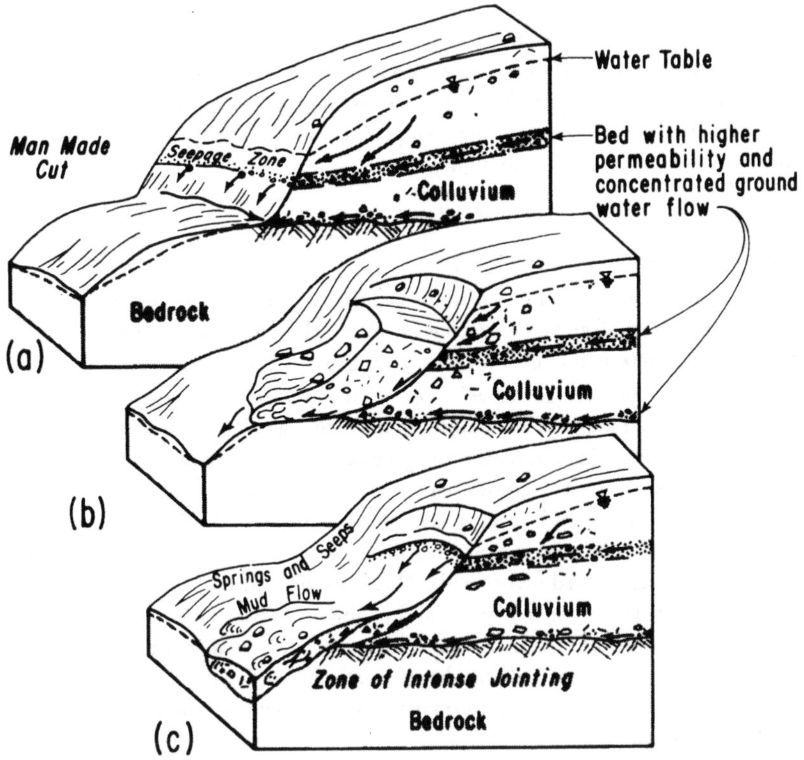

Figure 3. Slope failure by piping (a), (b) and (c). Erosion of fines from a thin aquifer is initiated by ground water discharging to the new cut (a). In (b), piping continued until the slope fails by slumping. With prolonged ground-water discharge and surface runoff (c), the slumped mass may give rise to a mud flow and the exposed slope is subjected to a new cycle. In (d) piping is initiated within a three-foot thick sand bed located between two glacial tiils. The piping tubes and small gullies eroded by discharging ground water are obvious. In time the 18-foot till bank above will slump.

after a slope failure has been initiated. This sequence tends to be the most economic from a highway maintenance viewpoint.

Less excusable are slope failures that result when highway cuts are excavated in landslide topography that was not recognized as such when planning and designing the highway. Although slopes may have been stabilized after a series of slides were complete, they may be reactivated by removing toes of slides that acted as buttresses to overlying slide material, by adding to driving forces by placing fills on the upper parts of slides, and/or by changing the surface-water and ground-water regimen.

A good example of slide topography is shown in Figure 5 taken along Swift Current Creek in south-central Saskatchewan. Many new slope failures have developed adjacent to highways and railroads constructed through formerly stable landslide topography.

* * * * * * *

Turbidity Increase

Increases in the sediment load in ground water and surface water may result during and after highway construction. This may cause changes in the physical characteristics of surface- and ground-water reservoirs as well. Reduction in the storage capacity of surface-water reservoirs is the most obvious change. Fine-grained sediments may change the infiltration capacity and permeability characteristics of ground-water reservoirs, may damage pump and distribution facilities where suspended silt, sand, and clay are encountered in wells and springs, and may require costly filtration. Similar problems may result in water supplies derived from lakes and rivers containing increased sediment loads. Not only does this cause greater water development costs, inconveniences, and problems for water supply operators, but also siltation of channels and reservoirs may reduce the infiltration capacity of streambeds, reduce the recreational and wild-life value of the water supply, as well as cause increased flood damage, channel migration and similar problems.

Scheidt (1967) indicates that sediment caused by erosion of soil exposed during construction is one of the most serious detriments to the environment caused by highways. Damage may occur in the form of: added regrading costs to the contractor; downstream damages to wildlife, streams, rivers, reservoirs, navigation channels, and; damage to adjacent landowners where water and sediments trespass on their property.

* * * * * * *

The Interstate Commission on the Potomac River Basin (1963), reports that 2,500,000 tons of sediment per year are dumped into the tidal estuary each year, which requires annual dredging to keep navigation channels open. Erosion from highway construction is believed to contribute considerably to this sediment load. Similar conclusions undoubtedly can be drawn from highway construction elsewhere. Sediment loads, however, are expected to be highly variable depending upon soils, vegetation, climate, geology, construction and erosion control practices employed.

* * * * * * *

Sediments may be derived from fresh road cuts, embankments, borrow pits, etc. Sediment problems are most serious when surfaces are freshly exposed prior to seeding and emplacing pavements. However, increased discharges and redistributed surface runoff from pavement and embankment areas may promote new erosion of vulnerable soils. This serves to aggravate turbidity and siltation problems.

Highways have been constructed in advance of water-supply reservoirs in some watersheds to avoid these problems (Lewistown, Pennsylvania) and highways have been postponed or rerouted in other areas where water-supply reservoirs already existed (the Keystone Shortway near DuBois, Pennsylvania).

Suspended sediments may be encountered in wells and springs under certain field circumstances. This problem, which is not as well known as turbidity problems in surface water supplies, may arise because the filtration media (grass sod, forest litter, soil profile, etc.) and the hydrologic balance are upset by excavation. Where coarse-grained sand and gravel (Fig. 6a), cavities in carbonate bedrock (Fig. 6b and d), and fissures, joints and other openings in other bedrock types (Fig. 6c) are exposed adjacent to excavations, they provide avenues into which turbid surface water can migrate. Where suspended particles are small compared to the size of voids, they are free to migrate with ground water without being filtered out. This is especially a problem for well-sorted coarse gravel deposits and bedrock containing extensive openings. In mixed fine-grained and well-graded granular deposits a filter cake quickly develops as suspended grains are filtered from the percolating water.

Turbid water is known to have traveled vertically from a road cut through more than 50 feet of dry glacial gravel and horizontally through more than 2,000 feet of saturated gravel to springs serving Milford, Pennsylvania, indicating travel distances that are possible for entrained sediment. Damage to the Milford water supply will exceed several hundred thousand dollars before a final solution is provided. Where open conduits occur, as is common in carbonate rocks, and ground-water flow rates are 10's of feet or more per hour, turbid waters derived from excavations may travel a mile or more with

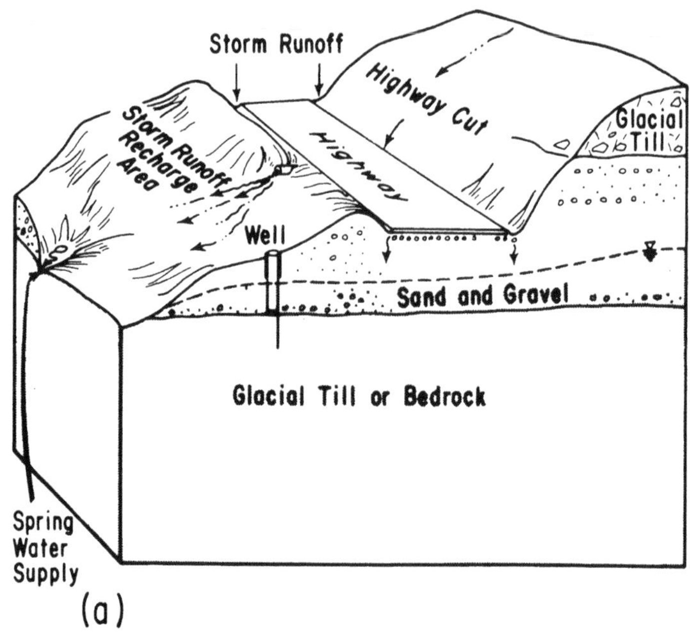
Storm Runoff
Highway Cut
Glacial Till
Storm Runoff Recharge Area
Highway
Well
Sand and Gravel
Glacial Till or Bedrock
Spring Water Supply

(a)

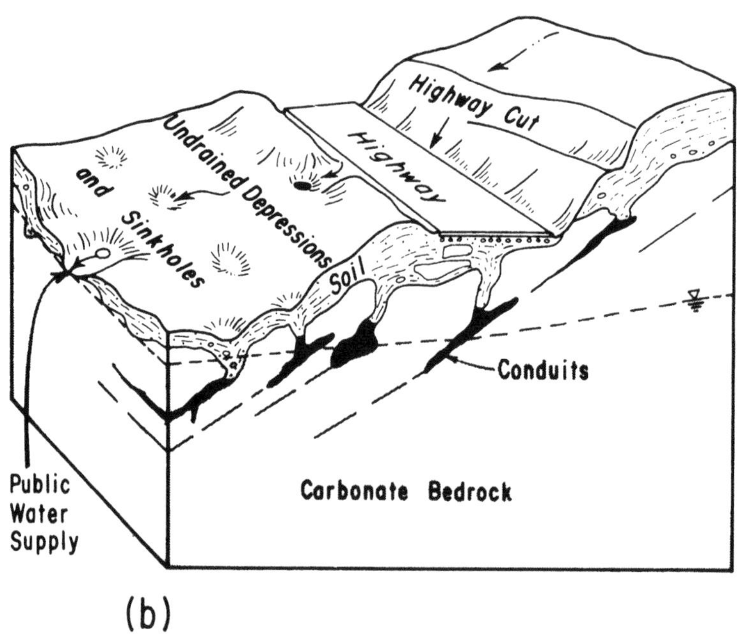
Highway Cut
Highway
Undrained Depressions
and
Sinkholes
Soil
Conduits
Public Water Supply
Carbonate Bedrock

(b)

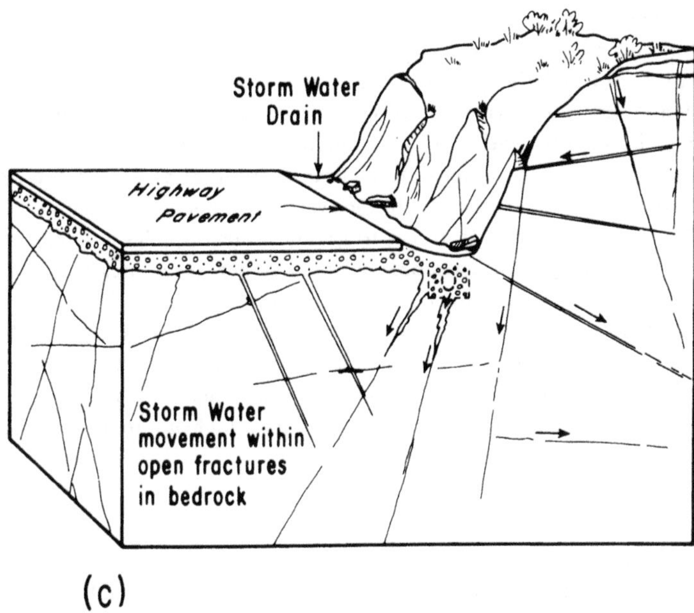

Figure 6. Conditions favoring turbidity and roadside pollution of aquifers in sand and gravel (a), carbonate bedrock (b) and fractured rock (c). Storm runoff diverted by highway drainage facilities concentrates recharging waters carrying pollutants may erode sediment within intergranular and fracture openings.

little, if any, filtration or reduction in suspended solids. Turbidity problems are most likely to occur: where the water table is shallow, and road cuts extend near to or into the top of bedrock; where open fissures and cavities are uncovered by the excavation, and; where sediment laiden surface water is free to enter open sink holes and exposed gravel downslope from construction.

Turbidity problems may be intermittent and may occur during periods of increased recharge and runoff during and after construction. Once fine-grained sediments have been added to openings and have not yet stabilized, they are free to be remobilized and transported by soil-water and ground water as flow volumes and rates periodically increase.

Turbidity may result by changing the ground-water flow regimen. Runoff from highways may be redistributed and be added to the ground-water reservoir in new areas and in increased amounts. As flow volumes and velocities increase or flow becomes turbulent, unstable silt and clay particles contained in granular deposits and in fissures and cavities may become mobilized and transported by ground water. Soils that were formerly stable can pipe into underlying solution cavities when subjected to increased flows, be winnowed from coarse-grained granular deposits, or eroded from walls and floors of conduits developed in carbonate aquifers. Turbidity resulting from the latter cause can reappear from time to time after construction during periods of extensive recharge until the system becomes readjusted.

This potential turbidity threat to ground-water supplies is harder to assess than are many of the other problems noted earlier. It requires a knowledge of the hydrologic and geologic framework in each area which is normally incomplete, hence an element of risk is always involved. Adequate filtration facilities may be provided during and after construction, an alternate water supply may be developed as a stand-by in advance of construction or the supply may be replaced entirely. An alternate highway route might also be considered.

* * * * * * *

Acid Drainage

Mine drainage derived from the oxidation of iron disulfide (FeS_2) or pyrite-bearing shale and coal beds is a problem of major concern in many states. Strip and deep coal mining activities are responsible for the breakup of rock strata which are susceptible to acid production in the presence of water and oxygen. Increased surface areas of broken strata and the accessibility provided for oxygen by mining greatly increase the rate of acid production over that of the undisturbed strata.

* * * * * * *

Strata exposed along road cuts, and thoroughly disaggregated fills containing pyrite-bearing strata are equally suitable environments that promote acid drainage problems comparable to those created by coal mining operations.

* * * * * * *

In addition, mine drainage may be redistributed to the road cut which now serves as a ground-water drain, hence causing pollution of adjacent wells and tributary watersheds.

The potential pollution threat derived from pyrite-bearing strata should not be overlooked, particularly where receiving bodies of water afford little if any dilution or neutralization, and where there is direct use of these waters in water-supply systems.

* * * * * * *

FINAL STATEMENT

* * * * * * *

Impact of highways on land-use patterns, land values, and the general economy of a region have been investigated in some detail in recent years. Establishing criteria to aid in future highway locations and land-use planning and zoning has been a prime or long-ranged objective of these studies. Hydrogeologic factors have often been of less concern in highway design and planning than have geologic and economic factors. Only recently have many highway projects required extensive excavations, often below the water table. Frequently, major highways were located in more remote areas to minimize land purchase and condemnation expenses, hence, have not caused noticeable damage to local water users. In metropolitan areas highways commonly transect regions served by municipal sewage and water supply facilities. Hydrology has often not been an important consideration. However, as the competition for open space becomes more intense, and population densities increase, the impact of highways on the hydrologic regimen will cause increasing concern. Potential hydrologic problems can be minimized if during the planning stages of highway projects reasonable alternative solutions are available, thereby reducing the damage to adjacent landowners, avoiding costly delays and litigation. Also, it is only proper that real and total damage to private interests be appraised and compensated when acquiring land for public interest. However, as the interaction of man's activities on his environment becomes better understood and the impact of these activities on his environment and his own interests are recognized, he will no longer overlook actions of others that he considers to be an infringement on his rights. Alteration of the hydrologic regimen by highways may be but another such infringement in the future that he may no longer tolerate without adequate compensation. Our increased knowledge should allow us to predict more of these changes in advance of construction.

Based on judgment, experience, and investigation, some of these changes will be judged as insignificant in some regions, others more significant. However, at this juncture, detailed investigation of some of these possible changes in the hydrologic regimen is warranted. This may lead to an expansion of the variety of people employed and training required in the future for those engaged in highway design and planning.

Specific research projects may consider the influence of highway construction on sediment production, changes in the regimen of surface creeks and rivers, and erosion under a variety of topographic, pedalogic, geologic, and hydrologic settings. Problems created by increased turbidity in surface-water and ground-water reservoirs and conditions that favor sediment production should be investigated as well as the influence of highways on ground-water and surface-water flow and ground-water recharge.

* * * * * * *

REFERENCES

Bartltrop, D. and Killala, N.J.P. 1967. Faecal excretion of lead by children: Lancet. v. 2, p. 1017.

Boswell, V.R. 1952. Residues, soils and plants: *The Year Book of Agriculture for 1952* edited by A. Stefferud, p. 284-297. U.S. Gov. Printing Office.

Boswell, V.R. 1953. Insecticides and the soil: Atlantic Naturalist, v. 8, p. 246-251.

Bowers, G.N. 1970a. The effect of highways on public drinking water supplies in Connecticut; memo to concerned individuals and officials of responsible agencies: by the Director–Clinical Chemistry, Dept. of Pathology, Hartford Hospital, Hartford, Conn. 3 p.

Bowers, G.N. 1970b. Atmospheric lead pollution: Hartford Hospital Bull. v. 25, n. 2, p. 70-78.

Bureau of Public Roads. 1966. Instructional memorandum No. 20-3-66: U.S. Dept. of Commerce, June 7.

Caruccio, F.T., and Parizek, R.R. 1967. An evaluation of factors influencing acid mine drainage production from various strata of the Allegheny Group and the ground water interactions in selected areas of western Pennsulvania: Coal Res. Board, Commonwealth of Penn. Spec. Rept. SR-65, 213 p.

Caruccio, F.T. 1968. An evaluation of factors affecting acid mine drainage production and ground water interactions in selected areas of western Pennsylvania: Second Symp. on Coal Mine Drainage Research, Mellon Inst., Pittsburgh, Pa.

Dansereau, H.K. 1964. Selected attitudes toward highway change: Monroeville and Blairsville compared: Inst. for Research on Land and Water Resources, The Pennsylvania State Univ., Res. Rept. 2, 27 p.

Dowler, C.C., Sand, P.F., and Robinson, E.L. 1963. The effect of soil type on pre-planting soil-incorporated herbicides for witchweed control: Weeds, v. 11, p. 276-279.

Dutcher, R.M., Jones, E.B., Lovell, H.L., Parizek, R.R., and Stefanko, R. 1966. Mine drainage Part 1: Abatement, disposal, treatment: Mineral Industries, v. 36, n. 3, December, The Pennsylvania State Univ., 8 p.

Fair, G.M., and Geyer, J.C. 1954. *Water supply and waste-water disposal;* John Wiley and Sons, Inc., New York.

Ferris, J.G. 1950. Quantitative methods for determining ground-water characteristics for drainage design: Agr. Eng. v. 31, n. 6, p. 285-291.

Ferris, J.G., Knowles, D.B., Brown, R.H., and Stallman, R.W. 1962. Theory of aquifer tests: U.S. Geol. Survey Water-Supply Paper 1536-E, 174 p.

Finley, S., Exploring Lake Barcroft with dragline and ten-wheeled truck: Small Sanitary District, Fairfax County, Virginia.

Foster, A.C., Boswell, V.R., Chisholm, R.D., Carter, R.H., Gilpin, G.L., Pepper, B.B., Anderson, W.S., and Gieger, M. 1956. Some effects of insecticide spray accumulations in soils on crop plants: U.S. Dept. Agr. Tech. Bull. 1149, 36 p.

French, D.W. 1959. Boulevard trees are damaged by salt applied to streets: Minnesota Farm and Home Sci., v. 16, n. 9, p. 22-23.

Frey, J.C., *et al.* 1960. Planned vs. Unregulated development in a suburban community: Agr. Economics and Rural Sociology Rept. 23, The Pennsylvania State Univ.

Garin, A.N., and Foster, G.W. 1940. Effect of soil erosion on the costs of public water supply in the North Carolina Piedmont: U.S. Dept. Agr., Soil Conserv. Serv. SCS-EC-1.

Grace, A.B. 1966. Severance damage studies: Right-of-Way, v. 13, n. 1, p. 55-58.

Hibben, C.R. 1962. Investigation of sugar maple decline in New York woodlands: N.Y. Nursery Notes, Cornell, Dec.

Industrial Hygiene Section of the American Public Health Association. 1964. Occupational lead exposure and lead poisoning: New York: Am. Public Health Assoc.

Interstate Commission on the Potomac River Basin. 1963. Preliminary study of sediment sources and transport in the Potomac River Basin, Washington, D.C.

John Hopkins University. 1966. Report on Patuxent River Basin, Maryland: John Hopkins Univ., Baltimore Md. June

Karadi, G., Krizek, R.J., and Elnaggar, H. 1968. Unsteady seepage flow between fully-penetrating trenches: Jour. Hydrol. v. 6, p. 417-430.

Kehoe, R.A. 1964. Metabolism of lead under abnormal conditions: Arch. Environ. Health, v. 8, p. 235-243.

Korkish, J., and Hazan, I. 1965. Anion exchange separations in hydrobromic acid-organic solvent media: Anal. Chem., v. 37, p. 707-710.

LaCasse, N.L., and Rich, A.E. 1964. Maple decline in New Hampshire: Phytopathology, v. 54, Sept., p. 1071-1075.

Lagerwerff, J.V. and Specht, A.W. 1970. Contamination of roadside soil and vegetation with cadmium, nickel, lead and zinc: Environmental Sci. and Tech., v. 4, n. 7, p. 583-586.

Lierboe, R.T. 1967. Interrelation of highway and reservoir location: Jour. of Highway Division, ASCE v. 93, n. HW1, Proc. Paper 5186. p. 1-6.

Mariolo, J.R., and Dansereau, K. 1965. Planning, zoning, and interchange protection: a report of leadership attitudes: Inst. for Research on Land and Water Resources, The Pennsylvania State Univ., Res. Publication No. 45, 22 p.

McCarty, P.L., and King, P.H. 1966. Movement of pesticides in soils: Proc. of the 21st Industrial Waste Conference, Part I, Purdue Univ., p. 156-171.

McGough, B.C. 1968. A treatise on highway impact studies: Right-of-Way, v. 15, n. 3, p. 40-45.

Morrow, P.E., and Casarett, L.J. 1961. *Inhaled particles and vapors:* Oxford, England: Pergamon Press, p. 167.

Neill, C.R. 1964. A review of bridge engineers: Canadian Good Roads Assoc. Tech. Pub. No. 23, Res. Council of Alberta, Contribution No. 281, 37 p.

Neuman, S.P., and Witherspoon, P.A. 1970. Finite element methods of analyzing steady seepage with a free surface: Water Resources Research, v. 6, n. 3, p. 889-897.

Page, A.L., and Ganje, T.J. 1970. Accumulations of lead in soils for regions of high and low motor vehicle traffic density: Environ. Sci. and Tech., v. 4, p. 140.

Parizek, R.R. 1970. Land use problems in Pennsylvania's ground-water discharge areas in soils (accepted for publication), Penn. Geol. Survey.

Phillips, W.M. 1959. Residual herbicidal activity of some chloro-substituted benzoic acids in soil: Weeds, v. 7, p. 284-294.

Robinson, E., *et al.* 1963. Variation in atmospheric lead concentrations and type with particle size: SRI Project Pa-4211, Menlo Park, Calif.: Stanford Research Institute, p. 40.

Robinson, E., and Ludwig, F.L. 1964. Size distribution of atmospheric lead aerosols: SRI Project Pa-4788, Menlo Park, Calif.: Stanford Research Institute, p. 22.

Rockey, M.B., and Frey, J.C. 1964. Farms and new highways: problems and adjustments: Inst. for Research on Land and Water Resources, The Pennsylvania State Univ., Res. Rept. 1, 29 p.

Salt Institute. 1962. Technical analysis of salt (NaCl) for ice and snow removal: Salt Institute, Chicago, Illinois, 59 p.

Salt Institute. 1966. Storing road de-icing salt: Salt Institute, Chicago, Illinois, 16 p.

Sauer, G. 1967. On damages by de-icing salts to plantings along the Federal highways: Nachrichtenblatt des Deutschen Pflanzenschutzdientes (Verlag Eugen Ulmer, Stuttgart, Germany), v. 19, n. 6.

Savini, J., and Kammerer, J.C. 1961. Urban growth and the water regimen: U.S. Geol. Survey Water-Supply Paper 1591-A, p. A1-A43.

Sheets, T.J. 1967. The extent and seriousness of pesticide build-up in soils: *Agriculture and the Quality of our Environment*, edited by N.C. Brady, Am. Assoc. Advancement of Science Pub. 85, p. 311-330.

Sinclair Refining Co. 1964. Product information Bull. P-74.

Smith, W.T. 1963. Ground-water control for highways: California Division of Highways, Highway Research (Abstracts), v. 33, n. 12, 43 Annual Meeting, Jan. 13-17, 1964, Washington, D.C., p. 84.

Strong, F.C. 1944. A study of calcium chloride injury to roadside trees: Michigan Agr. Expt. Sta. Quart. Bull. v. 27, p. 209-224.

Scheidt, M.E. 1967. Environmental effects of highways: Jour. Sanitary Engineering Div. Proc. of Amer. Soc. Civil Engineers. v. 93, n. SA5, p. 17-25.

Terzaghi, K. 1950. Mechanism of landslides: in *Application of geology to engineering practice*, Berkley Volume, Geol. Soc. Amer., p. 83-123.

Terzaghi, K., and Peck, R.B. 1948. *Soil mechanics in engineering practice:* John Wiley and Sons, New York, 566 p.

Thomas, H.V., Milmore, B.K., Heidbreder, G.A., and Kogan, B.A. 1967. Blood leads of persons living near freeways: Arch. Environ. Health, v. 15, p. 695-702.

Van Tuyl, D.W. 1951. Ground water for air conditioning in Pittsburgh, Pennsylvania: Penn. Geol. Surv. 4th Series, Bull. W10, 34 p.

Vermont Dept. Forests and Parks. 1961. Sugar maple decline: Tree Pest Inf. No. 2, Pest Control Div. Vermont Dept. Forests and Parks, Montpelier, Vermont.

Walton, W.C. 1963. Estimating the infiltration rate of a streambed by aquifer-test analysis: Int. Assoc. Scientific Hydrol., Pub. No. 63, Berkeley meeting, p. 409-420.

Weiner, D.J. 1966. Unpublished report, R.A. Taft Sanitary Eng. Center, Cincinnati, Ohio, Aug.

Williams, R. 1967. The influence of ground-water flow systems on pavement stability in highway cuts: Ground Water, v. 5, n. 4, p. 23-26.

Wolman, M.G. 1964. Problems posed by sediments derived from construction activities in Maryland: Maryland Water Pollution Control Comm., Jan.

Working Group on Lead Contamination. 1965. Survey of lead in the atmosphere of three urban communities: Public Health Service Pub. No. 999-AP-12.

Reprinted from *Smithsonian Institution Report*, 273–301 (1950)

8

PERMAFROST [1]

By Robert F. Black
Geologist, United States Geological Survey

[With 12 plates]

The term "permafrost" was proposed and defined by Muller (1945). A longer but more correct phrase is "perennially frozen ground" (Taber, 1943a). The difficulties of the current terminology are presented by Bryan (1946a, 1946b), who proposed a new set of terms. These are discussed by representative geologists and engineers (Bryan, 1948). Such terms as cryopedology, congeliturbation, congelifraction, and cryoplanation have been accepted by some geologists (Denny and Sticht, unpublished manuscript; Judson, 1949; Cailleux, 1948; Troll, 1948) in order to attempt standardization of the terms regarding perennially frozen ground and frost action. The term permafrost has been widely adopted by agencies of the United States Government, by private organizations, and by scientists and laymen alike. Its use is continued here because it is simple, euphonious, and easily understood by all.

Extent.—Much of northern Asia and northern North America contains permafrost (fig. 1) (Jenness, 1949; Sumgin, 1947; Muller, 1945; Obruchev, 1945; Troll, 1944; Taber, 1943a; Cressey, 1939; and others). The areal subdivision of permafrost into continuous, discontinuous, and sporadic bodies is already possible on a small scale for much of Asia, but as yet for only part of North America. Refinements in delineations of these zones are being made each year. The southern margin of permafrost is known only approximately, and additional isolated bodies are being discovered as more detailed work is undertaken. The southern margin of permafrost has receded northward within the last century (Obruchev, 1946).

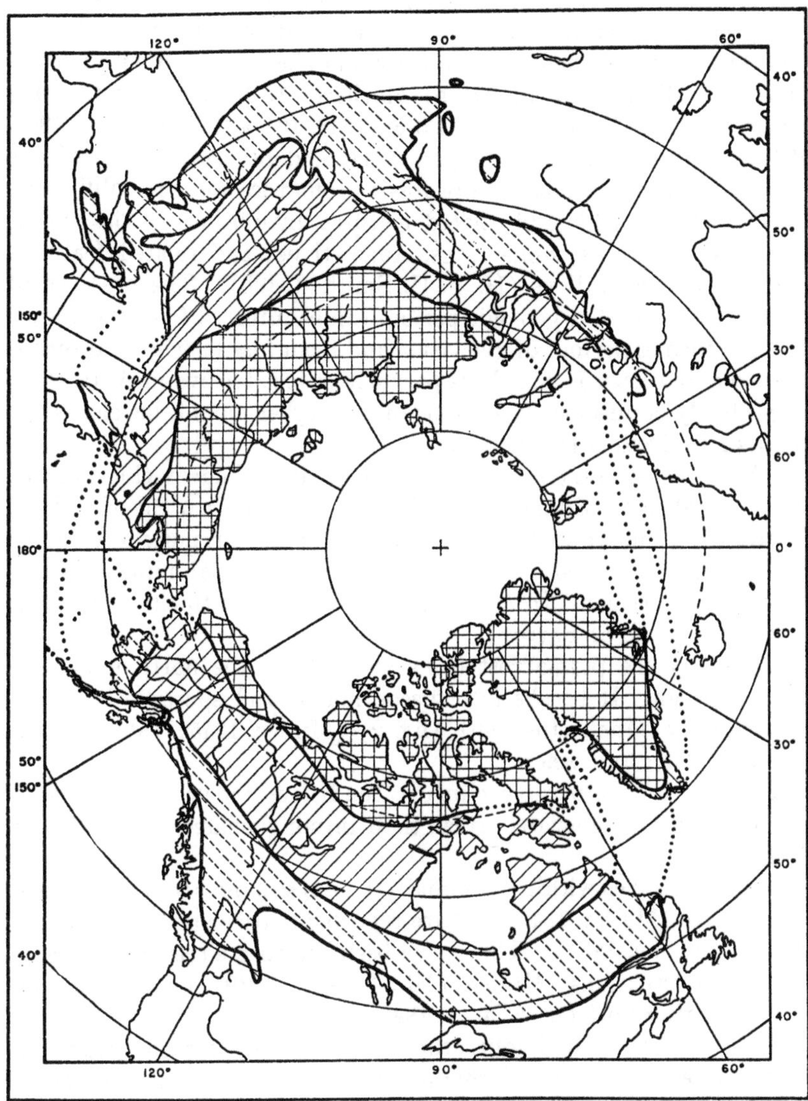

FIGURE 1.—Areal distribution of permafrost in the Northern Hemisphere.

Double hatching: Approximate extent of continuous permafrost. Ground temperature at a depth of 30 to 50 feet generally below —5° C. Diagonal hatching: Approximate extent of discontinuous permafrost. Ground temperature in permafrost at a depth of 30 to 50 feet generally between —5° and —1° C. Dotted diagonal hatching: Approximate extent of sporadic permafrost. Ground temperature in permafrost at a depth of 30 to 50 feet generally above —1° C. Reliability: Eurasia, good; Alaska, fair; all other, poor. (Eurasia after Sumgin and Petrovsky, 1940, courtesy of I. V. Poiré.)

Permafrost is absent or thin under some of the existing glaciers, and it may be absent in areas recently exhumed from ice cover.

A greater extent of permafrost in the recent geologic past is known by inference from phenomena now found to be associated with permafrost (H. T. U. Smith, 1949b; Horberg, 1949; Richmond, 1949; Schafer, 1949; Cailleux, 1948; Poser, 1948, 1947a, 1947b; Troll, 1947, 1944; Zeuner, 1945; Weinberger, 1944, and others). Some of the more important phenomena are fossil ground-ice wedges, solifluction deposits, block fields and related features, involutions in the unconsolidated sediments, stone rings, stone stripes and related features, and asymmetric valleys (H. T. U. Smith, 1949b). The presence of permafrost in earlier geologic periods can be inferred from the known facts of former periods of glaciation and from fossil periglacial forms.

In the Southern Hemisphere permafrost is extensive in Antarctica. It probably occurs locally in some of the higher mountains elsewhere, but its actual extent is unknown.

Thickness.—Permafrost attains its greatest known thickness of about 2,000 feet (620 meters) at Nordvik in northern Siberia (I. V. Poiré, oral communication). Werenskiold (1923) reports a thickness of 320 meters (1,050 feet) in the Sveagruvan coal mine in Lowe Sound, Spitsbergen. In Alaska its greatest known thickness is about 1,000 feet, south of Barrow.

Generally the permafrost thins abruptly to the north under the Arctic Ocean. It breaks into discontinuous and sporadic bodies as it gradually thins to the south (fig. 2) (Muller, 1945; Taber, 1943a; Cressey, 1939; and others).

In areas of comparable climatic conditions today, permafrost is much thinner in glaciated areas than in nonglaciated areas (Taber, 1943a).

Unfrozen zones within perennially frozen ground are common near the surface (Muller, 1945) and are reported to occur at depth (Taber, 1943a; Cressey, 1939). They have been interpreted as indicators of climatic fluctuations (Muller, 1945; Cressey, 1939), or as permeable water-bearing horizons (Taber, 1943a).

Temperature.—The temperature of perennially frozen ground below the depth of seasonal change (level of zero annual amplitude) (Muller, 1945) ranges from slightly less than 0° C. to about −12° C. (I. V. Poiré, oral communication). In Alaska the minimum temperature recorded to date is −9.6° C. at a depth of 100 to 200 feet in a well about 40 miles southwest of Barrow (J. H. Swartz, 1948, written communication). Representative temperature profiles in areas of (1) continuous permafrost are shown in figure 3, *a*; of (2) discontinuous permafrost, figure 3, *b*; and of (3) sporadic bodies of permafrost, figure 3, *c*.

Temperature gradients from the base of permafrost up to the depth of minimum temperature vary from place to place and from time to time. In 1947–48 four wells in northern Alaska had gradients between 120 and 215 feet per degree centigrade (data of J. H. Swartz, G. R. MacCarthy, and R. F. Black).

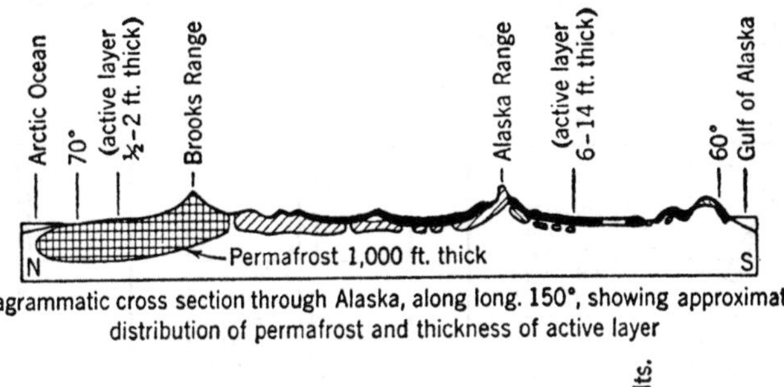

Diagrammatic cross section through Alaska, along long. 150°, showing approximate distribution of permafrost and thickness of active layer

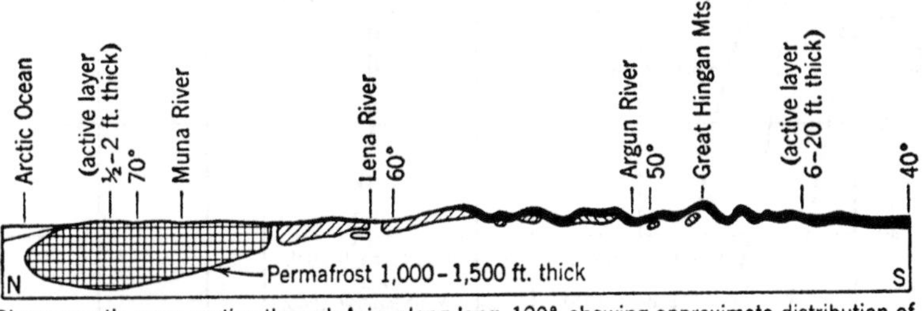

Diagrammatic cross section through Asia, along long. 120°, showing approximate distribution of permafrost and thickness of active layer. (Modified from unpublished cross section by I.V. Poire'.)

Active layer — Discontinuous permafrost — Continuous permafrost — Sporadic permafrost

FIGURE 2.—Representative cross sections of permafrost areas in Alaska and Asia.

The shape of a temperature curve indicates pergelation or depergelation (aggradation or degradation of permafrost) (Muller, 1945; Taber, 1943a). Some deep temperature profiles have been considered by Russian workers to reflect climatic fluctuations in the recent geologic past. No known comprehensive mathematical approach has been attempted to interpret past climates from these profiles, although it seems feasible. Some of the effects of Pleistocene climatic variations on geothermal gradients are discussed by Birch (1948) and Ingersoll et al. (1948).

Character.—Permafrost is defined as a temperature phenomenon, and it may encompass any type of natural or artificial material, whether organic or inorganic. Generally permafrost consists of variable thicknesses of perennially frozen surficial unconsolidated materials, bedrock, and ice. Physical, chemical, or organic composition, degree of induration, texture, structure, water content, and the like range widely and are limited only by the extremes of nature or the caprice of mankind. For example perennially frozen mammals, rodents, bacteria, artifacts, beds of sand and silt, lenses of ice, beds of peat, and varied junk piles, such as kitchen-middens, mine dumps, and ships' refuse heaps are individual items that collectively can be lumped under the term permafrost.

Ground perennially below freezing but containing no ice has been called "dry permafrost" (Muller, 1945).

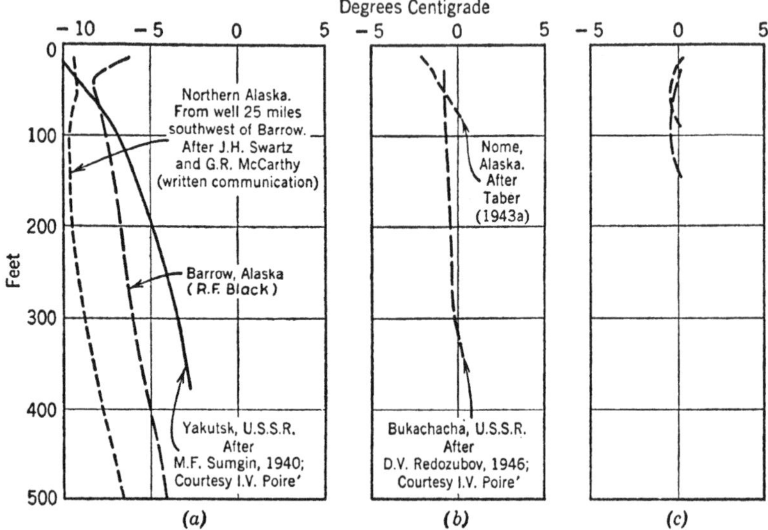

FIGURE 3.—(*a*) Representative temperature profiles in areas of continuous permafrost. (*b*) Representative temperature profiles in areas of discontinuous permafrost. (*c*) Hypothetical temperature profiles in areas of sporadic permafrost.

Permafrost composed largely of ice is abundant particularly in poorly drained fine-grained materials (pls. 1, 2, and 3). The ice occurs as thin films, grains, fillings, veinlets, large horizontal sheets, large vertical wedge-shaped masses, and irregular masses of all sizes. Many masses of clear ice are arranged in geometric patterns near the surface, that is, polygonal ground (pl. 4) and honeycomb structure. The ice may be clear, colorless, yellow, or brown. In many places it contains numerous oriented or unoriented air bubbles (pl. 5, fig. 1), and silt, clay, or organic materials. Size, shape, and orientation of the ice crystals differ widely (pl. 5, fig. 2). Discordant structures in sediments around large masses of ice are evidences of growth (Taber, 1943a; Leffingwell, 1919).

Relation to terrain features.—In the continuous zone of permafrost the upper limit (permafrost table, Muller, 1945) is generally within a few inches to 2 feet of the surface. Large lakes and a few large rivers lie in thawed areas slightly larger than the basins they occupy

Upper: Left, irregular settling of Kougarok Railroad on frozen tundra about 7 miles north of Nome, Alaska; photographed August 10, 1948. Right, caving and irregular settling in gravel fill over thawing ice wedges in runway at Umiat, Alaska; photographed August 30, 1946.

Lower: Left, road icing, about 4 feet thick, on Slana-Tok Cut-off about 58.7 miles north of Slana in east-central Alaska; photographed February 27, 1946. Right, earth mounds 4 to 8 feet high produced by stripping of vegetation in farming and subsequent thawing of ice wedge polygons near College, Alaska; photographed September 19, 1948.

1. Bridge piling on the Alaska Railroad broken by downhill creep in mud on top of permafrost in the Nenana River Gorge, central Alaska. Photographed September 24, 1948.

2. Slump in gravel on street in Barrow, where steam line produced thawing of permafrost. Photographed May 12, 1950.

1. Mud flow on top of permafrost along the Alaska Railroad resulting in track settling, near Moody, Alaska. Photographed June 22, 1948.

2. Settling cracks in foundation wall of the U. S. Post Office, Nome, Alaska, after thawing of permafrost under the building. Photographed August 10, 1948.

(Black and Barksdale, 1949; Muller, 1945). Well-drained coarse-grained materials may thaw annually to a depth of 6 feet. Poorly drained fine-grained materials protected from solar radiation and insulated with moss and other vegetation may thaw annually to a depth of only 4 inches.

In the discontinuous zone permafrost is absent under most major rivers and lakes. It may be absent in the tops of some well-drained low hills. Seasonal thaw (active layer, Muller, 1945) penetrates 1 foot to 10 feet, depending on insulation, insolation, drainage, and type of material.

Sporadic bodies of permafrost may be relics below the active layer or may be forming in favorable situations in poorly drained fine-grained materials on north-facing slopes. In the zone of sporadic permafrost the active layer may or may not reach the permafrost table, and it ranges between 2 and 14 feet in thickness.

Generally the depth of thaw is at a minimum in northern latitudes and increases to the south. It is at a minimum in peat or highly organic sediments and increases successively in clay, silt, and sand to a maximum in gravelly ground or exposed bedrock. It is less at high altitudes than at low altitudes; less in poorly drained ground than in dry well-drained ground; at a minimum under certain types of tundra and increases successively in thickness under areas of bog shrubs, black spruce, larch, white spruce, birch, aspen, and poplar to a maximum under tall pines. It is less in areas of heavy snowfall; less in areas with cloudy summers; and less on north-facing slopes (Muller, 1945; Troll, 1944; Taber, 1943a; and others).

Works of man commonly upset the natural thermal equilibrium and may tend to destroy permafrost or to aid in its formation. Most roads, runways, and other structures on the surface of or in the ground generally have lower permafrost tables than undisturbed natural areas adjacent to them. Structures above the ground and insulated from the ground protect the surface from solar radiation and commonly produce higher permafrost tables.

Origin.—The origin of perennially frozen ground is discussed by Jenness (1949), Muller (1945), Zeuner (1945), Taber (1943a), Cressey (1939), Nikiforoff (1932), Leffingwell (1919), and others. Generally it can be stated that most sporadic bodies of permafrost are relics of colder climates. Discontinuous bodies of permafrost are largely relics, but under favorable conditions may grow in size, and new deposits are being perennially frozen. In areas of continuous permafrost, heat is being dissipated actively from the surface of the earth to the atmosphere, and new deltas, bars, landslides, mine tailings, and other deposits are being pergelated (incorporated in the permafrost) (Bryan, 1946a).

922758—51——19

Local surface evidences indicate that heat, in some places at least, is being absorbed at the base of permafrost faster than it is being dissipated at the surface (Hopkins, 1949; Young, 1918). Hence the cold reserve is being lessened, and the thickness of permafrost is decreasing from the base upward.

The mean annual air temperature required to produce permafrost undoubtedly varies many degrees because of local conditions. Generally it is given as 30° to 24° F.; theoretically permafrost can form above 32° F. (Theis, unpublished manuscript), and apparently it is doing so locally in parts of southwest Alaska where poor drainage, abundant vegetation, cloudy summers, and low insolation are found (S. Abrahamson, oral communication, and Ernest H. Muller, written communication).

The relative effects of past climates have been inferred qualitatively through a study of present temperature profiles and indirectly through a study of past deposits, pollen analysis, vegetal changes, structural soils, and blockfields.

The origin of large, clear ice masses in the permafrost is a special problem in itself. Numerous theories are extant, and one or more may apply to a particular mass of ice (Taber, 1943a; Leffingwell, 1919; and others).

* * * * * * *

ENGINEERING SIGNIFICANCES

In Alaska during World War II the difficulties encountered by our armed forces in obtaining permanent water supplies and in constructing runways, roads, and buildings in permafrost areas focused attention on permafrost as nothing else could (Wilson, 1948; Jaillite, 1947; Barnes, 1946; Taber, 1943b). Only then did most people realize that in Russia similar difficulties with railroads, roads, bridges, houses, and factories had impeded colonization and development of the north for decades. Now with the recent progress in aviation, and because of the strategic importance of the north, active construction and settlement for military and civilian personnel must increase, and the problems of permafrost must be solved.

Fortunately we can draw on the vast experience of the Soviet Union. Their engineers have shown that it is—

> . . . a losing battle to fight the forces of frozen ground simply by using stronger materials or by resorting to more rigid designs. On the other hand, the same experience has demonstrated that satisfactory results can be achieved and are allowed for in the design in such a manner that they appreciably minimize or completely neutralize and eliminate the destructive effect of frost action . . . Once the frozen ground problems are understood and correctly evaluated, their successful solution is for the most part a matter of common sense whereby the frost forces are utilized to play the hand of the engineer

and not against it. . . . it is worth noting that in Soviet Russia since about 1938 all governmental organizations, municipalities, and cooperative societies are required to make a thorough survey of the permafrost conditions according to a prescribed plan before any structure may be erected in the permafrost region. [Muller, 1945, pp. 1–2, 85–86.]

Specifically we must think of permafrost in construction of buildings, roads, bridges, runways, railroads, dams, and reservoirs, in problems of water supply, sewage disposal, telephone lines, drainage, excavation, ground storage, and in many other ways. Permafrost can be used as a construction material or as a base for construction, but steps must be taken to insure its stability. Otherwise it must be destroyed and appropriate steps taken to prevent it from returning.

* * * * * * *

PRACTICAL APPLICATION AND SOLUTION OF THE PROBLEMS

In a permafrost area, it is imperative that the engineer have a complete understanding of the extent, thickness, temperature, and character of the permafrost and its relation to its environment before construction of any buildings, towers, roads, bridges, runways, railroads, dams, reservoirs, telephone lines, utilidors, drainage ditches and pipes, facilities for sewage disposal, establishments for ground-water supply, excavations, foundation piles, or other structures. The practical importance of the temperatures of permafrost cannot be overemphasized. A knowledge of whether permafrost is actively expanding, or the cold reserve is increasing, is stabilized, or is being destroyed is essential in any engineering problem. Past experience has amply demonstrated that low cost or high cost, success or failure, is commonly based on a complete understanding of the problems to be encountered. Once the conditions are evaluated, proper precautions can be taken with some assurance of success.

Muller (1945) and Liverovsky and Morosov (1941) give comprehensive outlines of general and detailed permafrost surveys as adapted to various engineering projects. These outlines include instructions for the planning of the surveys, method of operation, and data to be collected. Rarely does the geologist or engineer on a job encounter "cut and dried" situations, and it is obvious that discretion must be exercised in modifying the outlines to meet the situation at hand.

In reconnaissance or preliminary survey to select the best site for construction in an unknown area, it is recommended that the approach be one of unraveling the natural history of the area. Basically the procedure is to identify each land form or terrain unit and determine its geologic history in detail. Topography, character and distribution of materials, permafrost, vegetation, hydrology, and climate are studied and compared with known areas. Then inferences, deductions, extrapolations, or interpretations can be made with reliability commensurate with the type, quality, and quantity of original data.

Thus the solution of the problems depends primarily on a complete understanding of the thermal regime of the permafrost and active layer. No factor can be eliminated, but all must be considered in a quantitative way. It is understandable that disagreement exists on the mean annual air temperature needed to produce permafrost. Few, if any, areas actually have identical conditions of climate, geology, and vegetation; hence, how can they be compared directly on the basis of climate alone? Without doubt the mean annual temperature required to produce permafrost depends on many factors and varies at least several degrees with variations in these factors. For practical purposes, however, units (terrain units) in the same climate or in similar climates may be separated on the basis of geology and vegetation. Thus there is a basis for extrapolating known conditions into unknown areas.

The advantages of aerial reconnaissance and study of aerial photographs for preliminary site selection are manifold. Aerial photographs in the hands of experienced geologists, soils engineers, and botanists can supply sufficient data to determine the best routes for roads and railroads, the best airfield sites, and data on water supply, construction materials, permafrost, trafficability conditions, camouflage, and other problems. Such an approach has been used with success by the Geological Survey and other organizations and individuals (Black and Barksdale, 1949; Wallace, 1948; Woods et al., 1948; Pryor, 1947).

Emphasis is placed on the great need for expansion of long-term applied and basic research projects as outlined by Jaillite (1947) and referred to by Muller (1945) for a clearer understanding and evaluation of the problems.

Recognition and prediction.—Recognition and prediction of permafrost go hand in hand in a permafrost survey. If natural exposures of permafrost are not available along cut banks of rivers, lakes, or oceans, it is necessary to dig test pits or drill holes in places to obtain undisturbed samples for laboratory tests and to determine the character of the permafrost.

Surface features can be used with considerable degree of accuracy to predict permafrost conditions if the origin of the surface forms are clearly understood. Vegetation alone is not the solution, but it can be used with other factors to provide data on surficial materials, surface water, character and distribution of the permafrost, and particularly on the depth of the active layer (Denny and Raup, unpublished manuscript; Stone, 1948; Muller, 1945; Taber, 1943a). Cave-in or thermokarst lakes (pl. 8, fig. 1), thaw sinks (Hopkins, 1949; Black and Barksdale, 1949; Wallace, 1948; Muller, 1945), and ground-ice mounds (Sharp, 1942a) are particularly good indicators of fine-grained materials containing much ground ice. Polygonal ground can be used with remarkable accuracy also if the type of polygonal ground and its origin are clearly known. Numerous types of *strukturboden*, polygonal ground, and related forms have been described and their origins discussed (Wittmann, 1950; Richmond, 1949; Cailleux, 1948; Washburn, 1947; Troll, 1944; Sharp, 1942b: Högbom, 1914). The type of ice-wedge polygon described by Leffingwell (1919) (pl. 4) can be delimited from others on the basis of surface expression. The author's work in northern Alaska (1945 to present) reveals that the polygons go through a cycle that can be described as youth, maturity, and old age—from flat surface with cracks to low-centered polygons and, finally, to high-centered polygons. Size and shape of polygons, widths and depths of troughs or cracks, presence or absence of ridges adjacent to the troughs, type of vegetation, and other factors all provide clues to the size-grade of surficial materials and the amount of ice in the ground. Frost mounds, frost blisters, icings, gullies, and many other surficial features can be used with reliability if all factors are considered and are carefully weighed by the experienced observer.

Geophysical methods of locating permafrost have given some promise (Sumgin and Petrovsky, 1947; Enenstein, 1947; Swartz and Shepard, 1946; Muller, 1945; Joestings, 1941). (See p. 282.) Various temperature-measuring and recording devices are employed. Augers and other mechanical means of getting at the permafrost are used (Muller, 1945, and others).

Construction.—Two types of construction methods are used in permafrost areas (Muller, 1945). In one, the passive method, the frozen-ground conditions are undisturbed or provided with additional insulation, so that the heat from the structure will not cause thawing of the underlying ground and weaken its stability. In the other method, the active method, the frozen ground is thawed prior to construction, and steps are taken to keep it thawed or to remove it and to use materials not subject to heaving and settling as a result of frost action. A preliminary examination, of course, is necessary to determine which procedure is more practicable or feasible.

Permafrost can be used as a construction material (if stress or load does not exceed plastic or elastic limit), removed before construction, or controlled outside the actual construction area. Muller (1945) has shown that it is best to distinguish (*a*) continuous areas of permafrost from (*b*) discontinuous areas and from (*c*) sporadic bodies. Russian engineers recommend that in (*a*) only the passive method of construction be used; in (*b*) or (*c*) either the passive or active method can be used, depending on thickness and temperature of the permafrost. Detailed information and references on the construction of buildings, roads, bridges, runways, reservoirs, airfields, and other engineering projects (pls. 9, 10, 11, and 12) are presented by Huttl (1948); Hardy and D'Appolonia (1946); Corps of Engineers (1946, 1945);

Zhukov (1946); Muller (1945); Richardson (1944); and others. Refinements of the techniques and data on Alaskan research projects (Wilson, 1948; Jaillite, 1947; Barnes, 1946) are contained largely in unpublished reports of various federal agencies.

Eager and Pryor (1945) have shown that road icings (pl. 10, fig. 3) are more common in areas of permafrost than elsewhere. They, Tchekotillo (1946), and Taber (1943b) discuss the phenomena of icings, classify them, and describe various methods used to prevent or alleviate icing.

One of the major factors to consider in permafrost is its water content. Methods of predicting by moisture diagrams (epures) the amount of settling of buildings on thawing permafrost are presented by Fedosov (1942). Anderson (1942) describes soil moisture conditions and methods of measuring the temperature at which soil moisture freezes.

Emphasis should be placed again on the fact that permafrost is a temperature phenomenon that occurs naturally in the earth. If man disturbs the thermal regime knowingly or unknowingly, he must suffer the consequences. Every effort should be made to control the thermal regime, to promote pergelation or depergelation as desired. Generally the former is difficult near the southern margin of permafrost. If the existing climate is not cold enough to insure that the permafrost remain frozen, serious consideration should be given to artificial freezing in those places where permafrost must be utilized as a construction material. Techniques that were used at Grand Coulee Dam (Legget, 1939) or on Hess Creek (Huttl, 1948) can be modified to fit the situation. It should be borne in mind that the refrigerating equipment need be run only for a matter of hours during the summer after the ground has been refrozen and vegetation or other means of natural insulation have been employed. Bad slides on roads and railroads, settling under expensive buildings, loosening of the foundations of dams, bridges, towers, and the like probably can be treated by refreezing artificially at less cost than by any other method. In fact the day is probably not far off when airfields of Pycrete (Perutz, 1948) or similar material will be built in the Arctic where no construction materials are available.

Where seasonal frost (active layer) is involved in construction, the engineer is referred to the annotated bibliography of the Highway Research Board (1948) and to such reports as that of the Corps of Engineers (1945, 1946, 1947).

Water supply.—Throughout permafrost areas one of the major problems is a satisfactory source of large amounts of water. Problems encountered in keeping the water liquid during storage and distribution or in its purification are beyond the scope of this report.

Small amounts of water can be obtained generally from melted ice or snow. However, a large, satisfactory, annual water supply in areas of continuous permafrost is to be found only in deep lakes or large rivers that do not freeze to the bottom. Even then the water tends to have considerable mineral hardness and organic content. It is generally not economical to drill through 1,000 to 2,000 feet of permafrost to tap ground-water reservoirs beneath, although artesian supplies have been obtained under 700 feet of permafrost (Dementiev and Tumel, 1946) and under 1,500 feet of permafrost (Obruchev, 1946).

In areas of discontinuous permafrost, large annual ground-water supplies are more common either in perched zones on top of permafrost or in nonfrozen zones within or below the permafrost (Cederstrom, 1948; Péwé, 1948b).

Annual water supply in areas of sporadic permafrost normally is a problem only to individual householders and presents only a little more difficulty than finding water in comparable areas in temperate zones.

Surface water as an alternate to ground water can be retained by earthen dams in areas of permafrost (Huttl, 1948).

Throughout the Arctic, however, the quality of water is commonly poorer than in temperate regions. Hardness, principally in the form of calcium and magnesium carbonate and iron or manganese, is common. Organic impurities and sulfur are abundant. In many places ground water and surface water have been polluted by man or organisms.

Muller (1945) presents a detailed discussion of sources of water and the engineering problems in permafrost areas of distributing the water. Joestings (1941) describes a partially successful method of locating water-bearing formations in permafrost with resistivity methods.

Sewage disposal.—Sewage disposal for large camps in areas of continuous permafrost is a most difficult problem. Wastes should be dumped into the sea, as no safe place exists on the land for their disposal in a raw state. As chemical reaction is retarded by cold temperatures, natural decomposition and purification through aeration do not take place readily. Large streams that have some water in them the year around are few and should not be contaminated. Promiscuous dumping of sewage will lead within a few years to serious pollution of the few deep lakes and other areas of annual surface-water supply. Burning is costly. As yet no really satisfactory solution is known to the writer. In discontinuous and sporadic permafrost zones, streams are larger and can handle sewage more easily, yet even there sewage disposal still remains in places one of the most important problems.

Agriculture.—Permafrost as a cold reserve has a deleterious effect on the growth of plants. However, as an impervious horizon it tends to keep precipitation in the upper soil horizons, and in thawing provides water from melting ground ice. Both deleterious and beneficial effects are negligible after 1 or 2 years of cultivation, as the permafrost table thaws, in that length of time, beyond the reach of roots of most annual plants (Gasser, 1948).

Farming in permafrost areas that have much ground ice, however, can lead to a considerable loss in time and money. Sub-Arctic farming can be done only where a sufficient growing season is available for plants to mature in the short summers. Such areas are in the discontinuous or sporadic zones of permafrost. If the land is cleared of its natural insulating cover of vegetation, the permafrost thaws. Over a period of 2 to 3 years, large cave-in lakes have developed in Siberia (I. V. Poiré, oral communication), and pits and mounds have formed in Alaska (pl. 10, fig. 4) (Péwé, 1948a, 1949; Rockie, 1942). The best solution is to select farm lands in those areas free of permafrost or free of large ground-ice masses (Tziplenkin, 1944).

Mining.—In Alaska, placer miners particularly, and lode miners to a lesser extent, have utilized permafrost or destroyed it as necessary since it was first encountered. Particularly in placer mining, frozen ground has been the factor that has made many operations uneconomical (Wimmler, 1927).

In the early part of the century, when gold was being mined so profitably at Dawson, Fairbanks, Nome, and other places in northern North America, it was common for miners to sink shafts more than 100 feet through frozen muck to the gold-bearing gravels (P. S. Smith, unpublished manuscript). These shafts were sunk by steam jetting or by thawing with fires or hot rocks. If the muck around the shafts or over the gravels thawed, the mines had to be abandoned.

Now, with the advent of dredges, such ground is thawed, generally with cold water, one or more years in advance of operations. In the technique used holes are drilled in or through the permafrost at regular intervals of possibly 10 to 30 feet, depending on the depth and types of material, and cold water is forced through the permafrost into underlying permeable foundations or out to the surface through other holes. Hot water and steam, formerly used, are uneconomical and inefficient. Where thick deposits of overburden cover placers, they are removed commonly by hydraulicking. Summer thaw facilitates the process (Patty, 1945).

Permafrost is commonly welcomed by the miners in lode mining, as it means dry working conditions. Its effect on mining operations other than maintaining cold temperatures in the mine is negligible unless it contains aquifers. Because of cold temperatures, sealing such aquifers with cement is difficult, and other techniques must be used as the situation demands.

Some well drilling in permafrost requires modifications of existing techniques and more careful planning for possible exigencies (Fagin, 1947). Difficulty may be encountered in getting proper foundations for the rig. In rotary drilling, difficulty may be experienced in keeping drilling muds at the proper temperature, in finding adequate water supplies, or in finding proper local material for drilling muds. In shallow holes particularly, the tools will "freeze in" after a few hours of idleness. In many places refreezing of permafrost around cased holes produces pressures great enough to collapse most casing. Cementing of casings is costly and very difficult, as ordinary concrete will not set in subfreezing temperatures. Deep wells below the permafrost may encounter high temperatures (100° to 150° F.), and the hot drilling muds on returning to the surface thaw the permafrost around the casing and create a settling hazard in the foundation of the rig and also a disposal problem. In some foundations refrigerating equipment must be used to prevent settling.

Permafrost also may act as a trap for oil or even have oil reservoirs within it. The cold temperature adversely affects asphalt-base types particularly and cuts down yields. Production difficulties and costs go up (Fagin, 1947).

Refrigeration and storage.—Natural cold-storage excavations are used widely in areas of permafrost. They are most satisfactory in continuous or discontinuous zones. Permafrost should not be above 30° F.; if it is, extreme care in ventilation and insulation must be used. Properly constructed and ventilated storerooms will keep meat and other products frozen for years. Detailed plans and characteristics required for different cold-storage rooms are described by Chekotillo (1946).

Trafficability.—In the Arctic and sub-Arctic most travel overland is done in winter, as muskegs, swamps, and hummocky tundra make summer travel exceedingly difficult (Navy Department, 1948–49; Fagin, 1947). Tracked vehicles or sleds are the only practical types. Wheeled vehicles are unsatisfactory, as most of the area is without roads.

Permafrost aids travel when it is within a few inches of the surface. It permits travel of D8 caterpillar tractors and heavier equipment directly on the permafrost. Sleds weighing many tons can be pulled over the permafrost with ease after the vegetal mat has been removed by an angle-bulldozer. Polygonal ground, frost blisters, pingos, and small, deeply incised thaw streams (commonly called "beaded" streams), rivers, and lakes create natural hazards to travel.

In areas of discontinuous and sporadic permafrost, seasonal thaw is commonly 6 to 10 feet deep, and overland travel in summer can be accomplished in many places only with amphibious vehicles such as the weasel or LVT. Foot travel and horse travel are very slow and laborious in many places because of swampy land surfaces and necessity for making numerous detours around sloughs, rivers, and lakes.

Military operations.—Permafrost alters military operations through its effects on construction of airbases, roads, railroads, revetments, buildings, and other engineering projects; through its effects on trafficability, water supply, sewage disposal, excavations, underground storage, camouflage, explosives, planting of mines, and other more indirect ways (Edwards, 1949; Navy Department, 1948–49). Military operations commonly require extreme speed in construction, procuring of water supply, or movement of men and material. Unfortunately it is not always humanly possible to exercise such speed (Fagin, 1947). Large excavations require natural thawing, aided possibly by sprinkling (Huttl, 1948), to proceed ahead of the earth movers. Conversely, seasonal thaw may be so deep as to prevent the movement of heavy equipment over swampy ground until freeze-up. Or, similarly, it may be necessary in a heavy building to steam-jet piles into permafrost and allow them to freeze in place before loading them. These tasks take time, and proper planning is a prerequisite for efficient operation.

Camouflage is a problem on the tundra. Little relief or change in vegetation is available. Tracks of heavy vehicles or paths stand out in marked contrast for years. It is easy to see in aerial photographs footpaths and dog-sled trails abandoned 10 years or more ago.

Mortar and shell fire, land mines, shaped charges, and other explosives undoubtedly respond to changes in the character of permafrost, but no data are available to the author.

* * * * * * *

REFERENCES

ANDERSON, ALFRED B. C., ET AL.
1942. Soil-moisture conditions and phenomena in frozen soils. Trans. Amer. Geophys. Union, 1942, pt. 2, pp. 356–371.

BARNES, LYNN C.
1946. Permafrost, a challenge to engineers. Military Eng., vol. 38, No. 243, pp. 9–11.

BIRCH, FRANCIS.
1948. The effects of Pleistocene climatic variations upon geothermal gradients. Amer. Journ. Sci., vol. 246, No. 12, pp. 729–760.

BLACK, R. F., and BARKSDALE, W. L.
1949. Oriented lakes of northern Alaska. Journ. Geol., vol. 57, No. 2, pp. 105–118.

BRYAN, KIRK.
1946a. Cryopedology, the study of frozen ground and intensive frost-action with suggestions on nomenclature. Amer. Journ. Sci., vol. 244, No. 9, pp. 622–642.
1946b. Permanently frozen ground. Military Eng., vol. 38, No. 246, p. 168.
1948. The study of permanently frozen ground and intensive frost-action. Military Eng., vol. 40, No. 273, pp. 304–308; discussion, pp. 305–308.
1949. The geologic implications of cryopedology. Journ. Geol., vol. 57, No. 2, pp. 101–104.

BRYAN, KIRK, and ALBRITTON, C. C., JR.
1943. Soil phenomena as evidence of climatic changes. Amer. Journ. Sci., vol. 241, No. 8, pp. 469–490.

CAILLEUX, ANDRÉ.
*1948. Études de cryopédologie. Expéd. Arct., Sec. Sci. Natur., pp. 1–67.

CEDERSTROM, D. J.
[1948]. Ground-water data for Fairbanks, Alaska. Manuscript on open file, U. S. Geological Survey, Washington, D. C.

CHANEY, R. W., and MASON, H. L.
1936. A Pleistocene flora from Fairbanks, Alaska. Amer. Mus. Nat. Hist. Nov. No. 887, 17 pp.

CHEKOTILLO, A. M.
1946. The underground storage places in the permanently frozen ground. Priroda, vol. 11, pp. 27–32. (Translation by E. A. Golomshtok in Stefansson Library.)

CONRAD, V.
1946. Polygon nets and their physical development. Amer. Journ. Sci., vol. 244, No. 4, pp. 277–296.

CORPS OF ENGINEERS.
*1945. Construction of runways, roads, and buildings on permanently frozen ground. U. S. War Dept. Techn. Bull. TB 5–255–3, 64 pp.
1946. Airfield pavement design—construction of airfields on permanently frozen ground. Eng. Man. for War Dept. Construction, pt. 12, ch. 7, 20 pp. U. S. War Dept.
*1947a. Frost investigation, 1945–1946—Comprehensive report and nine appendices. New England Division, CE, U. S. War Dept., mimeographed.
*1947b. Report on frost investigation, 1944–1945, and 15 appendices. New England Division, CE, U. S. War Dept., mimeographed.

CRESSEY, G. B.
1939. Frozen ground in Siberia. Journ. Geol., vol. 47, pp. 472–488.

DEMENTIEV, A. I., and TUMEL, V. F.
1946. Civil engineering in frozen soil, U. S. S. R. Canadian Geogr. Journ., vol. 32, No. 1, pp. 32–33.

DENNY, C. S., and RAUP, H. M.
(Unpublished manuscript.) Notes on the interpretation of aerial photographs along the Alaska Military Highway.

*Articles that are comprehensive in scope or that contain extensive bibliographies are marked by asterisk.

922758—51——20

DENNY, C. S., and STICHT, J.
(Unpublished manuscript.) Geology of the Alaska Highway.

EAGER, WILLIAM L., and PRYOR, WILLIAM T.
*1945. Ice formation on the Alaska Highway. Public Roads, vol. 24, No. 3, pp. 55–74.

EDWARDS, N. B.
*1949. Combat in the Arctic. Infantry Journ., January, pp. 4–8.

ENENSTEIN, B. S.
1947. The results of electrometric investigations carried out by means of direct current on permanently frozen soils. (In Russian.) Inst. Merzlotovedeniia Trudy, vol. 5, pp. 38–86. Geophysical Abstract No. 10080, U. S. Geol. Surv. Bull. 959–B, pp. 126–127, 1948.

FAGIN, K. MARSHALL.
*1947. Petroleum development in Alaska. (In four parts.) Petrol. Eng., August, pp. 43–59; September, pp. 150–170; October, pp. 180–193; December, pp. 57–69.

FEDOSOV, A. E.
1942. Forecasting of the settling of buildings after the thawing of permanently frozen ground, by the method of moisture diagrams. Prognoz osadok soorusheniy na ottaiviushchey merzlote (Metod v lazhnostnyh epiur). Issledovanie vechnoy mersloty v Iakutskoy respeblike Akad. Nauk U. S. S. R., Inst. Merzlotovedeniia, No. 1, pp. 52–85. Moscow-Leningrad. (Condensed by I. V. Poiré.)
1944. Predictions of settling of structures caused by the thawing of the permafrost. (In Russian.) Inst. Merzlotovedeniia Trudy, vol. 4, pp. 93–124, 1944. Geophysical Abstract No. 10406, U. S. Geol. Surv. Bull. 959-C, p. 238, 1948.

FLINT, R. F.
1947. Glacial geology and the Pleistocene epoch. 589 pp. New York.

GASSER, G. W.
1948. Agriculture in Alaska. Arctic, vol. 1, No. 2, pp. 75–83.

GATTY, O., FLEMING, W. L. S., and EDMONDS, J. M.
1942. Some types of polygonal surface markings in Spitzbergen. Amer. Journ. Sci., vol. 240, pp. 81–92.

GIDDINGS, J. L.
1938. Buried wood from Fairbanks, Alaska. Tree-Ring Bull., 1938, pp. 3–6.

GRIGGS, ROBERT F.
1936. The vegetation of the Katmai District. Ecology, vol. 17, No. 3, pp. 380–417.

HARDY, R. M., and D'APPOLONIA, E.
1946. Permanently frozen ground and foundation design, Parts 1 and 2. Eng. Journ. Canada, vol. 29, No. 1, pp. 1–11. (Reviewed by W. H. Ward, in Journ. Glaciol., vol. 1, No. 2, pp. 80–81, July 1947.)

HIBBEN, F. C.
1941. Archeological aspects of Alaskan muck deposits. New Mexico Anthropologist, vol. 5, No. 4, pp. 151–157.

HIGHWAY RESEARCH BOARD.
*1948. Bibliography on frost action in soils, annotated. Nat. Res. Counc., Highway Res. Board, Bibliography 3, 57 pp.

HÖGBOM, B.
*1914. Über die geologische Bedeutung des Frostes. Geol. Inst. Uppsala Bull., vol. 12, pp. 257–389.

HOPKINS, DAVID M.
1949. Thaw lakes and thaw sinks in the Imuruk Lake area, Seward Peninsula, Alaska. Journ. Geol., vol. 57, No. 2, pp. 119–131.

HORBERG, LELAND.
1949. A possible fossil ice wedge in Bureau County, Illinois. Journ. Geol., vol. 57, No. 2, pp. 132–136.

HUTTL, JOHN B.
1948. Building an earth-fill dam in Arctic placer territory. Eng. and Min. Journ., vol. 149, No. 7, pp. 90–92.

INGERSOLL, L. R., ZOBELL, O. J., and INGERSOLL, A. C.
1948. Heat conduction, with engineering and geological applications. 278 pp. New York.

JAILLITE, W. MARKS.
*1947. Permafrost research area. Military Eng., vol. 39, No. 263, pp. 375–379.

JENNESS, J. L.
1949. Permafrost in Canada. Arctic, vol. 2, No. 1, pp. 13–27. (See also later correspondence by J. D. Bateman in Arctic, vol. 2, No. 3, pp. 203–204.)

JOESTINGS, H. R.
1941. Magnetometer and direct-current resistivity studies in Alaska. Amer. Inst. Mech. Eng. Techn. Publ. 1284, 20 pp.

JUDSON, SHELDON.
1949. Rock-fragment slopes caused by past frost action in the Jura Mountains (Ain), France. Journ. Geol., vol. 57, No. 2, pp. 137–142.

KALIAEV, A. V.
1947. Anabiosis under conditions of frozen ground. Mikrokiologiya, vol. 16, No. 2.

LANE, ALFRED C.
1946. Northern climatic variations affecting geothermal initial. Canadian Inst. Min. and Metal. Bull. 411, pp. 397–402.

LEFFINGWELL, E. DE K.
*1919. The Canning River region, northern Alaska. U. S. Geol. Surv. Prof. Pap. 109, 251 pp.

LEGGET, R. F.
1939. Geology and engineering. 650 pp. New York.

LIVEROVSKY, A. V., and MOROZOV, K. D.
1941. Construction under permafrost conditions. 244 pp., 151 figs. Leningrad-Moscow. (Abstracted by E. A. Golomshtok, Stefansson Library.)

MOSLEY, A.
1937. Frozen ground in the sub-Arctic region and its biological significance. Scottish Geogr. Mag., vol. 53, No. 4, pp. 266–270.

MULLER, SIEMON W.
*1945. Permafrost or permanently frozen ground and related engineering problems. U. S. Geol. Surv. Spec. Rep., Strategic Engineering Study 62, 2d ed., Military Intelligence Div., Office Chief of Engineers, U. S. Army. (Also lithoprinted, Ann Arbor, Mich., 1947, 231 pp.)

MULLIS, IRA B.
1930. Illustrations of frost and ice phenomena. Public Roads, vol. 11, No. 4, pp. 61–68.

NAVY DEPARTMENT, BUREAU OF YARDS AND DOCKS.
*1948–49. Cold-weather engineering, chs. 1–5, 109 pp.

NIKIFOROFF, C.
1932. The perpetually frozen subsoil of Siberia. Soil Sci., vol. 26, pp. 61–78.

NORDALE, A. M.
1947. Valuation of dredging ground in the sub-Arctic. Trans. Canadian Inst. Min. and Metall., vol. 50, pp. 487–496; Canadian Min. and Metall. Bull. 425, September.

OBRUCHEV, V.
1945. Eternal frost: the frozen soil of northern Russia and Siberia. Nat. Rev., vol. 124, pp. 220–227.
1946. The fifteenth anniversary of the ground frost study of the Academy of the U. S. S. R. Priroda, No. 5. (Translated by Stefansson Library.)

ORSTRAND, C. E. VAN.
1939. Observed temperatures in the earth's crust. Ch. 6, Physics of the earth; ch. 7, Internal constitution of the earth.

PATTY, E. N.
1945. Placer mining in the sub-Arctic. Western Miner, vol. 18, No. 4, pp. 44–49.

PERUTZ, M. F.
1948. A description of the iceberg aircraft carrier and the bearing of the mechanical properties of frozen wood pulp upon some problems of glacier flow. Journ. Glaciol., vol. 1, No. 3, pp. 95–102.

PÉWÉ, TROY L.
1948a. Origin of the Mima Mounds. Sci. Month., vol. 66, No. 4, pp. 293–296.
[1948b]. Ground-water data for Fairbanks, Alaska. Manuscript on open file, U. S. Geological Survey, Washington, D. C.
1949. Preliminary report of permafrost investigations in the Dunbar area, Alaska. U. S. Geol. Surv. Circ. 42, 3 pp.

POSER, HANS.
1947a. Dauerfrostboden und Temperatur verhaltnisse während der Würm-Eiszeit im nicht vereisten Mittel- und Westeuropa. Naturwiss., Jahrg. 34, pp. 1–9.
1947b. Aufbautiefe und Frostzerrung im boden Mitteleuropas während der Würm-Eiszeit. Naturwiss., Jahrg. 34, pp. 232–238, 262–267.
*1948. Boden- und Klimaterhältnisse im Mittel- und Westeuropa während der Würm-Eiszeit. Erdkunde, vol. 2, Nos. 1–3, pp. 53–68. (Reviewed by Ernst Antevs, Journ. Geol., vol. 57, No. 6, pp. 621–622, 1949.)

PRYOR, W. T.
*1947. Aerial surveying on the Alaska Highway, 1942. Public Roads, vol. 24, No. 11, pp. 275–290.

RAUP, H. M.
1941. Botanical problems in Boreal America. Bot. Rev., vol. 7, pp. 147–248.
1947. The botany of southwestern Mackenzie. Sargentia 6.

RICHARDSON, H. W.
1944. Construction in the tundra. Eng. News, vol. 132, pp. 956–961.

RICHMOND, GERALD M.
1949. Stone nets, stone stripes, and soil stripes in the Wind River Mountains, Wyoming. Journ. Geol., vol. 57, No. 2, pp. 143–153.

ROCKIE, W. A.
1942. Pitting on Alaska farm lands: a new erosion problem. Geogr. Rev., vol. 32, pp. 128–134.

SCHAFER, J. P.

1949. Some periglacial features in central Montana. Journ. Geol., vol. 57, No. 2, pp. 154–174.

SHANNON, W. L., and WELLS, W. A.

1947. Tests for thermal diffusivity of granular materials. Presented at the 50th annual meeting of the American Society for Testing Materials, June 16–20.

SHARP, R. P.

1942a. Ground-ice mounds in tundra. Geogr. Rev., vol. 32, No. 3, pp. 417–423.

*1942b. Soil structures in the St. Elias Range, Yukon Territory. Journ. Geomorph., vol. 5, pp. 274–301.

SHELESNYAK, M. C.

1948. History of the Arctic Research Laboratory, Point Barrow, Alaska. Arctic, vol. 1, No. 2, pp. 97–106.

SMITH, H. T. U.

1949a. Periglacial features in the driftless area of southern Wisconsin. Journ. Geol., vol. 57, No. 2, pp. 196–215.

*1949b. Physical effects of Pleistocene climatic changes in nonglaciated areas—eolian phenomena, frost action, and stream terracing. Bull. Geol. Soc. Amer., vol. 60, pp. 1485–1516.

SMITH, P. S.

(Unpublished manuscript.) Permanent ground frost in Alaska. On file with U. S. Geological Survey, Washington, D. C.

SMITH, W. O.

1939. Thermal conductivities in moist soils. Proc. Soil Sci. Soc. Amer., vol. 4, pp. 32–40.

1942. The thermal conductivity of dry soil. Soil Sci., vol. 53, No. 6, pp. 435–459.

STECHE, H.

*1933. Beitrage zur Frage der Strukturboden. Berichte Vorhandl. Sächsischen Akad. Wiss. Leipzig, math.-phys. Kl., vol. 8, pp. 193–272.

STONE, KIRK.

1948. Aerial photographic interpretation of natural vegetation in the Anchorage area, Alaska. Geogr. Rev., vol. 38, No. 3, pp. 465, 474.

SUMGIN, M. I.

*1947. Eternal ground frost in the U.S.S.R. 2d ed., rev. (not seen). Vladivostok.

SUMGIN, M. I., and PETROVSKY, A. A.

1947. The importance of electrical methods for the study of permanently frozen ground. (In Russian.) Inst. Merzlotovedeniia Trudy, vol. 5, pp. 15–17. Geophysical Abstract 10089, U. S. Geol. Surv. Bull. 959–B, p. 130, 1948.

SWARTZ, J. H., and SHEPARD, E. R.

1946. Report on a preliminary investigation of the possible application of geophysical methods to the studies of permafrost problems in Alaska. Ozalid Rep., U. S. Bur. Mines.

TABER, STEPHEN.

*1930a. The mechanics of frost heaving. Journ. Geol., vol. 38, pp. 303–317.

*1930b. Freezing and thawing of soils as factors in the destruction of road pavements. Public Roads, vol. 11, pp. 113–132.

*1943a. Perennially frozen ground in Alaska—its origin and history. Bull. Geol. Soc. Amer., vol. 54, pp. 1433–1548.

1943b. Some problems of road construction and maintenance in Alaska. Public Roads, vol. 23, No. 9, pp. 247–251, July–September.

TCHEKOTILLO, A.

1946. Solving the problem of "Nalyeds" in permafrost origins. Eng. News-Rec., No. 28, pp. 62–65.

THEIS, CHARLES V.

(Unpublished manuscript.) Thermal processes related to the formation of permafrost.

TOLMACHOFF, I. P.

1929. The carcasses of the mammoth and rhinoceros found in the frozen ground of Siberia. Trans. Amer. Philos. Soc., vol. 23, pt. 1, pp. 12–14.

TREMAYNE, MARIE.

1948. Bibliography of Arctic research. Arctic, vol. 1, No. 2, pp. 84–86.

TROLL, CARL.

*1944. Strukturboden, Solifluktion und Frostklimate der Erde. Geol. Rundsch., vol. 34, pp. 545–694.

1947. Die Formen der Solifluktion und die periglaziale Bodenabtragung. Erdkunde, vol. 1, pp. 162–175.

1948. Der subnivale oder periglaziale Zyklus der Denudation. Erdkunde, vol. 2, pp. 1–21.

TUCK, RALPH.

1940. Origin of the muck-silt deposits at Fairbanks, Alaska. Bull. Geol. Soc. Amer., vol. 51, No. 9, pp. 1295–1310.

TZIPLENKIN, E. I.

1944. Permafrost and its influence on agriculture. Trudy, Obruchev Inst. Permafrostology, vol. 4, pp. 230–255. Moscow-Leningrad.

WAHRHAFTIG, CLYDE.

1949. The frost-moved rubbles of Jumbo Dome and their significance in the Pleistocene chronology of Alaska. Journ. Geol., vol. 57, No. 2, pp. 216–231.

WALLACE, R. E.

1948. Cave-in lakes in the Nebesna, Chisana, and Tanana River Valleys, eastern Alaska. Journ. Geol., vol. 56, No. 3, pp. 171–181.

WASHBURN, A. L.

1947. Reconnaissance geology of portions of Victoria Island and adjacent regions, Arctic Canada. Geol. Soc. Amer. Mem. 22, 142 pp.

WEINBERG, B. P.

1940. Studies on eternally frozen ground and on freezing of soil. Trans. Amer. Geophys. Union, vol. 21, pp. 770–777.

WEINBERGER, L.

1944. Frostspalten und Froststrukturen in Schottern bei Leipzig. Geol. Rundsch., vol. 34, pp. 539–544.

WERENSKIOLD, W.

1923. Frozen soil in Spitzbergen. Abstract, Month. Weather Rev., vol. 51, p. 210.

WILKERSON, A. S.

1932. Some frozen deposits in the gold fields of interior Alaska. Amer. Mus. Nat. Hist. Nov. No. 525, 22 pp.

WILSON, WALTER K., JR.

1948. The problem of permafrost. Military Eng., vol. 40, No. 270, pp. 162–164.

WIMMLER, N. L.

1927. Placer-mining methods and costs in Alaska. U. S. Bur. Mines Bull. 259, pp. 37–40.

WITTMANN, OTTO.

*1950. Diluvialprofile mit periglazialen Erscheinungen von Chateau de Jeurre zwischen Etampes und Etrechy (Seine und Oise). Neues Jahrb. Geol. und Paläontol., Monatshefte, No. 3, pp. 65–79.

WOODS, K. B., ET AL.

1948. Use of aerial photographs in the correlation between permafrost and soils. Military Eng., vol. 40, pp. 497–499.

YOUNG, JACOB W.

1918. Ground frost in Alaska. Eng. and Min. Journ., vol. 105, No. 7, pp. 338–339.

ZEUNER, F. E.

*1945. The Pleistocene period—its climate, chronology, and faunal successions. 322 pp. Ray Society, London.

*1946. Dating the past. 444 pp. London.

ZHUKOV, V. F.

*1946. The earthworks during the laying of foundations in the permafrost region. Obruchev Inst. Permafrostology, pp. 3–130. Moscow-Leningrad. (Translated by Stefansson Library.)

Reprinted from *Amer. J. Sci.*, **258**, 161–176 (1960)

9

American Journal of Science

MARCH 1960

ENTRENCHMENT OF THE WILLOW DRAINAGE DITCH, HARRISON COUNTY, IOWA*

RAYMOND B. DANIELS
Soil Conservation Service, U. S. Department of Agriculture, Ames, Iowa

ABSTRACT. The changes in the Willow Drainage Ditch since 1919-1920 have been reconstructed by comparison of the original profile of the ditch, historical records, and a survey of the ditch in 1958. Since construction, the drainage ditch has filled where the ditch was constructed on the Missouri River Válley. The drainage ditch in the Willow River Valley has entrenched, however, and becomes progressively deeper upstream, attaining a maximum depth of 42 feet at the Monona-Harrison County line.

The drainage ditch in the Willow River Valley apparently deepened by channel scour and headward movement of knickpoints, but in a number of entrenchments rather than a single entrenchment followed by stabilization. Once a knickpoint has passed a point in the drainage ditch, stabilization of the channel does not necessarily follow but channel scour may deepen the ditch more than passage of the knickpoint.

The filling of the lower part of the drainage ditch probably is caused by the sharp decrease in gradient of the ditch, and by coinciding periods of high water in the Boyer Drainage Ditch or the Missouri River, to which the Willow is affluent. The entrenchment in the Willow River Valley probably was produced by a constructed increase in gradient of the drainage ditch as contrasted to that of the original river gradient in the same area.

The entrenchment of the Willow Drainage Ditch has been responsible for much of, but cannot explain all of, the entrenchment of its tributary streams.

INTRODUCTION

The rivers in Harrison County, Iowa, commonly flooded their valleys during the early part of this century. To alleviate the flooding, the river systems were modified by construction of drainage ditches throughout the lower reaches of the rivers. Since construction, many of the drainage ditches have filled in the lower reaches, and at the same time have deepened and widened considerably in the upstream reaches.

The Willow Drainage Ditch, for example, has filled where the ditch was built in the Missouri River flood plain, but in the Willow River Valley it has deepened. Although the changes in the Willow Drainage Ditch are more severe than many of the other drainage ditches in Harrison County, these changes afford an excellent example of the method of stream entrenchment in the area. Also, the entrenchment of the Willow Drainage Ditch is important in explaining the entrenchment of its tributary streams.

The Willow River heads in southwestern Crawford County, Iowa, and flows in a southwesterly direction through southeastern Monona County. In Monona County the Willow River is diverted into the Willow Drainage Ditch two miles north of the Monona-Harrison County line. The drainage ditch ex-

* Joint contribution of Soil Survey Investigations, Soil Conservation Service, U. S. Dept. of Agriculture; and the Iowa Agricultural and Home Economics Experiment Station. Journal Paper No. J-3600 of the Iowa Agricturał Experiment Station, Ames, Iowa. Project No. 1250.

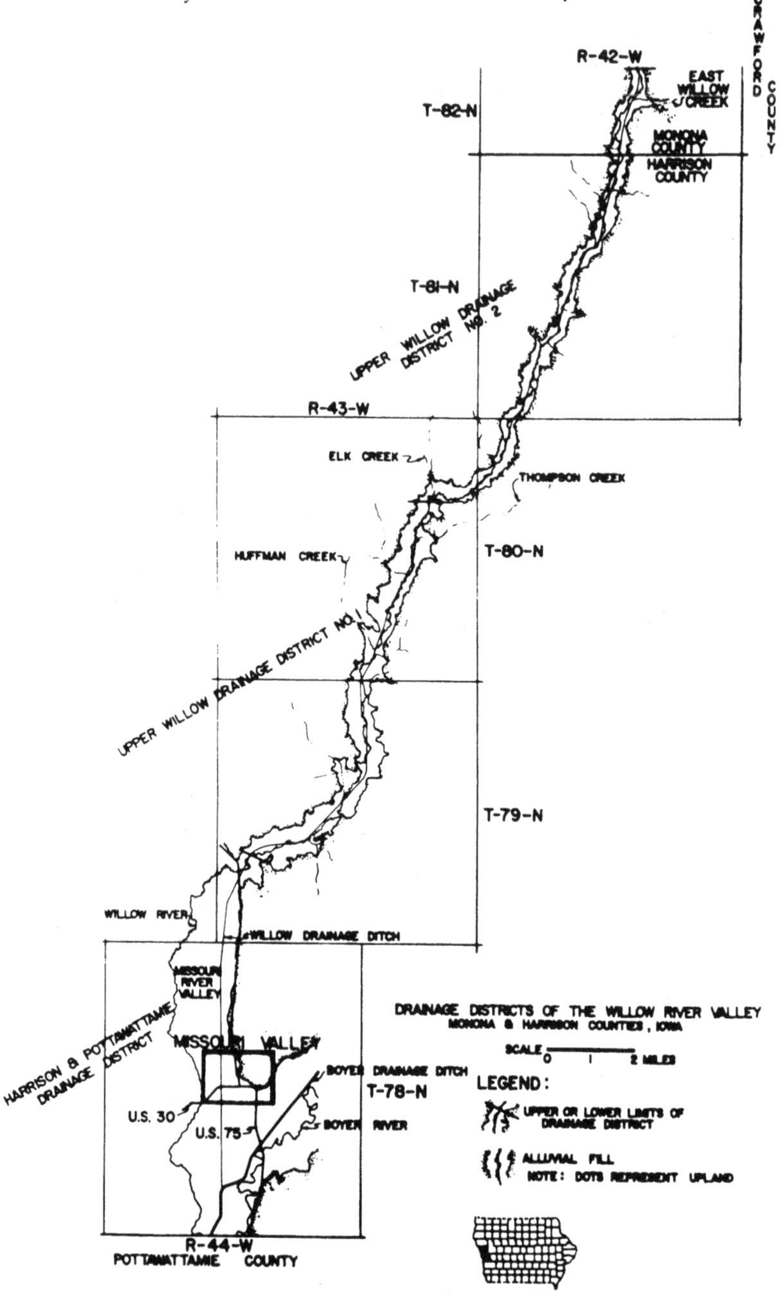

Fig. 1. Location of Willow Drainage Ditch and the former channel of the Willow River. The boundaries of the drainage districts do not necessarily coincide with the limits of the alluvial fill in the Willow River Valley.

tends from southern Monona County to southwestern Harrison County where it joins the Boyer Drainage Ditch south of Missouri Valley, Iowa (fig. 1). The Willow Drainage Ditch is under control of the Harrison and Pottawattamie Drainage District and Upper Willow Drainage Districts No. 1 and 2 (fig. 1). The Willow River, upstream from the head of Upper Willow Drainage District No. 2, has not been straightened except in minor areas near bridges.

CHARACTERISTICS OF THE WILLOW RIVER AND DRAINAGE DITCH

Willow River Prior to Straightening

Prior to straightening, the Willow River in Harrison and Monona Counties was a small stream that meandered considerably before flowing into the Missouri River Valley (fig. 1). The sinuous course of the Willow is shown when river miles are compared to valley miles. The stream length in Harrison County was approximately 26.3 miles, but the valley length is approximately 20.2 miles. By comparison, the Boyer River in Harrison County, Iowa, was approximately 60 miles in length, and the Boyer River Valley 28 miles in length (Smith, 1888, p. 21).

Only sparse information is available on the width and depth of the river prior to 1900. Sales (1853, p. 244) recorded that in 1853 the channel of the Willow River in Monona County was 6 to 7 feet wide, 1.5 feet deep, and that the stream was flowing about 6 feet below the flood plain.

In 1916-1918 the Willow River in Upper Willow Drainage Districts No. 1 and 2 flowed 10 to 12 feet below the flood plain level. The bank-to-bank width of the stream ranged from a maximum of about 100 feet to a minimum of about 60 feet.[1] No information could be found on the width and depth of the Willow River where it flowed into the Missouri River Valley.

The gradient of the Willow River in Upper Willow Drainage District No. 1 averaged 5.2 feet per mile but ranged from a minimum of 1.7 feet per mile to a maximum of 7.9 feet per mile. In Upper Willow Drainage District No. 2 the gradient of the river averaged 7.5 feet per mile, and ranged from 4.6 to 12.2 feet per mile (fig. 2). The wide range in gradients of the river in relatively short distances appears to be related to the entry of the larger tributaries into the river system. For example, the gradient of the channel upstream from Elk Creek (fig. 2, mile 15) is 6.4 feet per mile, but immediately below Elk Creek the gradient flattens to 1.7 feet per mile. Other examples of the flattening of the gradient below a tributary mouth are Thompson Creek, Huffman Creek, and tributaries A, D, and E. The gradient of the Willow River downstream from the mouths of tributaries B, C, G, H, and South Willow Creek decreases slightly or is the same as the gradient upstream. An exception is the steepening of the channel downstream from tributary F (fig. 2).

The flattening of the gradient of the Willow River immediately below the entry of a tributary into the river may have been due to some structural control, but glacial till or bedrock are not known to be exposed in the channel of the drainage ditch. Generally, the sediments exposed in the drainage ditch to

[1] The width and depth of the Willow River were measured at the points that the Drainage Ditch crossed the river as shown by the plats and profiles of the drainage ditch on file at the Drainage Clerk's Office, Court House, Harrison County, Iowa.

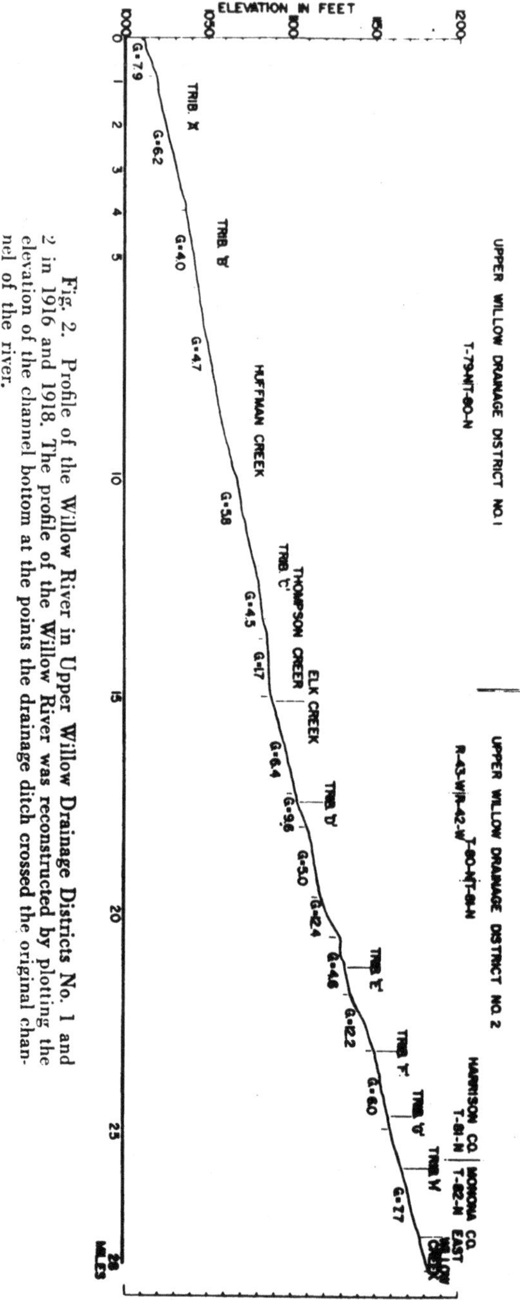

Fig. 2. Profile of the Willow River in Upper Willow Drainage Districts No. 1 and 2 in 1916 and 1918. The profile of the Willow River was reconstructed by plotting the elevation of the channel bottom at the points the drainage ditch crossed the original channel of the river.

depths considerably below the depths of the channel bottom of the original Willow River are dominantly silts with only sparse sand and gravel lenses. Thus structural control does not appear to have been responsible for the majority of the inflection points in the Willow River.

An alternative explanation for the flattening of the gradient below the tributaries is that the larger tributaries were overloading the Willow River with sediment and aggrading the channel of the Willow for a short distance downstream.

Price, in a survey of the Willow River in Drainage District No. 1 in 1916, indicated that the river was filling (Harrison County, Iowa, Drainage Record No. 7, p. 604[2]). Price suggested the filling was caused by the meandering of the river.

The Willow River frequently flooded its valley to a depth of several feet as early as 1851, and the valley was, in general, considered unfit for cultivation (Anderson, 1851a, p. 190; 1951b, p. 168). In the early part of this century when the Willow River and its tributaries overflowed, they damaged crops on the inundated area. Local residents frequently disagree on the amount of the damage caused by flooding, but historical records indicate that the flooding frequently prevented planting or killed growing crops (Harrison County, Iowa, Drainage Record No. 7, p. 604, 1104). The damage from flooding was especially serious where the Willow River flowed in the Missouri River Valley (Harrison County, Iowa, Drainage Record No. 3, p. 43).

Characteristics of Constructed Drainage Channel

To alleviate the flooding by the Willow River, a drainage ditch was constructed from Monona County, Iowa, to the Boyer Drainage Ditch (fig. 1). Construction of the Willow Drainage Ditch in the Harrison and Pottawattamie Drainage District was started on June 20, 1906, and completed late in 1908 (Harrison County, Iowa, Drainage Record No. 3, p. 171, 198).

The dimensions of the drainage ditch in the Harrison and Pottawattamie Drainage District were an 18-foot bottom, 1:1 side slopes, and a depth of 15 feet from the top of the berm to the bottom of the channel[3]. The top of the berm was approximately 3 feet above the natural ground level. The total length of the drainage ditch was 6.6 miles. Total fall from the head to the mouth of the ditch was 13.5 feet, or an average of 2.04 feet per mile.

The construction work on the drainage ditch in the Upper Willow Drainage District No. 1 started October 1, 1916, and was completed on October 16, 1919. The ditch was cut with a bottom width of 12 feet, 1:1 side slopes, and 15-foot berms. A strip of land 140 feet wide was condemned as right-of-way for the drainage ditch. Average depth of cut was approximately 15 feet, but varied somewhat according to local conditions (fig. 3). The gradient of the ditch was 7.66 feet per mile throughout the drainage district (fig. 3). Total length of the ditch was slightly over 10.2 miles (Harrison County, Iowa, Drainage Record No. 7, p. 605, 683, 711).

[2] On file at the Drainage Clerk's Office, Court House, Harrison County, Iowa.

[3] The dimensions of the drainage ditch were obtained from the profile of the ditch on file in the Drainage Clerk's Office, Court House, Harrison County, Iowa.

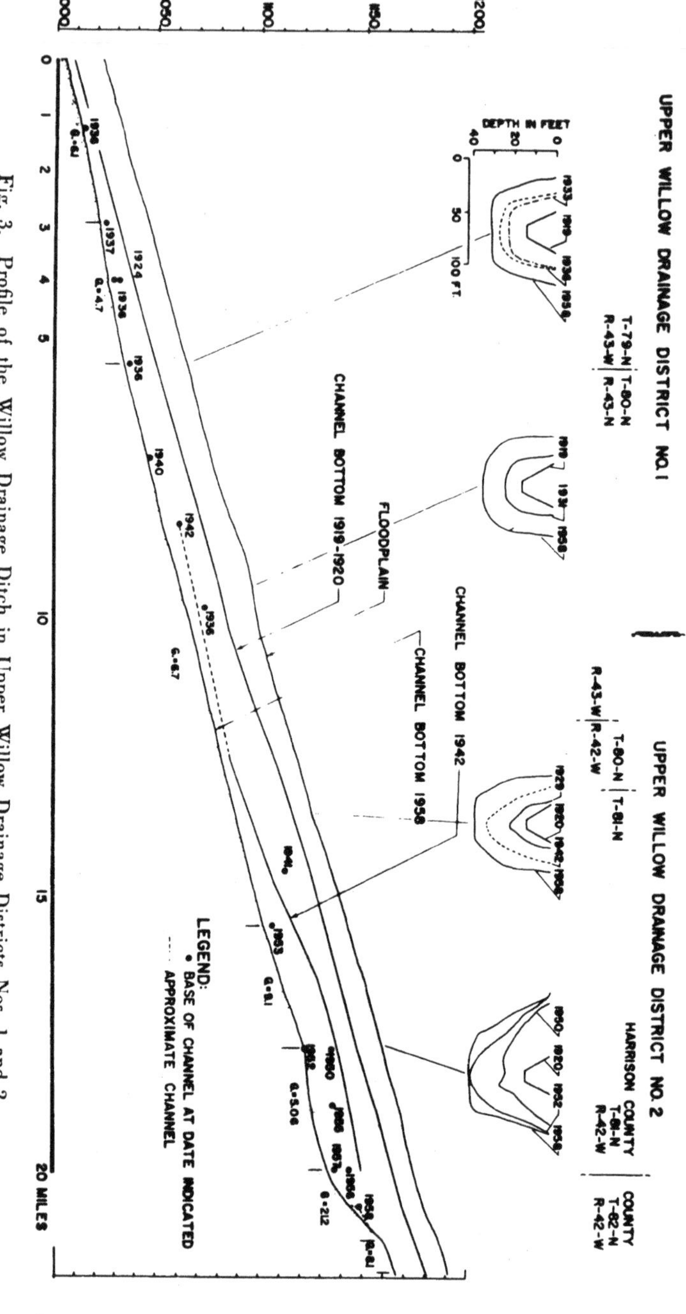

Fig. 3. Profile of the Willow Drainage Ditch in Upper Willow Drainage Districts Nos. 1 and 2.

The drainage ditch in the Upper Willow Drainage District No. 2 was constructed between April 1, 1919, and June 1, 1920. The bottom width of the ditch was 12, 10, and 8 feet, with the width becoming less in the upper part of the district. Side slopes of 1:1 and a berm width of 15 feet were required in all segments of the drainage district. Depth of cut averaged 11 feet throughout the district (Harrison County, Iowa, Drainage Record No. 7, p. 1106, 1140). The gradient of the ditch was variable, ranging from 12.14 to 8.45 feet per mile (fig. 3).

In conjunction with construction of the drainage ditch in the Upper Willow Drainage Districts, most of the larger tributary streams also were straightened. An example of the amount of work on tributaries of the Willow is shown by the costs in the Upper Willow Drainage District No. 1. The estimated cost of the total project in drainage district no. 1 was $85,000, of which $49,630 was for construction of the main channel and the balance for the tributaries (Harrison County, Iowa, Drainage Record No. 7, p. 689-690).

Changes in the Willow Drainage Ditch Since Construction

The Willow Drainage Ditch in the Harrison and Pottawattamie Drainage District (fig. 1) has been subject to filling since construction. At low water level the water flows 6 to 12 inches deep, and meanders over a bed of silt. The ditch was cleaned in 1916 and again in 1941. Prior to cleaning, the bottom of the ditch had filled with silt, and many of the drainage outlets were below the channel bottom (Harrison County, Iowa, Drainage Record No. X, p. 469, 475, 510). In 1954 Smith (Harrison County, Iowa, Drainage Record No. XI, p. 834) reported that " . . . Heavy sedimentation has filled up the bottom of the ditch greatly reducing its capacity. In places the bottom of the channel is only 1 or 2 feet lower than natural ground level." Local residents believed that the Missouri River flood waters in April, 1952, had backed up the Boyer and Willow Drainage Ditches, causing considerable silting in the channel of the Willow Drainage Ditch and impeding drainage (Harrison County, Iowa, Drainage Record No. XI, p. 856). Inspection of the drainage ditch by the author in the fall of 1958 showed that the bottom of the ditch was 4 to 10 feet below natural ground level, and that numerous drainage outlets were partially or completely buried by the silt in the channel bottom.

In direct contrast to the filling of the channel of the Willow Drainage Ditch in the Harrison and Pottawattamie Drainage District is the entrenchment of the drainage ditch in the Upper Willow Drainage Districts.

Exact records on the changes of the Willow Drainage Ditch in the Upper Willow Drainage Districts prior to 1924 are not available, but interviews with local residents indicate that the ditch started to deepen and widen almost immediately. The earliest accurate record available is a measurement at a bridge[4] in drainage district no. 1 (fig. 3, mile 4.1) in 1924. The drainage ditch had deepened 6 feet at this point in approximately 5 years. Other records prior to 1931 are not available, but a farmer (oral communication, Mr. Earl Thompson, Woodbine, Iowa, 1957) living in the Willow Drainage Ditch area stated that by 1930 this ditch in the vicinity of Thompson Creek (fig. 1) had widened to

[4] Bridge records were obtained from county and state highway official records.

a top width of about 60 to 70 feet, and had deepened to 20 to 25 feet below the flood plain level.

The Willow Drainage Ditch by 1931 had entrenched approximately 8 feet at mile 9.4, and had widened from an original width of 42 feet to about 57 feet (fig. 3, cross section mile 9.4). The changes in the ditch at mile 5.6 were similar (fig. 3). In both cases the cross section of the channel was distinctly U-shaped, and not the trapezoidal shape of the original channel.

County bridge records show that by 1936-37 the Willow Drainage Ditch had entrenched to about its present depth from the lower end of district no. 1 upstream to mile 5.6 (fig. 3). Upstream from this point in the drainage ditch the records prior to 1940 or 1942 are not available because the bridges had been replaced. When the depth of the ditch in 1940 at mile 7.2 is compared to the depth at mile 8.3 in 1942, however, a distinct difference is noted (fig. 3). Observations by Mr. Burkholder (oral communication, Mr. Verne Burkholder, Logan, Iowa, 1957) suggested that the ditch deepened slightly but widened considerably in the vicinity of mile 8.3 between 1941 and 1943. The sharp increase in gradient between miles 7.2 and 8.3 also suggests the drainage ditch was actively cutting in this segment between 1940 and 1942.

The Soil Conservation Service, U. S. Department of Agriculture, conducted an Open Drainage Ditch Reconnaissance Survey of the Willow Drainage Ditch in Harrison County in 1942. Records are available for parts of the drainage ditch. In the lower end of the survey the drainage ditch had deepened about 22 feet since 1920 (fig. 3, mile 12.5); upstream the ditch shallowed considerably and at mile 17.7 had increased the depth of its channel 11 feet since 1920 (fig. 3).

Ileff (1942, p. 31-41) had recorded a number of knickpoints on the tributaries of the Willow Drainage Ditch in 1942 from mile 12.5 upstream to mile 17.7 (fig. 3). He also recorded that the bottom of the ditch was badly broken and checked in this segment, and many pronounced riffles with intervening stretches of slower water were present. Apparently the drainage ditch was actively eroding between mile 12.5 and 17.7 just prior to 1942. This contrasts to the relatively stable channel bottom of the drainage ditch in most of drainage district no. 1 during the same period.

The channel of the Willow Drainage Ditch in 1958 is distinctly different than the original channel of the ditch in 1919-1920. The lower end of the drainage ditch in Upper Willow Drainage District No. 1 is 4 feet deeper and about 40 feet wider than the original channel. The ditch becomes progressively deeper and wider upstream, and has a maximum depth at the Monona-Harrison County line in Drainage District No. 2 (fig. 3, mile 19.6). At the Monona-Harrison County line the ditch has increased from an original depth of 11 feet in 1920 to a depth of 42 feet in 1958. The 1920 top width of 30 feet has increased to 110 to 120 feet.

Accurate records on the rate of deepening of the Willow Drainage Ditch are not available, but a series of cross sections of the ditch at a bridge site on the north line of Sec. 16, T. 81N., R. 42W., Harrison County, Iowa, illustrates the rapidity of entrenchment (fig. 3, mile 17.7). Between 1950 and 1952 the ditch had deepened 15 feet and widened about 20 feet. Since 1952 the bottom

PLATE 1

A. Willow Drainage Ditch downstream from the mouth of Thompson Creek. The ditch at this point is 36 feet deep and 110-120 feet wide.

B. Knickpoint in the Willow Drainage Ditch, September, 1957. Note the action of the water at the plunge pool.

of the ditch has been relatively stable, but some adjustment in the steepness of the banks has taken place.

The walls of the drainage ditch throughout the lower 17.7 miles of the ditch in the Upper Willow Drainage Districts are vertical, or nearly so, and the channel has a distinct U-shape (pl. 1A). Upstream from mile 17.7 the distinctive U shape of the ditch is modified by abundant slump blocks, and the cross sections of the channel are very irregular.

The drainage ditch has been cut into a compact silt, and only sparse sands occur throughout its length. The base of the stream, however, is flowing on silt, sand, and gravel, 1 to 3 feet thick, that are constantly being shifted by the stream. Above the Monona-Harrison County line the water is flowing directly on a compact silt, and only in the pools or quiet stretches is a thin silt, sand, or gravel bed present.

The gradients of the drainage ditch in 1958 also differ from the gradients of the original ditch. In the upper Willow Drainage District No. 1 a uniform gradient of 7.66 feet per mile was constructed in the drainage ditch (fig. 3). The gradients in 1958 ranged from 4.7 to 6.7 feet per mile. In Upper Willow Drainage District No. 2 the original gradients of the drainage ditch ranged from 7.04 to 10.56 feet per mile, but in 1958 ranged from 5.1 to 21.2 feet per mile. The gradients of the ditch in 1958 are, in general, less than those of the original channel, but somewhat more than the original Willow River channel (fig. 2).

ENTRENCHMENT

Method

Recorded changes in the Willow Drainage Ditch in the vicinity of the Monona-Harrison County line since 1953 permit an analysis of the method of entrenchment of the ditch.

A knickpoint was present in the Willow Drainage Ditch slightly above a bridge on the north line of Sec. 9, T.81N., R.42W., Harrison County, Iowa, in 1953 (fig. 3, mile 18.8). The knickpoint moved upstream, and in November, 1956, was less than 100 yards north of the Monona-Harrison County line (oral communication, Mr. Al Schafer, Dunlap, Iowa, 1957). Between November, 1956, and July 9, 1957, the headward movement of the knickpoint was approximately one-half mile.

The knickpoint was first seen by the author on July 9, 1957, and had a vertical height of 1.5 feet. Between July 9, 1957, and April 27, 1958, the knickpoint moved upstream 85-90 feet and the vertical height increased to 3 feet (plate 1B). From April 27 to May 1, 1958, the knickpoint moved upstream about 600 feet; it was then essentially stationary until July 1, 1958, but between July 1 and August 15 moved upstream 1400 feet before disappearing into a series of riffles.

The rapid movement of the knickpoint between April 27-May 1 and July 1-August 15, 1958, may be related to periods of high water in the drainage ditch because rainfall exceeding 1 inch occurred in the drainage basin during these periods. The movement of the knickpoint could not be attributed to cutting in more friable materials, such as sands, because the sediments were compact silts throughout the area of cutting.

During the periods of slow headward movement in 1957 and 1958, the knickpoint was characterized by a vertical wall (pl. 1B). As the face of the knickpoint was undercut, shear planes developed in the unsupported sediments, the undercut sediments slumped, and a vertical face was reestablished. These

PLATE 2

A. Action of water at the plunge pool of a knickpoint. Note the wave action produced.

B. Mass movement subsequent to passage of a knickpoint. Two distinct and one indistinct slump blocks are visible in the upper right of the photo. The surface of the blocks represent former flood plain levels.

processes were responsible for slow headward migration. The action of the water flowing over the knickpoint was similar to that described by Ireland, et al. (1939, p. 50, fig. 20). Some of the water always flowed down the face of

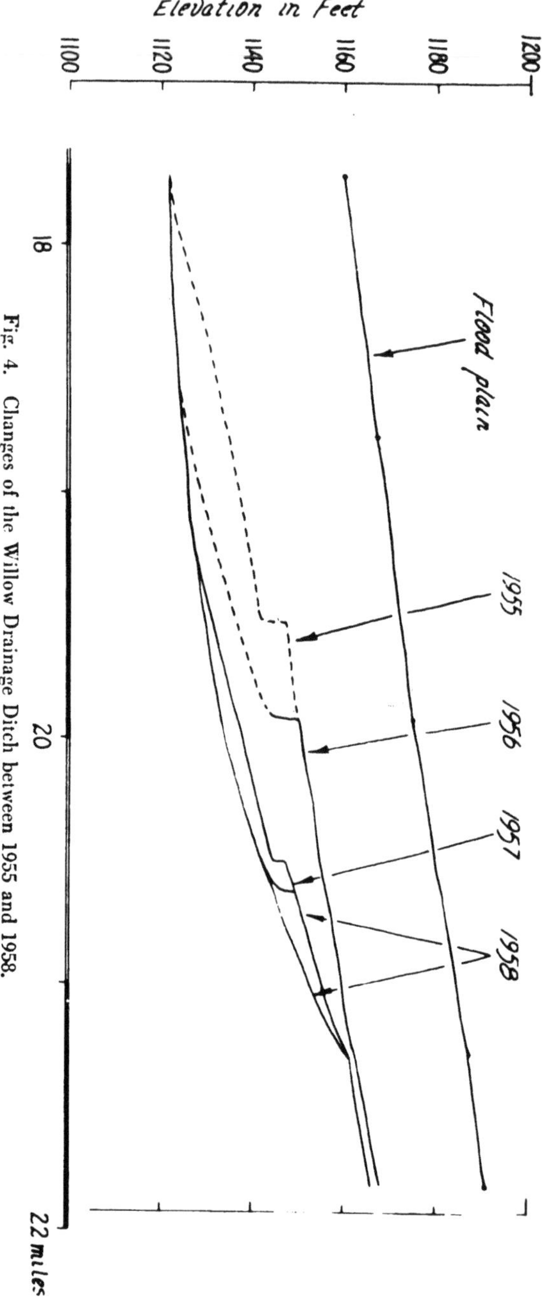

Fig. 4. Changes of the Willow Drainage Ditch between 1955 and 1958.

the knickpoint, even when undercut, but the percentage depended upon the stream discharge. A plunge pool 2 to 6 feet downstream from the base of the knickpoint always was present, and a swirling or wave action (pl. 2A) was produced in the plunge pool by the falling water. The wave action (pl. 2A) undoubtedly assisted in undercutting at all stages of the knickpoint development. Seepage water at the base of the knickpoint possibly was as important in undercutting as the back trickle and wave action of the plunge pool. The part of the knickpoint that undercut was usually at or below the seep line, and the saturated sediments possibly were less resistant to erosion than the drier overlying material. Thus, the water flowing down the face of the knickpoint, the swirling or wave action at the base, and the saturated sediments at the base probably contributed to the undercutting and slow upstream movement of the knickpoint.

Once a knickpoint has moved upstream from a point in the drainage ditch, several adjustments in the channel take place. The passage of a knickpoint normally exposes a seep line above the channel bottom, and mass movement of the saturated material is common (pl. 2B). The mass movement of the saturated material weakens the overlying sediments, and may be largely responsible for the formation of slump blocks. Slump blocks up to 30 feet in width have been measured downstream from an active knickpoint in the Willow Drainage Ditch, and as many as three distinctly separate slump blocks have been observed. The surface of each slump block was the surface of the original flood plain. In most cases the slump blocks have rotated backwards toward the valley walls. The blocks apparently move slowly downward toward the stream bottom and eventually are removed by the stream unless stabilization of the slumped mass occurs.

Deepening of the Willow Drainage Ditch has not been entirely by headward movement of knickpoints. For example, between 1955 and 1956 the knickpoint in the drainage ditch moved upstream approximately one-third mile (fig. 4), but the channel continued to entrench downstream from the knickpoint. Little change occurred in the channel upstream. In June, 1957, heavy rains in the drainage basin resulted in a large flow of water in the ditch and deepened the channel considerably between mile 19 and 21 (fig. 4). Upstream from mile 21 little change in the channel occurred. Between July, 1957, and August, 1958, the knickpoint continued to advance with concurrent deepening of the channel downstream, but little change in the channel of the drainage ditch upstream from the knickpoint was measured (fig. 4). Thus, the deepening of the drainage ditch by the passage of a knickpoint is minor in comparison to that produced by channel scour subsequent to passage of the knickpoint.

The changes in the Willow Drainage Ditch since 1936 show that entrenchment of the drainage ditch essentially is a process of upstream migration of the eroding surface. The channel upstream from the eroding surface, while subject to some channel scour during periods of heavy flow, is relatively stable. The process of erosion of the Willow Drainage Ditch is not unlike that detailed by Ruhe (in press) for the process of pedimentation in southwestern Iowa.

Cause

Since construction of the Willow Drainage Ditch, two changes have taken place. The ditch has filled on the Missouri bottom, but has entrenched in the Willow River Valley. The gradient of the ditch in the Missouri River Valley is about 2 feet per mile. and immediately upstream in the Willow River Valley 6.1 feet per mile (fig. 3). The abrupt change of gradient of the Willow Drainage Ditch where it enters the Missouri River Valley probably results in a decrease in velocity of the water with the subsequent deposition of part of the suspended sediment and aggradation of the channel., The deposition of sediment is undoubtedly increased if the period of high water in the Willow Drainage Ditch coincides with the period of high water in either the Boyer Drainage Ditch or the Missouri River because the water backs up in the Willow Drainage Ditch under these conditions. It is doubtful if much sediment is deposited during low water because during these periods the ditch carries one foot or less of water.

The Willow Drainage Ditch in the Willow River Valley was cut at a higher gradient than that of the Willow River in the same segments. To illustrate, the average gradient of the drainage ditch was 7.66 and 8.48 feet per mile in Upper Willow Drainage Districts Nos. 1 and 2, respectively. The average gradient of the Willow River in the same segments was 5.18 and 7.50 feet per mile. The increase in the average gradient of the drainage ditch over the Willow River was about 1-2 feet per mile.

The smooth, straight channel and increased gradient of the drainage ditch must have increased the velocity of the water, at least initially, and undoubtedly were responsible for the entrenchment of the ditch.

Effects

The main reason for straightening the Willow River was to prevent damage to crops by flooding (Harrison County, Iowa, Drainage Record No. 7, p. 604, 1104). Interviews with local residents indicate the last flood on the Willow Drainage Ditch in the vicinity of Thompson Creek (fig. 1) was about 1925 (oral communication, Mr. Earl Thompson, Woodbine, Iowa, 1957). At the upper end of the drainage ditch the last flood was in 1942 (oral communication, Mr. M. Bendict, Dunlap, Iowa, 1958). An example of the water-carrying capacity of the drainage ditch is shown by the absence of flooding in Upper Willow Drainage Districts during a period of heavy rainfall in June, 1957. In a 24-hour period a series of rains equalling or exceeding 7 inches failed to fill the drainage ditch more than half full (oral communication, Mr. Earl Thompson, Woodbine, Iowa, 1957). During the same rains, however, flooding did occur at or slightly below the lower end of Upper Willow Drainage District No. 1.

The absence of flooding in most of the Upper Willow Drainage Districts for a number of years is interesting in light of a report filed by the commissioners and engineer that surveyed the Willow River in Upper Willow Drainage District No. 2 in 1918 (Harrison County, Iowa, Drainage Record No. 7, p. 1106). "Judged from other straightened channels, for instance the Boyer [Drainage Ditch], this proposed straightened channel should wash out in a couple of years sufficiently to carry the maximum floods." It is doubtful, however, if the commissioners and engineer visualized a channel similar to the Willow Drainage Ditch in 1958.

Influence on Tributaries

Accompanying the entrenchment of the Willow Drainage Ditch has been the entrenchment of the tributaries of the ditch. Thompson Creek (fig. 1), for example, has entrenched at its mouth accordant with the channel of the Willow Ditch. The Trench of the creek deepens upstream for about 2 miles before shallowing where till is exposed in the channel bottom. In its headwaters Thompson Creek is cut to a depth below the adjacent flood plain of about 40 feet. The entrenchment of Thompson Creek is typical of the larger tributaries of the Willow Drainage Ditch. The entrenchment in the mouths of tributaries such as Thompson Creek can be related directly to the entrenchment of the Willow Drainage Ditch.

Shimek (1909, p. 298) described a number of creeks in Harrison County that had entrenched by 1909, which was 10 to 11 years before construction of the drainage ditch was completed in the Upper Willow Drainage Districts. Thus, not all of the entrenchment of the tributaries can be related to the entrenchment of the Willow Drainage Ditch.

SUMMARY

Since construction of the Willow Drainage Ditch, the ditch has filled in the Missouri River Valley and simultaneously entrenched in the Willow River Valley. The method of entrenchment of the drainage ditch essentially has been a process of headward encroachment of an erosional front on a channel that is relatively stable.

The cause of filling of the drainage ditch in the Missouri River Valley probably is the sharp decrease in the gradient of the ditch in this segment compared to the gradient of the ditch in the Willow River Valley. Concurrent flooding in the Boyer Drainage Ditch, or the Missouri River, causes water to back up in the Willow Drainage Ditch and also results in the deposition of sediment in the Willow Ditch.

The entrenchment of the drainage ditch in the Willow River Valley probably was caused by the increase in gradient of the constructed channel over that of the Willow River.

Entrenchment of the Willow Drainage Ditch is responsible for much of the deep entrenchment of its tributary streams.

REFERENCES

Anderson, Alexander, 1851a, Subdivision of Township No. 81 North of Range No. 42 West of the 5th Principal Meridian, State of Iowa: State Land Office, Office of Secretary of State, Des Moines, Iowa, v. 275, p. 129-190.

——, 1851b, Subdivision of Township No. 80 North of Range No. 43 West of the 5th Principal Meridian, State of Iowa: State Land Office, Office of Secretary of State, Des Moines, Iowa, v. 280, p. 107-168.

Ileff, Earl A., 1942, Harrison County, Iowa: Soil Conserv. Dist. Eng. Field Book No. 9, p. 31-41.

Ireland, H. A., Sharpe, C. F. S., and Eargle, D. H., 1939, Principles of gully erosion in the Piedmont of South Carolina: U. S. Dept. Agriculture Tech. Bull. no. 633, 142 p.

Ruhe, Robert V., (in press), Geomorphology of parts of the Greenfield Quadrangle, Adair County, Iowa: U. S. Dept. Agriculture Tech. Bull.

Sales, David J., 1853, Subdivision of Township 82 North of Range 42 West of the 5th Principal Meridian, State of Iowa: State Land Office, Office of Secretary of State, Des Moines, Iowa, v. 275, p. 193-244.

Shimek, B., 1909, Geology of Harrison and Monona Counties: Iowa Geol. Survey, v. 20, 485 p.

Smith, Joseph H., 1888, History of Harrison County, Iowa: Des Moines, Iowa, Iowa Printing Co., 491 p.

Editor's Comments on Papers 10 Through 14

10 Askochensky: *Basic Trends and Methods of Water Control in the Arid Zones of the Soviet Union*

11 England: *Problems of Irrigated Areas*

12 Poland and Davis: *Subsidence of the Land Surface in the Tulare-Wasco (Delano) and Los Banos–Kettleman City Area, San Joaquin Valley, California*

13 Hamilton and Meehan: *Ground Rupture in the Baldwin Hills*

14 Curry: *Soil Destruction Associated with Forest Management and Prospects for Recovery in Geologic Time.*

Terrain Changes Caused by Resource Development

Raw materials are vital and essential ingredients in today's civilized world. Since the industrial revolution, economic resources have been mined from the earth at an accelerated pace, and with the rapid increase in population during the twentieth century, there has been an almost exponential growth in man's use of such metallic and nonmetallic ores as iron, copper, lead, zinc, coal, limestone, sand and gravel, and water. Many of these are obtained at or near the earth's surface and their withdrawal has resulted in man's deliberate defacing of his own habitat. Most papers in this section illustrate damages that occur to the land–water ecosystem when man develops and uses materials of the earth's surface. Water is of such fundamental importance to the geomorphologist that most articles selected dramatize results that can occur when it is mismanaged. Although nearly all of man's activities that involve terrain considerations, even agricultural pursuits, can be viewed as resource development, this section stresses water and timber use to show how their extraction upsets the balance of nature and devastates the terrain.

There is a vast, indeed overwhelming literature on water resources. Much of this information is in government publications, both national and international. The first paper, by A. N. Askochensky, is a good example of the sort of information that may be found in UNESCO reports, a valuable source of world-wide hydrologic data. This paper dealing with a situation in the U.S.S.R., provides an international flavor for water use, both surface and groundwater, and gives a good perspective on methods that are used there to prevent land abuse. See also article by Vendrov (1964) on water problems in Siberia, and by Smith (1970) for irrigation in the Middle East.

It is a maxim that whenever man uses and exploits water, he changes it in some way. Except for discussion of salinization, the problems of water quality and pollution lie beyond the scope of this volume. Wolman (1971) provides a thoughtful review of surface waters in this context, and Strahler (1972) provides a good case history for a specific area, Cape Cod, in the use of groundwater. Wolman indicates the problems that are encountered when attempts are made to evaluate river quality but emphasizes that ". . . many rivers of the United States . . . are not as they were 70 years ago" (p. 916). Strahler demonstrates how important it is to ". . . preserve the high quality of ground water and the water level of the fresh-water ponds, requires strict attention to known basic principles of ground water movement and recharge" (p. 21). A good summary of water resources in the United States occurs in a volume of the Water Resources Council (1968). This book presents a comprehensive synthesis and discusses a large array of basic data, present water use, projected demands in the future, present problems, and likely problems of the future. Many of the categories that are classified are divided into urban and nonurban use, including industrial, municipal, irrigation, and rural domestic. It also contains sections on wilderness and free-flowing rivers. A further distinction must be made since water-use examples in this volume are largely restricted to local usage, rather than exotic usage, as discussed in Volume II—where an urbanized area imports water from some distant region.

The article by H. N. England occurs in a collection of excellent papers dealing with many different aspects of water development in Australia. Case studies by England show the large range of problems that can occur with water use in irrigated areas. One of the problems, that of salinization of soils, is not new to the twentieth century and was discussed in Volume I (p. 137–145). The problem occurs in semiarid climates of all countries in which the irrigated waters contain large amounts of dissolved solids. Several irrigation projects in Pakistan have failed because of excessive salts. It is a problem in the southwestern United States and is the focus of a dispute between the United States and Mexico, because irrigation practices in the United States have caused deterioration of the quality of Colorado River waters flowing into Mexico. In the Phoenix area, 300,000 tons of salt is irrigated into the soils annually (Halpenny et al., 1952), mostly from surface waters. However, the use of poor-quality groundwater is also creating problems in Arizona. With the rapid decline in water tables, deeper waters are being used, and since these waters have usually been in contact with sedimentary minerals of the valley fill longer than is the case with shallower waters, they contain excessive amounts of dissolved materials. Aschmann (1966) reports:

> In the last few year, an acute problem of crop failure has arisen due to the excessive salinity of irrigation water. Its immediate cause is an expansion of cultivation in the Wellton area of the Gila Valley, based on the exploitation of almost brackish well water. To produce crops with this water, it is necessary to over-irrigate, thereby preventing salt accumulation in the fields where it is applied. The runoff enters the Gila River and thence the Colorado just before it enters Mexico (p. 261).

Whereas salt encrustation of soils during irrigation is an age-old problem, the paper by J. F. Poland and G. H. Davis describes what is largely a twentieth-century phenomenon. When groundwater is mined, it creates a chain reaction of events that leads to a variety of damages and costs: (1) a decline in the water table; (2)

a lowering of water level requiring more energy to bring the water to the surface, thus increasing costs and making greater demands for energy (which also strains the environment); (3) wells must be drilled deeper, increasing costs; (4) deeper waters are usually poorer in quality and their treatment becomes expensive and when used in irrigation they can cause salinity problems; (in addition, mineral incrustation can occur in the well casing, causing reduced efficiency and ultimate failure of the casing); and (5) dehydration of earth materials, formerly buoyed by hydraulic pressure, causes compaction and collapse to occur between grains. This feeds up into the system, causing cracks and subsidence to occur on the earth's surface. It is this feature that is discussed in the next two articles. Man-induced land subsidence also occurs in urban areas, such as Venice and Mexico City, and can be caused by a variety of ways resulting from utilization of earth resources.

The topographic changes that result from fluid extraction are generally the same, whether it involves petroleum or water—reduction of fluid pressure and increase in effective stress (grain-to-grain load) borne by the reservoir skeleton. Although other subsidence articles have been written, the Poland and Davis paper is the best of the early publications on subsidence by groundwater mining, and thoroughly documents the scale of this phenomenon. In a more recent paper (1969) Poland and Davis report on amounts of subsidence in many different regions: 3–4 m at Tokyo and Osaka, Japan; up to 7.5 m in Mexico City; 1–4 ft in the Galveston–Houston area of the Texas Coast; 1.25 ft in the Denver area; 3.6 ft in an area 50 miles north of Tucson, Arizona; about 3 ft in Las Vegas, Nevada; 5 ft in the Santa Clara Valley south of San Francisco. They show that a maximum subsidence of 16 ft from 1943–1959 occurred south of the Mendota region (see Fig. 3).

The literature contains several examples of land subsidence in areas of petroleum extraction. The Goose Creek oil field in Harris County, Texas, was the first subsidence due to fluid withdrawal described in the literature (Minor, 1925). This oil field was started in 1917 and by 1925 maximum subsidence was more than 3 ft and affected an area 2.5 miles long and 1.5 miles wide and conformed to the area that contained the oil-producing wells. At Lake Maracaibo, Venezuela, oil was discovered in 1917, but large-scale production did not start until about 1926. Greatest subsidence is about 11 ft, which occurred mostly from 1926 to 1954. The damages are almost exclusively within the producing stratum or movements within the producing interval (Kennedy, 1961).

There are other areas where subsidence is occurring that may be due to multiple causes. At Nigata, Japan, large quantities of saline groundwaters containing methane gas are pumped, and by 1959 the annual subsidence rate here was 1.6 ft/year. Methane gas is produced from the Po Delta and by 1959 the rate of sinkage was 25 cm/year. The sinking of Venice, according to some authorities, is due to large groundwater withdrawals in the coastal region, but this is a complex region with other contributing factors, which include the rise in sea level, tectonic adjustment, and compaction under normal overburden. Land subsidence was noticed in 1940–1941 in the Wilmington oil field in the harbor area of Los Angeles and Long Beach, California. By August 1962, the subsidence was 27 ft at its center and included a 25-mile area where subsidence was 2 ft or more. According to Poland and Davis, this vertical settlement is the greatest known subsidence attributed to fluid withdrawal. Corrective measures have largely stopped this subsidence at the present time.

D. H. Hamilton and R. L. Meehan in their article show another dimension of subsidence—that of man-induced hazards. Withdrawal of fluids, mostly petroleum, from the Inglewood oil field in the Baldwin Hills area of California produced subsidence of 9.7 ft from 1917–1963. The earth movements that caused failure in the structure impounding the reservoir is attributed to the chain of events that started with general land subsidence, creating weaknesses in earth materials that were then activated when injection fluids were introduced into the system in order to recover more petroleum. The fluid-drive pressure was sufficient to lubricate the fault surfaces that triggered the ground ruptures. A recent series of earthquakes in the Denver area has been attributed by many (although not all geologists agree) to injection of disposal waste waters. In March 1962, a deep well (3671 m) at the Rocky Mountain Arsenal was first used for disposing of chemical waste waters. Soon afterward, earthquakes began in the Denver region—the first in the area since 1882. From April to November 1962, a series of 700 earthquakes occurred up to a magnitude of 4.3 on the Richter scale. The general reason given is that the fluid pressure from the liquids acted as a lubricant to old fault traces (the Rubey and Hubbard principle), reactivating them because the pressure reduced the frictional resistance (Evans, 1966; Healy et al., 1968). Underground nuclear explosions also trigger earthquakes as shown by Emiliani (1969); in Nevada significant earthquake activity occurs for at least 32 hours after explosions and at distances up to 860 km from the sites.

Subsidence is caused by man in a variety of other ways, such as drainage of water from organic deposits. Peat lands underlie about 450 square miles of the Sacramento–San Joaquin delta and form one of the largest recent organic deposits in the western United States. The delta islands were all above sea level in the early 1900s but are now more than 15 ft below sea level. For example, Mildred Island subsided 9.3 ft from 1922 to 1955 during drainage (Weir, 1950). Causes of the subsidence have been attributed to several factors, including oxidation from dewatering to permit their cultivation. Lofgren (1969) presents data to show that subsidence also results from application of surface water (hydrocompaction). There are extensive alluvial fans above the water table along the south and west parts of the San Joaquin Valley. These lands are used for agricultural purposes but have subsided 5 to 15 ft from the application of irrigation waters. A number of serious problems result from this land deformation, such as sunken irrigation ditches, undulating fields with damaged canals that constantly need re-leveling, sheared well casing, broken pipes, and displaced roads and transmission towers. Comparable damage has occurred in Arizona, Montana, Washington, and Wyoming. Subsidence also occurs in loess soils in the Missouri River Basin, as well as in Europe and Asia, where features as much as 2.5 m of relief have been created.

The last article in this section is by R. R. Curry, who shows destruction of land by the highly controversial method of clear-cutting of forests. Lumber companies, the federal government, and conservationists cannot agree on the best methods of logging. There is a wealth of information that shows the importance of forests for protecting soil from erosion, and Curry and others have amassed significant data indicating that clear-cutting can lead to serious terrain damage. Orme and Bailey (1970) compared Monroe and Volfe Canyons, two small drainage basins in the San Gabriel Mountains of California. In Monroe watershed, 17 hectares were cleared of woodlands and 57 hectares of chaparral was converted to grass. No man-made disturbance was observed in the Volfe watershed. In reaches of the Monroe watershed under observation during 1963–1969, 2200 m^3 of material was lost, which was eight times the material lost in comparable reaches of Volfe Canyon, where erosion was inhibited by the vegetative cover.

Proponents of clear-cutting—the removal of all trees in an area as contrasted with selected cutting—favor the method because of lower costs, whereas opponents believe the land–water ecosystem is severely jeopardized, leading to accelerated erosion, soil exhaustion, and siltation. The Curry article presents a one-sided persuasive thesis that clear-cutting does cause increased erosion and nutrient depletion of soils. McCaskill (1966) describes clear-cutting experience in New Zealand in the following terms:

> The timber resources of North Westland were regarded as a mine to be exploited rather than as a crop to be tended. There was no thought of perpetuating the native forests by selective cutting and controlled regeneration . . . Clear cutting followed by the burning of the slash was an acceptable policy. It was based on the assumption that the timber resources were almost limitless . . . But the terrace and morainic gravel soils of North Westland are bog podzols . . . where downward percolation of water in the soil is checked by multiple iron pans . . . After logging operations, the surface drainage channels become blocked with fallen trees, transpiration through the foliage is reduced with the removal of the forest canopy and rising groundwater converts the cut-over forest into bogs (p. 281).

The work of Kotschy (1964) in Austria showed that regeneration of trees in clear-cut areas was only one-half that in other tree-selection systems. Other effects of clear-cutting include increased density in roads and logging trails, resulting in increased erosion and downhill mass wasting and aggradation. In the H. J. Andrews experimental forest in Oregon, 65 percent of the landmass failures were caused by roads and road building, and in Idaho along the South Fork of the Salmon River, 90 percent of slope failures occurring after a storm period were associated with roads (Gray, 1969). Curry raises an interesting question about forest fires and indicates there is absence of appreciable soil loss after a fire. A contrary study by Doehring (1968) in the San Dimas Experimental Forest in California showed that forest areas burned out after a fire had 20 times the normal erosion rate the first year after the fire, and from the second to seventh year movement of material downslope was five times the average, and at the end of the period gravity-induced erosion was still abnormally high.

Bormann et al. (1967) summarize the effects of clear-cutting upon the ecosystem, claiming that it produces (1) reduction of transpiration and thus increases the amount of water passing through the system; (2) reduction in root surfaces able to remove nutrients from leaching waters; (3) removal of nutrients from forest products; (4) addition to the organic substrate available for immediate mineralization; and (5) a microclimate favorable for rapid mineralization.

Surface mining is one of the most obvious destroyers of the landscape because it is a calculated and deliberate operation by man to consciously destroy surface characteristics of the earth. Surface mining operations can be grouped into five kinds: open-pit mining, strip mining, auger mining, dredging, and hydraulic mining (see Gilbert, Vol. I, this series). Probably the best known open-pit iron mines are those of the Mesabi Range, extending intermittently more than 150 miles with individual mines that are more than 3 miles wide and benches hundreds of feet deep. Copper is another metal mined by this method in the western United States (e. g., Bingham, Utah, and Bisbee, Arizona) and in many other parts of the world.

Strip mining can be divided into two categories, area stripping and contour stripping. The ruggedness of the terrain determines which method is used. Coal and phosphate are materials commonly mined in this manner. More and more coal is being mined by surface methods, rather than underground methods, because the huge machines now used make it cheaper (proponents of the method also point out it is safer for the miners). For example, some of the equipment currently used consists of giant shovels which can remove as much as 300 tons per scoop, are 12 stories high, weigh 15,000 tons, and can take 50 bites per hour and load 160 railroad cars in a single day.

Perhaps no single other activity by man is wrecking his environment in so dramatic a fashion as strip mining; not least is the chain-reaction effect it produces on the land, waters, and ecology of the terrain. Typical damage that results includes:

1. *Destruction of original vegetative cover.* This disrupts the oxygen–carbon dioxide cycle and removes cover and food sources for wildlife.

2. *Erosion.* Studies in Kentucky have shown that under natural conditions only about 27 tons/square mile of soil is eroded from the land, whereas in hillsides that have been strip-mined, more than 27,000 tons/square mile were eroded.

3. *Siltation and flooding.* Eroded material from the hillsides collects in streams, clogs the channels, and produce higher flood crests and damages during the storm period. Reservoirs are affected also.

4. *Leaching and chemical pollution.* Because many new and fresh rock surfaces are exposed to atmospheric weathering and erosion, mineral constituents within the coal, such as pyrite and marcasite, are decomposed and produce sulfuric acids. The acid washes into the surface waters of the region, affecting flora and fauna, and even gets into the groundwater and ruins crops, causing soils to become sterile.

5. *Aesthetic pollution.* The rock holes and cuts produced by strip mining deface what originally was often beautiful country. Expecially is this so in Appalachia, where the mountains, hills, and valleys form an elegant, well-ordered pattern of rugged and picturesque terrain.

By 1965 strip mining and other surface operations in the United States had damaged more than 3.2 million acres, with an additional 320,000 acres disturbed by supporting mining roads and exploration activities. The full range of surface disruptions has caused losses in fish and wildlife in 2.0 million acres (U.S. Department of the Interior, 1967).

To avoid topic duplication, facsimile articles concerning mineral and rock mining are not reproduced here. Instead, the reader is referred to Volume I, where Gilbert discusses the terrain ravages created by hydraulic gold mining, and to Part III of this volume, where coal, gold, and iron mines are considered for examples of what can be done to rehabilitate damaged environments.

Reprinted from *The Problems of the Arid Zone, Proc. Paris Symp.*, 401–410 (1962)

BASIC TRENDS AND METHODS OF WATER CONTROL IN THE ARID ZONES OF THE SOVIET UNION

10

by

A. N. Askochensky

The Lenin Academy of Agricultural Sciences, Moscow

As far as the extent of its water resources is concerned, the vast territory of the Soviet Union, with its great variety of climatic zones, can be divided into two sections, one with inadequate water reserves and the other with surplus water. The salient feature of the first is that evaporation exceeds atmospheric precipitation, whereas in the other precipitation exceeds evaporation. The purpose of this report is to give a short summary of the methods and practices applied in the Soviet Union to rationalize the problem of water supplies in areas which are permanently or temporarily subject to drought.

We shall first consider the areas where there is a constant, annually recurring, water shortage. The main factor in these areas is that the rainfall distribution does not favour agriculture, as the maximum amount falls in the winter months and the minimum in the summer or crop-growing months. It must be stressed that such climatic conditions are fundamentally different from those of the monsoon areas of the eastern and southern regions of Asia, for instance, where maximum and minimum rainfall is locally variable. In the past, the particular features of the arid regions of the Soviet Union were responsible for the use of canal irrigation as a basis for arable farming and flood irrigation as a basis for animal husbandry.

If the problem of the development of irrigation and drainage is examined from a historical point of view, it can be seen that there is a direct connexion between this development and changes in social structure and successive forms of government. Within the narrow framework of this report, however, this interesting digression must be restricted to a brief description of the development of canal and flood irrigation immediately prior to the creation of the Soviet Government.

The arid regions are, for the most part, situated in the present-day Central Asian Republics (Uzbekistan, Kirghizia, Tadjikstan and Turkmenistan), the Transcaucasian Republics (Azerbaijan and Armenia) and also in a large part of Kazakhstan and Georgia.

In the period before the Revolution, the farms in these regions chiefly consisted of privately owned or rented smallholdings. This explains why irrigation at that time was haphazard in character, a fact which is to be seen from the dense network of small irrigation canals with each individual canal obtaining its water supply from isolated springs or from larger canals, although the latter were very few in number. The irrigated areas had a curious, mosaic-like appearance since they consisted of small fields cultivated by hand with the aid of draught animals and very primitive agricultural implements. This type of farming was entirely dependent on water, a factor which very often had serious consequences resulting in periods of flood and intense drought, in calamities on a national scale.

Animal husbandry, which was mainly nomadic, was in no better position. The migration routes connecting summer and winter grazing grounds were provided with water-holes or wells which were maintained by individual owners of large herds or by communities of a family or tribal character. Different tracts of the winter and summer pastures were also distributed among such people in accordance with a tradition which was tantamount to law. These tracts of land were used for primitive agriculture and were provided with stables, wells and stocks of feedstuff. Shortages of water and feedstuff often led to disputes between neighbouring herdsmen. Animal husbandry was more dependent on water than was arable farming; in dry years, the wells and pastures dried up and in winter the ground was often covered with a thin crust of ice or with thick snow which deprived the animals of food and led to the total depletion of herds and to the ruin of stock-farmers who could only deal with the problem by primitive means. One of the methods used was to collect the melting snow or rainwater in reservoirs in the spring, usually by making use of natural hollows in the ground. Sometimes, artificial reservoirs were created by erecting low earth dykes or dams to block wide saddles or old river beds. In the absence of underground water close to the surface, use was made of wells filled with rain or melted snow. Sometimes dome-shaped structures were built over these wells to reduce water losses through evaporation or to protect them from dust infiltration. In Central Asia these structures were called *sardobas*. Flood waters were used to improve the grass on pasture lands in cases where it proved easy to divert the water to flood low-lying land which was either enclosed by the natural topography or by small earth dykes. In the latter case, it was possible to use the flooded areas not only for watering and grazing cattle along the reservoir banks, but also for sowing plants for food and animal feedstuffs on strips of dried-out ground near the reservoirs or by creating small irrigation canals. It is easy to see that animal husbandry was more dependent on water and was more vulnerable than irrigated arable farming.

In regions where drought was an intermittent and not an annual occurrence, agriculture could not follow a settled pattern and, in critical periods, was able to maintain its level only by building up emergency and temporary stocks of feedstuffs and food. We shall call these regions 'semi-arid'; they include the southern areas of the Ukrainian and Russian Republics, the vast territory west of the Caspian Sea between the Terek, Kuma, Volga and Ural rivers, the middle and lower stretches of the Volga, the west Siberian regions and the northern and western parts of Kazakhstan. With a very few exceptions, the agriculture of these regions was traditionally dependent on natural precipitations; only comparatively small areas were irrigated for vegetable, horticultural and industrial crops and these occupied an insignificant place in agriculture as a whole. Ponds and small reservoirs, where flood and rain water were collected, were created for stock-farming, in addition to some wells. In other cases, isolated attempts were made to improve grazing by letting flood water flow on to the meadows and pasture (so-called 'silt' irrigation). In the same way as in the arid zone, stock-farming was subject to periodical crises through shortage of feedstuff and drinking water, or because of unfavourable winter conditions such as the formation of ice-crust on pastures, or snowfalls and storms, resulting in the widespread loss of livestock.

This was the agricultural and water supply situation in the arid and semi-arid zones in the period immediately before the formation of the Soviet Government. During the forty-two years of Soviet constructive activity, the situation in these zones has been set on an entirely new footing. It has, in fact, been radically altered and replaced by a new agricultural and water supply system based on socialist principles. Since the question of water supply is not an independent branch of the economy but is subservient to agriculture, its development is closely related to the progress achieved in the latter sphere. To enable development in water policy to be understood and appreciated, some light must first be thrown on the main changes in agriculture during the Soviet period.

As it is not our aim to give an exhaustive survey of all the changes in agriculture during this period, we shall merely single out those relating to water control: in the first place—and this is typical not only of agriculture, but also of the economy of the Soviet Union as a whole—its development has been based on sound governmental planning; secondly, two types of large agricultural organization have been created, the collective farms and the State farms; thirdly, through mechanization and electrification, agricultural production has now been industrialized; fourthly and lastly, in order to use the sources of production to their best advantage, agriculture has been divided into zones, each of which has different natural features.

We shall now consider the changes that have occurred in the water policy, firstly in the arid zones and then in the semi-arid zones, following the method we used to describe the pre-Soviet phase.

A vast work programme was required to bring the water policy of the arid zones into line with governmental planning; it involved the study of climatic, soil, hydrological, hydrogeological and economic conditions in the various regions of the zone, in so far as such conditions had a bearing on water requirements. The historical aspects of water policy formulation and the scientific bases for a complete revision of policy also had to be studied. The results of this research were then used to work out reconstruction and development schemes for the water policy of each river basin. The most distinctive feature of these schemes is their comprehensive character, the need to use the water resources in the interests of all branches of the national economy making a fully co-ordinated solution essential. In order to prepare projects for these schemes and also to supervise the actual implementation of water policy in the different republics, territories and regions in the arid zone, a well-knit system of governmental water control boards was devised. These bodies deal with all matters relating to operation, design, planning and scientific research for water control. The main role of these organizations was to create the water control proper by centralizing all the production concerns in charge of irrigation and the reclamation of saline land in zones requiring irrigation. To do this, vast projects were required, calling for the construction of large canals to replace the great number of small irrigation canals which had previously supplied a host of privately owned properties or groups of small farms. Based on master irrigation canals equipped with control and distribution devices, vast irrigation networks were created on an industrial scale to supply water to the consumers, i.e. the collective and State farms, on a properly planned basis.

Each farm receives water in accordance with a prearranged schedule and distributes it according to agricultural needs. Water requirements vary for different types of crop in different climatic and soil conditions. However, on the basis of general observations which appear to be typical for individual natural micro-regions, it is possible to work out average annual irrigation quotas, standard quantities for a single watering, and optimum watering schemes. The use of water on the collective and State farms is therefore scientifically planned. In this connexion, the State and collective farms receive any necessary help from the water control boards.

In industrializing agricultural production, large-scale projects of great technical complexity were required in order to adapt field characteristics to mechanized agricultural equipment. Modern agricultural methods obviously could not be applied to the diminutive fields with their irregular and random shapes and terraced lay-outs required by the flood-type irrigation which was so prevalent in the past. This state of affairs had, therefore, to be changed and fields of larger dimensions created. These were laid out in accordance with the slope of the ground so that the flood-type irrigation, formerly used for arable farming, could be transformed into irrigation through channels or through strips protected by earth embankments. Inside the enlarged fields themselves, trees, roads, ditches, or any other objects which might hamper mechanized equipment, had to be eliminated. The scale of the earthworks required to carry out the different stages of these projects can easily be imagined. Fields 0.25 to 0.5 hectares in area or even smaller were replaced by fields of 5, 10 or 20 hectares and more. As a general rule, these enlarged fields are oblong in shape with their long sides running parallel to the slope of the ground, i.e. in the direction of the irrigation channels and strips. Flood irrigation over horizontal areas was retained solely for rice growing.

It must not be thought that the work of transforming the old irrigation systems has been completed. The main task in the first stage of this work was to increase the size of the plots and fields requiring irrigation; in the second stage, the aim is to complete the levelling-off of the enlarged areas so as to distribute the water more regularly, while reducing the water consumption quotas, and also to increase the efficiency of the irrigation works by extending the irrigation channels and strips. In regions where the top-soils are barely productive and the sub-soil microstructure irregular, the surface area is being levelled off in successive stages while appropriate measures are being taken to restore the impaired fertility of the soil—all of which takes time.

The progress achieved in setting up agricultural production on a regional basis can be seen clearly in the case of cotton which is well known to be particularly responsive to climatic conditions. Cotton growing is centred in the Uzbek, Tadjik, Turkmen, Kazakh (southern regions), Kirgiz and Azerbaijan Republics (with the exception of the mountainous regions). Here the cotton crop is rotated with alfalfa and maize to a certain extent. Fruit trees, vegetables and a number of plants for industrial uses are grown on the irrigated land. As a general rule, these Republics are supplied with corn from regions specializing in the cultivation of cereals, whereas before the socialist period, a large part of the irrigated land in the cotton-growing regions was given up to the cultivation of corn and other cereals.

So radical a change in agricultural policy caused a sharp rise in water consumption and the transfer of its maximum use from the spring to the summer months. The rivers of Central Asia and Transcaucasia reach top water-level in the spring with the water from the melting snow, or in the spring and summer from melting snow and glaciers. A large-scale programme was therefore required to redistribute the stocks both in space and in time. This assumed concrete form in the construction of reservoirs and canals connecting abundant water sources with those lacking in supplies. Two such canals are the Great Fergansky Canal, some 350 kilometres long, which feeds the small shallow rivers of the southern parts of the Fergansky valley in the Uzbek Republic and the Leninabad region in the Tadjik Republic, with additional water from the River Narin, and the Karakum Canal at present under construction, the first section of which, some 400 kilometres in length, has already linked the River Amu-Daria, with its abundant water supply, to the shallow, smaller River Murgab in the Turkmen Republic.

Everything that has already been said about remodelling the old irrigation systems applies equally well to the large-scale projects undertaken during the Soviet period to irrigate the virgin lands.

During the Soviet period, the irrigated areas have almost tripled and now cover 12 million hectares.

Significant changes have also taken place in the water policy of the arid zone with regard to animal husbandry. This sector of agriculture has also been rationally planned; it has been organized into large socialist farms, supplied with modern equipment and sub-divided into natural zones. The new increased demands on the water supply system made all element of spontaneity in the provision of water and feedstuff for livestock out of the question and postulated an appreciable increase in the size of stock-raising farms. In view of these demands, a new technical basis was essential.

Large-scale hydro-geological research was undertaken everywhere in an endeavour to discover the extent of underground water stocks. In the desert areas of Central Asia and Kazakhstan, artesian basins were located, giving rise to an extensive well-boring programme. These new wells are used both for watering herds and for setting up small irrigated cases which enable emergency stocks of feedstuffs to be distributed throughout the grazing areas. In cases where the water in the artesian wells does not rise to the surface, widespread use is made of stationary or mobile mechanical pumps using oil or wind energy.

Water is brought from rivers or irrigation canals into the heart of the desert whenever it is possible and economically worthwhile to do so. Water can, for instance, be diverted without any great difficulty from the Kizil-Ordinsky Dam on the River Sir-Daria through an old river bed into the Kizil-Kum Desert, where it irrigates an enormous grazing area. The Kara-Kum Canal which crosses the Kara-Kum Salt Desert for more than 200 kilometres can irrigate the adjacent grazing areas, which do not possess any local water supplies.

Even as simple an operation as collecting and storing rainwater in natural or artificial reservoirs can be more effective if mechanically assisted—and some rain does fall even in desert conditions. Experience has shown that if the water-collecting area is increased by means of simple installations such as channels and dams, and if the reflecting surface of such reservoirs is reduced and their depth increased by means of embankments, the duration and efficiency of storage can be substantially increased. In the same way, the mineral content of the stored water can be reduced and in certain cases even salt lakes can be purified.

Passing on to the semi-arid zone we may note that the above remarks also generally apply to areas where irrigation-aided arable and stock farming are faced with difficulties through periodical drought. This general survey of water control in the Soviet Union should therefore be completed by a study of the questions which are relevant to this zone alone.

It must first be noted that the semi-arid zone covers a vast area of our country, and that irrigation occupies a relatively unimportant place in it. The fundamental law of agriculture in this zone is that as much water as possible must be collected and stored in the soil and that atmospheric moisture must be used economically. All types of agricultural activity, worked out by scientific and practical methods in the light of natural conditions in different parts of this zone, are governed by this law. Thus the so-called 'dry' agriculture in this zone, assisted by powerful scientific and technical methods and drought-resistant agricultural plant breeds, is able to withstand drought and very critical periods with a minimum of loss. However, future prospects for this zone hinge on the implementation of large-scale irrigation projects. The real prerequisite for these vast undertakings is the creation of basic hydro-schemes on such rivers as the Volga, Kama, Don and others. These hydro-schemes, which are at present used for producing hydro-electricity and as a means of transport, are fundamental to future irrigation planning. The question of how the change-over from dry farming to irrigated farming can be brought about in the vast area of the semi-arid zone, covering tens of millions of hectares, is one of practical interest for the immediate future.

Some methods have already taken shape with the creation of water-supply and irrigation systems. The basic idea of such systems is to construct head canals which can later serve as a basis for large-scale irrigation networks. In this way, as wide an area as possible of the semi-arid zone can be covered, thereby largely solving the problem of its water policy. In the first stage, basin irrigation will naturally predominate over standard irrigation, the volume of which will be determined in each stage by the actual demand and by the availability of water. The subsequent development of water-supply and irrigation systems will be shown by the gradual increase in the specific volume of irrigation within previously flood-irrigated areas. We shall give some examples of these water-supply and irrigation systems.

The Inguletzky system, covering a total area of more than 300 thousand hectares, has been created in the southern part of the Ukrainian Republic. A pumping station on the Inguletz River lifts as much as 30 cubic metres of water per second to a height of 65 metres into the main irrigation canal, from which all the natural ravines and hollows are filled. As a result, there is now a reservoir system in what was formerly a completely waterless area. The total irrigated area is about 60,000 hectares divided into individual tracts of land throughout the region, so that the irrigated area for each farm is about 500-1,000 hectares. Consequently, in spite of the relative instability of 'dry' agriculture in this zone, where dry summers recur once every three years, each farm can now always have feedstuff reserves available for its livestock or even an ensured income from cultivating such valuable produce as sugar beet, vegetables and maize on the irrigated land.

The Krasnoznamensky water-supply and irrigation system, at present being constructed in the Southern Ukraine, is a similar arrangement and is fed with water by gravity from the permanent Kakhovski hydroschemes on the River Dnieper. The North Crimean Canal will be constructed from the Krasnoznamensky Canal to drain off the northern and central regions of the Crimean peninsula and to provide them with selective irrigation.

In the southern regions of the Russian Republic, the Nevinnomyski canal has been constructed; it carries water from a dam on the River Kuban to the small River Egorlyk, on which the Troitzkaya Dam has been built. The 200-kilometre Pravo-Egorlykski drainage irrigation system starts from this dam. Water will also be taken from the Kuban for the Kuban-Kalausky drainage and irrigation canal, at present under construction, by using the periodically dry river bed of the Kalaus. The water problems of a vast area of the Stavropol steppe which is used as pasture ground for thin-fleeced sheepbreeding, will be solved by these systems. As a result, water and a regular supply of feedstuffs will be ensured.

The canal connecting the Terek and Kuma rivers and the dams on these two rivers were constructed for the same purpose. It is planned to extend this supply and irrigation canal from the dam on the Kuma farther north, with the aim of solving the water problem of the three million hectares of low-lying land near the Caspian Sea which are also used as pasture grounds for thin-fleeced sheep.

In this way, the drainage and irrigation systems can contribute towards solving not only the problems of arable farming but also those of animal husbandry. In the former case the intensive type of irrigated agriculture is being contemplated, but in the second, irrigation may take on simpler forms such as those of silt or moisture-fed irrigation, fostering the cultivation of feedstuffs and improving the hay crops and grazing in temporarily flooded dry valleys, former river-beds, water meadows and natural depressions such as saddles and shallow enclosed depressions.

To complete this survey of the water policy for the arid and semi-arid zones of the Soviet Union, we must consider the question of the extent to which governmental and public funds are used for these undertakings.

In the first stages of the creation of farms of the socialist type, when they did not have their own funds, the cost of hydro-technical irrigation work was mainly borne by governmental funds and the collective farms affected often contributed to the work with their own labour; because of the shortage of earthmoving equipment, the less complex work of constructing earth embankments and canals was done by manual labour. In some cases, such as the construction of the Great Fergansky Canal, the labour contribution from the collective farms was of the highest importance.

Subsequent developments are characterized on the one hand by the growing prosperity of the collective farms, which began to dispose of appreciable funds and, on the other, by the powerful development of Soviet industry, which meant that manual labour was no longer needed for hydro-schemes or for any other sector of the building industry. Under present-day conditions, the situation has been completely transformed. Collective farms co-operate by investing large sums from their own funds or from those of the Land Reclamation Fund in irrigation or drainage works. This enables the government contribution to be reduced, so that it can be directed towards other projects such as those for irrigating and developing the virgin lands where there are no collective farms and where agricultural activity of a predominantly State-farm type must first be organized.

Thus government and public interests are combined in solving the very important nation-wide problem of creating surplus agricultural produce.

Having completed this short survey of the basic trends in water policy, the second part of this report is devoted to describing the methods used to implement the policy, still dealing, as before, with irrigation and water supply in the arid and semi-arid zones of the USSR.

Irrigation in the Soviet Union chiefly consists of free-flow networks, planned so as to supply each community with water from rivers, canals and reservoirs. As can be seen from irrigation works all over the world, the commonest features of such networks are the appreciable water losses caused by seepage because the water is transported through unlined earthbank canals, the silting-up of canals because of the high proportion of sediment in the water, the rise of the water table and the saline impregnation of irrigated land in areas where the natural run-off of groundwater is inadequate. These features are all to be found in irrigation works in the Soviet Union. The methods used in our country to combat these adverse phenomena should therefore first be described.

Water losses through river banks vary with the over-all surface area through which the seepage occurs, and with the magnitude of the seepage coefficient, which depends on the porosity of the ground. The effectiveness of a given irrigation system, in other words its efficiency, can be defined as the ratio of the amount of water delivered to the fields to the amount of water taken in at the irrigation source.

As a result of the change-over from small privately owned farms requiring extensive water distribution, to large-scale collective and State-type farms irrigated by larger, and therefore fewer, canals, water is now distributed by means of concentrated flows, with the result that the over-all surface area of the banks has been substantially reduced. Consequently, water losses due to seepage have also been reduced and the efficiency of the irrigation networks increased. Proof of the effectiveness of this method is provided by the several million hectares on which irrigation has been remodelled. Enlarging the canals forming the irrigation networks, while decreasing their number, and supplying water through these canals by means of concentrated flows, can be recommended as one of the simplest methods of enhancing the efficiency of irrigation networks.

By increasing their discharge, the transport capacity of the enlarged canals is also increased. This means that the silting-up of small canals supplying water over a large area can be appreciably reduced by increasing the discharge and velocity of the water flowing through such canals. In this way, the reconstruction of irrigation networks first led to fundamental changes in agriculture and also had the favourable effect of reducing the annual amount of work required to clear the canals of sediment, with a corresponding saving on operating expenses and reduction in the cost of irrigated water.

The two results mentioned above are closely connected; the fact is that muddy water flowing through canals clogs the surface area of the bed, thereby decreasing the seepage coefficient, i.e. reducing the water losses. If the surface area of the bed and the banks remains clogged, the over-all water losses are correspondingly lower. Since the canals become silted-up, however, the sediment has to be cleaned out once a year, which involves removing the consolidated superficial layer; consequently, water losses due to seepage rise sharply in canals which have been cleaned out until they become clogged up again.

It is therefore not difficult to see that by passing concentrated flows with their increased volume, the water flowing through enlarged canals may be wholly or partially desilted and may heighten the effectiveness of the irrigation network while at the same time reducing maintenance costs. It is perfectly natural that the method described above should have been extensively applied not only to remodel existing irrigation networks, but also to construct new ones.

In practice, however, it is only possible to increase efficiency by 0.5 to 0.7 by such methods. If the irrigation networks are to become even more effective, other measures are required, aiming at eliminating water losses due to seepage or at least reducing them to a minimum. Appropriate scientific research has been carried out along these lines and its results have been practically applied.

The world-wide method of lining canal banks with massive or prefabricated concrete slabs does not call for any special comment. The first type of lining requires a system of machinery, working along the canals on the principle of a continuous production line. The second type requires the manufacture of ready-made slabs, their transport to the place where they are to be used and the mechanical equipment to set them in place in the bed. Both types therefore need adequate mechanization. A more expensive material—reinforced concrete—is used to reduce the thickness of the slabs and to make them simpler in construction. In other instances, reinforced concrete chutes are laid down on the canal-bed or on special reinforced concrete supports.

This technical solution undoubtedly has great merit and sharply decreases water losses due to seepage; with the chutes on supports, in fact, seepage is eliminated completely. It has not been widely used chiefly for economic reasons such as the shortage of cement, steel, special equipment and means of transport, and above all, because of its high cost.

Concrete and reinforced concrete linings are predominantly used for irrigation works in the Soviet Union in the following cases: in sections of canals passing through high impervious ground such as faulted rock, shingle and sand; on sloping sections (to prevent breaches and landslides); along canal sections with steep slopes to prevent the bank from being washed away; in sections where there are deep excavations to reduce the volume of work, and so on. For instance, reinforced concrete chutes were used for irrigation works in the Georgian and Armenian Republics under very unfavourable topographical conditions. The use of concrete and reinforced concrete can thus be justified in each individual case; its wholesale use is still not widespread, however.

Attempts have been made to find cheaper types of material; asphalt and bitumen have been used for the purpose, but did not prove successful. Local materials such as brick and stone were unsuitable for our conditions, because the cost of labour was too high.

It must be pointed out that, in many cases, the use of any sort of rigid lining has to be ruled out for engineering and geological reasons—one such example was the first irrigation period in the virgin lands, where irrigation had previously never been applied. The fact is that there are regions where the soil reacts to irrigation and gives rise to a series of volumetric deformations, manifesting themselves in the appearance of cracks, craters and the widespread and irregular sinking of the ground. At times, such deformations may assume a dynamic character, whereas at others, they may develop slowly over a period of two or three years or even more. In most cases, both types of phenomenon are to be observed. Generally speaking, these soils are of the loess type, with poor cohesive qualities and contain salts which are readily soluble; they are often macro-porous, aeolian and deluvial-proluvial-type soils with a low density ratio. Under these conditions, great care must be taken to carry out the irrigation works in successive stages and, as far as possible, to set up the regulating installations and main dykes, and to line the canals, if this is necessary, during the period when the soil is becoming relatively stabilized. If the installations or dykes have to be constructed before soil stabilization occurs, their foundations should be thoroughly saturated, thereby provoking the required deformations, so that the soil can be prepared to take the load in the shortest and at the most suitable time. However, provision must be made in the structures themselves for a means of preserving their strength and stability, while making allowance for possible irregular precipitations.

The above remarks on the use of concrete and reinforced concrete as a material for lining canals clearly explain why every country at present carrying out irrigation schemes is attempting to find other, simpler and economically preferable, solutions. This question has been given sufficient attention in technical literature; it formed the subject of a discussion at the Third International Congress on Irrigation and Drainage held in San Francisco in May 1957, during which various ways of combating seepage from canals, most of them of an experimental nature, were mentioned. Chemical means being contemplated include the treatment of the banks with sodium chloride and the grouting of different compounds with the aim of cementing the ground and reducing porosity. Other methods have been proposed; these include thermal processes, consisting of burning the ground, and biochemical methods, where deep, almost impervious and gleyed blankets are made with organic matter. The congress also received information on mechanical methods of consolidating clayey and loamy soils without the need for imported materials.

The last-mentioned method is of the greatest importance for the future as it entails very little expense and is reliable and highly effective. It was first studied in theory and was then used experimentally before being verified under actual working conditions. Special machinery for consolidating soil in canals has been designed and is being mass-produced. It is being used on irrigation works in the Uzbek and Azerbaijan Republics and in other regions.

We would be right to anticipate other recommendations from the chemical industry, which is producing synthetic materials (polymer products) such as plastics, polyethylene and so on. In the Soviet Union, the United States and other countries, experiments are being conducted on the use of these materials in the struggle against canal seepage. In our age of vast discoveries and rapid developments in physics and chemistry, the problem of creating irrigation systems with higher efficiency should find a complete and economically feasible solution in the near future.

We shall examine the question of how to prevent sediment from being deposited in the canals in two parts, dealing separately with the methods used against deposited sediment and with those used against sediment in suspension.

We shall naturally not consider instances where the water is supplied from reservoirs where both liquid and solid flows are stored. However, if the water is supplied direct from low-head dams or from undammed intakes, appropriate measures have to be taken to prevent sediment from being deposited in the canals.

Sediment deposits containing a high proportion of shingle, pebbles, gravel and coarse-grained sand are a normal occurrence in mountainous areas or in the foothills of such areas. The aim is to prevent as much as possible of the sediment from settling in the canals and to leave it in the rivers. Several different ways of solving this problem are known: in some cases, special silt excluders are incorporated in the dams and are washed out and cleaned either hydraulically or mechanically; in other cases, the headwaters are used to hold back the sediment, which is periodically or continually washed away into the tailwater through surface or submerged flushing outlets (evacuating tunnels). There is an even greater variety in respect of such details as the use of sediment control gratings (trashracks and grilles), raising the sill of intake installations and so on.

Soviet research and design organizations have carried out a vast amount of work in this field, and their results could form the subject of a special report. From all the different alternatives, however, we should like to single out one solution which has been conceived and put into practice in the Soviet Union. This is connected with the theory of the transverse circulation of currents, from which the natural relationship between liquid and solid phases in flow dynamics has been determined. By artificially agitating the transverse circulation, it was found possible to control the state of the sediment in the transformed current, diverting it to the sides of the intake from the centre. A number of dams have been designed by this method; mention should be made of the first of them to be completed—the Kampir-Ravatsky and Sar-Kurgansky dams in the Fergan Valley in the Uzbek Republic, which gave rise to the type of dam now known as the 'Fergansky' type. The irrigation canals branching out from these dams are almost all free of coarse-grained sediment (not more than 5 per cent of the total quantity). The finer sediment in suspension passes through the canals.

The problem is more complex in the case of undammed intakes in the middle and lower reaches of rivers with unstable banks. The correct choice of site for the intake is particularly important in such cases. Allowance must be made for topographical and geological conditions. When setting-out and stabilizing the headworks and diverting the canal from the river, sufficient head must be provided to control the irrigated area; hydrological conditions, such as the stability of the banks, flow parameters and the configuration of the bed, must also be considered. The last condition is related to the natural transverse circulation processes occurring in the current. If the installations are set out on the convex side of the current (on the curved bank), more satisfactory conditions will be obtained for the intake, since the sediment flow is directed at the opposite bank, i.e., towards the convex outline of the bank.

In a number of cases it is advisable to strengthen the transverse circulation in front of the intake, and special floating, jet-orientated devices can be placed in front of the installations in the river-bank for this purpose. This system was designed and verified on the Amu-Darya and other rivers. It enables sandy sediment deposits to be appreciably reduced in canals with undammed intakes. In some cases, optimum conditions can be obtained by special levelling-off of the river banks at the intake site.

It is to be noted that each head intake problem has to be solved on its own merits and that the solutions chosen must be corroborated and amended by model tests, usually at full-scale. Model techniques for intake installations are studied in detail in our scientific and research institutions.

The ultimate aim of the methods used to prevent canals from being silted-up by suspended sediment is to make them entirely silt-free; if this aim is to be achieved, the canals must have an adequate transport capacity; designs for silt-free canals are prepared in our scientific research institutions. The surplus sediment can be held back in the silt excluders—theories for which have also been examined—thus enabling silt-free irrigation networks to be accurately designed.

A better technical and economic solution to the problem of deposited and suspended sediment is to be found in the enlarged irrigation networks. The new agriculture and water policy described above has therefore provided hydrotechnology with means of extending the improved systems.

One of the most pressing problems is that of dealing with the salt impregnation of irrigated land, which occurs when the groundwater level is high and when the soluble salt content rises sharply with the capillary flow at groundwater level. Theoretical work on this problem is at present being carried out in all parts of the world and still further efforts will be called for in coming years. The importance of this problem has increased with the widespread development of irrigation and with the accompanying large-scale saline impregnation of land in the vast majority of irrigated areas. The methods used in the Soviet Union to deal with this adverse factor are therefore of interest.

We must first point out that the above comments do not apply to regions which are not threatened by salt impregnation because their natural conditions are satisfactory. Such conditions include the shallow bedding of the draining soil, the existence of natural drains (deep canyons to gently sloping saddles) resulting from water or wind erosion, adequately developed slopes ensuring that the groundwater can flow off from the irrigated area and so on. In these areas, which are satisfactory from the standpoint of land reclamation, the water regime of the soil can only be regularized by means of the downward flow of the irrigated water—by which the transient stocks of moisture in the soil are replenished.

The case of high groundwater levels is somewhat different, since their upward capillary flow can contribute to feeding plant life. Under certain conditions, this additional factor may play an important part in maintaining the water equilibrium of the soil and may appreciably lower water requirements for crops in the growing season; this is very valuable in cases where water resources are deficient.

The irrigation of land threatened by salt impregnation can be dealt with in one of two ways. First, when the groundwater level is rather low and there is still a gap between it and the soil moisture, plant life is fed in the first way, i.e., by the downward flow. In the second case, when the gap resulting from the rise in the groundwater level disappears, the plants are fed in the second way.

In the first case, the problem of keeping the groundwater level stable for irrigation purposes may arise. The special features of each project are taken into account when measures to meet such a situation are instituted. A number of such preventive measures are given below.

For instance, agricultural crops must be rationally selected and sited. One of the fundamental factors to be considered is whether the plants need large or small quantities of moisture; plants such as rice, for example, which need a great deal of moisture, can only grow in limited quantities and be planted in places within the irrigated area which are safe as far as reclamation and hydrology are concerned.

Annual and specific irrigation quotas must strictly conform to plant needs and the irrigation must not exceed the water-holding capacity of the soil.

The operation of irrigation canals must be strictly limited to the period when the irrigated water is actually required; wherever possible it is desirable for the irrigation canals to be shut off and the water allowed to flow away, so that the water equilibrium of the region is not disturbed through having absorbed excess seepage water; the public water supply in these cases should be based on other sources and not on the water supplied through the irrigation canals.

When irrigation networks are being designed, the desirability of enlarging and extending the canals, as well as of reducing their number, should be borne in mind; this factor is also related to the danger of disturbing the water equilibrium.

In regions where water and labour for agriculture are in plentiful supply, it may be temporarily advisable to reduce the land utilization factor to some extent, by rationally selecting and siting the irrigated tracts of land within the entire irrigation network in accordance with hydrogeological conditions.

In view of the shortages pointed out above and of similar measures taken to organize and develop agriculture, greater attention should be paid to the problem of seepage in canals. The most reliable and economic methods should be selected, and should be used on a scale justified by technical and economic considerations. Such methods were outlined above.

The reclamation measures adopted should be based on the principle that the intake and discharge of water is balanced throughout the irrigated area. This should ensure that the basic groundwater regime remains stable. If the groundwater level continues to rise after all the preventive measures have been taken, it becomes essential to change over from the first reclamation method to the second, unless, as a last resort, vertical drainage is carried out on a very limited scale.

This method has been tested in the Soviet Union under varying hydrological and geological conditions. As a result, two criteria for its technical advisability have been formulated; there must be a fairly deep layer of water-bearing soil with good filtering qualities, i.e. with a high discharge, and, in addition, this layer must not be separated by an impervious boundary from the higher lying soil which acts as a protection for the ground layer. In such cases, vertical drainage may prove to be technically efficient; local conditions determine whether it is economically sound.

In changing over from the first method of reclamation to the second, such measures as the choice of crop and its site, the limiting of irrigation quotas, the curtailment of the period during which the irrigation canals operate and the application of rationally designed irrigation networks are all still of importance. They should, however, be supplemented by horizontal drainage, which is of capital importance and has already proved successful.

The construction of drainage channels, organically linked to the irrigation network as a whole, is a universally recognized method of combating water-logging and soil impregnation of the soil. It is a well-known fact that, because of the downward seepage flow, drainage can be used to wash out salt-ridden soils, to evacuate mineral-impregnated groundwater and to supply a larger than usual amount of water having no detrimental effect on plant life.

Horizontal drainage may be of the open-air type, which it is comparatively easy to construct but considerably more difficult to maintain in operation. Covered or tube-type horizontal drainage costs more to construct, but is undoubtedly more advantageous to operate. Open drainage channels were used in the first stages of irrigation schemes, but we have now turned to covered-type drainage as being a more progressive, reliable and profitable method of reclaiming salt-impregnated soils.

In an endeavour to increase the speed at which the salt soils are washed out, the permanent drainage system is often supplemented by temporary, ploughed drainage channels—usually of the open type; these channels are only about one metre deep, but are spaced at fairly frequent intervals.

Thus, the methods used to increase the efficiency of irrigation networks, to decrease silting and to prevent and eliminate water-logging and salt impregnation of the soil, are all based on socialist planning for agricultural and water policy, by which they gain in effectiveness. The methods do not follow a set pattern, but vary with the natural conditions of each region. They also undergo modifications over a period of time, as progress is made in Soviet agriculture, science and economics. Economic developments are well illustrated by the fundamental changes which have taken place in water policies during the Soviet period.

Heavy industry is at the root of the economy of the Soviet Union and was completely remodelled during the early years of Soviet rule. As heavy industry developed, agricultural and hydroproduction were also industrialized. Although in the twenties and thirties, the earthworks for remodelling and extending irrigation networks were largely constructed with manual labour, since the end of the second world war the number of hydro-schemes which have been mechanically equipped has grown rapidly. In the course of the last ten years, mechanization of earthworks for irrigation and water-supply networks has been increasingly introduced and has accounted in recent years for 95 per cent of the over-all volume of earthworks under construction and from 85 to 90 per cent of those in use.

We need not give a detailed description of all the equipment used in the construction of hydro-schemes, such as excavating shovels, scrapers, bulldozers, dredgers and so on. It should be noted, however, that it has been our practice at every stage to concentrate on hydromechanical methods.

When construction equipment was in short supply, widespread use was made of hydromechanical methods in a simple but highly effective form, as will be seen from the following examples: the evacuating capacity of the current was used to construct canals; the transporting capacity of the current was used to carry earth from excavations or quarries to earthfill embankments; pumps were used to increase water velocities where natural slopes were inadequate; blasting was used to loosen the earth so that it would be washed out more effectively.

The hydraulic fill method was used for constructing earthfill embankments (earth dams, weirs, road embankments, approach works for bridges, etc.), the water being held back by small earth dykes placed on the surface of the gradually rising embankment; by using this method, it was no longer necessary for the mass of the embankment to be mechanically consolidated, since a type of soil which rapidly loses its cohesive structure was employed.

Blasting was sometimes also used for such specific problems as making deep excavations by means of disruptive explosions, constructing cofferdams to close off steep river banks by means of limited explosions and so on. All these methods can be successfully used in hydro-schemes when sufficient mechanized means are not available. As hydro-systems become more highly mechanized, the above methods are gradually being supplanted by improved methods which will give a much greater degree of efficiency. In the transition period, the most elementary forms of hydromechanical equipment were used in combination with construction equipment; for example, earth from excavations was transported vertically by manually-loaded conveyors and dumped by means of hydraulic pumps. Still other combinations were used, in all of which a fundamental part was played by the flow of water.

As hydromechanization developed, we began to use hydraulic earth excavators in excavations and quarries, and to introduce dredging on a large scale, with dredgers of various sizes. The manufacture and widespread application of small-gauge dredgers, adapted to the construction and maintenance of medium-sized irrigation canals, is particularly noteworthy. Powerful dredgers are widely used for the silting of earth dams, for filling-in river beds and constructing larger canals. In this connexion, it is worth mentioning the experiment made during the construction of the Kara-Kumsky Canal in the Turkmen Republic. Standard excavating equipment was first used to drive a pilot canal, supplying a small amount of water, from the river towards the middle of the desert; it was then followed by a series of pontoon dredgers, which widened the canal banks to the design dimensions.

Earthworks are particularly important in the creation of hydro-schemes. It is generally recognized, however, that problems arise in constructing hydrotechnical installations because of their varying features and because installations of comparatively small dimensions are dispersed over a wide area. In order to solve this problem on an industrial basis, extensive research was devoted to typical designs which would be suitable for all kinds of irrigation installation. For this reason it was decided to prefabricate a minimum number of typical units and parts for such structures. Factories were set up in areas where hydro-schemes were being carried out to manufacture typical concrete and reinforced concrete units and parts based on these designs. Typical locks, hoist equipment, steel built-in parts and other metal parts are prefabricated in modern, mechanized factories. These hydrotechnical installations are then erected on the actual site from the parts prefabricated in the factory.

Mechanization of operating conditions should also be noted—especially with regard to the irrigation flow. The main feature of our system is to manufacture and use highly efficient machinery to feed the water from the canals or from the pressure conduits; a tractor of 54 to 80 H.P. serves as the basis for such machinery; a double-cantilever water-sprinkling device with an over-all span of 100 metres (i.e. the width of strip irrigated as the machine passes) is mounted on the tractor. This type of machine can supply irrigated water to 100 to 200 hectares per season. Other types of machinery have also been designed and produced; these may have either medium or long-range nozzles, or prefabricated pipes, equipped with short-span nozzles. They can be fed by pumping stations which are either mobile (on the banks or on pontoons) or stationary.

Everything that has been said about irrigation also applies to the simpler task of supplying water and feedstuffs for grazing. Methods of constructing water-supply schemes were initially quite elementary, but have now developed both technically and economically.

For instance, in addition to the mechanically-driven shaft wells with a depth of 20 to 30 metres, bore wells of 100-200 metres or more are now widely used. The water is raised mechanically by means which vary according to the discharge of the wells and bore-holes, to the height through which the water has to be raised, and to the power supply. Side by side with stand-type pumps, electrically-driven immersed pumps have been widely used. For wells where the discharge is particularly small, very simple belt-type hoists—the water being raised by an endlessly moving belt—have proved of value. Solid fuel and the wind are generally the chief energy sources. Experiments are being made on the applications of solar energy.

Areas where feedstuffs are grown are irrigated by the simplest possible method—that of flooding (silt irrigation); the top-water levels of the spring are generally used for the purpose. Experience has shown that shallow, multilayer flooding is preferable to deep flooding. The basic installations for silt irrigation consist of low earth dykes and weirs, which hold back the water collected. The building and maintenance of such structures present no problem.

The irrigation of small tracts of land from artesian wells follows the general trend, i.e. every effort is made to ensure that the best possible and most economic use is made of the water which, in waterless areas, is an expensive commodity.

The storage of rain and flood water in small reservoirs and ponds is widely practised. These reservoirs are used for such widely differing purposes as supplying water and irrigation, breeding fish and wild-fowl, etc. Very simple earth dams are usually built for these reservoirs and may or may not have spillway installations. In the latter case, the volume of the reservoir and the height of the dam are designed for a storage calculated on the wettest years.

To conclude this report on the basic trends and methods applied in Soviet irrigation practice, we should like to single out a number of important problems which should be solved in collaboration with the countries also concerned with such problems: (a) methods of preventing reservoirs from being silted; (b) methods of preventing water-losses in tanks and small reservoirs; (c) the prospection of useful supplies of subterranean water, organizing the control for their use, and methods of artificially replenishing supplies of subterranean water; (d) the purification of mineral-impregnated, surface or subterranean water; (e) methods of combating water-logging and salt impregnation of soils in irrigated areas, the selection of rational reclamation schemes, and methods of hastening the reclamation of salt-marshes and deserts; (f) the use of solar energy in waterless deserts and steppes.

Reprinted from *Water Resources Use and Management,* E. S. Hills, ed., 399–418 (1964)

C11 11

PROBLEMS OF IRRIGATED AREAS

by H. N. ENGLAND*

Summary

The classic problems of large-scale irrigation projects are those presented by rising water-tables and soil salinity—problems created by disturbances of the natural hydrological balance.

Development of shallow water-tables and soil salinity and means of controlling or minimizing their ill effects are discussed generally and in relation to the Australian landscape, particularly that of the riverine plain of south-eastern Australia where major irrigation development has so far been concentrated.

The peculiar features of the Australian landscape—resulting from comparative lack of recent orogenic activity—are responsible for the peculiar character of the Australian irrigation scene, as compared with the rest of the world, and peculiar methods of control of soil salinity without general (sub-surface) drainage.

The conclusion is advanced that the potential for further major project development lies in schemes of a similar character in similar situations—low intensity irrigation without general drainage, on areas of flood-plain deposition in semi-arid regions. The importance of high quality irrigation water—another peculiar feature of the Australian landscape—and especially of rain in the control of soil salinity under shallow, saline water-table conditions is emphasized: under arid conditions drainage is essential.

Water-table and salinity problems are to a large degree disabilities of scale; to this degree isolated small or individual irrigation projects are free of them.

Introduction

Irrigation disturbs the natural hydrological equilibrium of the locality. It is equivalent to an increase in precipitation and clearly will cause increases in other local hydrological processes such as evapotranspiration, run-off, infiltration and deep percolation. A new hydrological equilibrium will eventually be attained.

This paper deals with the problems which arise from the changes in the hydrological environment and with the physical measures aimed at establishing a new equilibrium without ill effects or with minimum ill effects on soil and crop.

It deals especially with the problems presented by shallow water-tables and salinity—the classical problems of irrigation schemes—but not with the politico-social problems, such as resettlement, which arise from physical or economic inability to solve them. Neither does it deal with the effects of

* Water Conservation and Irrigation Commission, Sydney.

shallow water-tables and salinity on soil or plant except insofar as they affect drainage and reclamation measures.

There is no attempt to present a conspectus of Australian irrigation problems region by region. The major problems are outlined and are illustrated by examples from regions more familiar to the author—the riverine plain of south-eastern Australia, especially the Murrumbidgee Irrigation Areas.

Certain peculiar features of the Australian environment bound up with the comparatively recent geological history of this continent intensify some of the problems but ease the solution of others.

The Australian irrigation environment

The Australian landscape is old and broken down in comparison with other continents. This continent largely escaped the intense mountain-building phase that took place on an otherwise world-wide scale in early Quaternary times.

The period of active erosion and of deposition of coarse-grained, unleached materials has long passed. In general Australian rivers in their lower courses meander through extensive flood-plains built up by slow deposition of fine-grained materials already leached and broken down by secular decay. The soils formed from these materials are generally senescent and of comparatively low fertility.

Australia's geological history is not only responsible for its limited water resources, surface and sub-surface, but also for the very limited occurrence of ideal conditions for irrigation—youthful, undifferentiated, permeable, fertile soils, underlain by more permeable material with natural drainage outfalls of high capacity—and with a surface topography suitable for effective control of water supplies and of natural and irrigation run-off. Generally our developed and potential irrigation land is lacking in these attributes.

Our erratic rainfall not only adds to the difficulties of water conservation and the design of irrigation ventures but also the problems with which this paper is concerned.

Because of its climate, particularly its erratic rainfall and semi-insular temperature range, Australia is essentially a pastoral country with pastoral attitudes. Dry-land arable farming generally has a pastoral basis, although the reasons for this extend beyond our pastoral predilections. Irrigation is also predominantly pastoral in character; apart from horticulture and other intensive production such as tobacco, our irrigation has been in general either wholly pastoral or pastoral with incidental cropping. Certain factors responsible for the pastoral basis of dry-land agriculture, such as restoration of soil structure and nitrogen levels, apply with greater force to our irrigation agriculture. In addition, as is shown later, there are certain conditions under which permanent pasture or long pasture courses may be necessary for salinity control.

This form of salinity control demands irrigation water of low salinity. Fortunately in this respect the Australian environment, with its weathered mountain catchments and low ground-water contributions to river-flow, favours irrigation.

The predominantly pastoral character of Australian irrigation is unique.

Elsewhere irrigation is principally used for arable farming or rice-growing. The pastoral character of Australian irrigation is a major weakness economically and there is widespread interest in the development of high-return crops.

Another peculiar feature of the Australian irrigation scene is the use in southern Australia of annual pastures that are irrigated (and grow) only in the cooler months. It is not only the pasture that is unique: in other parts of the world where non-perennial irrigation is practised, the irrigation season is generally the summer.

Development of high water-tables

Irrigation introduces two new sources of deep percolation and ground-water accessions: from channel seepage, and application of irrigation water in excess of crop requirements. In addition accessions from precipitation—of which forms other than rain can be neglected in this context—are increased.

Ground-water accessions from rainfall

Most Australian irrigation is in semi-arid zones with a high variability of rainfall and where rainfall in very wet years may be substantially greater than even dry-year importations of irrigation water. Even in dry years and in semi-desert climates rainfall can result in substantial increases in ground-water accessions as soil-moisture in irrigated land will generally be higher than under natural conditions. Thus at times, particularly when rainfall follows irrigation, infiltration will be greater than is necessary to meet crop moisture requirements.

Substantial losses of horticultural plantings have occurred in the Murrumbidgee Valley (New South Wales) and the Goulburn Valley (Victoria) in wet years such as 1931, 1939 and 1956, but these losses were due only in part to shallow water-tables; a substantial proportion, probably the greater proportion, was due to superficial waterlogging and local flooding.

Fig. 1 shows the relationship of rainfall to the occurrence of shallow water-tables in parts of the Mirrool Irrigation Area, New South Wales. Most of the areas to which Fig. 1 applies are underlain by considerable thicknesses of dense clays; water-tables are perched and fluctuate according to the intensity of irrigation and rainfall. The results presented are derived from water-table surveys which have been carried out each winter in all horticultural land.

The greater frequency of shallow water-tables when high rainfall has occurred in the previous six months is apparent in Fig. 1. It is also noticeable that the net effects of dry and wet seasons are dependent on pre-existing ground-water conditions.

Periods of high rainfall will also affect general ground-water systems although the effects may not be so direct, either in time or magnitude, as in the case of localized perched water-tables. This is illustrated in Fig. 2, which compares rainfall with the behaviour of an observation well in a semi-confined aquifer. The aquifer, consisting of coarse sand and gravel, occurs between the depths of 36 ft. and 52 ft. at the location of Well 101, and is part of a system underlying a large proportion of Yanco Irrigation Area. Prior to irrigation this system was dry, but has since become surcharged.

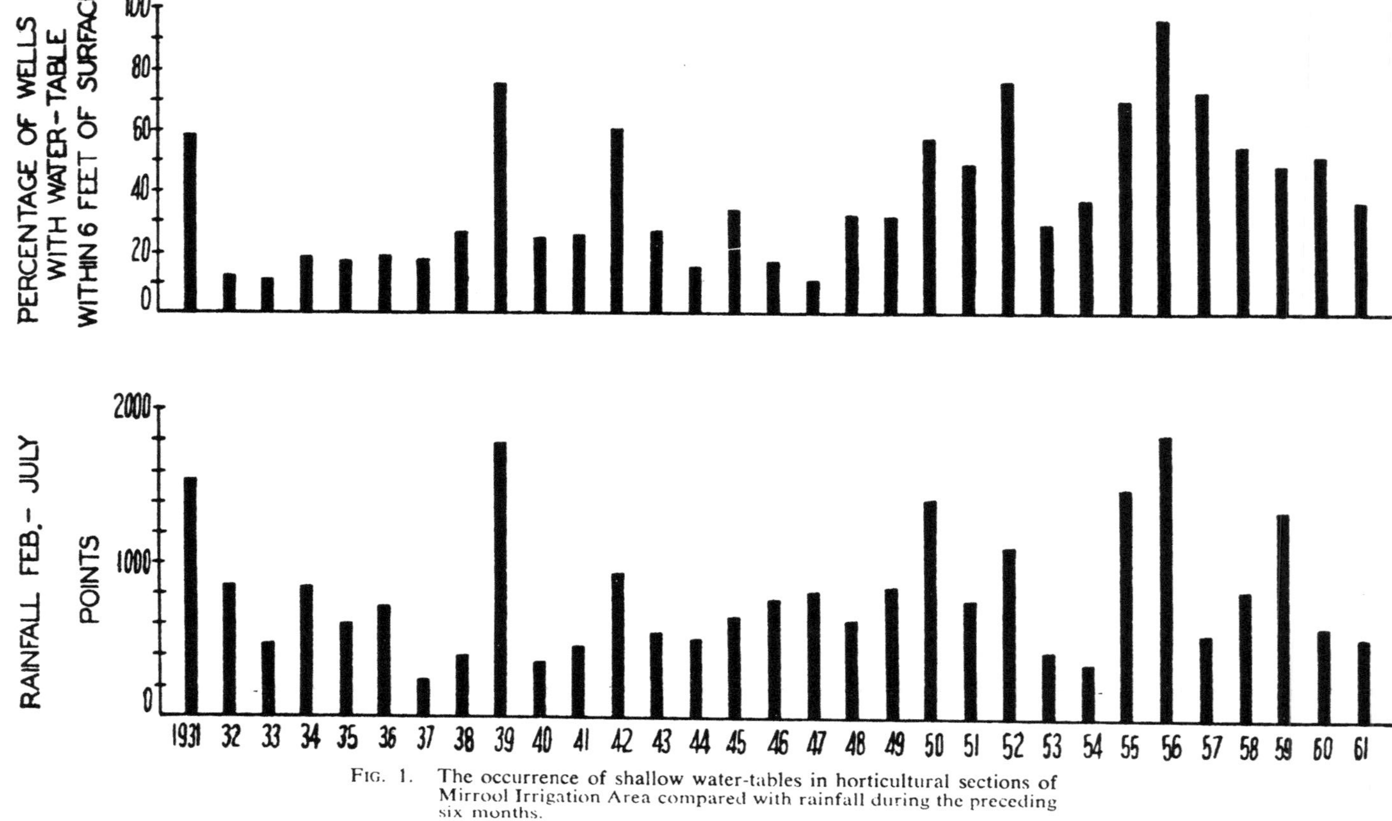

FIG. 1. The occurrence of shallow water-tables in horticultural sections of Mirrool Irrigation Area compared with rainfall during the preceding six months.

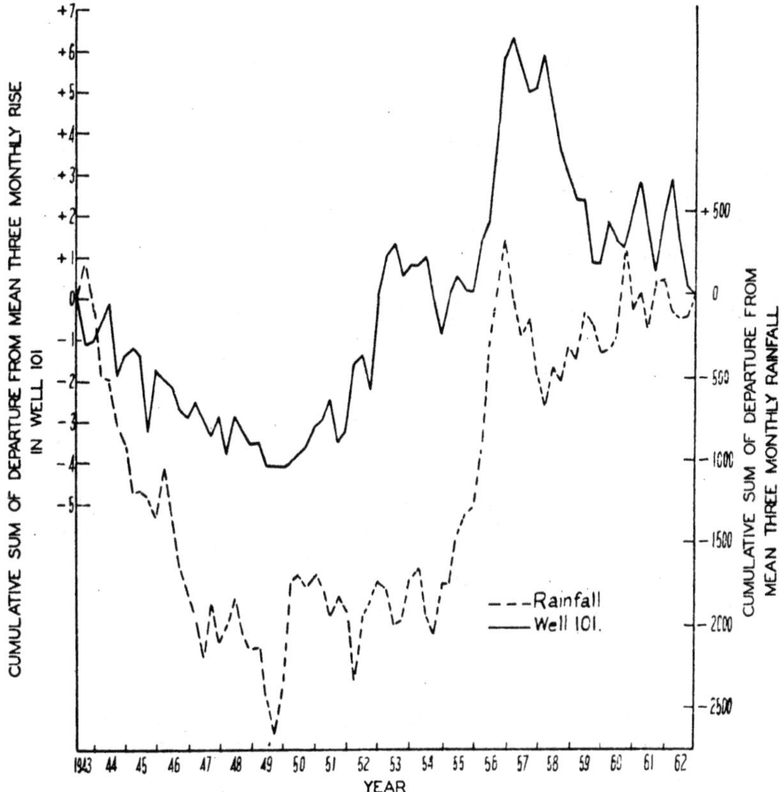

FIG. 2. Residual mass curves for rainfall and pressure level in semi-confined aquifer.

Ground-water accessions from channel seepage

In many overseas irrigation schemes channel seepage is regarded as the major contribution to the ground-water systems. Losses of 50 per cent and more between diversion and farm delivery points are not uncommon; evaporation and transpiration from emergent and marginal weeds can account for only a small proportion of these losses.

Generally in Australia losses from channel seepage are low by overseas standards, for the reason that our soils and underlying sediments are generally of comparatively low permeability.

The effect of channel seepage on marginal water-tables can be readily demonstrated, but measurement of flows in the marginal materials is difficult and expensive.

Ground-water accessions from irrigation

From the plant point of view irrigation is efficient when it does no more than satisfy plant requirements: ostensibly, the objective of irrigation, apart from the special case of rice irrigation, is to satisfy, but no more than satisfy, the crop's moisture requirements by periodic additions of moisture to the soil.

In practice this cannot be achieved, neither by the traditional surface flow methods of application nor by the modern development of sprinkler irrigation.

For surface flow methods there is systematic variation in penetration down the irrigation run, and for sprinkler irrigation there are several sources of systematic variation in the rate or the pattern of application, apart from haphazard variation in pattern due to wind. Unless the whole or part of the crop is consistently under-irrigated, some water will infiltrate beyond the root zone and eventually appear as ground-water.

The special case of rice irrigation, in which water is ponded for long periods, deserves mention. Here soil tensions play but a transient role in infiltration, and the rate and volume of deep percolation may be governed by deeper-seated conditions instead of soil characteristics. For example, the consumption of water for rice-growing is consistently higher in those parts of Yanco Irrigation Area underlain by a shallow aquifer system (within 30 ft. of the land surface) than in other parts where these aquifers are absent and fine-textured sediments extend to much greater depths. Mean consumptions in the two different environments for three recent seasons are given in Table 1. Although the shallow aquifers are now surcharged there apparently is still sufficient underdrainage to permit significant deep percolation during the period of inundation of the rice crop. In earlier years the differences in consumption between the two environments were greater.

TABLE 1

MEAN CONSUMPTION OF WATER FOR RICE-GROWING IN DIFFERENT HYDROGEOLOGICAL ENVIRONMENTS IN YANCO IRRIGATION AREA

Environment	*Mean water consumption (acre-ft./acre)* 1958/9	1959/60	1960/1
Land underlain by shallow aquifers	5·62	6·56	5·94
	*	†	*
Land not underlain by shallow aquifers	5·13	5·61	5·39

* Difference significant at 1% level.
† Difference significant at 0.1% level.

Ground-water behaviour

Increased deep percolation from these three sources will obviously disturb the pre-existent ground-water equilibrium. It will result in higher ground-water levels and, where deep percolation is impeded, the development of new perched ground-water systems.

The conditions regarded as ideal for irrigation, permeable soils and more permeable underlying materials, favour large ground-water accessions from all sources—rainfall, channel seepage and irrigation. Less favourable, i.e. less permeable, soil conditions are often accompanied by still less permeable underlying materials; although deep percolation may be limited this is achieved at the expense of perched water-tables.

Modern engineering has permitted far more drastic interference with the natural equilibrium than was possible in earlier times when irrigation seldom extended beyond control of seasonal flooding. It permits not only perennial irrigation, but perennial irrigation remote from rivers, and large accessions to ground-water systems remote from their natural outfalls.

It is hardly to be expected that natural drainage could cope safely with ground-water accessions from the modern, large perennial irrigation scheme. Probably in no case has an overall ground-water equilibrium been established; for remote outfalls this could require something approaching the geological time-scale.

In general the modern, large irrigation scheme is marked not only by the development in places of shallow water-tables but by continual rise in ground-water levels in extensive ground-water systems and extension of high water-table areas, and by attempts to cope, locally, with these extensions.

Webster (1957) has given several examples of rises in ground-water levels following irrigation in some of the older irrigation districts in northern Victoria, and has stated that in the newer Murray Valley Irrigation District, where the ground-water level was 80–100 ft. below the surface prior to irrigation, "permanent water tables are rising, under irrigation in the intensively developed areas, at the rate of three feet per year".

Specific examples of the development of, and rise in ground-water systems in the Murrumbidgee Irrigation Areas are given in Figs. 3 and 4. Fig. 3

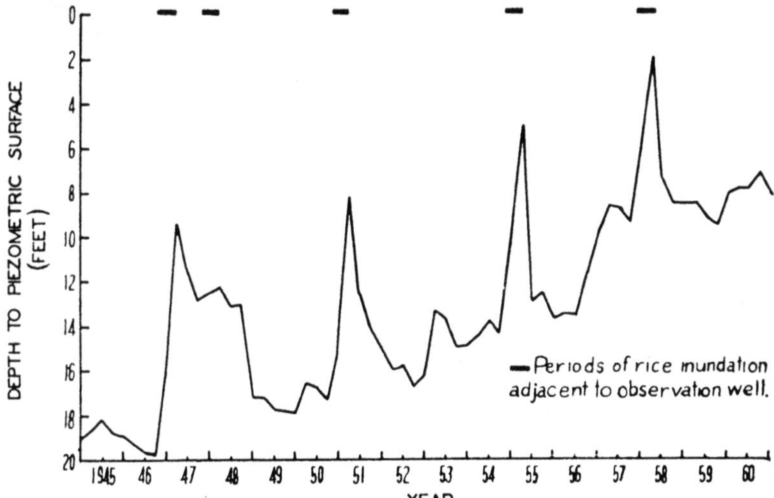

FIG. 3. Continuous rise in pressure in a semi-confined aquifer in Yanco Irrigation Area and effects of rice inundation adjacent to observation well.

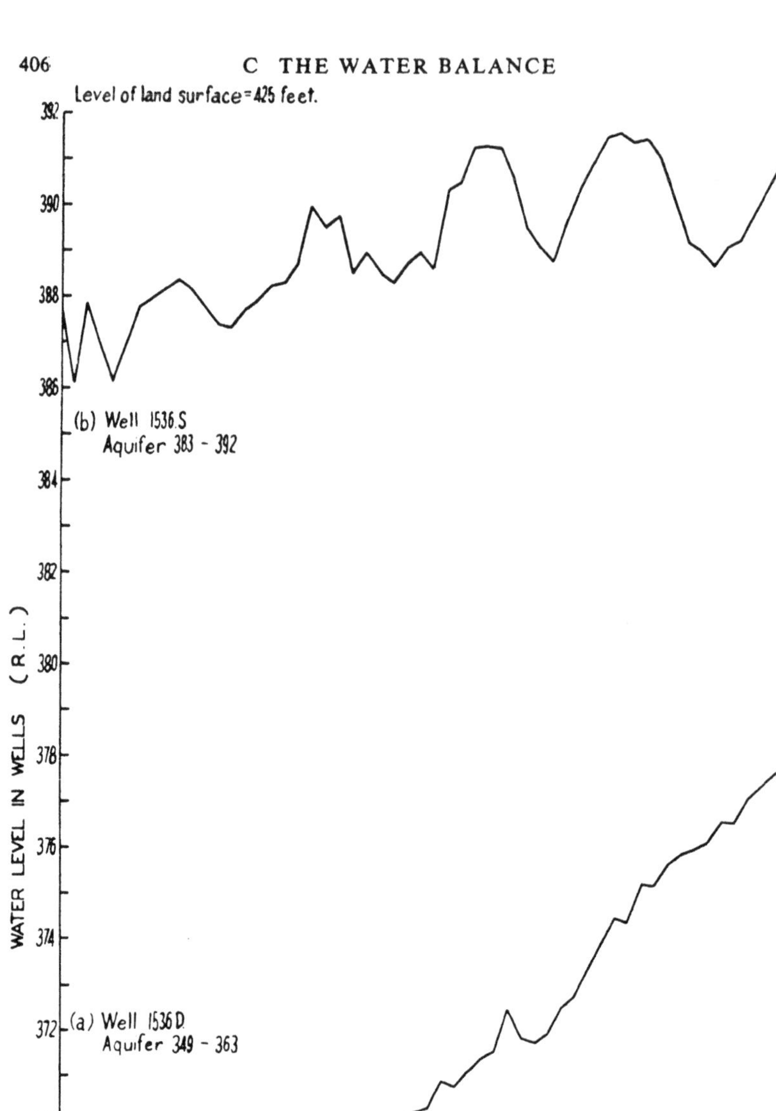

FIG. 4. Hydrographs of two observation wells at the same location showing (a) steady rise in pressure in deeper aquifer, (b) rise and fluctuations of a water-table in previously dry shallow aquifer.

shows the hydrograph of an observation well placed in a six-foot-thick sand stratum which occurs at a depth of 30 ft. in one of the older parts of Yanco Irrigation Area. This aquifer was surcharged at the time observations were first made, but has shown a continuous increase in pressure with time. Superimposed on the continuous rise are very sharp rises which occurred each time a rice crop was grown at a short distance from the observation well.

Wells 1536S and 1536D (Fig. 4) are situated at the same location in a recent extension to Yanco Irrigation Area where irrigation commenced in 1956–7. The shallow aquifer (Well 1536S) was previously dry, but a water-table developed in it within a very short period after the commencement of irrigation. The aquifer is gradually being filled and the water-table shows seasonal fluctuations caused by irrigation. Pressure in the deeper aquifer (Well 1536D) is increasing at a more or less constant rate without the seasonal irrigation effects. The rise is attributed to accessions from the shallow aquifer together with slow lateral movement from adjoining areas which have been irrigated for a much longer period and which are underlain by the same system.

Nevertheless, not all ground-water accessions are undesirable. As discussed later, salinity control demands a deep percolation component either from rain or from irrigation in excess of crop requirements. However it follows that leaching of salts should be as uniform as possible in pattern; that is, channel seepage is wholly undesirable but ground-water recharge from rain—without irregularities from flooding—or from relatively uniform over-irrigation, may at least in part be beneficial.

Soil salinity

Movement of water into and through the soil and underlying materials will obviously result in some redistribution of their virgin soluble-salt content. However leaching is seldom completely effective; effective leaching could be expected only in homogeneous materials, completely devoid of structure on any scale.

Local surface concentrations of salt sometimes appear in freshly irrigated or channelled land by lateral unsaturated movement without the formation of a water-table, but usually this is only a transient phase.

As a result of downward leaching, ground-water bodies built up in irrigated areas as new perched systems, or as superimposed parts of existing systems, usually have some significant content of soluble salts from this source.

In addition, the soluble salts contained in irrigation (and rain) water are concentrated by evapotranspiration and will accumulate at and near the soil surface unless leached downward by irrigation or rain in excess of plant requirements. This is a further source of salts to the ground-waters.

Should a saline water-table rise to within capillary range of the surface, surface accumulations of salts will result.

The salinity patterns that so develop are usually highly irregular in concentration in space and in time, because of irregularity of field conditions and the incidence of a complex of variable, interacting and opposing factors. The main factors governing the dynamics of salt movement under shallow, saline water-table conditions are: the salinity of the ground-water, the depth to the

o

water-table, the potential thickness of the capillary fringe, the capillary conductivity of the soil above the water-table, the rate of evapotranspiration, and the leaching effects of rain or irrigation. At times and in places the net result will be upward movement of water and salt, and at other times and places the flow will be reversed.

An effect of practical significance is that, for a number of reasons, once surface salt accumulation reaches substantial levels the process tends to accelerate and the affected area to enlarge.

However there are conditions, apart from leaching, under which upward movement of water and salt virtually ceases—when the rate of evapotranspiration exceeds the rate of capillary rise. The critical depth of water-table for this situation will obviously vary with soil characteristics and with the rate of withdrawal of moisture from the soil by evapotranspiration, i.e. with meteorological and crop conditions at any particular time.

The factors governing the dynamics of water movement (and thus of salt movement) and the critical water-table concept are discussed at this Symposium by Philip, and Slatyer and Denmead.

The following generalizations conform to theoretical considerations and field observation:

(a) the critical depth to the water-table is greater for coarse-textured soils (i.e. soils with a high capillary conductivity) than in fine-textured soils;
(b) under similar saline water-table conditions salination occurs more rapidly in coarse-textured soils than fine-textured soils;
(c) salination occurs more readily in bare soils than those with a crop cover;
(d) at high water-table levels (above the critical depth range) surface salt accumulation takes place at all seasons of the year and more rapidly in summer;
(e) at lower water-table levels (but still within the critical depth range) the danger period is in cooler weather.

Although the balance is in favour of the fine-textured soils (albeit this is of no significance where water-tables are very high), the coarse-textured soils have the advantage for the reverse process: they are more readily leached and drained.

In the Murrumbidgee Irrigation Areas, where salinity is not a particularly serious problem, obvious signs of surface salt accumulation have mainly developed in the following situations:

(a) towards the bottoms of slopes—usually stony or gravelly hill-slopes, but also the slopes of "soil mounds" remote from the stony hills—presumably as a result of down-slope seepage;

(b) marginally to channels and intensively irrigated areas, especially on land which is somewhat higher in elevation and is not irrigated; such occurrences are more frequent under local perched water-table conditions;

(c) sporadically over areas not irrigated, or seldom irrigated and underlain by surcharged shallow aquifers—within about 30 ft. of the surface—and a general ground-water system associated therewith;

(d) more generally over low-lying areas, irrigated or not irrigated, underlain by surcharged shallow aquifers and an associated general ground-water system; in these situations the water-table is, of course, closer to the surface than in the unaffected adjoining lands.

These occurrences of obvious salt accumulation are not related to original soil salinity levels; indeed salt shows up more readily in the coarser-textured soils whose salt content is generally lower than that of the fine-textured soils.

Further, there is a suggestion that high concentrations of salt at the surface seldom occur without concentration by lateral movement. West and Howard (1953) did not obtain any appreciable salt accumulation in isolated soil columns with a maintained shallow water-table, using a soil in which salt easily accumulates under field conditions. However, except in some of the situations described above, shallow water-tables—within 2 to 3 ft. of the surface—rarely persist for long periods in the Murrumbidgee Irrigation Areas.

To describe the converse position in the Murrumbidgee Irrigation Areas: obvious signs of surface salinity do not usually occur where relief is relatively unbroken, and underlying materials are slowly permeable and probably progressively less permeable to considerable depths. The specific reasons for this are not known and in any case cannot be simple. Ground-water behaviour obviously plays an important part; there is some evidence that in most of these situations there is still appreciable natural vertical drainage.

The salinity patterns that have developed in other parts of the riverine plain of south-eastern Australia are not inconsistent with those in the Murrumbidgee Irrigation Areas. In the Kerang-Cohuna locality in northern Victoria where there is a long-standing salinity problem, and in adjoining parts of Wakool Irrigation District and Tullakool Irrigation Area in southern New South Wales where a salinity problem is developing, most of the severe salting appears to be associated with shallow aquifers. However in both regions salinity is more severe and widespread than in the Murrumbidgee Irrigation Areas: virgin-soil salinity levels were probably higher and for some reason water-tables persist closer to the surface.

In the Sunraysia District of north-western Victoria and nearby parts of New South Wales and South Australia, where mallee lands with an irregular wind-sculptured relief are irrigated for horticulture, local perched water-tables develop rapidly, and salinity, particularly from down-slope and marginal seepage, was a widespread problem before tile drainage was fairly generally installed.

Apart from these spectacular accumulations from persistent, shallow saline water-tables there will be opportunity from time to time for surface accumulations at lower concentrations. The danger periods for such accumulations appear to be immediately following a temporary rise in the water-table from rain or irrigation, or the onset of cool conditions, when capillary rise can reach the surface from an otherwise safe water-table level.

Fig. 5 shows the change which occurred in the salt profile of an unirrigated (roadside) brown soil of heavy texture near Leeton, following the abnormally wet winter and spring of 1956. At April 18, 1956, the water-table was at a depth of 38 in. From May to October a total of over 17 in. of rain fell, and the water-table rose at times to the surface and did not recede below a depth of 25 in. Samples taken throughout and at the close of this period showed salt distributions not significantly different from that of April 15. Ground-water salinity remained relatively constant throughout the period, and apparently the rain prevented any surface salt accumulation.

After the cessation of the wet period, the water-table fell and by January 16, 1957, was at a depth of 46 in. The surface accumulation of salt, as shown for that date, apparently occurred while the water-table was receding, but was still at shallow depths. It will be noted that the final surface concentration of salt, although probably innocuous for most crops, was achieved quite rapidly in spite of the comparatively low capillary conductivity of this fine-textured soil.

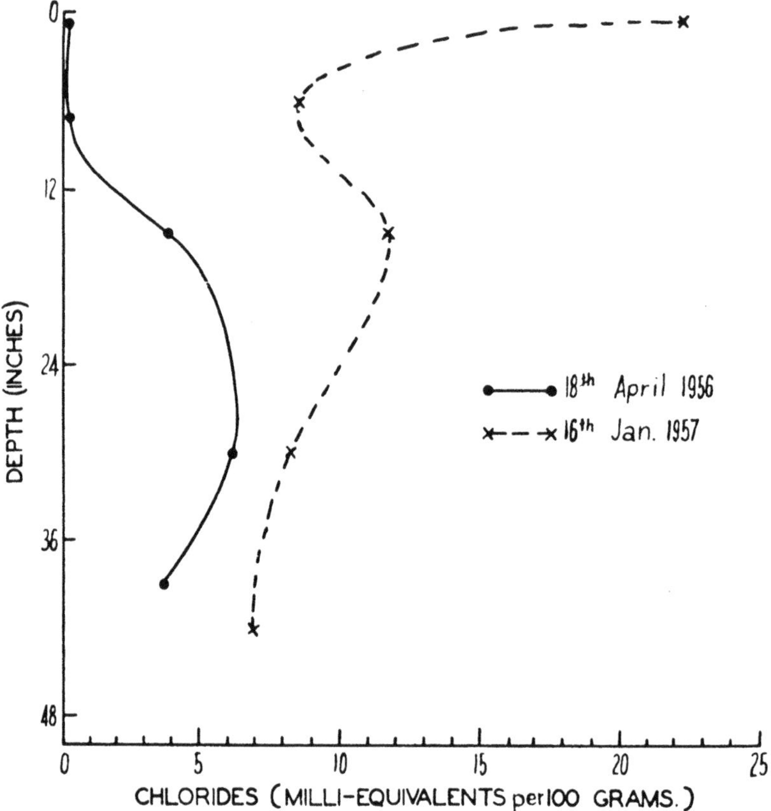

FIG. 5. Salt distributions in a brown soil of heavy texture showing accumulation at surface following a water-table rise during a wet period.

In the Sunraysia District in dry winters, i.e. in the absence of rain or irrigation, there are recurring concentrations of salt in the root zone sufficient to affect citrus plantings (Benton and Fraser, 1946). It seems likely that this results from capillary rise of saline water from water-tables that under normal conditions are at safe levels.

Salinity of irrigation water

In the Murrumbidgee and Murray Valleys from which these illustrations have been taken, irrigation water makes only minor contributions to superficial accumulations of soil salinity. This is for three reasons:

(a) the soluble salt content of Murray and Murrumbidgee waters is very low, about 70 to 80 p.p.m.—salinities of waters of major rivers in other parts of the world are much higher, generally of the next order;

(b) because it is only a semi-arid region irrigation does not have to provide full water requirements;

(c) rain provides significant leaching effects.

Without leaching from rain or excess irrigation it is only a matter of time before the salt content of the root zone reaches dangerous levels from importations alone.

The time required for high concentrations of salt may be quite short. Assuming irrigation water with a salinity of 600 p.p.m. (by no means a high figure by world standards) and a water requirement of 4 ft. to be wholly supplied by irrigation, simple calculation will show that salts imported in twenty years will result in a salt content of 1 per cent in the surface 3 ft. of soil. Since salt is normally irregularly disposed, much higher concentrations could occur in places and in a shorter time.

Where irrigation water has a substantial soluble-salt content, the leaching process is of vital importance; however the leaching component necessary to maintain soil fertility and crop production is not merely a simple function of the concentration of soluble salts in the irrigation water.

Various workers at the United States Salinity Laboratory (Richards, 1954) have shown that the quality of irrigation waters is a complex matter involving both concentration and composition of soluble salts (including toxic elements), a number of soil characteristics, drainage, crop, and management factors. In particular they have shown that under certain conditions the leaching component should reach high proportions of the water added to the soil.

Since this leaching component can be supplied either by rain or applied irrigation water, the following statement can be made:

With increasing aridity there is need for—higher standards of irrigation water, as regards both concentration and composition of soluble salts; higher standards of soils, as regards exchangeable base status and internal drainability; higher leaching components of irrigation applications; and higher standards of drainage, natural or artificial.

Ground-water and salinity control

Although control of soil salinity is conventionally exercised through control of ground-water by drainage, there is some incompatibility between the measures to be taken; as previously pointed out, a leaching component of some magnitude is necessary for salinity control and may reach a high proportion of the applied water. Reclamation of saline soils will demand a higher leaching component, and it is in this respect only that reclamation measures are discussed in this paper.

Control of ground-water may be exercised by reducing accessions at source or by (sub-surface) drainage. Reduction of ground-water accessions can be achieved by surface drainage, by reduction of channel seepage and by improved irrigation efficiency.

Surface drainage

McCutchan has dealt with the need for surface drainage of irrigated lands, more particularly to avoid flooding from storm rains and to dispose of surplus irrigation water. Rapid removal of surplus water from these sources will, of course, also reduce ground-water accessions.

In some Australian irrigation schemes, particularly the older ones, lack of surface drainage or inadequate surface drainage has been a substantial factor in the development of high water-tables, and is still a factor in their maintenance, but in some cases, such as in the Sunraysia District, provision of surface drainage is impracticable in places because of wind-sculptured relief.

In general, provision of surface drainage was not envisaged in partial irrigation schemes because of their low original irrigation intensity but the need—and the difficulties of installation—have become apparent with increasing intensity.

Channel lining

Channel seepage, and thus ground-water accessions, may be reduced by a variety of means, including lining with artificial materials such as cement concrete, with uncompacted clay, or by consolidating native or imported earth materials.

Irrigation efficiency

Ground-water accessions should be reduced to the minimum requirement for salinity control. This is difficult and perhaps impracticable of achievement by surface flow methods when relief and soil disposition are irregular as in parts of the Sunraysia District.

Drainage

The conventional means of controlling high water-tables to avoid damage to susceptible crops, of controlling surface salt accumulations to avoid damage to both soil and crops, and of reclaiming saline land, is by (sub-surface) drainage.

The main criteria of drainage—the minimum depth to which the water-table is to be reduced and the minimum rate of this recession—are governed by a complex of factors, but, these factors being constant, can differ with the aim of drainage—whether control of waterlogging with incidental control of salinity, control of salinity with incidental control of waterlogging, or reclamation of saline land.

As instances, protection of crops highly susceptible to waterlogging in the root zone requires a high rate of recession of the water-table, but if the salinity risk is slight or absent, the minimum depth to which the water-table is to be reduced may be less than if the salinity risk is high. However, if salinity control is the main consideration, i.e. the crop is not particularly susceptible to waterlogging in the root zone, the minimum rate of recession could be reduced. Reclamation may require a higher rate of recession than is necessary for prevention of salinity, either absolutely or to speed up the reclamation processes.

The answers to the questions of whether drainage is physically practicable and of the particular type of drainage measures that might be considered are dependent on the hydraulic properties of the soil to be drained, particularly its hydraulic conductivity, and on all those factors governing ground-water behaviour in the system of which the soil to be drained forms a part. The physical environment to a considerable extent determines the particular type of drainage to be employed, or at least limits the choice. However, all drainage measures are expensive and the choice of whether or not to drain, and what type of drainage to employ, are ultimately economic problems.

Tile drainage and field drainage. These are the commonest forms of drainage. In Australia, the use of tile drainage is virtually confined to horticultural lands, and field drainage is not used, although deeper parts of surface drainage systems may have marginal effects on high water-tables. Virtually, the only limitation (and this in the main is economic) on the use of tile drainage or field drainage is that the soil at and around the level of the drains should be reasonably permeable.

There are slight differences in water-table behaviour between field drainage and tile drainage, mainly those concerned with entry into and flow through the tile lines.

Design of tile drainage systems has been greatly facilitated in recent years by development of means of measuring hydraulic conductivity of the soil *in situ*, e.g. Maasland and Haskew (1957).

Field drainage requires stable soils so as to avoid waste of land, but under the best conditions waste of land and maintenance costs are considerable; a further disadvantage is the limitation imposed on the use of agricultural machinery. For these reasons it has been replaced by tile drainage in most advanced countries.

The main regions where tile drainage is used in Australia are the Sunraysia District—mainly for control of salinity—and the Murrumbidgee Irrigation Areas—mainly for control of shallow water-tables, especially shallow water-tables resulting from heavy or persistent rains.

Mole drainage. Since there is usually some salinity risk with irrigation, and the depth of mole drainage is generally too shallow for salinity control, this form of drainage has little application to irrigation conditions.

Tube-well drainage. In this form of drainage, water is pumped from underlying aquifers and vertical drainage of the soil follows reduction of pressure in the aquifer.

Since operational costs in pumping water from considerable depths are high, the aim is to provide the greatest area of pressure reduction—and of vertical drainage—per unit volume of water pumped. There are thus two conflicting requirements: a wide radius of pressure relief per unit volume pumped can be achieved only in an aquifer with a high degree of confinement, which precludes rapid vertical drainage of the soil; and, conversely, under water-table conditions or with only a minor degree of confinement of the aquifer, large volumes of water are pumped relative to the area of influence. The conditions for low-cost tube-well drainage rarely occur—a degree of confinement which will permit a wide radius of influence on the one hand, and reasonably rapid adjustment of phreatic levels to piezometric levels on the other hand.

Further, unless aquifer conditions are more or less uniform, location of drainage points by test drilling and test pumping is an expensive procedure. Also, under these conditions, the pattern of drainage points must be advanced experimentally; although test-pumping techniques yield otherwise sufficient information on aquifer characteristics, they do not disclose the shape of the area to be drained.

For these reasons, tube-well drainage has not been commonly used. In Australia, it has been used successfully by the Water Conservation and Irrigation Commission in the Murrumbidgee Irrigation Areas for drainage of some horticultural lands, commencing with an experimental installation in 1945, and recently by the State Rivers and Water Supply Commission for reclamation of saline lands in the Central Gippsland Irrigation District (Webster, priv. comm.).

An extensive programme of tube-well drainage—apparently under water-table conditions—is contemplated by the Pakistan Government in the Indus Valley (West Pakistan Water & Power Dev. Auth., 1961).

In some irrigated regions of south-western U.S.A., effective drainage is provided by tube-well systems whose function is to supplement surface water supplies, although in some cases the original purpose was drainage. In some places, the ground-water thus used is not the irrigation recharge but is drawn from deeper levels where salinity is lower.

Weeping-well drainage. This is a special case of tube-well drainage in which systems of tube-wells tapping a surcharged aquifer discharge freely into deeply-excavated open drains. Weeping-well drainage is being used in conjunction with tube-well drainage in the Central Gippsland Irrigation District (Long and Webster, 1960).

Drainage into underlying aquifers. In its simplest form, this is another special case of tube-well drainage in which water from a high-level aquifer is permitted to discharge into a low-level aquifer by means of a tube-well screened at both levels. It may also be used to discharge the effluent from other forms of drainage, such as tile drainage. Obviously, it requires special hydrogeological conditions, including a high discharge capacity of the low-level aquifer.

It was used to some extent in the Mildura district (Victoria) until pressures rose to high levels in the underlying aquifers (Lyon and Tisdall, 1962). It is being used successfully for discharge of tile-drain effluent in the Waikerie district (South Australia) where highly porous low-level aquifers are cut into the deeply incised gorge of the River Murray.

Disposal of drainage waters

Disposal of drainage waters, which are invariably more saline than the original irrigation waters, presents a problem. Their re-use, undiluted or diluted in the supply system, or by return to the river, demands still higher standards of salinity control.

Salinity control without drainage

In parts of northern Victoria, especially the irrigation districts around Cohuna and Kerang, permanent pastures are being successfully grown without drainage in spite of the presence of persistent saline water-tables at very shallow depths—2 ft. to 3 ft. or even less from the surface. Lyon and Tisdall (1942) stated as long ago as 1942 that, "in the Kerang district, improvements in the farm practices have now demonstrated that reclamation may be secured by improvements in agrostology, and irrigation methods". Morgan and Garland (1954) also have shown at Swan Hill that it is possible to reclaim soils which have become intensely salinized, by the establishment of permanent pastures following several annual crops; this was done without lowering the water-table and with a reduction in salinity of only the top six inches of soil.

It appears that many fields of permanent pasture have been successfully maintained for thirty years or so, despite the shallow saline water-tables. On appearances, and especially as reclamation has been achieved without drainage, this form of land-use seems to be permanent.

It is possible that the water balance and the salt balance are assisted by some natural under-drainage. Also, it is noteworthy that areas of intensely salinized lands—usually in low-lying situations—and even swamps, occur throughout the high water-table regions and the possibility arises that these may assist in maintaining the balances in the irrigated lands: that is, there is a possibility that salinity control in parts may be achieved only by the sacrifice of other parts, which in general are less suitable or unsuitable for irrigation.

In any case it is clear that the low salinity of the irrigation water and the substantial winter rains are major factors in salinity control, even if salinity balance is achieved on only a local scale and not throughout the groundwater province.

The disadvantage of this form of salinity control is that land-use is rigid or at best limited. In situations where the water-table is within the critical range for long periods of the year, the permanent pasture cover must be maintained; exploitation of built-up fertility by arable cropping and grading or reconstruction of faulty irrigation layouts are out of the question.

There could also be situations in which the water-table is at critical levels only in the cooler parts of the year. Here it would not be essential to maintain the permanent pasture cover and winter crops and pastures could be grown, but not irrigated arable summer crops. In general this would be a lower form of land-use than grazing of permanent pastures, and probably would be attended by some surface salt accumulation requiring periodic reclamation.

There is only one irrigated summer crop that can be grown safely with a saline water-table within capillary range of the surface, and that is rice—the traditional reclamation crop.

In the Murrumbidgee Irrigation Areas, the Commonwealth Scientific and Industrial Research Organization and the Water Conservation and Irrigation

Commission are collaborating in a research project aimed at devising for lands underlain by persistent, shallow saline water-tables, systems of land-use involving periodic rice-growing in rotation with winter-growing pastures and crops, and of defining the environmental conditions under which these systems may be practised without undue salinity risk. Here again, essential factors are low salinity of irrigation water and substantial winter rainfall. Others that may be necessary for permanence are some measure of natural under-drainage and sacrifice of low-lying (generally unsuitable) lands.

Neither of these systems is a deliberately designed salinity control measure. Dairying on permanent pasture, or mainly permanent pasture, is the main form of land-use in the Cohuna–Kerang region of northern Victoria. In the Murrumbidgee Irrigation Areas, where the rice-pasture rotation is standard practice (apart from horticultural farms), experience to date suggests that there are good prospects of salinity control under this system and the main objectives of the research project are to devise modifications of the system to suit particular environmental conditions.

There are good reasons why similar systems have not been evolved in other parts of the world. Apart from the requirements of irrigation water of low salinity and a semi-arid climate with some effective winter rainfall, pastoral irrigation is rarely practised elsewhere because of its comparatively low returns. Further, these forms of salinity control, especially the rice-pasture rotation, depend on low capillary conductivities of the soils, and in many parts of the world such soils are regarded as unsuitable for irrigation, in part because of their hydraulic properties, and in part because of their relative unsuitability for arable farming.

Although this paper is largely concerned with ground-water and salinity problems, it is deemed advisable to describe the conditions under which so far no serious salinity problems have arisen in Australian irrigation schemes, in spite of greatly impeded vertical drainage.

The conditions are: soils of varying surface texture, but with fine-textured subsoils underlain to very considerable depths by dense clays of apparently decreasing permeability, and at deep levels by aquifers with a high degree of confinement. These conditions apply over much of the Murrumbidgee Irrigation Areas and apparently over much of the riverine plain of south-eastern Australia where most of our irrigation is located.

In this environment in the Murrumbidgee Irrigation Areas, local high-level perched water-tables are readily developed with intense irrigation or heavy rainfall. With less intense irrigation and the absence of heavy rains, the ground-water dissipates. The processes involved in dissipation have not been followed, but the volume of (free) water involved is slight, as the drainable porosity of these dense sediments is very low. There has been some rise in pressure levels in the deep-seated aquifers, but observations are very limited and it is not known whether accessions to the aquifer are on a general or a local scale.

It may be of significance that there is no opportunity for concentration of salt by lateral movement of ground-water: because of the low permeability of the subsoils and deeper sediments, movement of water and salt must be largely vertical. Also, the water-table cannot be maintained from remote intakes.

It remains to be seen whether freedom from surface salinity and persistent high water-tables will be permanent, but on appearances there are prospects of permanence.

The hydrology of this type of environment needs study but not merely because of existing irrigation projects. The conditions are typical of mature flood-plains and are probably typical of most of this country's potential for major irrigation projects.

Individual and small irrigation projects

So far discussion has been concerned mainly with the problems of major irrigation projects, in Australia usually Government-controlled schemes, but in some cases co-operatively controlled. However, a considerable proportion of Australia's irrigation is in the form of individual or small group projects, usually by pumping from rivers under some form of licence from the water-supply authority.

Many of the problems discussed are problems of scale. Obviously, the greater the area under irrigation in any ground-water province, the greater the tax upon the discharge capacity of the ground-water system.

Small isolated irrigation projects are relatively free from these disabilities of scale. Generally they are free from ground-water and salinity problems except under conditions conducive to development of local perched water-tables, down-slope and marginal seepage. Further, they are often underlain by free-draining aquifers connected with the river from which they draw their supplies.

Conclusions

There are good reasons why Australia's major irrigation projects have been developed as comparatively low intensity irrigation schemes, largely pastoral in character, in areas of flood-plain deposition in semi-arid climatic zones and without widespread drainage.

The environment generally and the low irrigation intensity minimize soil salinity troubles, but as shallow water-tables develop, the range of land-use, particularly arable cropping, is restricted.

The permanence of this type of project has yet to be demonstrated, but so has the permanence of major irrigation projects in river valleys where the physical conditions will permit, and eventually demand, general drainage—with the increasing problem of disposal of saline drainage waters.

Our potential for further major irrigation projects seems to be largely restricted to schemes of a similar character in similar geomorphological situations in semi-arid regions where high-quality water is available. In such projects drainage may be essential only for particular crops, such as horticultural ones, or for special soil and ground-water conditions.

In arid climates high standards of drainage, natural or artificial, are essential.

Ground-water and salinity problems are largely problems of scale. Isolated small-scale and individual irrigation ventures largely avoid these problems.

Acknowledgments

The invaluable assistance of the author's colleague, Mr. S. E. Flint, in the preparation of this paper is gratefully acknowledged. The author is also indebted to the State Rivers and Water Supply Commission and its staff for information gained from time to time in the field on water-table and salinity problems in the irrigation districts of northern Victoria.

REFERENCES

Benton, R. J., and Fraser, L. R. (1946). Unpubl. rept., N.S.W. Dept. of Agric.

Long, A. A., and Webster, A. (1960). "Surface and subsurface drainage investigations in the Central Gippsland Irrigation District Victoria", *Proc. 4th Cong. Int. Comm. Irrig. & Drainage*, 11.97–11.113.

Lyon, A. V., and Tisdall, A. L. (1942). *Production of dried grapes in Murray Valley irrigation settlements: 2. Irrigation, drainage and reclamation*, Coun. Sci. Industr. Res., Aust., Bull. No. 149.

Maasland, M., and Haskew, H. C. (1957). "The auger hole method of measuring the hydraulic conductivity of soil and its application to tile drainage problems", *Proc. 3rd Cong. Int. Comm. Irrig. & Drainage*, 8.69–8.114.

Morgan, A., and Garland, K. R. (1954). "Salt-land reclamation on the Swan Hill flats", *J. Agric. Vict.*, **52**: 268–70.

Richards, L. A. (ed.) (1954). *Diagnosis and improvement of saline and alkali soils*, U.S. Dept. Agric. Hdbk. No. 60.

Webster, A. (1957). "Drainage in the riverine plain of Northern Victoria, Australia, with special reference to ground-water hydrology", *Proc. 3rd Cong. Int. Comm. Irrig. & Drainage*, 10.27–10.42.

West, E. S., and Howard, A. (1953). "Movement of salts in isolated soil columns", *Aust. J. Agric. Res.*, **4**: 82–7.

West Pakistan Water and Power Development Authority (1961). "Programme for water-logging and salinity control in the irrigated areas of West Pakistan".

Reprinted from *Trans., Amer. Geophys. Union*, **37**, 287–296 (1956)

Subsidence of the Land Surface in the Tulare-Wasco (Delano) and Los Banos-Kettleman City Area, San Joaquin Valley, California

12

J. F. Poland and G. H. Davis

Abstract—Releveling of bench marks in 1953 and 1954 by the U. S. Coast and Geodetic Survey indicates that subsidence of the land surface has now exceeded ten feet in two areas of the San Joaquin Valley. In the Tulare-Wasco (or Delano) area of Tulare County, subsidence which was as much as five feet in 1940 now has about doubled. The maximum rate of subsidence in recent years has been about 0.8 foot a year. In the Los Banos-Kettleman City area of western Fresno County, major subsidence extends from Ora Loma on the north beyond Huron on the south, a distance of 70 miles or more. The maximum rate there approaches one foot a year. Plots of subsidence against decline in artesian pressure suggest that pressure decline is a major cause of the subsidence. Compaction of the soil after irrigation is known to have caused substantial local subsidence in the Los Banos-Kettleman City area, and tectonic adjustment and other causes also may have contributed to the subsidence.

Introduction

Occurrence of land subsidence—For many years substantial subsidence of the land surface has been occurring at several places in California, in other parts of the United States, and in foreign lands. In California, subsiding areas that have received the most attention are the San Jose area in the Santa Clara Valley [*Tolman* and *Poland*, 1940] and the Wilmington area of Los Angeles Harbor [*Gilluly* and *Grant*, 1949], but the subsidence observed for the longest time (since 1922) has been in the peat lands of the Sacramento-San Joaquin Delta [*Weir*, 1950]. The principal areas of known subsidence in California are shown on Figure 1. Elsewhere, possibly the best known subsidences are in the Houston-Galveston area of Texas [*Winslow* and *Doyel*, 1954] and in the Mexico City area [*Zeevaert*, 1949]. A lesser subsidence has been described in the Lake Mead area [*Longwell*, 1954], and local substantial subsidence in the Heart Mountain Division of the Shoshone Irrigation Project in Wyoming [Swenson, in preparation].

Subsidences in the vicinity of many oil fields also have been described. Of these, the Wilmington subsidence, those in the Goose Creek and Sour Lake oil fields in Texas, and the one in the Lake Maracaibo area of Venezuela probably are the best known. In addition, minor subsidences have been noted at many California oil fields, such as Signal Hill, Huntington Beach, and Santa Fe Springs.

Land subsidence in the San Joaquin Valley—Subsidence of the land surface in the San Joaquin Valley was first described in 1941 by *Ingerson* [1941] who reported briefly on the subsidence in the Delano (Tulare–Wasco) area. Since that time subsidence has been observed in at least two additional areas: The Los Banos–Kettleman City area and the Arvin–Wheeler Ridge–Maricopa area. Information on the magnitude of these subsidences is available chiefly through first-order leveling and releveling of the U. S. Coast and Geodetic Survey, but supplementary leveling is available from several other agencies, especially the United States Geological Survey, the Bureau of Reclamation, and the California Division of Water Resources. Control is best in the Tulare–Wasco area and poorest in the Arvin–Wheeler Ridge–Maricopa area. These principal areas, in each of which several feet of subsidence is known to have occurred, are outlined on Figure 1.

This paper presents information on the status of subsidence in the Tulare–Wasco (Delano) and Los Banos–Kettleman City areas, the two areas of greatest subsidence, as of 1954. In each of these areas, subsidence is known to have exceeded ten feet. Figures 2 and 3 show the extent of subsidence in the Tulare–Wasco area from 1948 to 1954 and in the Los Banos–Kettleman City area from 1943 to 1953. These subsidences occur in irrigated areas traversed by costly canals of large capacity and low gradient, and additional trunkline and distribution canals are proposed. The extent, magnitude, and rate of subsidence and amount of future subsidence (or recovery of land surface) are of paramount importance in planning construction or repair of such structures.

Subsidence in the Tulare – Wasco (Delano) Area

General features—The Tulare – Wasco area is in the southeastern part of the San Joaquin Valley, chiefly in Tulare but extending south into Kern County. The town of Pixley in the northern part of the subsidence area is roughly equidistant from

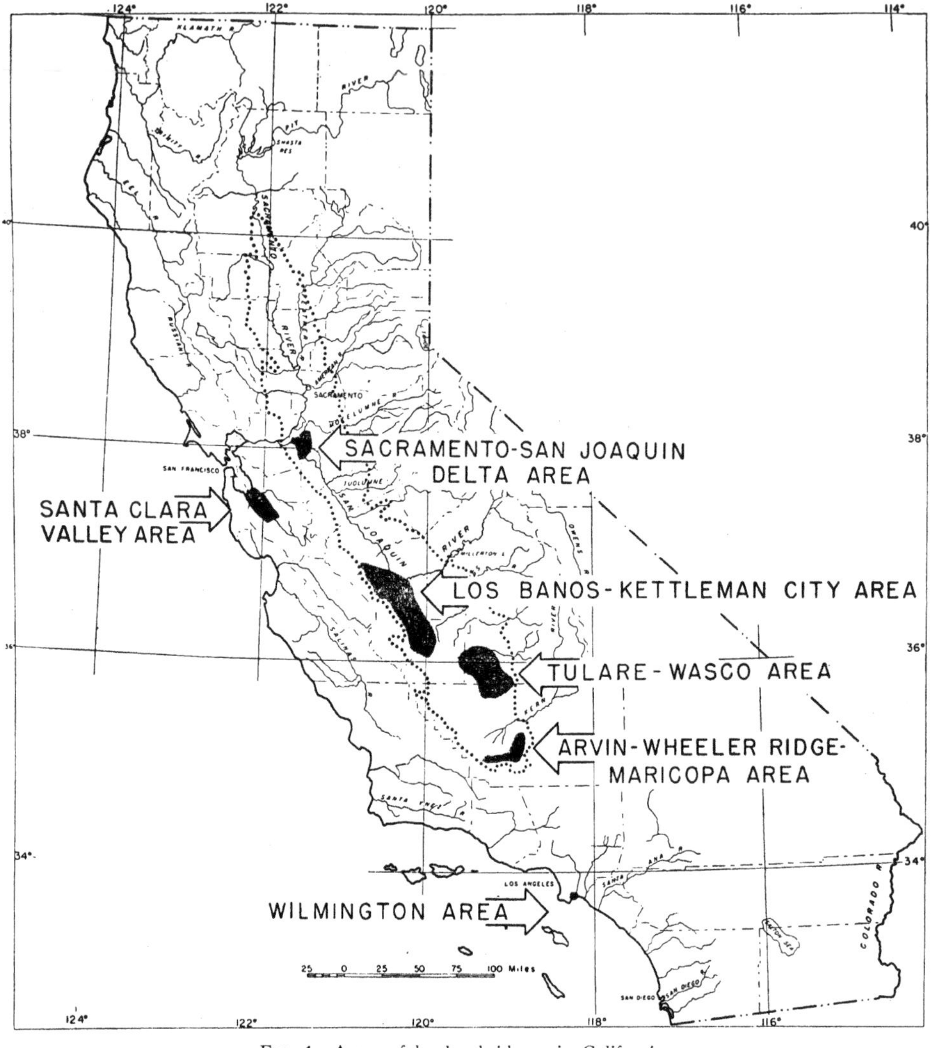

FIG. 1 – Areas of land subsidence in California

Fresno and Bakersfield, and the area of subsidence is nearly bisected by U. S. Highway 99 and the Southern Pacific railroad. Delano is the largest city within the area of substantial subsidence. The area is agricultural and much of it is irrigated, until 1950 almost wholly from wells but since then in part by water imported from the San Joaquin River through the Friant – Kern Canal.

Wells have been used for irrigation in Tulare and Kern Counties since about 1885 [*Mendenhall* and others, 1916], and much of the subsiding area is irrigated by water from wells at the present time. Because of increased pumping demand and the deficient local supply, water levels have declined for many years. In spite of the scope of the ground-water development, however, the geology and hydrology of the Tulare – Wasco area are not well known.

In a field examination of wells in 1905–07 *Mendenhall* and others [1916, pl. 1] found that the area of flowing artesian wells extended east from Tulare Lake about to a line through Tulare, Pixley, and Earlimart, about two miles west of Delano and three miles west of Wasco. Thus, even in 1905 it was known that an extensive area of confined water underlay the western part of both counties. Recent work by the Bureau of Reclamation and others has shown that the principal confining bed is a diato-

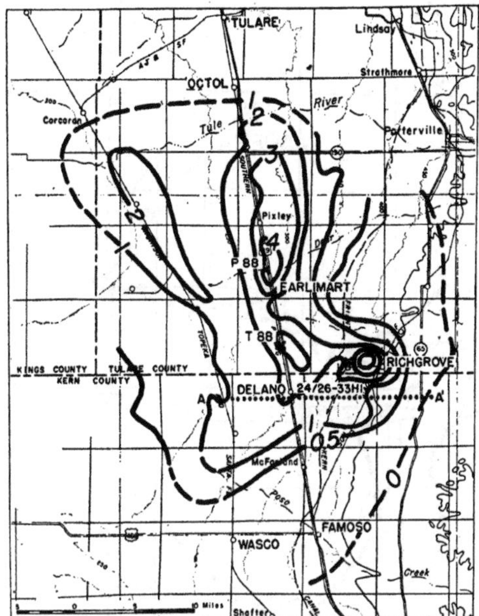

FIG. 2 – Land subsidence, Tulare-Wasco (Delano) area, 1948–54 (lines of equal subsidence in feet)

maceous clay at a depth of several hundred feet below the land surface. This clay, which is the principal confining bed beneath several thousand square miles of the San Joaquin Valley, separates unconfined and semi-confined water above from confined water beneath; it extends about a mile east of Earlimart and Delano. The separation of water bodies extends farther east, however, possibly owing to the presence of other confining beds in the section tapped by water wells.

Water levels in the Tulare – Wasco area have declined substantially from 1905 to date. For example, at Delano the 1905 water-level altitude was approximately 285 ft above sea level. In 1921 it was about 275 ft, but by 1952 the water table had declined to about 190 ft and the pressure head in deeper zones was at about 160 ft. At Richgrove in 1905 the altitude of the water level was about 340 ft, in 1921 it was about 325 ft, and in 1952 it was about 110 ft above sea level. Thus, there had been an overall decline of head in confined aquifers of about 125 ft at Delano and 230 ft at Richgrove, very largely since 1921.

Subsidence—Subsidence of the land surface in the Tulare – Wasco area was first described in print by *Ingerson* [1941, p. 40–42], who noted that in 1935 I. H. Althouse, Consulting Engineer, Porterville called attention to the possibility of a definite subsidence of land surface in the vicinity of Delano. The map prepared by Ingerson indicating lines of equal subsidence from 1902 to 1940 showed the one-foot subsidence contour passing through Pixley and McFarland and a maximum subsidence of five ft half a mile east of U. S. Highway 99 (Southern Pacific R. R.) and three miles north of Delano. Profiles of subsidence and of depth to ground water along the Southern Pacific R. R. were presented.

Precise leveling in the area was carried out first in 1902 by the U. S. Geological Survey. A first-order line was established in 1930–31 by the U. S. Coast and Geodetic Survey along the Southern Pacific R. R., and this line and others were releveled by the same agency in 1940 and 1943. Because these relevelings and levels run by other agencies showed continuing subsidence, at the request of several agencies the U. S. Coast and Geodetic Survey in 1948 established a level net designed to give reasonably close control in the area of known subsidence. This control net was releveled between December 1953 and March 1954. Subsidence of the land surface as indicated by changes of elevation of bench marks from 1948 to 1954 is shown by lines of equal subsidence on figure 2. The area enclosed by the one-ft subsidence line approximates 450 sq mi. The subsidence during the six-year period ranged from nothing south of Famoso and along the eastern margin of the valley to more than four ft between Earlimart and Pixley and in an area west of Richgrove. Figure 2 shows also the location of profiles, bench marks, and wells to be discussed.

Changes in elevation of bench marks along the Southern Pacific R. R. from Octol to Famoso for the period 1902 to 1954 are shown on Figure 4. Levelings in 1902, 1931, 1940–43, 1948, and 1954 indicate that the maximum subsidence along the line has been 10.2 ft since 1902 at bench mark T 88, about four miles north of Delano, although the maximum subsidence from 1948 to 1954 was recorded at bench mark P 88, between Pixley and Earlimart. Subsidence at P 88 totaled 4.7 ft during the six-year period, indicating an average rate of about 0.8 ft per year.

Subsidence of the land surface and change in head in the confined aquifers along line A-A′ (Fig. 2) are compared on Figure 5. The lower profiles show the altitude of the piezometric surface of the confined water along most of line A-A′ in 1948, 1952, and 1954. Comparison of the profile for 1952 with that for 1948 indicates decline in head all along the profile, but comparison of the profile for 1954 with that for 1952 shows substantial recovery

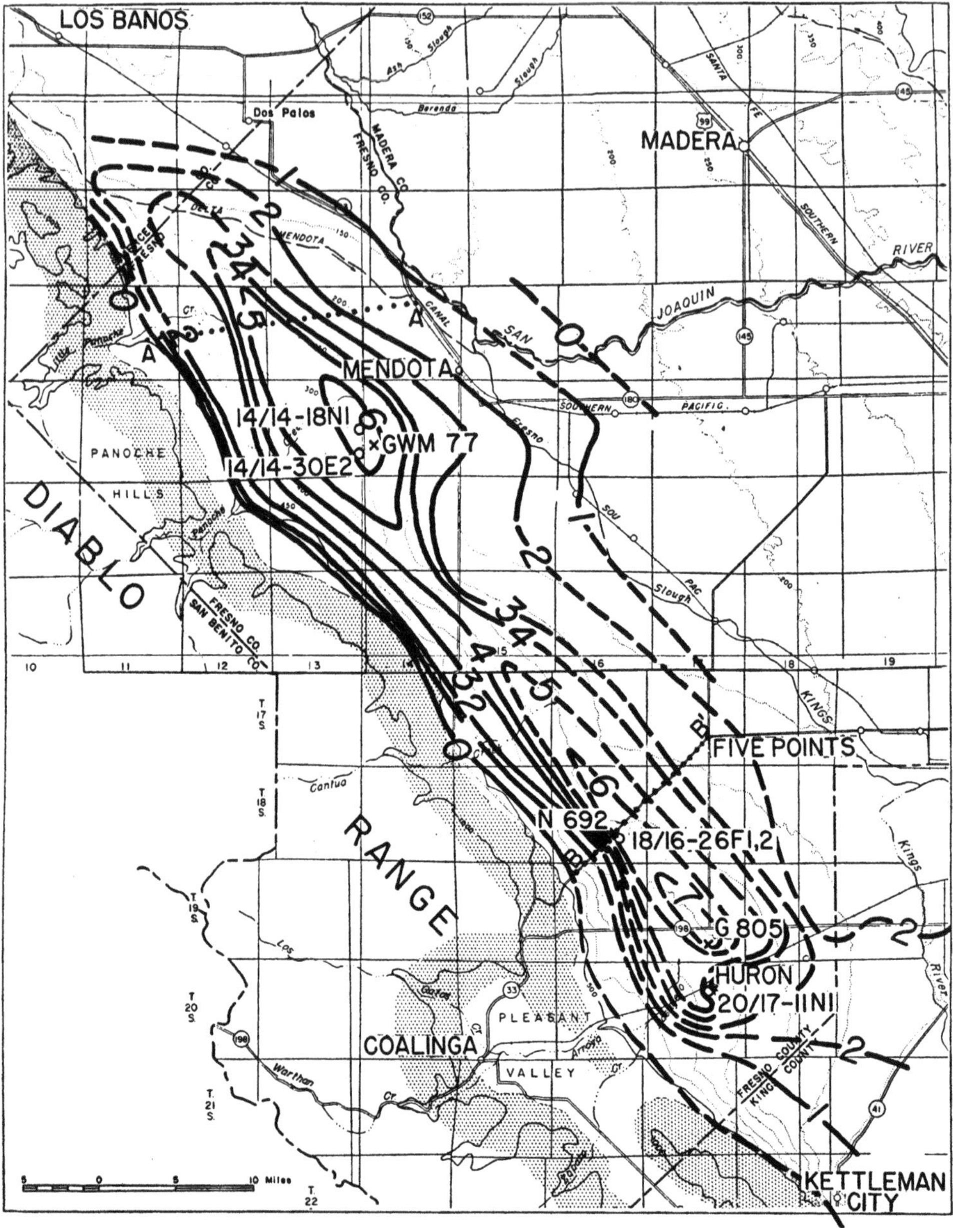

FIG. 3 – Land subsidence in the Los Banos-Kettleman City area, 1943–53 (lines of equal subsidence in feet)

in the area east of Delano. This recovery, which evidently is related to importation of new surface-water supplies to the area via the Friant-Kern Canal, even appears as a net rise of about five ft for the period 1948–54 about 1½ miles east of Delano. The graphs of subsidence and change in head agree fairly closely in the vicinity of Delano, but near the eastern end of the profile of water-level change a marked discordance is evident. This decrease in the ratio of subsidence to decline in head is not fully understood. It may well be related to a change in the elastic or other physical properties of the sediments, or may represent a decrease in the degree of confinement to the east;

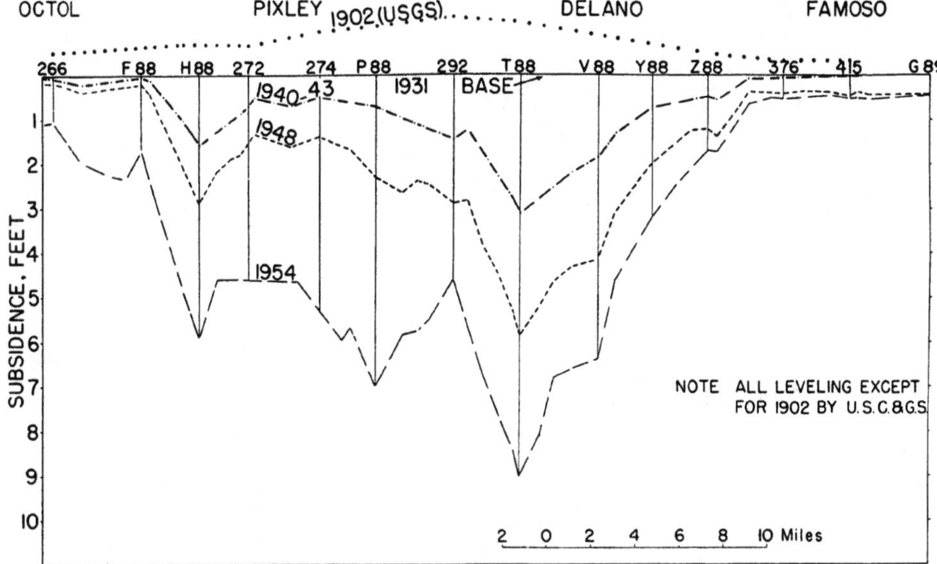

FIG. 4 – Profiles of land subsidence, Octol to Famoso, Calif., 1902–54

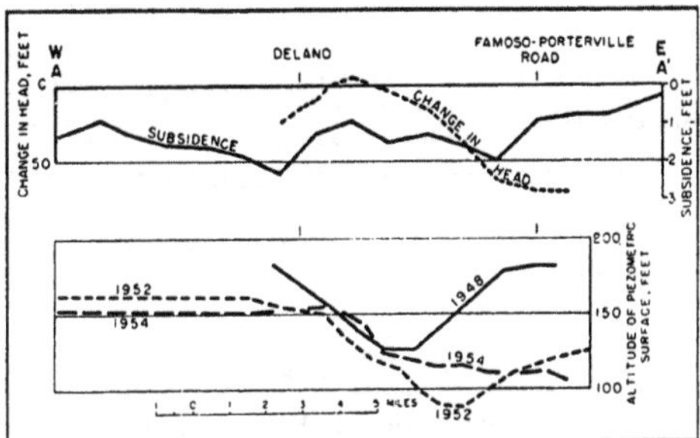

FIG. 5 – Profiles showing land subsidence and change in artesian head along Kernell Road south of Delano, 1948–54 (for location of line A–A′, see Fig. 2)

it may also represent a time lag of several years in adjustment to changes in head.

Figure 6 shows changes in altitude at bench marks T 88 and G 758 and change in pressure head in well 24/26-33H1 (see Fig. 2). The hydrograph of well 24/26-33H1, about five miles northeast of Delano shows seasonal fluctuations and the long-term water-level trend since 1930 in a typical well tapping the confined aquifers. It shows a continuing decline from 1930 to 1951, evidently the result of overdraft on local supplies of ground water. After the beginning of delivery of imported water by the Friant–Kern Canal (Fig. 6), however, the long-term trend began a pronounced rise which continued through 1954. Bench mark T 88 is six miles west of well 33H1. The graph for this bench mark, based on leveling in 1931, 1940, 1943, 1948, and 1954, shows good correlation with change in head in well 33H1 except after 1951, when the water level in the well began to rise but the bench mark continued to subside. The graph for bench mark G 758, only half a mile from the well, based on leveling in 1948, 1951, and 1954, however, indicates a lessening of the rate of subsidence between 1951 and 1954 as compared with 1948 to 1951, possibly as a result of recovery of artesian pressure after 1951.

SUBSIDENCE IN THE LOS BANOS–KETTLEMAN CITY AREA

General features—The Los Banos–Kettleman City area is in the west-central part of the San Joaquin Valley, almost wholly in Fresno County but extending north into Merced County and south into Kings County. The area is about 90 miles long and extends from the western edge of the valley about to Fresno Slough and the San Joaquin River, an average width of about 16 miles. Most of the area is irrigated by ground water pumped from approximately 1000 irrigation wells; only north of the Delta–Mendota Canal is surface-water irrigation extensive. Most of the ground-water development has occurred since 1940 and pumpage tripled from 1945 to 1953. In 1953 it was about 1,200,000 ac ft [*Davis* and others, 1954].

The deposits containing fresh water in this area

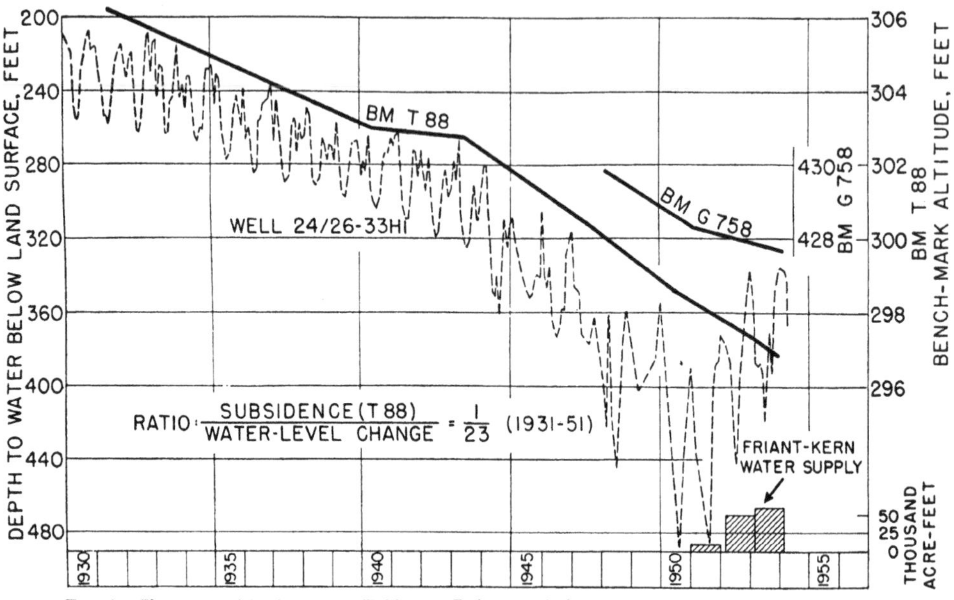

Fig. 6 – Change in altitude at BM T 88 near Delano and change in artesian head in nearby well

are of Recent to Pliocene age and extend to depths ranging from less than 1000 ft to at least 3000 ft below the land surface. They can be divided into (1) an upper unit of clay, silt, and sand, chiefly alluvial-fan deposits, that extends from the land surface to depths of 400 to 800 ft; (2) a middle unit of impervious diatomaceous clay 20 to 120 ft thick; and (3) a lower unit of clay, silt, and sand, in part of lacustrine origin, commonly 600 to 1500 ft thick but locally as much as 2500 ft thick, that extends down to beds containing saline water.

A body of semiconfined water occupies most of the upper unit; the water table at its top is ten to 300 ft below the land surface. The ground water in the lower unit, the lower water-bearing zone, is confined in most of the area by the diatomaceous clay. Probably 80 pct or more of the pumping draft is from this lower zone. The general positions of the upper and lower water-bearing zones and of the diatomaceous clay that separates them, as encountered along line A-A′ (Fig. 3), are shown on Figure 7.

The water table in the upper water-bearing zone has not changed greatly from its initial position. However, the heavy pumping draft has drawn down the pressure head in the lower water-bearing zone as much as 300 ft in the past 20 years; the decline has ranged from 150 ft near Mendota on the northeast to roughly 300 ft near Huron on the south. Since 1945 the average rate of yearly decline has been four to seven feet in the northern

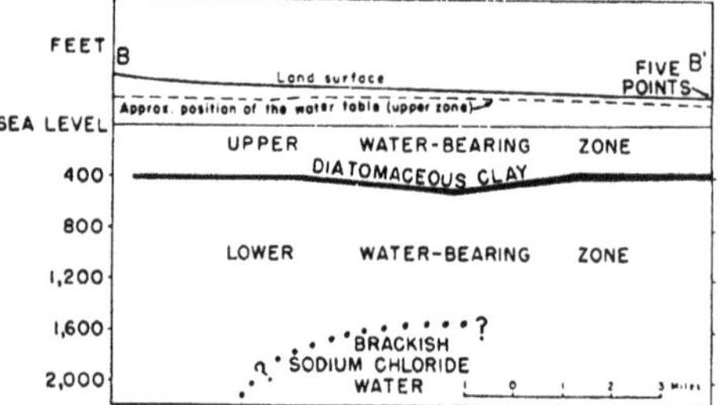

Fig. 7 – Geologic section B–B′ (for location of line B–B′, see Fig. 3)

part and 20 to 30 ft in the southern part of the area.

Subsidence—Subsidence of the land surface in the Los Banos–Kettleman City area was noted as early as 1943, when releveling by the U. S. Coast and Geodetic Survey of bench marks established in 1935 indicated changes in altitude of several feet in some areas. Subsequent levelings by the Coast and Geodetic Survey in 1947, 1953, and 1954 and by the Bureau of Reclamation, the Geological Survey, and the California Division of Water Resources have supplied the data for the map showing subsidence from 1943 to 1953 (Fig. 3). The land-surface subsidence is shown by generalized lines of equal subsidence based on leveling along widely separated lines. Approximately 1050 sq mi is enclosed within the one-foot subsidence line. Two areas of sharp subsidence in which the decline exceeded six feet during the

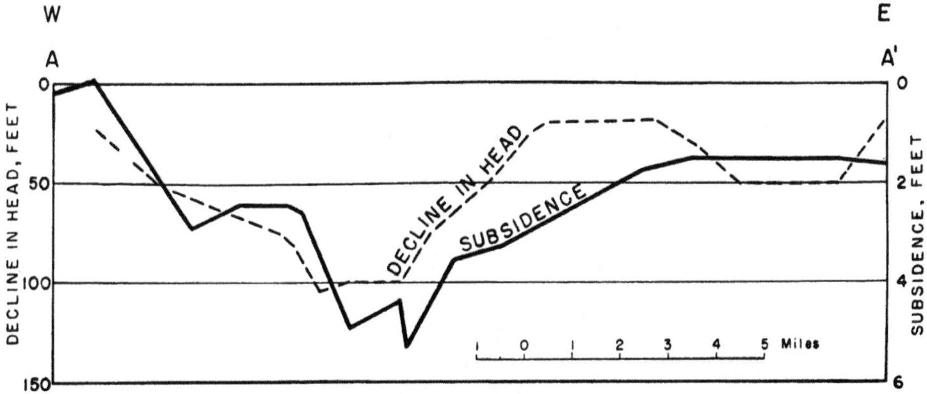

FIG. 8 – Profiles of land subsidence and decline in artesian head from Little Panoche Creek east to Benito, 1943–53 (for location of line A-A', see Fig. 3)

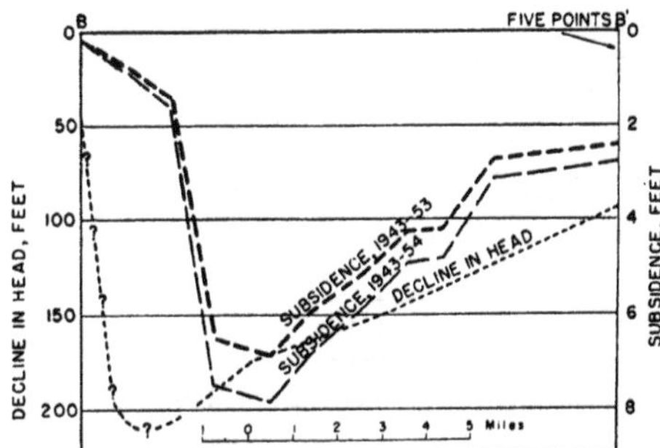

FIG. 9 – Profiles of land subsidence and decline in artesian head along Fresno-Coalinga Road southwest from Five Points, 1943–53

ten-year period 1943 to 1953 are shown on Figure 3, one in the southern part of the area centered about 4½ miles north of Huron, and the other in the northern part of the area centered about seven miles southwest of Mendota. The maximum subsidence recorded was at bench mark G 805, 2½ miles north of Huron, which declined about 7.5 ft between 1947 and 1954. (Unadjusted preliminary elevations, 1954, by U. S. C. and G. S. adjusted 0.8 foot downward to make previously stable area in Kettleman Hills stable with respect to valley area.) The maximum subsidence from 1943 to 1953 in the northern area occurred at bench mark GWM 77, about 7½ miles southwest of Mendota, which declined 6.8 ft. Bench mark 246 USGS, about four miles north of GWM 77, apparently subsided 5.7 ft between 1919 and 1943, thereby suggesting a total subsidence in the northern area of about 12 ft.

Profiles of land subsidence and decline in artesian head of the lower water-bearing zone along lines A-A' and B-B' of Figure 3, presented on Figures 8 and 9, respectively, show the relationship of subsidence to decline in head in the ten years 1943–53. The profiles along line A-A' (Fig. 8) indicate a rude correlation; the maxima on both profiles are about coincident, although the ratio of subsidence to decline changes along the profiles. Figure 9 shows subsidence from 1943 to 1954, based on precise leveling, and subsidence for the ten-year period 1943–53, extrapolated from the 1943–54 change on the basis of rate of subsidence at bench mark N 692 since 1947 (Fig. 10). The graphs of subsidence and decline in head correspond reasonably well except that the trough of maximum subsidence is about 2½ miles east of the trough of maximum decline in head. The pressure trough, however, has been migrating westward with recent development of land along the western edge of the valley and accompanying heavy draft on ground water, and the subsidence trough also may be migrating with a lag as indicated.

Changes in altitude of bench marks of maximum subsidence in the areas of heaviest subsidence and of pressure levels in nearby wells are illustrated on Figures 10, 11, and 12, respectively. Bench marks GWM 77 and N 692 have behaved similarly, both showing a pronounced increase in the rate of subsidence since 1947. The pressure head of the confined water in the lower water-bearing zone also has been declining at an accelerated rate in recent years as a result of rapid development of irrigation in the area since the close of World War II. This accelerated rate of decline is well shown by the composite hydrograph of well 18/16-26F1 and F2 (Fig. 10). The plot of bench mark G 805 (Fig. 12) is based on only two levelings and was included merely to show conditions in the area of greatest ten-year subsidence, but leveling to other bench

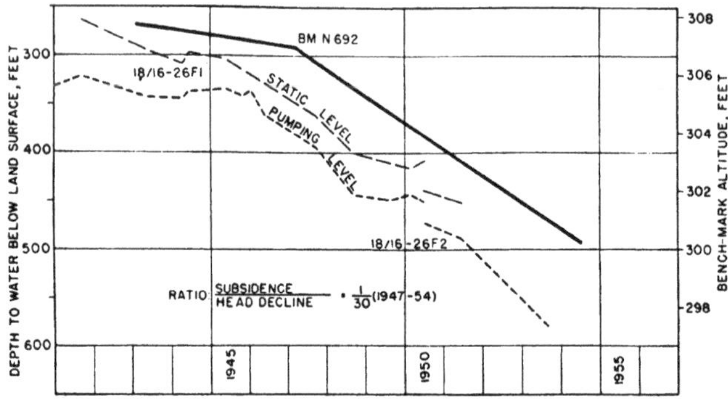

FIG. 10 – Change in altitude at bench mark nine miles southwest of Five Points and change in artesian head in nearby wells

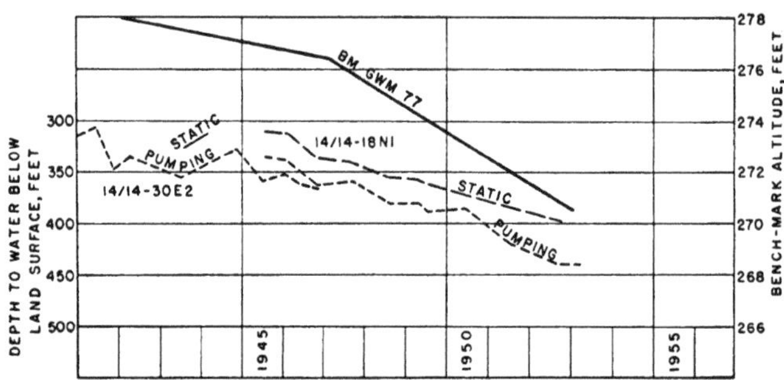

FIG. 11 – Change in altitude at bench mark southwest of Mendota and change in artesian head in nearby wells

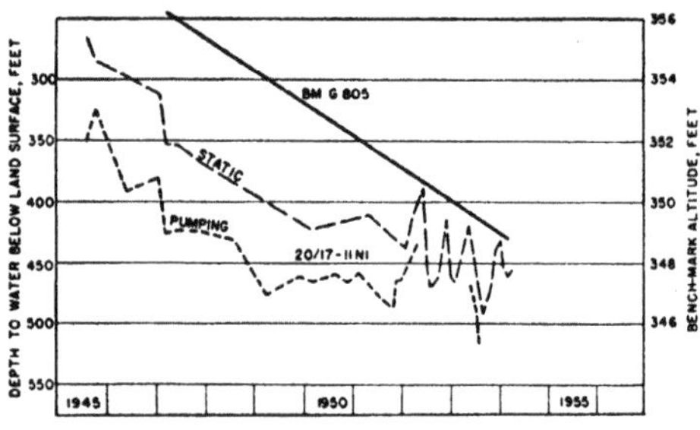

FIG. 12 – Change in altitude at bench mark near Huron and change in artesian head in nearby well

marks in the vicinity suggests that the rate of subsidence there also has been increasing in recent years. All three of the figures suggest a general correspondence between the trend of land-surface subsidence and pressure decline. The ratio of subsidence to pressure decline as shown on Figure 10 for the seven years 1947–54 was 1/30. In the Tulare–Wasco area the corresponding ratio shown on Figure 6 from 1931 to 1951 was 1/23.

CAUSES OF SUBSIDENCE

Subsidence of the land surface has been noted in many areas and ascribed to various causes. Possible causes are summarized in the following table and the likelihood of each cause being responsible for or a contributor to subsidence in the Los Banos – Kettleman City and Tulare–Wasco areas is indicated on Table 1.

In considering these possible causes, note: (1) There has been no general loading of the land surface with structures in the areas of subsidence. (2) The only widespread vibrations at or near land surface in the areas have been from land cultivation, from the drilling of water wells, and from infrequent earthquakes. Locally in the Tulare–Wasco area, vibrations from the passing of railroad trains have occurred many times each day. (3) There is good evidence that compaction due to application of irrigation water is a contributing cause. (4) It seems probable that solution is a contributing but minor cause. (5) Excessive and sustained drying does not ordinarily occur in irrigated areas. (6) Soils in these areas do not contain a high percentage of organic matter. (7) The water table has not declined appreciably in the Los Banos–Kettleman City area; in areas where it has declined and the confining beds are continuous and relatively impermeable, the resultant removal of load from deeper aquifers that

TABLE 1 – *Possible causes of land-surface subsidence and probability of their contributing to subsidence in the Los Banos - Kettleman City and Tulare - Wasco areas*

No.	Item	Unlikely	Possible	Probable
1	Loading at land surface	x		
2	Vibrations at or near land surface		x	
3	Compaction due to irrigation and farming			x
4	Solution due to irrigation		x	
5	Drying out and shrinking of deposits	x		
6	Oxidation of organic material	x		
7	Lowering of water table		x	
8	Decline of pressure head in confined aquifers			x
9	Decline of pressure in oil zones due to removal of oil and gas		x	
10	Tectonic movement		x	

are under pressure would tend to offset the effects of lowering the water table. (8) Decline of pressure head in confined aquifers has proved to be a major factor in several other areas, and available evidence suggests that it should be considered so in these areas. (9) Producing oil and gas fields are located in or adjacent to part of the subsidence areas, and decrease in fluid pressure due to removal of oil and gas would be a contributing cause. (10) Tectonic activity has been demonstrated to be responsible for at least a portion of the movement in the Arvin–Wheeler Ridge–Maricopa area, but whether it has contributed appreciably to subsidence in the two principal areas discussed in this paper is not yet known.

On the basis of presently available data it is concluded that the extensive drawdown of the confined water has been one of the primary causes if not the major cause of subsidence in the Los Banos–Kettleman City and Tulare–Wasco areas. Field reports of noticeable settlement of the land surface of as much as several feet in the vicinity of wells and ditches on newly irrigated land on the west side of the San Joaquin Valley suggest that, locally at least, compaction due to irrigation and farming is a major cause of settlement. However, no specific measurements of subsidence of this nature are yet available to assist in evaluating its relative importance in the regional subsidences here discussed. In addition, it is considered that six other possible causes may contribute to the overall land-surface subsidence.

In the Arvin–Wheeler Ridge–Maricopa area, tectonic movement has been demonstrated to be responsible for at least a part of the change in land-surface altitude, but very possibly decline in head in confined aquifers, compaction due to irrigation, and decrease in fluid pressures due to the removal of oil and gas also contribute to subsidence.

INTER-AGENCY COMMITTEE

The subsidence of the land surface in the San Joaquin Valley, especially in the two principal areas described in this paper, poses serious problems in connection with maintenance of present large-capacity low-gradient canals and construction of proposed additional canals, as well as in construction and maintenance of other engineering structures such as irrigation distribution systems, trunk pipelines, drainage and sewerage systems, power-transmission lines, highways, railroads, and buildings, and in various aspects of land development and use. The Bureau of Reclamation already has large canals in each of the two areas, and largely at the suggestion of that agency an Inter-Agency Committee of interested Federal and State agencies has been formed for the purpose of formulating and carrying out a program designed to furnish information on the extent, magnitude, rate, and causes of the land subsidences in the San Joaquin Valley, in order that planning and construction agencies may be able to estimate with fair accuracy the location, rate, and magnitude of future subsidence (or recovery) of the land surface.

REFERENCES

DAVIS, G. H., J. F. POLAND, AND OTHERS, Ground-water conditions in the Mendota-Huron area, Fresno and Kings Counties, Calif., *U. S. Geol. Survey Water-Supply Paper 1360*, in preparation; issued also in mimeographed form, 102 pp., 1954.

GILLULY, JAMES, AND U. S. GRANT, Subsidence in the Long Beach Harbor area, California, *Bull. Geol. Soc. Amer.*, **60**, no. 3, 461–530, 1949.

INGERSON, I. M., The hydrology of the southern San Joaquin Valley, California, and its relation to important water supplies, *Trans. Amer. Geophys. Union*, pt. 1, pp. 20–45, 1941.

LONGWELL, C. R., Interpretation of the leveling data in Lake Mead comprehensive survey of 1948–49, U. S. Geol. Survey open-file rept., v. I, 71–80, 1954.

MENDENHALL, W. C., R. B. DOLE, AND HERMAN STABLER, Ground water in San Joaquin Valley, California, *U. S. Geol. Survey Water-Supply Paper 398*, 310 pp., 1916.

SWENSON, F. A., Geology and ground-water supply of the Heart Mountain and Chapman Bench Division of the Shoshone Irrigation Project, Wyoming, *U. S. Geol. Survey Water-Supply Paper*, in preparation.

TOLMAN, C. F., AND J. F. POLAND, Ground-water, salt-water infiltration, and ground-surface recession in Santa Clara Valley, Santa Clara County, Calif., *Trans. Amer. Geophys. Union*, pt. 1, pp. 23–35, 1940.

WIER, W. W., Subsidence of peat lands of the Sacramento-San Joaquin Delta, California, *Hilgardia*, **20,** 37–56, 1950.

WINSLOW, A. G., AND W. W. DOYEL, Land-surface subsidence and its relation to the withdrawal of ground water in the Houston-Galveston region, Texas, *Econ. Geology*, **49,** 413–422, 1954.

ZEEVAERT, LEONARDO, An investigation of the engineering characteristics of the volcanic lacustrine clay deposit beneath Mexico City; Ph.D. thesis, Univ. Illinois, 234 pp., 1949.

U. S. Geological Survey, Sacramento, California

(Manuscript received January 17, 1956; presented at Regional Meeting in Berkeley, California, February 4, 1955; open for formal discussion until November 1, 1956.)

Reprinted from *Science,* **72,** 333–344 (1971)

23 April 1971, Volume 172, Number 3981 **SCIENCE**

13

Ground Rupture in the Baldwin Hills

Injection of fluids into the ground for oil recovery and waste disposal triggers surface faulting.

Douglas H. Hamilton and Richard L. Meehan

On the Saturday afternoon of 14 December 1963, water burst through the foundation and earth dam of the Baldwin Hills Reservoir, a hilltop water storage facility located in metropolitan Los Angeles. The contents of the reservoir, some 250 million gallons of treated water that had filled the artificial, 20-acre clay- and asphalt-lined basin to a depth of 70 feet, emptied within hours onto the communities below the Baldwin Hills, inundated a square mile of residences with mud and debris, and damaged or destroyed 277 homes *(1).* Fortunately for those in the path of the flood wave, indications of imminent failure had been observed by a reservoir caretaker several hours before the final breach occurred; even so, police evacuation teams had barely sufficient time to clear the area. Consequences of the disaster were minimal compared with what would have occurred if no warning had been provided, but they included five lives lost, $12 million in property damage, and loss of the reservoir itself.

The remains of the Baldwin Hills Reservoir stand empty today, the northern rim of the bowllike structure having been gashed from crest to foundation by the escaping water (see cover). A linear crack issuing from the base of this gap can be traced across the asphalt floor of the reservoir. It reappears as a slight buckling of road pavement on the far side of the reservoir basin and thence becomes a faint, discontinuous break in the ground surface, which trails off south of the reservoir into the brush-covered and excavation-scarred terrain of the Inglewood oil field.

Since 1963 geologists and engineers have been intensively investigating this crack and several similar ones nearby, all known to be surface expressions of deep, near-vertical faults of Pleistocene or greater age. For there is no doubt that the disaster occurred as a result of displacement along faults in the unconsolidated sediments that underlie the reservoir. These displacements led to rupture of the protective clay lining that covers the floor of the bowllike structure, which had been constructed in 1951. Ironically, the 10-foot-thick (3-meter) lining and its underdrain had been especially designed and constructed to isolate the soft, sandy foundation rock from leakage of the reservoir water, thus providing what was thought to be a margin of safety against *piping,* a process characterized by gradual development of eroded subsurface cavities and channels. What the designers evidently had not thought possible were offsets along one or more of the buried faults great enough to destroy the lining.

On the day after the failure, it was apparent that major offset had occurred along what was to become known as the "Reservoir fault," the west side of the fault having moved relatively downward with respect to the east side (Fig. 1). The offset was sufficient to crack the lining all the way across the floor of the reservoir, and surface displacement of as much as 6 inches (15 centimeters) was created. The crack line was punctuated by several ragged, cavelike sinkholes, which marked the points of water entry that had led, possibly over a period of hours or days, to the deterioration and eventual destruction of the reservoir foundation.

Responsibility for the disaster has never been formally fixed. Insurance carriers for the Los Angeles City Department of Water and Power, constructor, owner, and operator of the reservoir, promptly paid for flood damages. A report on the results of an investigation carried out by the State of California's Department of Water Resources *(2)* provided a taut narrative of events on the day of the disaster and a cool appraisal of failure as the result of an unfortunate combination of physical factors: "Sitting on the flank of the sensitive Newport-Inglewood fault system with its associated tectonic restlessness, at the rim of a rapidly depressing subsidence basin, on a foundation adversely influenced by water, this reservoir was called upon to do more than it was able to do."

Two lawsuits filed in 1966 by the city and its insurers against the oil companies active in the Inglewood oil field at the time of dam failure charged that the oil field operations had led to the events directly associated with breaching of the dam. These suits were settled out of court for nearly $3.9 million dollars, thus disposing of the immediate financial issues that arose from the ground rupturing beneath the reservoir. Origin of the ruptures is a question that remains unadjudicated.

Mr. Hamilton is a geologist and Mr. Meehan is a civil engineer; both are partners in the consulting firm of Earth Sciences Associates, Palo Alto, California.

Geologic Setting

The Baldwin Hills, site of the Inglewood oil field and the failed reservoir, form part of an interrupted chain of low hills that rise in striking contrast to the surrounding flat terrain of the Los Angeles basin (Fig. 2). Processes of both folding and faulting have contributed to the uplift of this chain (*3, 4*). In the Baldwin Hills, the primary anticlinal fold structure has been much modified by faulting, especially by lateral and dip-slip displacement along the Inglewood fault, which bisects the hills. Many unnamed subsidiary faults, apparently related genetically to the Inglewood fault, are present throughout the hills and are especially numerous in the vicinity of the reservoir.

Moody and Hill (*5*), among others, consider the Inglewood anticline to have developed as a drag fold during deformation of the thick, plastic sedimentary section overlying the Newport-Inglewood zone of basement right-lateral transcurrent faulting. The anticlinal folding began and continued while marine sedimentation was still going on, as evidenced by thinning and lateral discontinuities in the stratigraphy of the overlying section of late Tertiary and Quaternary strata. As movement continued along the Newport-Inglewood zone, the deformation exceeded the capacity of the overlying strata to adjust by folding, and failure was propagated through the section as normal and strike-slip faulting. The surface strain pattern associated with this tectonic deformation (*6*) is illustrated in Fig. 3, and the subsurface structure is shown in Fig. 4.

The near-parallel, north-striking faults that splay outward from the Inglewood fault south of the Baldwin Hills Reservoir (Fig. 3) were probably formed as an array of tear faults developed in response to strike-slip displacement along the dominant Inglewood fault. The Reservoir fault is a member of this family of steeply dipping faults, all of which intersect the Inglewood fault at depths of no more than a few thousand feet.

The apparent sense of displacement on these secondary faults is largely vertical. At the time of construction of the Baldwin Hills Reservoir, vertical stratigraphic separation of as much as 26 feet was noted at surface excavations across what became known as the Reservoir fault; lesser apparent vertical displacement was recorded for several related faults beneath the reservoir. Some

Fig. 1. Baldwin Hills Reservoir. Seepage through the ruptured reservoir lining (foreground) undermined and then breached the Baldwin Hills dam embankment (background). The relatively downward displacement of the ground surface to the left of the crack is clearly visible in the photograph. [Wide World Photo]

horizontal displacement on the Reservoir fault was inferred on the basis of observed striations on the fault surface. However, the significant vertical offset component that is evident along these faults clearly places them in the category of normal (gravity) faults.

Inglewood Oil Field

The Inglewood oil field occupies an irregularly oval area that extends diagonally across the trend of the hills along the axis of the faulted Inglewood anticline (Fig. 3). The field is localized by the anticlinal structure and by the distribution of its sedimentary strata. The extensive breaking associated with the Inglewood fault has disrupted the original fold structure, so that it now has the form of a series of slices translated successively downward toward an axial trough along the west side of the Inglewood fault (Fig. 4). The Inglewood fault forms a structural break between the east and west blocks of the field and effectively isolates the two producing blocks hydraulically from each other. Some subsidiary faults within both blocks are also fluid barriers, as was confirmed by changes in reservoir pressure with time and in response to both production and water flooding. The barrier effect of many subsidiary faults in the east block may be local, however, and may not extend the entire length of any given fault.

Although oil production has been primarily through solution gas drive, a peripheral water drive exists along the northeast margin of the east block and also along the west margin of the west block. The edgewater condition of the east block encroaches from the area of artesian formation water in the northerly extension of the permeable oil sand horizons, beyond the area of petroleum entrapment where the groundwater formerly existed at or near hydrostatic pressure.

Exploitation of the Oil Field

The Inglewood oil field was discovered in 1924. It was explored and developed so rapidly that its period of greatest yield occurred within the first 3 years of production. Although the fluid pressure in the oil field reservoir was at hydrostatic levels under preexploitation conditions (570 pounds per square inch in the east block), fluid pressures measured in wells declined through the years to about 50 pounds per square inch in the early 1950's (*2, 7, 8*). Castle and Yerkes (*4*) have suggested, however, that blocks of higher pressure may have persisted after the general pressure declined, owing to fault compartmentalization of the reservoir. In any case, production and development, mainly by "infill" drilling between wells, has continued steadily to the present.

A significant modification of the extraction program was started in 1954, when the Standard Oil Company initiated a pilot "water-flood" program of secondary recovery in its east block leases. The history and technical development of this program were well described by Oefelein and Walker in 1963 (*9*). They also presented detailed information on the structure and stratigraphy of the main producing horizon, known as the Vickers zone, of the Inglewood field east block. The pilot water flood involved the injection of brine in selected intervals from two arrays of four wells each, generally at pressure gradients in the range of 0.5 to 0.9 pound per square inch per foot of depth.

Results of the pilot program were sufficiently encouraging to prompt a full-scale program in 1957. The east block secondary recovery program has been expanded by increments since its inception, with a general pattern of increasing numbers of injector wells and increasing volume and pressure of injection (Fig. 5). Although the program has been successful, it has involved many technical difficulties, including maintenance problems in the wells (7). Problems such as loss of major amounts of injected fluid in narrow intervals, sudden increases in formation "take" with concomitant loss of injection pressure, casing failures, "breakthroughs" to producing wells, and surface leaks of injected fluid seem to have increased as greater numbers of injection wells, operating under increasing pressure, have been brought into service. Such problems frequently have occurred after significant raising of injection pressure in specific wells, and some have been severe enough to cause abandonment of the injection well. However, techniques of sealing damaged intervals and repairing wells have often been effective in restoring wells to injection after each episode of failure, and the overall injection program has been progressively expanded.

Subsidence and Ground Rupture in the Baldwin Hills

Although the phenomenon of "earth crack" ground rupturing may not have been definitively observed in the Baldwin Hills prior to 1957, the existence of a more widespread and larger-scale form of ground surface movement was recognized by engineers of the Los Angeles Department of Water and Power as early as 1943, when comparison of leveling surveys indicated that elevation changes were taking place in the area (*10*). The ground subsidence was

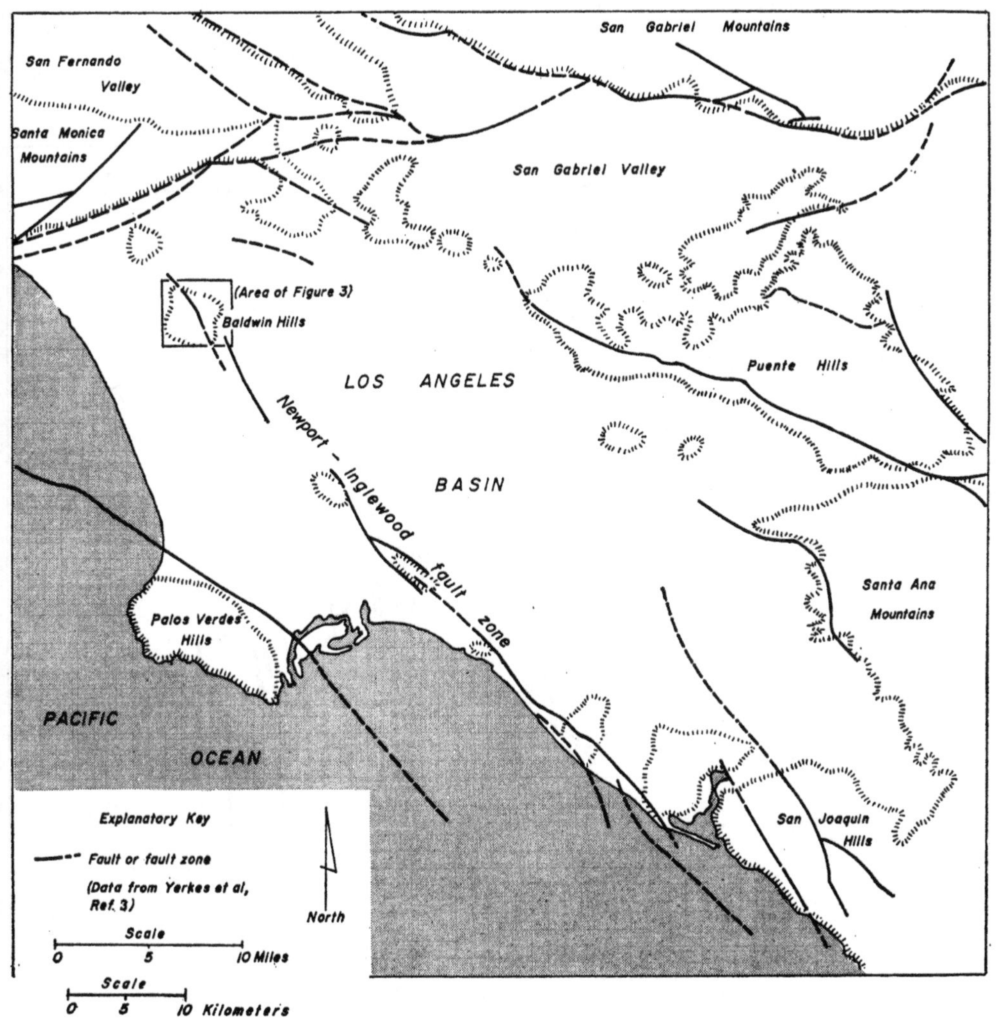

Fig. 2. Generalized location map, Los Angeles basin.

of concern to the department because of its plans to construct and operate a reservoir, but it was not known by the community at large. Not until 1955 were sufficient data available to establish that the changes in elevation defined a bowl-shaped area of subsidence (the outline of which appeared to coincide with the outline of the Inglewood oil field) (*10*). In 1957, 6 years after the reservoir had been put into operation, the beginning of surface cracking and faulting in the vicinity of the intersection of Stocker and La Brea streets, southeast of the reservoir, attracted the attention of public officials.

The spatial relations among patterns of ground deformation in the Baldwin Hills, preexisting geologic structural features, and the area of production in the Inglewood oil field are illustrated in Fig. 6. Table 1 summarizes the sequence of ground-rupturing events.

The principal features of contemporary ground subsidence and surface rupture are clearly established. They have been thoroughly documented by the U.S. Geological Survey (*4*), the Los Angeles Department of Water and Power (*10, 11*), the State of California Department of Water Resources (*2*), and by several consultants (*12, 13*) and other investigators. Some interpretations have differed slightly from others as to the exact pattern, magnitude, and history of this deformation. Explanations for the origin of the deformation have, in contrast, been sharply segregated into two categories: (i) response to lowering of fluid pressure, fluid withdrawal and injection, and related phenomena and activities associated with operation of the Inglewood oil field; and (ii) deformation largely or wholly of tectonic origin.

Tectonic theory. This theory is based on records of recent surface rupturing along known "active" faults and on indications that the sense of the contemporary movement appears to be indistinguishable from that of prior displacements. It is argued that vertical offset along the family of faults in the east block has occurred in recent geologic time as a result of lateral displacement along the Inglewood fault or as part of the process of anticlinal folding, or both, and that it may be reasonably assumed that the tectonic environment that created these structures, relatively recently and rapidly, persists today. The tectonic theory is offered by some geologists as an explanation of

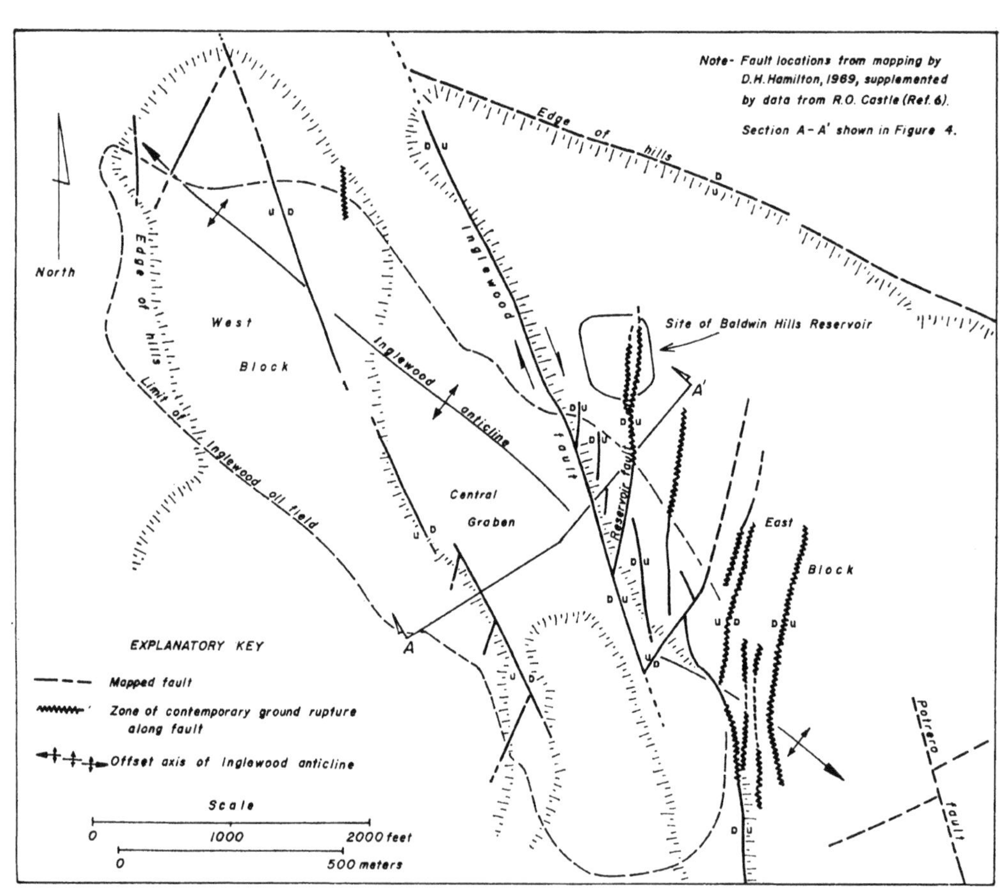

Fig. 3. Geologic map of the Baldwin Hills.

the subsidence in the Baldwin Hills as well as the cause of the ground rupture in the east block area.

The tectonic theory is open to objections of several kinds. First, and most important, displacement along subsidiary faults generally occurs in response to a relatively greater displacement along the principal fault, which in the Baldwin Hills is clearly the Inglewood fault. Yet relatively large offsets have occurred on the subsidiary faults in the east block, whereas no evidence has been produced to indicate related creep, rupture, or crustal strain that would give rise to displacement along the Inglewood fault or nearby major faults in the Baldwin Hills or adjacent area.

Continued upward warping of the Baldwin Hills might well lead to the observed normal faulting, but detailed analysis of survey leveling data by Hayes, Castle, Leps, and others indicates that little or no such regional uplift has occurred during the past half-century for which data are available. Further, the margins of a near-surface, upwarping, anticlinal fold should move outward, away from the fold axis, but surveys of four bench marks situated on the flanks of the anticline show the ground to be moving toward the anticlinal axis (and radially toward the center of the subsidence bowl) (Fig. 6). Moreover, offset of the folded formations by the Inglewood fault demonstrates that the major episodes of folding must have predated much of the faulting. This suggests that faulting has superseded anticlinal folding as the dominant style of tectonic deformation in the Baldwin Hills.

Finally, the shape of the subsidence bowl does not correspond to the shape of the "central graben," as would be expected if the subsidence were caused by further downfaulting.

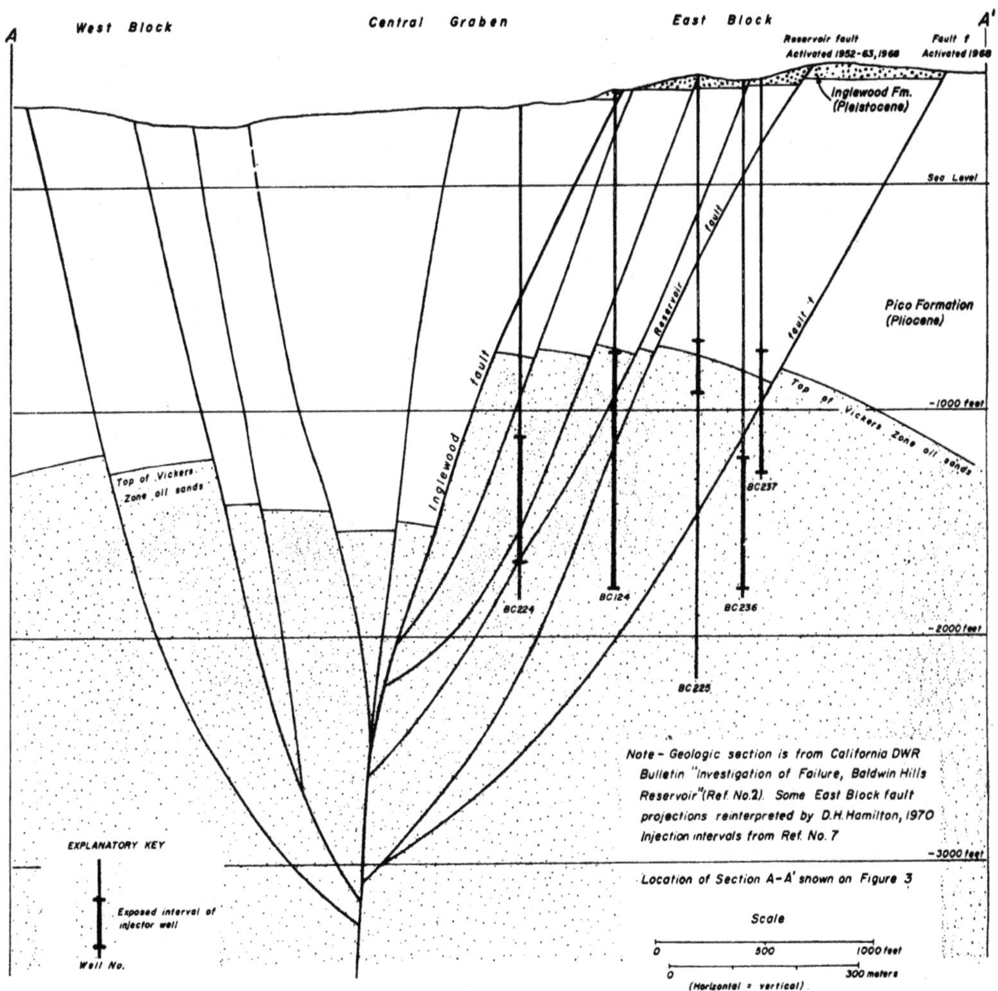

Fig. 4. Geologic structure section through the Baldwin Hills–Inglewood area and the Inglewood oil field.

Table 1. Sequence of ground rupturing along earth cracks in the Baldwin Hills, 1957–68. For location of cracks, see Fig. 6.

Crack No. or name	Occurred or first observed (date)	Length (feet)	Cumulative length (feet)
1	May 1957	2,600	2,600
3, 4	Jan. 1958	1,900	4,500
2	Mar. 1958	600	5,100
5a	Before 1961	400	5,500
5	Feb.–Mar. 1963	400	5,900
1a	1962–1963 (?)	800	6,700
8	Feb. 1963	400	7,100
Wilson's fault I*	Aug.–Dec. 1963	2,000	9,100
Wilson's fault V	Aug.–Dec. 1963	800	9,900
Wilson's fault I	1968	Recurrent movement (2,400)	
Wilson's fault V	1968	Recurrent movement	
Hamilton's fault f	1968	1,400	11,300

* Wilson's fault I is now identified as the Reservoir fault.

Subsidence theory. Advocates of the subsidence theory argue that the 6 feet or more of ground subsidence in the Inglewood field is clearly related in space and time to the net withdrawal of some 67,000 acre-feet (*14*) of oil, water, and sand from shallow producing horizons, a contention supported by theoretical mechanics and by analogous experience in other areas (*4*). Almost all the observed surface rupturing has occurred on the edge of the subsidence bowl; stretching on the rim of the sagging ground surface is an obvious and mechanically proved consequence of subsidence [demonstrated by Grant (*15*), Deere (*16*), Lee and Strauss (*17*), and others], and tension cracking of the near-surface materials is one result (illustrated diagrammatically in Fig. 7). Subsidence also exerts a downward drag on the sediments that rim the subsidence bowl, and, where faults or other established surfaces of weakness are present, accumulating elastic strain may be relieved by upward "popping" of the ground on the edge of the bowl.

Two objections might be leveled at the subsidence theory as an explanation of ground rupture in the east block. First, ground rupturing was not directly observed prior to 1957, and the only positive evidence of drag or tension-induced failure before 1957 is a record of continuous extension of the rupture caused by movement since 1952 of the Reservoir fault under the Baldwin Hills Reservoir inspection gallery. However, most of the ground subsidence between the early 1920's and 1963 had already occurred by 1957, and the rate of subsidence in the east block had slowed markedly in the late 1950's and early 1960's. Why then should intense surface faulting suddenly begin to occur as late as 1957?

The other possible objection to the subsidence explanation is the local distribution of ground rupturing. The downward drag and stretching presumably would be distributed around the rim of the subsidence area, but almost all the rupturing has occurred in one limited sector of the bowl circumference (Fig. 6).

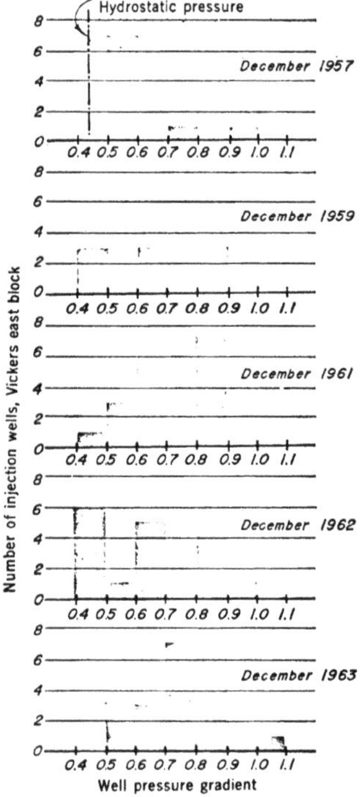

Fig. 5. Injection history for Vickers east block, 1957–63.

Influence of Secondary Recovery Operations

Recognition of these two serious objections, and further recognition that the procedure of exploitation in the Inglewood field changed from simple extraction to extraction combined with injection, has led to modification of the subsidence theory. As here presented, it eliminates the objections noted above, is supported by striking correlation of water-flood events and episodes of fault movement leading to ground rupture, and provides a satisfactory mechanical basis for the observed history of ground rupture. The revised theory may be referred to as the "subsidence fluid injection" theory. In substance, it specifies that injection of fluid into the ground under pressures only slightly greater than normal hydrostatic pressure will reduce and may, in areas subject to normal faulting, eliminate shearing resistance along potential failure planes. When this activity is carried out in ground earlier affected by faulting, and where additional stresses have been set up by differential subsidence, reduction in shear strength will give rise to movements that will be concentrated along preferred surfaces of weakness such as preexisting faults. According to this theory, fault activation and consequent ground rupturing should develop preferentially in an area being affected by fluid injection. This prediction is verified by the localization in space (Fig. 8) and time (Fig. 9) of earth-crack ground rupturing in the Baldwin Hills in and adjacent to the area of fluid injection for secondary recovery, which began almost simultaneously with the initiation of the "full-scale" water-flood program. The correlation of fluid injection events and surface rupturing is even more striking when it is examined in detail.

One of the first of the full-scale injector wells was placed in operation in May 1957. In that same month, a fault, the probable subsurface projection of which lies within or very near the injection interval of this well, was activated and became the first recognized earth crack in the east block area. Injection was started in 21 additional wells in that area between 1957 and 1963 (Figs. 5 and 8), during which time at least eight more faults were activated. The relation of injection intervals to the east block subsurface structure is illustrated by the geologic cross section in Fig. 4. This section passes through two of the east block faults, each of which had been related to surface rupturing by 1969.

The history of contemporary movement along the Reservoir fault (beneath the east side of the Baldwin Hills Reservoir basin) is known from monthly readings taken on strain gauges extending across a crack in the concrete inspection gallery beneath the reservoir and across the fault. The opening of the crack, which reflected displacement along the underlying fault, began in 1952 and was monitored until the time of reservoir failure in 1963. The history of crack activity (Fig. 10) shows three stages of progressively accelerating movement, upon which are superimposed individual jumps that reflect events of larger fault movement. These stages correspond to episodes of surface rebound or uplift, identified by Leps (*13*) from records of comparative leveling of bench marks at and south of the reservoir.

When the history of movement along the Reservoir fault is compared with the operational history of injector wells located in the vicinity of the fault's southerly subsurface projection, the evident influence of pressure fluid injection on fault activity can be recognized. The earliest stage of crack extension began prior to the initiation of the water-flood program and continued until just after the third of the first three injectors along the fault was brought into service. This stage may therefore represent ground movement attributable to differential subsidence with little or no effect of pressure injection. Then, in late 1957, a jump in the crack extension was followed by a doubling in the rate of extension and also by the first episode of surface rebound. This level of activity then continued for 3½ years, until 1961. At that time, three more injector wells were activated in the area. This increase in injection was followed several weeks later by another fault jump and a further increase—a tripling—in the rate of crack extension. Mid- and late 1963 was a period marked by intense activity in injection operations. Injection was started (or restarted) in four additional wells close to the reservoir, and severe and unusual operational problems were occurring in those wells (now nine) near the fault. Uncontrolled loss of fluid occurred in five injectors. One well appears to have been

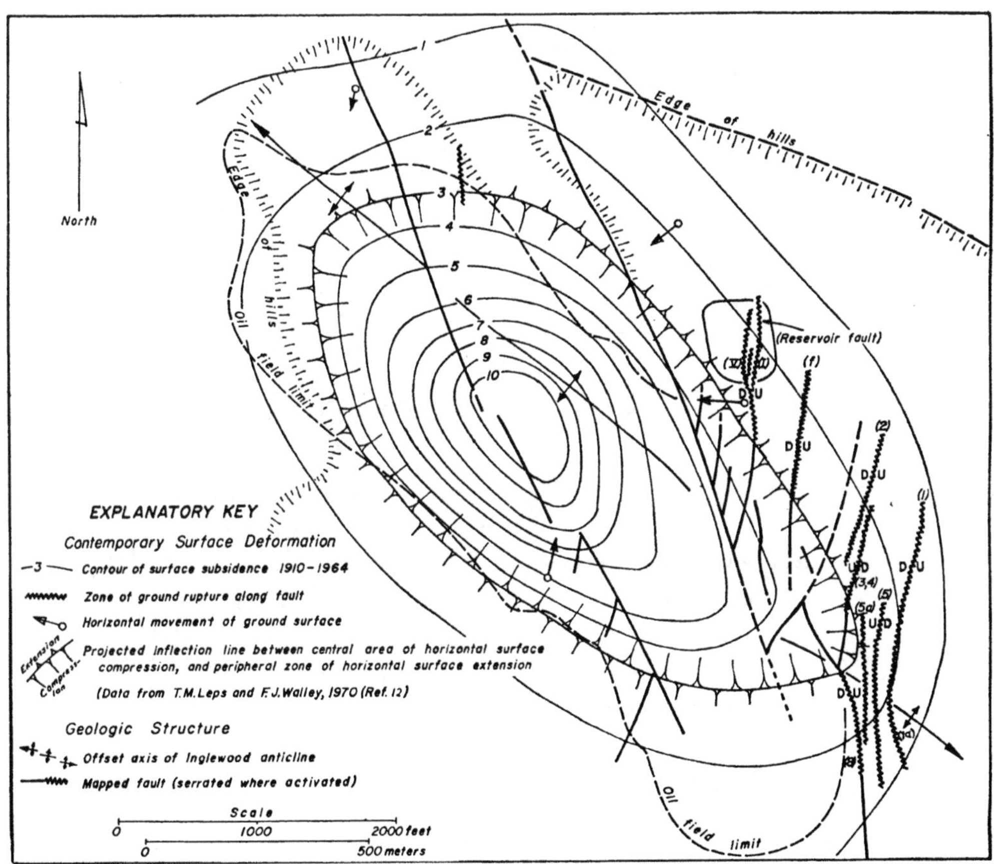

Fig. 6. Contemporary surface deformation, Baldwin Hills–Inglewood area.

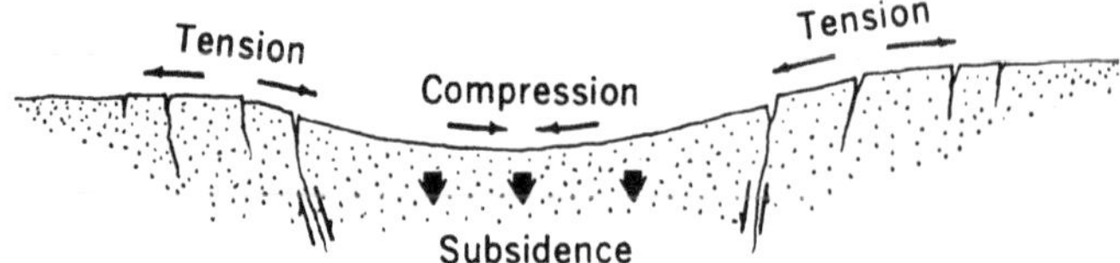

Fig. 7. Diagrammatic representation of fault activation attributed to subsidence.

pinched or sheared off at depth. In May, 6 months before the reservoir failure, brine was observed seeping from cracks in the ground surface south of the reservoir and on the trace of the Reservoir fault. Wellhead injection pressures during 1963 were erratic, but pressures in excess of 400 pounds per square inch were common; 400 pounds per square inch creates a gradient of 0.73 pound per square inch per foot at a depth of 1200 feet, the top of the injection interval. Late in 1963, an episode (or episodes) of movement occurred along the Reservoir fault, which culminated in failure of the reservoir. This fault movement, though apparently normal in sense of displacement, actually involved upward movement of the footwall block and was therefore distinctly related to the earlier episodes of surface rebound in the area.

In considering the possible cause-and-effect relations among injector activity, subsurface problems, and fault movements, it seems significant that all recorded episodes of fault movement since 1957 have occurred *after* one or more of the following: initiation of injection in nearby wells, increases of injection pressure, or problems such as dropping of fluid pressure concomitant with increases of fluid take, loss of fluid in narrow zones, and so on. The sequence of events suggests that the injection caused or contributed to the movement.

As is evident from Fig. 7, most of the ground cracking occurred in and beyond the outer margin of the injec-

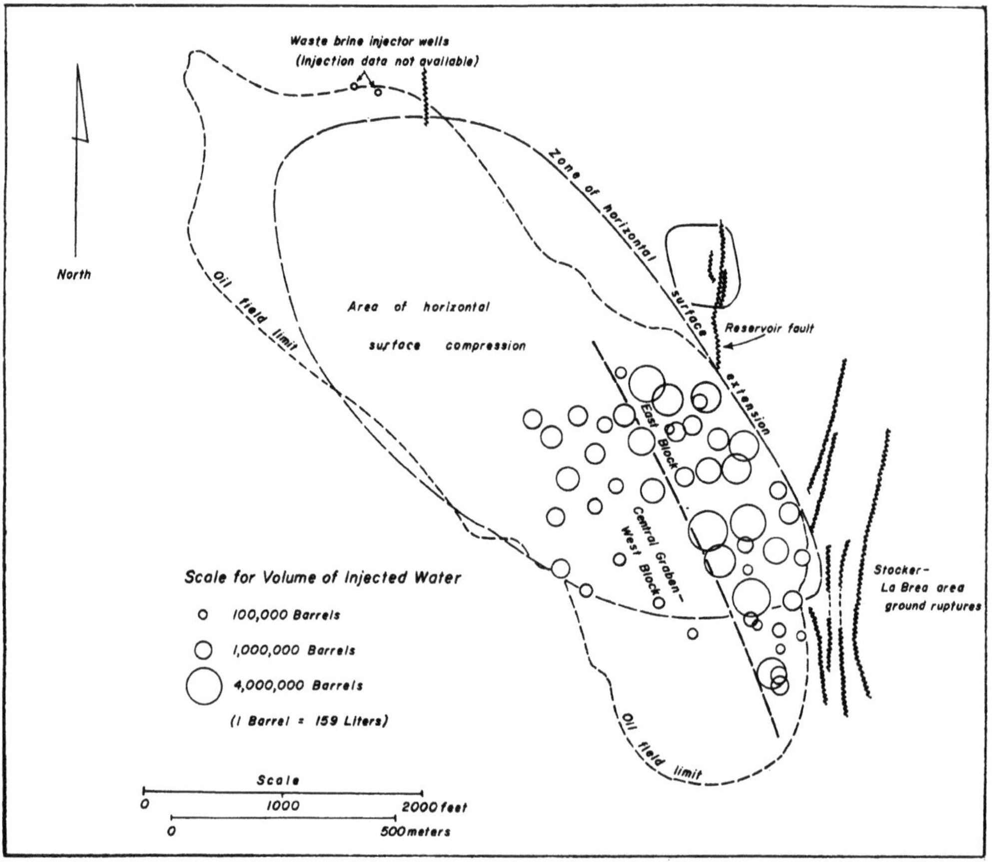

Fig. 8. Zones of ground rupture and cumulative amounts of injected water, Baldwin Hills–Inglewood oil field, through 1963.

tion area. The explanation for this condition lies in the three-dimensional geometry of the faults, and also in the relation of the area of surface rupture to the area of ground surface extension and compression induced by subsidence. Both projection of fault plane dips measured at the surface and reference to the cross section constructed from subsurface data (Fig. 4) indicate that most of the east block faults dip to the west. Consequently, fault planes in the immediate vicinity of injection intervals reach the ground surface, 800 to 1500 feet above, some hundreds of feet to the east. Faulting initiated at depth and propagated to the ground surface would therefore result in surface rupturing hundreds of feet to the east of the subsurface point of origin. Also, reference to the map of surface deformation (Fig. 6) shows that the surface rupture has occurred in the zone of ground surface extension but that it dies out as the faults enter the zone of surface compression. The surface compression evidently "clamps" the fault surfaces together, whereas the extension is associated with a pulling apart.

The only active ground rupturing known to us outside the east block area in the Baldwin Hills is a north-south zone of cracking at the north margin of the Hills (Fig. 8). This cracking reportedly began to develop about 10 years ago, well before the time of any secondary recovery operations in that vicinity. However, the rupturing is very near two fluid injection waste disposal wells, which probably have affected the strength of the subsurface in the same fashion as the pressure fluid injection for secondary recovery of oil.

Although water flooding was under way in 1970 throughout most of the Inglewood oil field, no ground rupturing other than that just noted has been recognized in the field's central area, or around its north, west, and south margins; it still continues, however, in the east block area. The absence of fault activation within other parts of the field can be explained by the "clamping" effect of surface compression in the ground there. The west margin of the field is probably in surface extension; unlike the faults in the east block, however, most of those in the west block die out upward at points beneath the ground surface. Consequently, even if fault activation has occurred at depth, it evidently has not propagated upward through the unfaulted shallower ground.

Fluid Injection Causes Faulting

The mechanical behavior of systems that consist of solid particles with fluid-filled voids has long been investigated experimentally by workers in the field of soil mechanics, who have derived mechanical principles that have been of great practical use in the design of civil engineering works. Changes in volume and in strength of soil and rock materials with fluid-filled voids have been shown to be fundamentally related to change in stress within the solid skeleton. An increase in stress in the solid skeleton produces a decrease in its volume, owing to elastic deformation of the skeleton and, in the case of uncemented assemblages, to closer packing of particles. For liquid-saturated systems, stress within the solid skeleton is generally expressed in terms of effective stress $\bar{\sigma}$ (*18*), given by the expression

$$\bar{\sigma} = \sigma - u \tag{1}$$

where σ is the total stress acting across some plane within the system and u is the fluid pressure in the region of the plane. This relation is more complex for a solid-liquid-gas system, but it holds approximately if the liquid and gas pressures are approximately equal. It follows from the effective stress principle that a decrease in fluid pressure produces a like increase in effective stress and therefore a decrease in volume of the solid skeleton.

The principle of effective stress provides a reliable theoretical basis for the prediction of settlements caused by changes in either total stresses or fluid pressures, if the compressibility of the solid phase is known from laboratory tests. No laboratory testing of the Inglewood oil sands has been performed, to our knowledge, but an a posteriori theoretical analysis of subsidence in the Inglewood oil field has been made by Castle and Yerkes (*4*), who used reasonable compressibility data for the oil sands. Their analysis indicates that increased effective stresses accompanying fluid withdrawal and decreasing fluid pressure will lead to subsidence of the same order of magnitude as that shown in Fig. 6.

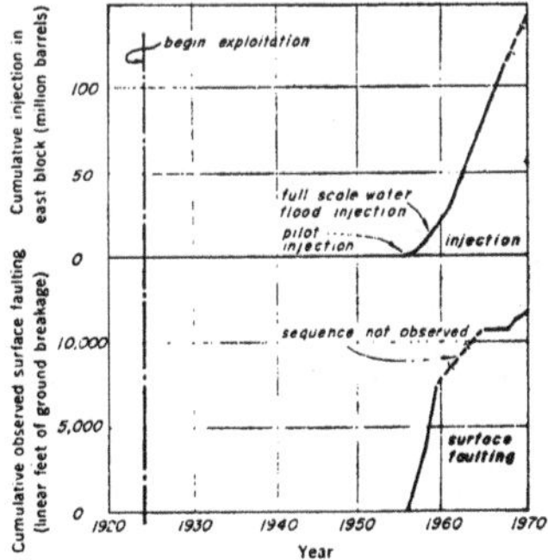

Fig. 9. Time relationship of ground rupturing and east block injection, Baldwin Hills and Inglewood oil field.

Subsurface fluid *injection* has been practiced less widely than fluid withdrawal, and theoretical and empirical understanding of the mechanical effects of high-pressure fluid injection has lagged accordingly. Such injection into the subsurface has many practical applications, among the most important of which is water (and steam) flooding of oil-producing zones, with the object of enhancing production by maintaining hydraulic pressure and flushing oil toward producing wells. This process is known as secondary recovery.

Secondary recovery is now widely practiced by the oil industry, and an advanced technology has been developed to service these operations for production enhancement. The economic feasibility of production in many older oil fields now depends on this technology.

The mechanical effects of high-pressure fluid injection on the volume of sediments may be described by the principle of effective stress in a fashion corresponding to that for withdrawal. It is apparent from Eq. 1 that an *increase* in fluid pressure, induced perhaps by pumping fluid into the ground at depth, would be accompanied by an equal decrease of the quantity described as the effective stress. This would result, in accordance with experimental results, in an expansion of the soil element subjected to the increase in fluid pressure. Pressure injection over a wide area might then be expected to result in a rise of the overlying ground surface, or, in the case of formations that had previously been compressed through reduction of fluid pressure, restoration of former higher pressures could result in rebound of the compressed sediments. Results of laboratory tests suggest that the amount of rebound may be roughly equal to or, under the less usual but predictable situation of rebound from virgin compression, substantially less than the compression accompanying a like decrease in fluid pressure.

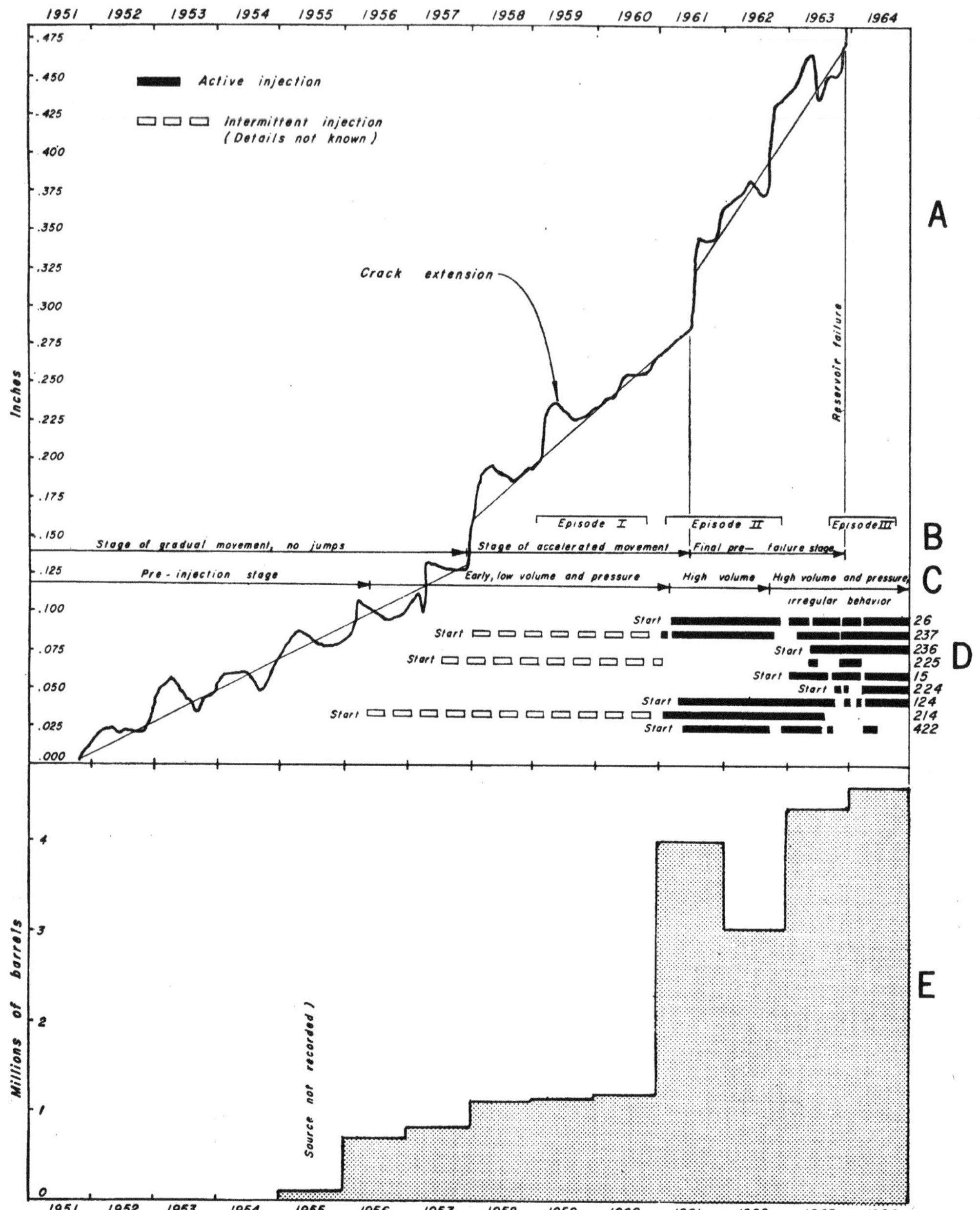

Fig. 10. Extension of Reservoir fault crack in inspection chamber of Baldwin Hills Reservoir as related in time to episodes of surface uplift, activity of injector wells, and volume of injected fluid near Reservoir fault. (A) Extension of crack over Reservoir fault in Baldwin Hills Reservoir inspection chamber (data from Los Angeles Department of Water and Power). (B) Episodes of surface uplift east of Reservoir fault (identified by T. M. Leps) and stages of crack extension activity. (C) Stages of injector well activity. (D) Times of operation of indicated injector wells located near the Reservoir fault. (E) Volume of fluid injected in wells located near the Reservoir fault.

Ground heave attributable to injection-induced volumetric expansion of substrata apparently has not generally given rise to secondary problems, perhaps because injection programs usually involve relatively small quantities or, not infrequently, are carried out in areas simultaneously or previously affected by subsidence and perhaps lacking structures such as reservoirs that could be seriously damaged. However, another side effect of high-pressure injection is the weakening of substrata that inevitably accompanies artificially induced increases in fluid pressure. This will occur in both cemented and uncemented rock materials; it can be more easily visualized in the case of uncemented sediments such as those in the Baldwin Hills.

The shear strength of unconsolidated sediments is mainly frictional and can be related to the quantity previously described as the effective stress. This can be shown experimentally by confining a sand specimen in a box (or flexible membrane) and then causing the specimen to fail by shearing. The shear stress on the failure plane at failure will be, for a typical sandy sediment, about 0.6 times the normal stress on the failure plane (Fig. 11A). This relation holds in a simple solid-fluid system, if the effective stress is taken as the normal stress. It can also be demonstrated theoretically and experimentally that the relation between major and minor principal stresses (effective stresses) at failure for the same material will be approximately as shown in Fig. 11B.

Subsurface stresses in relatively unconsolidated sediments under conditions of normal (gravity) faulting are approximately as shown on Fig. 11B. The major principal stress is vertical; expressed in terms of effective stress, it is equal to the pressure exerted by the weight of the overburden less any fluid pressure. The faulting is by definition a shear failure of the material, from which it may be deduced that the horizontal effective stress is at or near one-third the vertical effective stress.

Figure 12 illustrates the fluid pressure on a hypothetical system at a depth of 1600 feet—the depth of intensive injection activity in the east block—and initially in a state of incipient normal faulting. A geostatic pressure gradient of 1 pound per square inch per foot is assumed; thus, total vertical stress is 1600 pounds per square inch. The hydrostatic fluid pressure gradient is 0.43 pound per square inch per foot of depth, or 690 pounds per square inch at a depth of 1600 feet. From Eq. 1, effective vertical stress is then 1600 less 690, or 910 pounds per square inch. Horizontal stress includes a fluid stress component, equal to the assumed hydrostatic fluid pressure of 690 pounds per square inch at a depth of 1600 feet, and an effective stress component that, under the assumed state of limiting equilibrium, is about one-third of the vertical effective stress, or 303 pounds per square inch.

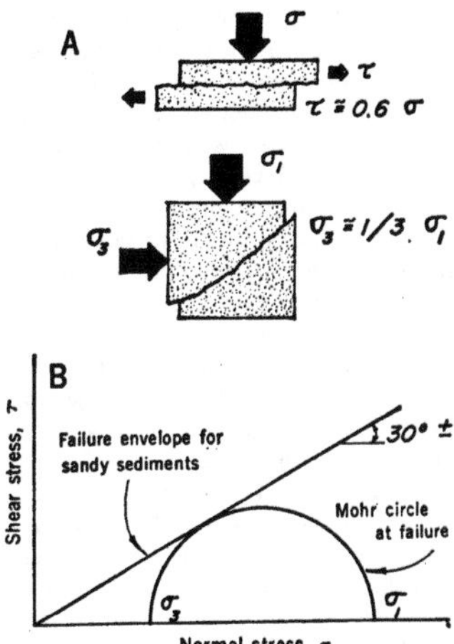

Fig. 11. Typical failure conditions in uncemented sediments: (A) as determined by laboratory tests; (B) Mohr failure criterion.

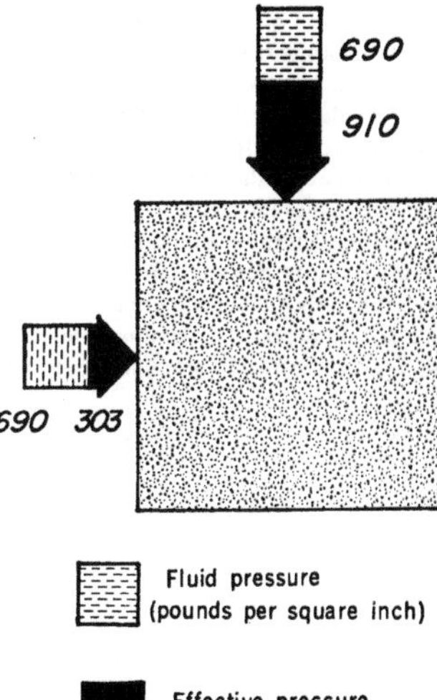

Fig. 12. Hypothetical natural stress conditions in element at a depth of 1600 feet. Normal faulting is imminent.

Reduction of fluid pressure increases both major and minor principal stresses, produces compression of the element, and, by reduction of the ratio of principal stresses, "backs away" from the critical state of shearing equilibrium.

Increase of fluid pressure has an opposite, destabilizing effect. If the fluid pressure is raised beyond the natural hydrostatic condition, the principal stress ratio will tend to increase, a condition not compatible with the previously defined state of limiting equilibrium; shearing will then occur and will continue until some readjustment of stresses restores equilibrium.

If normal faulting had occurred in the distant geologic past but a readjustment of stresses had since provided a margin of safety against reactivation of existing normal faults, and if subsidence were not contributing to disequilibrium, then the initial stress condition in the east block area might be somewhat different from the postulated condition of limiting equilibrium. However, unless a recent and major change in orientation or magnitude of principal stresses in the vicinity of the east block is postulated, it is reasonable to assume that the ratio of horizontal to vertical effective stresses across the recently active, normal fault planes would be no greater than about 0.5, a typical ratio obtaining in "normally consolidated" sediments under tectonically quiescent conditions. In this case the initial horizontal effective stress would be 455 pounds per square inch for the example in Fig. 12, and a fluid pressure increase from 690 to 915 pounds per square inch (gradient increase from 0.43 to 0.57) would be required to reestablish the limiting principal stress ratio of one-third; hence, faulting would be activated in the ground affected by such a pressure increase. It would thus appear that only a small increase in fluid pressure (a gradient of 0.6 or less) should be sufficient to trigger at least local shear failure in the east block, even if the material was not stressed to the point of failure by subsidence or active normal faulting.

The relation of fluid pressure to formation fracturing has been treated by Hubbert and Willis (*19*), who show that hydraulic fracturing will occur in tectonically relaxed regions at injection pressure gradients of less than 1. In the above example, the reduction of horizontal effective stress (455 pounds per square inch) to zero should result in fracturing along vertical or near-vertical planes and would occur at an injection pressure gradient of 0.62. As demonstrated above, this hypothetical fracture condition is statically inadmissible, given the shear failure criterion adopted for the material, without readjustment of stresses, which could arise only through shear deformation. The history of difficulties with injectors in the east block and comparison of injection gradients with fracture gradients in similar tectonic environments leave little doubt that fracturing could (and did) occur in the east block at injection gradients well below 1.0. Fracturing would and doubtless did result in extension of the effects of a single injector to points hundreds or thousands of feet away from the injection interval, thus exposing large volumes of sediment to the elevated injection pressures soon, or perhaps immediately, after the raising of pressure at the wellheads. Moreover, fracturing must have been accompanied by shear deformation, unless initial horizontal and vertical stresses in the injection zones were equal, which was certainly not the case in the recently faulted sediments comprising the east block.

In the Baldwin Hills east block area, injection pressure gradients of as much as 0.9 were employed, and relatively large bodies of ground, including an extensive reach of the Reservoir fault, were subjected to pressure gradients greater than 0.7. The history of casing failures in injection wells, sudden injection pressure losses, uncontrolled major entries of fluid at the tops of injection intervals (leading in at least one incident to discharge of injection fluid at the ground surface), all indicate that injection pressures were high enough to markedly reduce, or, in some areas, locally to eliminate the shear strength along preexisting fault surfaces and, also locally, to "fracture" the formation hydraulically and to create new fracture surfaces. The weakened zones must have been confined initially to the immediate vicinity of the injection wells, but subsequently, as new injectors were put into operation, relatively larger volumes of material became affected. In the case of the Reservoir fault, significant shear displacement probably had occurred along much of the break by mid-1963, and, at the end of that year, rupture propagating to the ground surface was of sufficient magnitude to shear the clay lining of the Baldwin Hills Reservoir and cause its failure.

Conclusions

That the earth-crack ground rupturing of the Baldwin Hills was genetically related to high-pressure injection of fluid into the previously faulted and subsidence-stressed subsurface seems established beyond reasonable doubt. The fault activation appears to be a near-surface manifestation of stress-relief faulting triggered by fluid injection, a mechanism identified as being responsible for the 1962–65 Denver earthquakes and for generation of small earthquakes at the Rangely oil field in western Colorado (*20*).

These examples of fault activation through the response of stressed ground to artificially induced increases in subsurface fluid pressure demonstrate some of the mechanically predictable consequences of injection of fluid into the ground, a practice that is becoming increasingly widespread not only in secondary oil recovery operations but as a means of industrial waste disposal and groundwater management.

Experience in the Baldwin Hills suggests that, although fluid injection operations may be carried out for beneficial purposes, the effects of such injection on the geologic fabric can be serious and far-reaching.

References and Notes

1. Metric equivalents: 1 gallon = 3.785 liters; 1 acre = 0.404 hectare; 1 foot = 0.3 meter; 1 square mile = 2.590 square kilometers.
2. "Investigation of Failure—Baldwin Hills Reservoir" (California Department of Water Resources, Sacramento, 1964).
3. R. F. Yerkes, T. H. McCulloh, J. E. Schoellhamer, J. G. Vedder, "Geology of the Los Angeles Basin—An Introduction," *U.S. Geol. Surv. Prof. Pap. 420-A* (1965).
4. R. O. Castle and R. F. Yerkes, "Recent Surface Movements in the Baldwin Hills, Los Angeles County, California," *U.S. Geol. Surv. Open File Rep.* (1969).
5. J. D. Moody and M. J. Hill, *Geol. Soc. Amer. Bull.* **67**, No. 9, 1207 (1956).
6. R. O. Castle, "Geologic Map of the Baldwin Hills Area, California," *U.S. Geol. Surv. Open File Rep.* (1960).
7. Unpublished 1969 operating data of the Inglewood oil field.
8. Metric unit: 1 pound per square inch = 0.0703 kilogram per square centimeter.
9. F. H. Oefelein and J. W. Walker, *J. Petrol. Technol.* **1964**, 509 (May 1964).
10. From unpublished data of S. A. Hayes (1943 and 1955) [cited in (*2*)].
11. From unpublished data of F. J. Walley (1963) [cited in (*2*)].
12. T. M. Leps and F. J. Walley, unpublished data.
13. T. M. Leps, unpublished reports and personal communications.
14. Metric unit: 1 acre-foot = 1233.5 cubic meters.
15. U. S. Grant, IV, "Subsidence of the Wilmington Oil Field, California," *Calif. Div. Mines Geol. Bull. 170* (1954), vol. 1, chap. 10, p. 19.
16. D. U. Deere, *Pa. State Univ. Miner. Ind. Exp. Sta. Bull.* **76**, 59 (1961).
17. K L. Lee and M. E. Strauss, paper presented at the International Symposium on Land Subsidence, Unesco, Tokyo, 1969.
18. K. Terzaghi, *Theoretical Soil Mechanics* (Wiley, New York, 1943).
19. M. K. Hubbert and D. G. Willis, *Trans. AIME* **210**, 153 (1957).
20. J. H. Healy, W. W. Rubey, D. T. Griggs, C. B. Rayleigh, *Science* **161**, 1301 (1968).

21. We thank T. M. Leps, R. H. Jahns, and others for suggestions and criticism.

Reprinted from *Ass. Southeastern Biologists Bull.*, **18,** 117–128 (1971)

Soils are seen as a dynamic nutrient reservoir formed by slow geochemical weathering over periods of tens of thousands to millions of years. Nutrient cation exchange capacities vary as a function of the original substrate in young soils, climate, soil hydrology, and vegetation. Land use practices such as forest cutting upset the nitrogen cycle within the soil which may permit bacterial oxidation of ammonia to nitrate with release of hydrogen ions. These H^+ ions displace biologically available stored soil nutrients and, in some cases, may render a soil unable to sustain any but primary successional species within a few decades after forest removal. Forest fire may kill the surface soil microorganisms by both heat and increase in pH and thus protect soil from natural nutrient depletion following fire. In addition, organic soil colloids are translocated into the superficial layers of soil during fire and further protect that soil from surface water erosion and clay-mineral nutrient leaching.

Major questions remain as to the effects of forest cutting and other land use practices such as repeated tilling and monocultural agriculture upon the nitrogen cycle and soil microfauna. Until some of these questions are answered, the advocacy of some federally approved and widely used logging practices is seriously questionable, and the consequences of continuation of present federal policies are grave.

14

Soil Destruction Associated with Forest Management and Prospects for Recovery in Geologic Time

ROBERT R. CURRY

Environmental Geology
University of Montana, Missoula 59801

I am a geologist and geologists generally deal with non-renewable resources. Forests are supposedly renewable resources but, in the sense of the lifespan of mankind, forest soils are only very slowly renewable and I believe that many of the forests which these soils sustain are essentially non-renewable. Much of the American landscape, upon which we rely for water, enjoyment, mental well-being, wood products and foods, I believe may be doomed to the same fate that befell the great forests of Dalmatia under the Greek and Roman axes, the Cedars of Lebanon, and the once productive areas of the English Midlands. I should like to address myself to a geologist's view of the development and maintenance of the world's soil resource and its conditions of equilibrium and disequilibrium. As an example of the biological crisis that faces much of this country as a result of short-sighted land use policies and lack of objective information on soils, I should like to discuss particularly the handling of our nation's forest resource. Much of this discussion will be directed toward management of the forest resource of western United States, but the principles apply to most attempts at management of primary biologic productivity from soil-covered lands including conventional agriculture and tree-farming as practiced in the Southeast.

In June, 1960, Congress passed the Multiple-Use Sustained-Yield Act (M.U.S.Y.A.) which, among other things, said that the National Forests should be maintained for multiple uses in *perpetuity*. Public debate, often rather one-sided, rages on about the merits of truck-mechanized logging in general as practiced in much of the U.S. including the Southeast and clearcut logging in particular. Most citizens are esthetically affronted by logged over land and intuitively feel that somehow the M.U.S.Y.A. is being interpreted to mean terminal logging. Such citizens frequently turn to the scientific community and a few renegade foresters to find valid proof that something is wrong with the forest practices that leave so much waste and ruin. Most technical criticism today is based upon erosion accompanying logging — there being some who believe that erosion is intolerably rapid while others claim some erosion is permissible and that the forests will not suffer as a consequence. Those poining out evidences of rapid erosion sometimes argue that, while gullying and soil loss in the forests may not damage future forest growth potential; the effects of the transplanted silt, sand, and clay upon adjacent streams bothers them. The facts are that in almost all measured cases, soils in logged areas are being lost faster than they are being formed and the sustained yield concept is being patently violated. I shall return to this point with specific details later.

Another argument sometimes offered by those offended by current forest removal practices — euphemistically called "management," is that actions in the once-forested areas upset the hydrologic regime of areas down valley from the cutover watersheds. Although those complaining may have a valid argument, the law is such that it offers little or no protection from hydrologic variables — generally termed "acts of god" by the English upon whom our legal precedents in this area are based. In many geographic areas, United States Forest Service and commercial logging company personnel can correctly argue that forest removal actually increases total water runoff from a watershed. We all know that we need more water so just because your fields are flooded in the spring and the irrigation ditch goes dry in the fall after forest excision, you shouldn't complain; someone downstream will be using that extra water, and we have made some more flood-control excuses for the Corps of Engineers. Also, it is agreed warm water is biologically more productive than that fishy cold water you had before. A Montana industrial planning director has gone so far as to suggest that westerners merely need to learn to relish catfish instead of native trout and they would then stop complaining about declining water quality.

In this paper, I wish to go into the data available on soil erosion and hydraulic variation associated with deforestation. I also want to summarize some recent work on soil and forest watershed chemistry that indicates that chemical reactions in the soil following logging in many areas cause the soils to be stripped of their nutrients so rapidly that soil sterility is virtually assured in a matter of two to, at maximum, six theoretical recyclings of forest growth — thus also fully negating the law as stated in the M.U.S.Y.A.

Let us first review some of the basic statistics regarding forests and forest production in the U.S. (derived largely from Statistical Abstracts of the United States, 1970).

In 1969, there were 600,000 lumber and wood products related jobs in the U.S. — 78,000 in the forests, 232,000 in sawmills, and the remainder in wood products manufacturing. The industry is losing people, offering 27,000 fewer jobs in '69 than in '60.

There are 0.76 million km^2 of National Forests, only 0.46 million of which actually possess merchantable trees — but this land produced a value of cut of 328 million dollars in 1969. Total gross receipts were actually only \$319 million — so someone did not pay up. The federal government allotted to the U.S.F.S., for fiscal year 1970's operation, only 31 million dollars — so you can see that they are quite dependent upon forest sales to exist as an organization — receiving only 12 million in income from grazing and special land use fees compared with 307 million for timber sales, of which most goes into the federal general fund. For comparison, Department of Interior had authorization for expenditure of \$967 million for the same year. Clearly, the U.S.F.S. funding is grossly inadequate in terms of the area of land for which they are responsible. Additional burdens are placed upon the agency by the fact that appropriated funds are earmarked for items such as roads and trails, fighting forest fires, preventing forest fires, timber sales, etc., and little opportunity is allowed for ecologically sound agency operation.

Of the total of 3.06 million km^2 of commercial forest lands in the United States, the mountain states and west coast including Alaska hold 1.45 million km^2 while the south Atlantic and Gulf Coast states, including Texas, hold 0.81 million km^2. The 29 percent of the U.S. acreage in the south produces 31 percent of the nation's annual board-foot production while the 47 percent of the nation's acreage in the west produces 60 percent of its annual cut. This is so despite the fact that southern forests are more productive in terms of board feet of timber grown per acre. The anomaly is simply due to the fact that old-growth forests are still being cut in the west but not in the south.

In 1969, there were 38 billion board feet of timber produced in the U.S. total while we "consumed" some 43 billion. Consumption has been essentially steady since 1950. We export 1 billion board feet, mostly to Asia. Last year when the President was calling for a crash program to supply the nation's timber and cut over the remaining forests, the Port Orford, Oregon, lumber dealers had such a glut of sawn timber with no buyers that they took sawed and billeted timber and chipped it to ship to Japan where the inscrutable Japanese use our wastes to make particle-board for their own consumption and can profitably ship the excess back to the U.S. for sale in the northwest.

Imports, largely from Canada, account for the 6 billion board feet deficit. Standing sawtimber in the U.S., including Alaska, totals 2 trillion board feet. We would, therefore, have 50 years of consumption at our present rate remaining if no new timber grew. What are the chances that it will grow?

Before we begin to discuss forests and their soils, we must define a soil in some workable fashion. I see a soil as a dynamic biogeochemical system of nutrient and water exchange. As a soil evolves, through physical and chemical weathering of rock and other parent materials at a given site, its ability to store and exchange upon biologic demand the essential nutrients for plant and animal growth increases with geologic time up to a maximum capacity, which it holds for several hundred thousand years, after which it slowly loses some of its nutrient exchange capacity. The nutrients exchanged are generally K, Na, Ca, Mg, & P with lesser amounts of the 14 so called micronutrients, like Fe, Mn, and Boron. The actual structures allowing this storage and exchange are clay minerals and organic soil colloids. Organic humus breaks down with time to a negatively charged complex which absorbs nutrient cations and keeps them from being readily leached away from the system by water as soon as they are released by chemical weathering of fresh mineral matter.

Clay minerals are complex alumino-silicate assemblages resulting from partial breakdown of some common rock-forming minerals like feldspars and micas. They are generally not present in any great amounts in unweathered bedrock or glacial deposits and should not be confused with "clay" in the sense that clay minerals have a specific chemical identity and physical structure, uniquely formed at or very near the earth's surface, whereas "clay" sometimes refers to a texture of clayey or silty particles which may or may not be composed of clay mineral particles.

The clay minerals are composed of sheet-like assemblages of Al, SiO_4 and OH^- molecules with spaces within the lattices and between the sheets to temporarily store cations like Ca, Mg, Al, K, etc., which can be released upon biologic demand or by leaching with acids whenever the small hydrogen ions squeeze into the lattice and bump the larger metallic cations out into solution. Plant root hairs release hydrogen ions to specifically accomplish their nutrient-gathering goal.

In summary then, soils are unique physico-chemical compounds which possess the ability to store and exchange nutrients. As bare mineral matter begins to accumulate a decomposing mat of humus and a few percent clay minerals, its potential nutrient storage budget increases and the kinds of plants it can support become more varied and may constitute a progressively larger biomass. So-called primary plan succession is generally nothing more than a reflection of soil development.

Chemical weathering of organic humus and minerals to form cation exchange media goes on at a rate which is a function of temperature and available moisture. In areas with summer drought and cold winter precipitation (like much of the mountain west), soils can form only very slowly. In western Montana, for instance, it has taken about 16,000 years to form a 30-60 cm. deep mantle of partly weathered mineral matter and small accumulated organic mat. Most of this soil is merely a thin oxidized zone on the surface of the earth. These mountain soils have at most about 9.8 kg. of cation exchange media per square meter of land area.

Organic matter is found in the top few centimeters, while the clays, once formed, are translocated to a clay accumulation zone about 15 cm. beneath the surface. Below that depth, little actual storage of nutrients is accommodated and plant roots generally find physical support and water as well as rather minute amounts of nutrients derived directly from the bedrock or oxidized soil by biologic release of hydrogen ions or from downward percolating water that gained its nutrient load from the overlying clays plus organic acids.

When you remove the soil, you remove the ability of young plants to become established if those plants are dependent upon a full and readily available nutrient supply after germination. If one compares soil to the earth's full breast, then the top 15 cm. is the teat. Different plant species have different requirements. The so-called pioneering species like lichens can directly release nutrient cations from fresh minerals by virtue of possessing a complex chemical arsenal of acids of various unique types and an ability to withstand great physiologic drought. Some conifers and many shrubs can reproduce on a bare soil with little or no humus so long as some base-cation-exchange clay minerals are present near the surface. Other conifers and hardwood forest species require deep clay accumulations for metallic cation storage plus water storage and much available humus for harboring water, microflora and fauna and essential nutrients like nitrogen.

One cannot readily "build" a soil. Laboratory clay mineral weathering is very time-consuming and natural distribution of clays is such that, although ubitiquous, there are not large minable deposits of clays of the types that hold most nutrients, like montmorillonite and vermiculite. Mere spreading of fertilizer on mineral soil or even incorporating it by tilling is fruitless and a criminal waste of energy. Without storage sites, the fertilizers merely dissociate and volatolize or leach away in 1-2 years, while many thousands of years are necessary for even the smallest accumulation of storage media. Leached fertilizers pollute streams and cause eutrophication by encouraging algal growth in those fertilized waters.

Erosion

Common erosion rates in tilled agricultural regions such as midcontinental U.S., average about 2-3 m./1000 years (compared with a pre-human erosion rate based upon geologic evidences of 5-10 cm. in the same areas) (Task Committee on Preparation of Sedimentation Manual, Committee on Sedimentation, Hydraulics Division, 1970; Walker, 1966; Ursic, 1965). Even in the warm, humid, long-stable southeastern U.S., soils seldom develop to much more than 3 m in depth and the young soils of glacial regions from Ohio northward possess generally less than 0.6 m of total exchangeable substrate materials. Clearly most U.S. agricultural regions will not sustain tilling for more than 300 years and many such hill regions in the southeast and the arid regions of Oklahoma and Texas are nearly depleated of their soils today and require annual expenditures of fertilizer and much energy just to make the most primitive plants grow.

For the U.S. as a whole, without the intervention of mankind, soil erosion goes on at a rate of about 3 cm per 1000 years (Judson & Ritter, 1964; Corbel, 1964). Thus, on the average, soils would have to form at a rate equal to or greater than 3 cm/1000 years or 3 mm/century to allow equilibrium. For the mid-temperate latitudes, this seems to be about the right order of magnitude. But averages do not tell the forest story. For montane western Montana about 0.7 m of soil, not all composed of materials with exchange capacity, has formed in 16,000 years for a rate of or 3.8 cm/1000 years absolute maximum rate of formation. True average rates may be one-half this figure. Local erosion, on specified watersheds, is as much as 15 cm/*100* years — or 162 kg/century removed from each square meter per century. If formed at a rate of 3.8 cm/1000 years and 0.7 m deep, we would have 400 years of soil left to be stripped. Since the 15 cm− rate, as measured from rates of siltation in streams and behind debris dams in small watersheds in logged areas, appears to have been in force for the last 50 or more years, we may have as little as only 350 years of forest soils left in some locales. Or perhaps enough time for three more cuts before sustained yield becomes a joke.

Erosion can be estimated from measurements of sediment carried in streams draining the areas (Ursic, 1965). Turbidity is not a measure of rate of erosion. There are three basic reasons for this:

1) Significant erosion can occur without measurable turbidity. By recent measure in a U.S.F.S. experimental forest, 100 kg/ha/yr suspended particulate matter can be carried off a small 160,000 m^2 (39 acre) deforested watershed without any measurable turbidity (Likens, et al, 1970). That is a lowering of 1.4 cm/1000 years or about one-half the rate of soil formation.

2) Erosion selectively removes the cation exchange media — the clays because they are light, fine and easily mobilized by water since they repel one another having like negative charges; and the organic matter since it lying upon the ground surface. Natural clear runoff contains about equal parts of mineral matter and organic matter.

3) Erosion can progress in deforested watersheds at great rates without the material actually reaching a stream for measurement. Current "good" management practice is to leave a vegetation buffer along streams to protect fish from water heating and stream siltation which deleteriously affect salmanoid species. The Federal Water Pollution Control Administration also recommends debris dams and the leaving of slash on the deforested hillsides to reduce sediment reaching water courses. This is all well and good but it does little to protect the soil — only the fish — and it even protects the fish only from silt without regard for the chemical quality or toxicity of the water. I will show shortly how this practice of leaving forested buffer strips along streams, while preserving spawning grounds a little longer, is assuring high fingerling mortality from chemical water pollution in that supposedly clear water draining "properly logged' watersheds is toxic to young fish near the logged regions.

When logging occurs in steep watersheds, erosion rates go up tremendously. In the Coast Ranges of Northern California, I and others have estimated regional rates of erosion based upon suspended sediment loads in streams to learn that, for instance, there is an average rate of lowering of 15 cm/100 years for the entire 7800 km^2 area of Eel River Basin (Wahrhaftig and Curry, 1967; Wallis, 1967). In this area of the redwood forests, soil is being lost 100 times faster than being formed for a predicted total soil loss in 400 years — much less time than it takes to grow a single second crop of the indigenous *Sequoia* on hillslopes.

Recent work by the U.S. Geological Survey and University of California personnel have shown our original estimates of rate of erosion to be far from correct. Our error was in assuming that the measured suspended sediment load of the river represented the total eroded materials. Later work measuring sediment load transported down the bed of the Eel River and tributaries showed the true erosion rate to be greater than our estimates. Since soil is largely carried as suspended particles, and since gullying supplying the bedload progressively exposes more topsoil to erosion as the gullies work headward, a value somewhat less than 400 years to total soil loss is probably most nearly correct.

Theoretically, areas with thick mature soils, much clay and organic matter, and a long geologic history of weathering and rapid soil formation should be the ideal places to log. In the U.S. such places are restricted to the Southeastern states (Georgia, N. Florida, Alabama, Louisiana, Texas, and the Carolinas) and coastal parts of the Pacific Northwest. Here soils are deep and should tolerate some erosion.

In the southeast, the deep red lateritic soils of the coastal plain are among the deepest and best developed in the United States. This is so because stable land surfaces have been exposed to warmth and summer rains for 2 to 10 million or more years to thoroughly weather and leach the top 6 or more meters of surface materials. Due to the high soil moisture and high ambient temperatures, as well as copious acid litter production and flat poorly drained topography, southeastern U.S. soil production would be higher than at any other U.S. site except parts of Hawaii. In the southeast rates of initial soil formation may be on the order of 6 cm/1000 years based upon soil development on elevated Pleistocene marine shorelines, but this rate of two or more times the national average drops to as little as 0.3 cm/1000 years after forming 3-6 m soil-depths in 1-2 million years. Thus, although initial rates of formation are high, limitations on effective depths of oxidation and hydrolysis ultimately limit the depth of soil development and further time merely permits the slow solution of the less soluable components of the soil and its nutrient loads, resulting in eventual breakdown of the clay mineral structures themselves. Thus, in the tropics and older soils of the southeast, organic soil colloids and the actual standing biomass comprise the major reservoir of nutrients. Areas subject to repeated fire or export of nutrient-containing biomass cannot long support a diverse flora and stands of low species diversity result, composed of species like slash pine and palmetto adapted to nutrient poor sites.

In the west, I have been recently looking at Oregon Coast Range soils in the Alsea and Unqua watersheds, especially those on old, flat, wave-cut marine terraces now elevated above the sea because erosion on flat areas would be expected to be low and rates of soil formation and past geologic erosion can be deduced from the ages of the terrace surfaces.

A cut through a 2-4 million year-old soil in Weyerhauser forest lands at about 300 m elevation between Eugene and Pt. Orford, shows a deep lateritic weathering profile reflecting virtual stability of this land surface for millions of years. Fossil evidence leads us to believe that this area has supported the present forest species for over 4 million years.

Nearby exposures of a younger forest podsol on top of the latosol suggest at least some periods of different soil-forming conditions during the cool wet periods of the glacial times when precipitation here may have reached 380 cm/annum (150″/annum) in comparison to today's 254 cm (100″). Absence of these younger soils or a thick humus layer on ridge tops suggest natural cumulative erosion of about 0.9 m (minimum) of soil in the last 500,000 to 1 million years (0.1 cm/1000 years). Today, following logging, these soils are gullying at an alarming rate. Ridgetop gullies do not normally occur in this country, even following severe fire (and I shall explain why this may be so later). Headward valley erosion is very slow and apparently confined to certain geologic episodes of changing seal level and stream regimes, coupled with sporadic landsliding to initiate the gullies. Healed gullies formed naturally before deforestation are not observed. Here in Oregon's Coast Ranges, erosion rates are likewise 100 to 1000 times the rate of soil formation and make a mockery of sustained yield concepts.

Thus by erosion alone, we are directly and surely turning parts of our western states into Dalmatias. It took 500 years to strip the soils from the 250-cm-rainfall-areas of the Dalmatian coast with more like 1000 years to denude the soils from Israel, Turkey, Greece, Italy, and Lebanon. Those poor Romans and their predecessors were without logging trucks and bulldozers so hadn't a chance to achieve our rates.

Floods and flooding accompany our deforestation, just as in the European case, but at least we haven't Florentine art treasures to lose. Shrubs, like nitrogen-fixing alders, the devils club of British Columbia and the Pacific Northwest; the chaparrel of California and the gorse of Oregon (introduced from deforested areas of Ireland and Scotland) are replacing our forests and may remain there for the thousands of years necessary to replace the cation exchange media. In the northern Rocky Mountains, it appears that about 5,000 years was necessary for the succession leading to our present forest ecosystems and this occurred at a significantly wetter and possibly slightly warmer more favorable soil forming time accompanying and immediately following our last deglaciation (see for instance Bryson, 1968). Today, it would probably take longer. One thousand years after final deforestation, the Mediterranian hillslopes have been negligible soil reformed. Progressive climatic change often renders forest ecosystems unstable after deforestation since inadequate soil moisture may be available for natural reproduction. Thus the coastal watersheds of California were deforested of redwoods *naturally* (probably by repeated fires) about 10,000 years ago and could not reproduce rapidly enough to save the soil from subsequent erosion so lost out to pioneering chaparrel. Relict stands, like those on the San Francisco peninsula, once cut are doomed since they are creating their own microenvironment and retaining their own soil. The same holds true of interior mountain west and parts of eastern U.S. Lodgepole pine scavenges up sites deforested of Douglas fir and Ponderosa, tolerates more soil erosion and less cation exchange capacity, and in some cases more summer drought. Weakened Ponderosa pine populations in Idaho and Montana, now made up dominantly of young individuals, are subject to more frequent and more highly virulent diseases than are mixed age stands. A viral disease is now sweeping the young Ponderosa of western Montana.

Thus erosional evidence gives us a finite life for our forests in all but the flat deep soil areas of the southeast where tree farms can be carried on like any other farming with full utilization of the materials and slow nutrient depletion offsetable by expensive and extensive fertilization.

Congress seems little concerned, and understandably so, about geologic rates of erosion — after all — M.U.S.Y.A. surely isn't expected to hold up for 400-1000 years, and anyway those responsible for its passage and its carrying out won't live to see the consequences. Do the Roman navies really care that Dalmatia is now scrub as a result of the construction of their warships?

Nutrient Depletion

Forest Service personnel and the general public have long noted that some logged areas do not recover, but instead become open grasslands or bald areas despite attempts to replant such areas (fig. 1). Frequently competition from understory species is blamed for the seedling failures and, upon occasion, published reports may speak of "climatic change" or periods of unusual summer aridity as being responsible. Other examples

Figure 1. — Forest plantation, Robbins Gulch near Darby, Montana. Area was planted with Douglas fir seedlings in 1966. This photo was taken in May, 1970. Note extensive growth of nitrogen-demanding thistles, indicative of upset of soil biogeochemical nitrogen cycle.

of regrowth failure occasionally quoted are insect or viral infestations selecting seedlings, ground-squirrel or gopher disturbance of the seedlings or seed, and small mammal or bird use of the artificially broadcast seed as food.

In the mountain west, probably the most common cause of the formation of such artificially deforested areas is slow progressive climatic change wherein increased summer aridity limits reproduction of tree species outside a forest canopy. However, this alone cannot explain why some areas, particularly those in more mesic clear-cut sites, seem to recover with such decreased vigor of growth of replanted species (Bolle, et al, 1970). Climatic change can logically be called upon to explain the uphill migration of the lower forest/grassland ecotone found in much of the arid-summer areas of the west but cannot well explain the lack of recovery or weak recovery of the higher elevation areas. Especially confusing is the role of fire. One can logically ask why a cut-

Figure 2. — Terraced clear-cut, Bitterroot Valley, western Montana. Both terracing and clear-cutting are practiced throughout much of western United States as a means of exposing shade-free bare mineral soil in order to promote monocultural forestry for second generation growth. At this spot, 20,000 years of accumulated soil and its stored nutrients were removed in a single season. Regrowth by pioneering species such as the Douglas Fir is slow and natural reseeding almost impossible.

over forest will not recover with vigor even when artificially replanted, yet a similar area may reseed readily with high survival after a fire. Since most western logging operations do not export much of the biomass from the forests but leave it as slash which may be burnt to recycle the nutrients, the answer to this enigmatic observation cannot be simply a question of recycling of above-ground nutrients.

Two kinds of recent research have suggested possible explanations for the problem of lack of forest recovery on logged watersheds. The work in question here is that done on the Hubbard Brook experimental watershed in New Hampshire and that done on water-repellency of forest soils following fire.

The Hubbard Brook study in particular has spurred similar research in the Pacific Northwest and Northern Rockies where clear-cutting practices (fig. 2) constitute the most frequent management decision of choice by the U.S. Forest Service and yet do not seem to be recovering according to the management manuals. The Hubbard Brook study has demonstrated that certain conditions lead to marked loss of productivity of forest soils through nutrient depletion following a single cutting. If similar biogeochemical upset is going on in the west where the bulk of the nation's forest reserves remain, the consequences could be grave indeed. Preliminary evidence suggests that the western forest soils are being depleted of their exchangable nutrients although the actual processes of nutrient loss have not yet been fully studied. A chief difficulty is that the Forest Service does not maintain adequate research personnel or facilities in the west and is at present completely unable administratively to approach the challenge of responsible management based upon scientifically prudent decisions.

In the Hubbard Brook study, an interdisciplinary team of foresters, biologists, and geologists from Yale, Dartmouth, the U.S. Forest Service, and the U.S. Geological Survey set up a rather drastic study in which, in the winter of 1965, a small 160,000 m^2 (39 acre) watershed near Hanover, New Hampshire in the U.S. Forest Services' Hubbard Brook experimental forest was clear cut, by hand, without any vehicles and with virtually no disturbance to trigger erosion (Likens, et al, 1970; Bormann, 1968; 1969). Immediately adjacent to this watershed were two similar-sized unlogged watersheds used as controls in the experiment. All of the trees, largely hardwoods and shrubs were cut to within 1.5 m of the ground and all were left in place — not exported from the ecosystem. The following spring, in 1966, a herbicide (Bromacil) of known chemical composition was carefully applied to suppress virtually all growth. The purpose of this was to see what would happen when the nutrient and hydrologic cycle was interrupted. The New England forest assemblage differs from those of the western states in that the species are largely deciduous and the soil has thus several cms of humus. The New Hampshire glaciated terrain of metamorphic rocks supported a 10-15,000 year old forest podsol soil of acid pH, somewhat different from those of the mountain states except the Cascades and

Oregon coast range, where one finds similar but better developed soils. Soil depths of Hubbard Brook are similar to those in the mountain west on similar aged materials but clays are kaolinitic rather than montimorillonite, and more litter is present to support a total greater ion exchange capacity, although less saturated with nutrients than found in western soils. Precipitation at Hubbard Brook is about 125 cm annum (49 in.) with ⅓ of it occurring in the summer.

As was expected after clear-cutting, annual runoff increased 30-40% over the values for the same uncut area. Summer runoff, without vegetative transpiration and interception, increased over 400%. And, of course, the runoff was warmer and the snow melted faster.

For the two springs following the cutting, there was no increase in turbidity and the drainage water ran clear and apparently clean. Analyses showed that, indeed, particulate matter had increased about 4-fold in the runoff, possibly due to increased raindrop splash erosion without protective forest canopy, but the water looked potable and drainages did not silt up.

Then Gene Likens, the aquatic ecogolist of the team, began to look at the runoff water chemistry. Total chemical input from precipitation, which contain significant numbers of ions in the polluted northeastern skies, as well as the compounds introduced in the herbicides, was accurately known (Fisher, et al, 1968; Likens, et al, 1970).

Outputs of ions, as measured by stream water chemistry for two years following cutting, increased dramatically. Even though there was more runoff to dilute it, average stream water concentration from the deforested watershed increased by:

417% for Ca
410% for Mg
1558% for K
177% for Na

and 56× for nitrate concentration (reaching levels of 82 ppm on occasional months — or 2× the Public Health standards for potable water without nitrogen poisoning) (McKee and Wolf, 1963, p. 224).

Ion flushing was going on with total dissolved nutrients being withdrawn at the rate of 97 metric tons/klm^2/year by the 2nd year or 8 times greater than would be expected for an undisturbed watershed.

Ninety seven metric tons/Klm^2 sound like quite a bit; but indeed, if one assumes a generous set of values for cation exchange capacity in soils of the mountain west of 20 milliequivalent/100 gms and 95% base-saturation of all storage sites and a full 30 cm of clay and organic rich soil, we could generously estimate that beneath western forests there would be something like 1700 metric tons of nutrient cations per sq. klm in storage. This is about the right order of magnitude for well-studied coniferus forests in similar young soils in the U.S.S.R. and Bavaria, and deciduous forests in Belgium — although those areas all have faster rates of soil formation (Duvigneaud and Denaeyer-DeSmet, 1970, p. 199).

So if nutrient leaching were to occur in the west as in the northeast after clear-cutting, and if the western soils did not erode at all and lay on perfectly flat ground, we might predict nutrient sterility 17.5 years after clear-cutting. Even if there were ten times the exchange capacity estimated, as for instance might be the case in western Washington, sterility might occur after only 175 years — i.e., before erosion strips the soil. But we do not know if the same processes are operating in the west as were found in the Hubbard Brook study, we do not know how long such leaching may progress after deforestation, and we do not know how much the forest understory, which was suppressed by herbicides at Hubbard Brook, may take up the biogeochemical slack and protect the soils against nutrient loss.

Likens has reported (personal communication, 1970) that analyses of runoff from old logged areas in New Hampshire still reveal higher than expected outputs of nutrient ions 10-20 years after logging. This might be due to the chemical nature of rainfall in that eastern area (pH 4.0 or less as weak nitric or sulfuric acid), which results from air pollution upwind in the vicinity of the southern and eastern Great Lakes. The answers to these and other fundamental questions must be answered before wise land-use decisions can be made based upon other than speculative information. To date, little or no effort is going into answering these questions from either federal, state, or private research agencies.

What apparently happens at Hubbard Brook is that as the spring sun warms the cut over area, the forest floor microflora convert organic nitrogen into ammonia. This ammonia is then rapidly oxidized at warm temperatures by bacteria (*Nitrosomonas* spp) to nitrite + H^+ ions. The nitrite is then further oxidized by other bacteria (*Nitrobacter* spp) to nitrate which runs out of the system in solution, while the hydrogen ions released in the first reaction displace the nutrient cations in the clays and organic colloids to flush the nutrients from the system with the spring runoff.

The soil bacteria multiply rapidly after clearcutting severs the nutrient cycle. Then the released cations enrich the streams and lakes as they flow seaward to cause local eutrophication, just as would be expected if one dumped fertilizers or phosphate detergents into a stream or lake. Soil bacteria may theoretically continue to function until all the sources of organic nitrogen are consumed or until vegetative growth recouples the nitrogen cycle. In the east, the litter layer and organic colloids supply the chief sources of organic nitrogen, as well as much of the cation exchange capacity. Regrowth of understory species is usually rapid in areas of natural disturbance or logging and one would expect that the nitrogen cycle would be completed but a few years after disturbance. In the west, however, where litter layers are weak or altogether absent, the largest nitrogen reservoir is to be found among the organic colloids within the mineral soil (Jenny, 1967, p. 31). This, coupled with the lesser acidity of western rainfall wherein most rains are weak carbonic rather than nitric or sulfuric acids, should theoretically make the possibilities for micro-organism initiated soil nutrient leaching less severe in the west, but the counteracting problem of summer drought and much slower resultant recovery of an understory or shrub layer may allow interruption of the nitrogen cycle and nutrient leaching for as long as 20 to 100 years, thus requiring nitrogen-fixing species such as Alder for long successional intervals after disturbance. A few years of leaching in the northeastern hardwood forests would remove only on the order of 10 to 20 percent of the total exchangable soil nutrient ions, but a 20 year leaching in the west could more than deplete the entire 15-25,000 year nutrient accumulation in the soil reservoir.

It is thus the interruption of this nitrogen cycle that flushes potentially available nutrients from the ecosystem. General studies of micro-nutrient cycling in forest ecosystems usually separate closed biologic nutrient cycles and open geochemical cycles and, prior to the recent Hubbard Brook study, the effects of biologic organisms on the geochemical cycle has not been fully appreciated. Nutrient losses from forest ecosystems with the occasional exception of calcium, have not been theoretically predicted and thus not looked for in previous studies (Leopold, 1970).

At Hubbard Brook, additional soil acidity resulted from the presence of SO_4 = (sulfate) in the polluted precipitation. In the west, we do not have anything like this except around large metropolitan areas, power plants, and smelters and in many such areas nutrient flushing has already gone on to such an extent that trees no longer grow to merchantible saw timber size and all that can be done with them is to pulp them for paper to liberate airborne sulfate from the pulp mills.

The single most pressing question raised by the Hubbard Brook study is is it happening in the mountain west? How do we measure it without this sort of destructive test cutting?

Three things suggest strongly that it is:

(1) Electrical conductivity of water draining clear cuts is 4-6 to even 10 times greater than before cutting — suggesting increasing dissolved ionic load.

(2) Eutrophication can be observed in streams and ponds below deforested hillsides with scummy streams and algal blooms directly correlated with percentage of the watershed cut in the previous two years.

(3) Trees do not grow as big as or as quickly as the forestry texts say they should or older foresters remember. Some clear cuts, without evident bare rock, just do not reproduce — and when replanted time and time again, the brush and alders seem to have a competitive advantage over trees.

Hard data is needed and is not being sought. President Nixon may soon again be calling for a doubling of forest cut to meet national needs. This increase must largely be met in the west as the southeast has quite accurately demonstrated that it can yield no more on a sustainable basis — even with massive fertilization. Forest Service personnel will be forced to act without knowledge and will, in my opinion, be committing pedological and ecological mayhem.

Forest Fires

One may well ask, perfectly logically, why all this does not happen after forest fires and how forest soils survive in the first place if they are so sensitive? I do not have a firm answer but have been perplexed by this problem for the last year and have the following theory:

Forest fires release amounts of nitrogen and other nutrients by taking it out of the standing crop and dropping it on the soil as ash. Hans Jenny (1962) has estimated that a California Ponderosa pine forest of 120′ trees contains 57 gm of nitrogen per m^2 above ground and 360 gm per m^2 below ground. Even the heaviest experimental burning did not reveal a measurable loss of nitrogen from the soil in Jenny's experiments. However, fires also temporarily kill off much of the soil microflora and fauna directly, and indirectly suppress recolonization of some through the effects of the fire ash on the soil pH. These soil organisms thus release their own nutrients to the system without forming and oxidizing the ammonia to release the hydrogen ions to flush the cations. Thus after a hot fire, a rich layer of nutrients lie on the surface in the form of alkaline ash and the soil nutrients remain fully stored and available for the next cycle of forest growth.

But why do soils not erode completely after repeated fires? We can see that southern California's fire-maintained chaparral ecosystems are virtually without soil clay mineral cation exchange media and are known to erode rapidly after each fire, so why does this also not happen in each of the Pacific northwest, mountain, and southeastern soils after burning?

I have observed that adjacent watersheds in southern Oregon's coast Ranges — one clear-cut by cable methods — not roads; and another burned at the same time without cutting, show marked differences in soil retention. It remains to be proven, but I think the reason for this may lie in the fact that hydrophobic organic colloid substances are translocated along a temperature gradient down into the soil from the litter layer during a fire. The hotter the fire, the more evaporation of organic matter in the little layer and the more precipitation of same in or on the mineral grains a few inches below the soil surface. This organic complex apparently causes the soil to repel surface water. USFS personnel have reproduced this effect in the laboratory and note that the organic-coated soil sheds water readily thus decreasing the capacity of the soil to absorb precipitation after a rain and increasing its overland flow yield to form severe flash floods in the river bottoms after fires but much less soil erosion than would be expected (DeBano, et al, 1969, 1970). Once established after a fire, grass and other vegetation seem to break down the organic cement and render the soil essentially the same as before the fire except that more organic matter capable of affording greater cation exchange capacity is found in the soil profile so the net result is an increase potential soil productivity and cation exchange capacity after a fire as compared with a marked decrease after forest extraction.

Conclusions

Present evidence suggests forest cutting, by any but the most conservative and careful methods, appears to defeat the soil-plant nutrient cycling and soil nutrient storage capacity, as well as increase erosion. Some western forests may have something less than 200 years of productive fertility remaining before permanent (in the sense of man's remaining timescale on earth) eradication of productivity for saw timber production. Means of chemically monitoring the effects of logging practices are not so complex that any reasonably competent citizen's organization should not be able to measure electrical conductivity and clay mineral and organic matter content of soils and their drainage water. I believe such semiquantitative citizens-data should be adequate to force full-scale soil stability investigation or reconsideration of land use, where appropriate, by agencies independent of the U.S. Forest Service, U.S. Department of Agriculture, and Soil Conservation Service, all of whom have had bias or technical limitation (Leopold, op. cit.). The matter is, in my opinion, so serious and potentially so completely damaging to our agricultural and forest lands that a full scale crash 4-year nationwide fact-finding effort should be undertaken now, administered by non-governmental and non-industry related agencies and personnel but working in cooperation with federal land-management agencies. A 4-5 year national forest cutting moratorium might further afford economic incentives to establish full scale paper recycling and waste-wood use to permit adoption of the better portions of European silvicultural practices upon re-opening our own public and private forests to carefully controlled selective cutting.

References Cited

Bolle, A. W., et al., 1970, *A University view of the Forest Service.* Prepared for the Committee on Interior and Insular Affairs, U.S. Senate, Doc. 91-115, Washington, 33 p.

Bormann, F. H., G. E. Likens, and J. S. Eaton, 1969, Biotic regulation of particulate and solution losses from a forest ecosystem. *BioScience* 19(7): 600-610.

Bormann, F. H., G. E. Likens, D. W. Fisher, and R. S. Pierce, 1968, Nutrient loss accelerated by forest clear-cutting of a forest ecosystem. *Science* 159; 882-884.

Bryson, Reid A., D. A. Baerreis, and W. M. Wendland, 1968, The character of late- and post-glacial climatic changes. *Rept. of the Symposium on Pleistocene and Recent environments of the Central Plains,* Oct. 25-26, 1968, 22 p.

Corbel, Jean, 1964, L'erosion terrestre, etude quantitative. *Annales de Geographie* 73(398): 385-412.

DeBano, L. F. and John Letey (eds), 1969, *Water-Repellent Soils,* Proc. of the Symp. on Water Repellent Soils, May 6-10, Univ. of Calif. at Riverside, 345 p.

DeBano, L. F., L. D. Mann, and D. A. Hamilton, 1970, Translocation of hydrophobic substances into soil by burning organic litter. *Soil Sci. Amer. Proc.* 34: 130-133.

Duvigneaud, P. and S. Denaeyer-De Smet, 1970, Biological cycling of minerals in temperate deciduous forests. In: *Analysis of Temperate Forest Ecosystems,* D. E. Reichle (ed), Springer-Verlag, Berlin, p. 199-225.

Fisher, D. W., A. W. Gambell, G. E. Likens, and F. H. Bormann, 1968, Atmospheric contributions to water quality of streams in the Hubbard Brook Experimental Forest, New Hampshire. *Water Resources.* 4(5): 1115-1126.

Jenny, Hans, 1967, Changes in soil nitrogen and organic matter under forest practices. In: *Man's Effect on the California Watershed,* Sect. 1, Rept. of the Subcommittee on Forest Practices and Watershed Management to the (California) Assembly Committee on Natural Resources. Committee Report, Assembly of the State of California, v. 25, Rept. 8, Jan. 2, 1967, Sacramento, p. 31-32.

———, 1962, *Soil Sci. Soc. America Proc.* 26: 200-202.

Judson, Sheldon, and D. F. Ritter, 1964, Rates of regional denudation in the United States. *Jour. Geophys. Res.* 69: 3395-3401.

Leopold, L. B., 1970, Hydrologic research on instrumented watersheds: UNESCO New Zealand Symposium on Experimental and Representative Research Basins, Wellington, N.Z., Dec. 1-14, 1970, 32 p. (Also available from the Office of the Regional Forester, U.S.F.S., Ogden, Utah)

Likens, Gene E., F. H. Bormann, N. M. Johnson, D. W. Fisher, and R. S. Pierce, 1970, Effects of forest cutting, and herbicide treatment on nutrient budgets in the Hubbard Brook watershed-ecosystem. *Ecol. Mono* 40: 23-47.

McKee, J. E. and H. W. Wolf (eds), 1963, *Water Quality Criteria,* 2nd ed. State of California Water Quality Control Board, Pub. 3-A, Sacramento, 548 p.

Task Committee on Preparation of Sedimentation Manual, Committee on Sedimentation of the Hydraulics Division, 1970, Chapt. IV, Sediment Sources and Sediment Yields. *Jour. Hydraulics Div., Amer. Soc. Civil Engineers* 7337 (HY 6): 1283-1329.

Ursic, S. J., 1965, Sediment yields from small watersheds under various land uses and forest covers. In: *Proc. Fed. Inter-Agency Sedimentation Conf.* 1963, p. 47-52 (USDA Misc. Pub. 970, Washington).

Wahrhaftig, Clyde, and R. R. Curry, 1967, Geologic implications of sediment discharge records from the northern Coast ranges, California. In: *Man's Effect on the California Watersheds,* Pt. III, Committee Rept. (Source Materials), C. R. Goldman (ed.), Inst. of Ecology, Univ. Calif. Davis, p. 35-58.

Walker, P. H., 1966, Postglacial erosion and environmental changes in Central Iowa. *Jour. Soil and Water Conservation* 21(1): 21-23.

Wallis, James R., 1967, *California's muddy streams — are they a problem we should solve?* Rept. of the Science and Public Policy Seminar, Kennedy School of Govt., Spring, 1967, 43 p.

Soil Conservation

II

Conservation means many different things to different people. To some it is simply the prevention of waste, and to others it is the preservation or nonuse of a resource. Still others conceive of it in a framework of maximum development with efficient and sustained yield for the benefit of most people. One definition of conservation is that it is an ". . . investment (1) in maintaining productive potential, (2) in decreasing the productivity deterioration or (3) in enhancing the productivity potential" (NAS-NRC, 1961). As used in the context of this series of volumes (Vol. I, p. 3) landscape conservation "... portrays constructive elements in man's attempts to live more harmoniously with his environment . . . (and) include(s) both preservation (protection) and reclamation." Prior to the twentieth century, man did very little in the way of attempting to restore damaged environments. Thus another aspect can now be added . . . that of rehabilitation or the act of repairing those areas that have undergone destruction. These extensions will be discussed in Part III. In Part II emphasis will be placed on the conservation movement and its special relationship where soils are involved. The reader is also referred to such foreign literature as Askew et al. (1970) and Coppock and Coleman (1970).

History of Conservation

Although some elements of conservation in the United States were present before 1900, there was no real conservation movement until the twentieth century. For example, the establishment of Yellowstone National Park in 1872 and Yosemite and Sequoia National Parks in 1890 were important steps. Another beginning was made in the Executive Order of 1891, setting aside the National Forest reserves, although these were not formulated into law with funding and administrative powers until

1933. Early leaders in alerting people to conservation problems were John Muir and Gifford Pinchot. Their beliefs were poles apart since Muir wanted to preserve the natural setting and Pinchot argued for reasonable use with benefits to the greatest number of people. His ideas are now anathema to present-day conservationists. For example, at various times Pinchot described conservation as

> Conservation is the use of natural resources for the greatest good of the greatest number for the longest time. Conservation implies both the development and the protection of resources, the one as much as the other. . . . The object of our forest policy is not to preserve the forests because they are beautiful . . . or because they are refuges for the wild creatures of the wilderness . . . but . . . the making of prosperous homes . . . Every other consideration comes as secondary (Herfindahl, 1961, pp. 2–4).

Pinchot in his position as chief of the Division of Forestry in 1898, which became the Forest Service of the U.S. Department of Agriculture in 1905, was very influential with President Theodore Roosevelt. Roosevelt's 1900 inaugural address contained some of Pinchot's *ideas,* and Pinchot was largely responsible for inspiring the famous Governor's Conference of 1908. This massive gathering of all governors or their representatives, members of the Congress and Supreme Court, scores of representatives from private organizations, and outstanding citizens and scientists, brought thoughts on conservation to the attention of the American public, and, to a lesser extent, to much of the world. The conference addressed itself to a broad spectrum of problems dealing with resources and their management. Soil erosion and conservation methods were discussed, but strangely these problems did not receive major attention. Instead, greater emphasis was placed on forest lands and other natural resources.

In 1911 H. H. Bennett (Simms, 1970) did a pioneer study of Fairfield County, South Carolina, which disclosed that 90,000 acres of formerly cultivated land had been so cut by gullies that it had to be classified as "rough gullied lands" and was useless. In addition, 46,000 acres of formerly rich bottomland had been converted into swampy meadow land because streams gorged with erosional products had lost their channel capacities and alluviated entire valleys. Unfortunately, it would be another two decades before such studies caught the attention of the public and government policy makers. It was the 1928 Bennett and Chapline article "Soil Erosion a National Menace" that finally ushered in a new era for soil conservation.

With this publication, the Congress began to show new concern, and Bennett won the interest of James P. Buchanan of Texas, whose committee in the House of Representatives began to hold hearings on soil erosion in 1928. This led to the passage of an amendment to the Agricultural Appropriations Bill for the 1930 fiscal year. This was a legislative breakthrough; the amendment stipulated that

> to enable the Secretary of Agriculture to make investigations, not otherwise provided for, of the causes of soil erosion and the possibility of increasing the absorption of rainfall by the soil of the United States, and to devise means to be employed in the preservation of soil, the prevention or control

> of destructive erosion and the conservation of rainfall by terracing or other means, independently or in cooperation with other branches of the Government, State agencies, counties, farm organizations, associations of business men, or individuals, $160,000 of which $40,000 shall be immediately available.

These were depression years, however, and although the Bureau of Chemistry and Soils, which Bennett headed, established some experimental procedures, it was not until another Roosevelt became President that the new conservation movement took form and substance. A series of events coincided to make soil-erosion programs feasible and necessary. The early 1930s were years of drought in the West. Many formerly plowed acres were abandoned after World War I, but with the lowering of farm prices additional acres were plowed since farmers had to grow more crops to receive the same amount of income. Along with deficient rainfall to wet down the fields came a period of years with abnormally high winds, which exceeded the threshold for movement of silt-size particles. This chain of events led to the creation of the "dust bowl" and triggered the infamous "black blizzards." Since many people were unemployed, the government saw the chance to unite these elements into programs that would combine work, stimulate the economy, and still perform the useful act of saving the lands. Thus, totally new directions were given to protection and repair of lands, and the government, for the first time, became deeply involved in conservation. These new programs provided manpower, as the Civilian Conservation Corps (CCC) and Works Project Administration (WPA), and new directions and leadership in conservation and reclamation projects occurred with the initiation of the Soil Conservation Service (SCS) and Tennessee Valley Authority (TVA).

On September 19, 1933, the Soil Erosion Service was established under the U.S. Department of Interior with a budget of $5,000,000 under authority of the National Industrial Recovery Act. Bennett headed the new agency, which had been conceived as a temporary measure, and within two years had established 40 large demonstration projects. The great dust storm of May 12, 1934, provided additional stimulus and received nationwide attention. This was the first dust storm in history big enough to retain its identity as it swept from the Great Plains to the Atlantic Ocean. It blotted out the sun and carried silt through windows of New York City skyscrapers. With such events and the prodding of Bennett, Secretary of Agriculture Henry Wallace was able to convince Secretary of Interior Harold Ickes that soil erosion was really an agricultural problem and that government expenditures would be duplicated less if such work were all handled by his department. On April 27, 1935, the Soil Conservation Act was passed, which committed the government to a long-range policy of national involvement in soil conservation. It states in part:

> It is hereby recognized that the wastage of soil and moisture resources on farm, grazing, and forest lands of the Nation, resulting from soil erosion, is a menace to the national welfare and it is hereby declared to be the policy of the Congress to provide permanently for the control and prevention of soil erosion and thereby to preserve the natural resources, control floods, prevent impairment of reservoirs, and maintain the navigability of rivers and harbors, protect public health, public lands, and relieve unemployment.

The following year Congress passed another important measure, which also had very far-reaching effects and set the stage for what would become an important power struggle between two different government agencies, the U.S. Department of Agriculture and the Corps of Engineers (U.S. Army). Although the first concepts for retardation of runoff and curbing of erosion by land-management practices were formulated as early as 1911, when Congress passed the Weeks Forest Purchase Act that authorized the Secretary of Agriculture to take certain steps for ". . . regulating the flow of navigable streams . . .", it was neglected, and implementation awaited the widespread program that was formulated in the Flood Control Act of 1936. This law formally established the important concept that flood control should consist of management of both the watershed and the river channel:

> It is hereby recognized that destructive floods upon the rivers of the United States, upsetting orderly processes and causing loss of life and property, including the erosion of lands, and impairing and obstructing navigation, highways, railroads, and other channels of commerce between the States, constitute a menace to national welfare; that it is the sense of Congress that flood control on navigable waters or their tributaries is a proper activity of the Federal Government in cooperation with states, their political subdivisions, and localities thereof . . . that the Federal Government should improve or participate in the improvement of navigable waters or their tributaries, including watersheds thereof, for flood control purposes if the benefits to whomsoever they may accrue are in excess of the estimated costs, and if the lives and social security of people are otherwise adversely affected.

This was the first federal law to recognize the need for runoff and water-flow retardation and soil-erosion prevention on watersheds as a principal means of flood prevention. Responsibility for upstream flood prevention was assigned to the Department of Agriculture. World War II delayed the beginning of operations under this law; however, beginning in 1937, the states passed laws that authorized farmers and ranchers to organize and govern soil conservation districts. There are now more than 2700 local districts throughout the country, dedicated wholly to planning and applying soil and water conservation measures to agricultural lands.

Since 1947, upstream planning and treatment with all measures needed to conserve soils and water and to reduce flood and sediment damage have been underway in many watersheds and cover more than 30 million acres. Another Federal action came in 1953, when Congress appropriated $5 million with which 60 pilot watershed projects were started. After the end of World War II in the mid-1940s, the power struggle between the Corps of Engineers and the Department of Agriculture came to a head. This stemmed from a conflict of jurisdiction concerning which agency would have primary responsibility for design of flood-control measures. It was also partly due to an influx of many engineers into the Department of Agriculture who wanted to expand their control systems in watershed design. The matter was finally resolved with the passage in 1954 of the Watershed Protection and Flood Prevention Act (Public Law 566, 83rd Congress). It provided for a new project-type approach to soil- and water-resource development, use, and conservation. Under this act, each project is a local undertaking with federal help. It has been subsequently amended several times by later Congresses. For example, as amended September 27, 1962, the act stated

> That erosion, floodwater, and sediment damages in the watersheds of the rivers and streams of the United States, causing loss of life and damage to property, constitute a menace to the national welfare; and that it is the sense of Congress that the Federal Government should cooperate with States and their political subdivisions, soil or water conservation districts, flood prevention or control districts, and other local public agencies for the purpose of preventing such damages and of furthering the conservation, development, utilization, and disposal of water and thereby of preserving and protecting the Nation's land and water resources.

A size constraint is placed upon structures which limits individual construction costs to $250,000 and the volume of water that is impounded cannot exceed 4000 acre-feet of total capacity (87th Congress, 76 Stat. 608). Within a 10-year period (1964), 100 projects had been completed, 447 had been authorized for construction but not completed, and 416 had been authorized for planning. Those completed or authorized for construction affected more than 30 million acres, those authorized for planning 33.5 million acres, with more than 1000 additional applications pending that involve 84.5 million acres.

Another thread in the matrix of conservation concerns lands that are in the public domain, which comprise nearly 180 million acres. Many of the public land laws between 1862 and passage of the Taylor Grazing Act of 1934 are still in existence. They are administered by the Bureau of Land Management of the Department of Interior instituted in July 16, 1946, when the Grazing Service and the General Land Office were consolidated under the Reorganization Act of December 20, 1945. The Taylor Act is a response to what the Secretary of Interior called: ". . . a vast empire over which there is at this time no adequate supervision or regulation, and which is rapidly becoming a no man's land through erosion and deterioration from unregulated use." The management of grazing districts seeks to establish a balance in grazing use, including game animals and conservation programs to reduce soil erosion and improve vegetation by reseeding, contour furrowing, soil pitting, controlling brush, and building detention dams and water spreading dikes, as well as insect and rodent control. The various states hold title to 80 million acres, wherein more than 30 million acres are set aside as wildlife preserves, 4.9 million acres as parks, and 5.1 million acres as forests, with related uses totaling another 19.3 million acres.

The Bureau of Reclamation, established by the Reclamation Act of June 17, 1902, is still another part of this total picture but will be discussed later. The final chapter to be told about the conservation movement starts in the mid-1960s with a great resurgence, a new awareness of ecology, and the total environment of humanity. A publication explosion is in progress at the present time, championing the rediscovery of man's habitat and its importance for his survival. There has also been the proliferation of new societies that have reached high levels of activism in their cause, which finally led to Environmental Policy Act of 1969. Some of these organizations that are currently involved in some measure with conservation matters are American

Farm Bureau Federation, National Farmers Union, National Grange, National Association of Soil Conservation Districts, Soil Conservation Society of America, Friends of the Land, The American Watershed Council, Izaak Walton League of America, The National Reclamation Association, The National Rivers and Harbors Congress, The American Water Works Association, Sierra Club, Council of Conservationists, American Forestry Association, National Parks Association, Audubon Society, and Wilderness Society.

Erosion-Control Methods

> To gain control over the soil is the greatest achievement of which mankind is capable. The organization of civilized societies is founded upon the measures taken to wrest control of the soil from wild Nature, and not until complete control has passed into human hands can a stable superstructure of what we call civilization be erected on the land (Jacks and Whyte, 1938, p. 1).

Conservation, preservation, reclamation, and rehabilitation become completely intertwined when methods of erosion prevention are considered. This section emphasizes geomorphic engineering practices that are used to stabilize soil and deter its loss, whereas Part III involves the much broader spectrum of landscape management and evaluation. Volume I (p. 391–454) of this series contains additional information on soil-erosion prevention prior to 1900. Many of the methods of the twentieth century are not new but mere refinements of older practices. However, with greatly improved and enlarged farm machinery and construction equipment (such as the bulldozer) the scale for undertaking preventative measures was broadened. Prior to 1900 there was little government assistance or intervention, and the dissemination of knowledge on soil conservation was minimal. Thus control methods were usually on a very local and small scale, and on a private or independent basis. Another difference manifesting in the twentieth century is to treat soil erosion as a scientific problem upon which research is needed; new developmental procedures were inaugurated by the experimental approach. Thus the need for data that would aid in discriminating the variables and their interrelationships was recognized. This was necessary before adequate and successful methods could be developed to arrest erosion.

In the United States, government involvement in these programs started in 1903 with field studies of hillside drainage by the Office of Experiment Stations of the U.S. Department of Agriculture. The purpose of these studies was to devise methods for reducing soil erosion. Hill and Chamberlin in the Governor's Conference of 1908 summarized some of the information on soil control methods:

> Our agricultural lands have been abused in two principal ways; first by single cropping, and second by neglecting fertilization . . . Rotation of crops and the use of fertilizers act as tonics upon the soil (Hill, p. 70–71).

> The key to the problem of soil conservation lies in due control of the water which falls on each acre. This water is an asset of great possible

> value. It should be computed by every acre-owner as a possible value, saved if turned where it will do good, lost if permitted to run away, doubly lost if it carries also soil and does destructive work below. A due portion of the rainfall should go through the soil to its bottom to promote soil-formation there; a due portion of this should go again up to the surface carrying solutions needed by the plants; a due portion should obviously go into the plants to nourish them; while still another portion should run off the surface, carrying away a little of the leached soil matter (Chamberlin, p. 79).

Chamberlin also discusses how this can be accomplished and methods that should be used: (1) deep tilth for soil granulation; (2) artificial underdrainage to prevent water-logging, promote granulation, and assist in absorption and transmission; (3) contour cultivation to arrest direct descent of water and distribute along slopes; (4) steep slopes require grassland strips, shrubland, and woodland alternating with plowed land, and (5) reservoirs at heads and in ravines to arrest storm floods.

In 1914 field investigations and experiments were conducted on terracing for erosion inhibition by the Bureau of Public Roads and the Bureau of Agricultural Engineering. The Forest Service was perhaps the first agency to measure in quantitative terms the amount of soil loss, which they did in 1915 in the Manti National Forest, Utah. The Agricultural Experiment Station of the University of Missouri was one of the early leaders in experimental work and by 1917 had developed a system of study plots where runoff and soil losses were measured after treatment of surfaces by different methods and by crop rotation. Other experiment plots were established at Spur, Texas, and Guthrie, Oklahoma, in 1926 to determine results of terracing on erosion control.

The work by Duley and Miller (1923) is typical of these early investigations. Their work showed that the character of rainfall and type of crop were very important in the amount of soil loss. The use of a cropping system was determined to be the most effective means for reducing erosion on rolling lands.

> From these experiments, it seems evident that farmers of the corn belt can do much toward reducing runoff and the disastrous effects of erosion by planning crop rotations in such a way that the land will be covered with a growing crop a very large portion of the time. The frequent use of clover and grass crops will aid materially in establishing such a cropping system (p. 45).

Grasovsky (1938) presents a summary of different methods used for soil erosion control in Nigeria, Algeria, Morocco, United States, Japan, Java, Malaya, Ceylon, and India. For example, in the extensive waste lands near Etawah, India, a forestation project was initiated in 1912 that is one of the earliest pieces of ravine reclamation carried out in semiarid regions. Reclamation of these lands included the planing off of the steep edges of ravines, and digging pits on the contours. Earth dams were installed in the deeper ravines both in and at their heads. Grazing was prohibited in the region, which was planted with 43 tree species. In Nigeria small-scale antierosion and experimental control sites were started in 1928. In the experimental site, all grazing was prohibited and firelines and ditches were cut to prevent fires from adjacent regions. Dams were constructed at the heads of gullies, and plantings were made on ridges and gully sides. In the United States, one of the most famous experiments was established in 1929 and is known as the San Dimas Forest and Range Experimental Station, California. It is about 17,000 acres in extent, and an entire network of instrumentation was established to monitor carefully all aspects of the water system and its interaction with the soil and hillslopes of the region.

The establishment of soil conservation districts is an innovation of the twentieth century. These range from small, local projects that are unipurpose to the giant multipurpose reclamation, conservation, and management programs, such as TVA (see Part III), that encompass thousands of square miles. Before discussing in detail the methods that have been used in erosion control, their effectiveness has been documented in the following typical cases. These examples demonstrate that man does have the ability to temper his destructive disposition and, when appropriate programs are instituted, to minimize soil losses and terrain disfigurement.

> In Brandywine Valley of Pennsylvania, a small band of farmers, devastated by the disastrous flood of 1942, got busy contouring, strip farming, tree farming, water holding, and planning guided runoff with the result that in 1955 (only thirteen years later), when thirteen inches of rain fell in ten days: "the damage was negligible, the flood waters almost clear, and the meadows were ready for grazing the next day . . . Most every factory in the valley was saved," related farmer J. W. Hershey (Rienow and Rienow, 1967, p. 77).

Albert and Spector (1955) show the cumulative results of what happened in one watershed where conservation practices were rather widely used. The Chattahoochee River in Georgia drains 928,000 acres. Soil conservation districts were organized starting in 1934, 50,000 acres of worn-out farmland were planted in trees, 60,000 acres of permanent pasture were established, 4000 miles of terraces were constructed, 52,000 acres were planted with annual cover crops, and good rotation practices were used for 60,000 acres. Other conservation practices were installed that included contour farming, construction of ponds, and planting of perennial grasses. The average annual turbidity declined steadily from 400 ppm in 1934 to 69 ppm in 1952.

Glymph and Storey (1967, p. 212–215) summarize from a variety of sources the effectiveness of different erosion-control methods:

1. Graded cropland terraces reduce erosion on fields at least 75 percent and in combination with crop rotation, mulching, minimum tillage, and soon, can reduce soil loss to practically nothing.

2. Converting croplands to good grasslands or pasture can reduce soil-erosion losses 90 percent and more.

3. In Texas Blacklands sediment yield from a 132-acre watershed with good conservation practices (including crop rotation, increased acreage of permanent grass, and graded cropland terraces) was only 12 percent that from an adjacent 176-acre watershed without conservation practices.

4. Sediment yield from soil erosion of the 88-acre Pine Tree Watershed in Tennessee was reduced 90 percent in the 5-year period following the program of check dams, rerimming, flowline staking, contour furrowing, and reforestation.

5. In loess hills of Mississippi, treating upland gullies with mulch, brush dams, grass seeding, tree plantings, and combinations reduced sediment yield 80 percent.

6. Streambank erosion-control methods in Buffalo Creek Watershed, New York, reduced suspended sediment loads 40 percent, while in adjacent untreated watersheds similar sediments increased 10 percent.

Editor's Comments on Papers 15 Through 20

Most of the articles in this section are from government publications or were written by government personnel. For a world-wide appraisal the reader is referred to United Nations publications (e. g., 1948). This is because the great majority of publications on erosion control occur in government publications and because their scientists are those who are specially assigned to study and deal with such problems. In the United States the Department of Agriculture developed an entire series of different publications for reporting their findings and recommendations for proper land use and inhibiting soil loss. The selections were chosen to represent ideas and the publications of the Department of Agriculture. Both generalized methods and specific detailed plans are provided in this series. Thus the first article, by H. G. Jepson, sets the stage and provides an overview of the variety of techniques in use to prevent gullies. He makes a point of treating the entire watershed area, realizing that the most important factor is runoff control, so it cannot channelize. He identifies and describes the diverse methods, which include (1) subsoiling; (2) contour furrows; (3) terracing; (4) earth fill; (5) diversion of runoff; (6) vegetation cover, natural, man-induced, and sodding; and (7) shrub check dams. These practices are in addition to effective farming methods that should include appropriate crop rotation, cover crops, strip cropping, and contour cultivation.

Whereas Jepson dealt with the entire spectrum of gully-control plans and devices, the E. C. Ramser article is one of the earliest and most complete articles in the twentieth century describing and illustrating general erosion-control methods. He shows that hillslope erosion occurs by two methods, *gullying* and *sheet wash.* To aid in their prevention, he recommends deep tillage with humus application, cover crops, contour plowing, pasturing and foresting, underdraining, and hillside ditches (see Vol. I). The greatest benefits derive when applied in conjunction with a system of terraces. Terraces are classified, and he describes their construction and maintenance.

The next two articles deal with soil erosion and its control in other parts of the world—China and Japan. In Japan, the Forest Act of 1897 was instrumental in starting the series of preventative measures that were undertaken in the early part of the twentieth century. Owing to the general steepness of the terrain, much of the major work in Japan has been aimed at torrent regulation, decreasing the amount of runoff into streams and decreasing the velocity of steep-graded channels. Serious erosion control in Japan really dates only from the beginning of the twentieth century, although cultivation practices did use terraces, but more to obtain land than as a check on erosion. It is fitting that a selected article should be written by one of the most influential American scientists, so important in developing U.S. practices and a legacy in the San Dimas experiment station which he founded—W. C. Lowdermilk (see also Vol. I). As pointed out in Part 1, there are always two facets to the destruction of lands: their erosion which strips away the soil and nutrients, and their ultimate deposition and siltation in some other locale. The article by S. Eliassen indicates this problem in China. He points out that the Yellow River transports 2.5 billion tons of soil per year; this is five times the amount transported by the Mississippi River. Although part of the cultivated hillslopes are terraced, these are subject to failures because of the unique properties of the loess soil which lead to piping and subsidence. He shows the status of the erosion problem in China up until the 1930s and mentions construction methods along with reforestation, that could be used to stop erosion. The area of land in China that required immediate attention was then about 130,000 square km, but ultimately it would be twice that. Since the siltation of river beds occupies space, any reduction of channel dimensions will automatically reduce the river's flood capacity. Although Eliassen does not illustrate methods at present in use, he provides insight into what needs doing and explains some of the benefits that would derive from appropriate land management practices to arrest soil loss. In the Soviet Union such articles as that by Sil'vestrov (1963) provide information on practices.

The short article by C. B. Brown provides a case history that emphasizes one aspect of the silt problem. Although it is vital to protect uplands from erosion, one cannot neglect what happens to the lowlands. He provides a study of what can be done where the local topography is suitable. Here erosion of uplands was contributing so much material that it made feasible a plan to create basins wherein the silt would be deposited and eventually become farmland. Crop yields in these newly reclaimed lands were greater than in adjacent lands (see also Kedar, Vol. I, for deliberate man-made soils for cultivation). Ruhe (1971) also describes increase in croplands in the Missouri River. Here less erosion was caused and new farm lands created simultaneously by careful design for altering the river course.

A related controversial issue which is now receiving a great deal of debate concerns man's alteration of rivers and channels in the belief that certain benefits will accrue. An unexpected result from stream straightening was illustrated in Part I. Thus the question of what is now called "channelization" needs continuing reappraisal as a conservation method (see also Daniels). So many aspects are interwoven into the fabric of channelization that it could lead to an entire string of related ideas that would include flood control management and flood plain design. Here we are only concerned with a brief review of one method currently in use and is included in numerous plans to reclaim land and prevent soil erosion. The question is whether it prevents or accelerates erosion. Many of the methods mentioned to this point have the approval of conservationists (especially when well designed) and are agreed upon as successful techniques. Channelization, however, is very argumentative and is opposed by many conservation groups.

The last article in this section, by C. J. Whitfield, differs from the others in several respects. Whereas the others emphasized erosion by water . . . sheet wash and fluvial processes . . . Whitfield discusses wind erosion and accretion of materials into sand dunes. He stresses two somewhat different yet overlapping concepts concerning the dune areas. First, he is interested in them from the viewpoint of reclamation. The areas they occupy consist both of lands previously uncultivated and of those lands that had formerly been farmed but which were later abandoned. Second, Whitfield shows the importance of stabilizing the dunes so that they do not migrate and ruin adjacent productive farmlands. This is one of the earliest articles that blends some of the ideas on dust storms of the 1930s and corrective measures that can be taken to place some of the areas in a stronger economic situation. Furthermore, the article unites conservation, reclamation, and rehabilitation, thus paving the way for expansion of these ideas in Part III.

Reprinted from *U.S.D.A. Farmers' Bull. 1813* (1939)

15

PREVENTION AND CONTROL OF GULLIES

By Hans G. Jepson, *assistant agricultural engineer, Engineering Division, Soil Conservation Service* [1]

CONTENTS

WHAT GULLIES DO

"SINCE THE ACHIEVEMENT of our independence, he is the greatest patriot who stops the most gullies." If this was true 150 years ago, when Patrick Henry made this statement, it is doubly true today. For gullies are now destroying land in every State. They have eroded many fields so badly that it has been necessary to discontinue the cultivation of areas that were good farm land only a few years ago. Year after year fields are abandoned as the old gullies take more of the land and new ones form.

When lands are gullied, the fertile soil is carried away, and unproductive soil may be deposited on rich bottom lands. Reservoirs and channels are also silted up, which then have to be dredged at great expense. Gullies that cannot be crossed readily by teams and farm machinery divide fields into smaller units and thus increase the cost of cultivation. As gullies tend to drain adjacent soil of its moisture, fields dry out much more rapidly near the gullies. This reduces crop yields on these fields. As these gullies become larger they branch out over the fields; and if they are allowed to develop unchecked, entire fields may have to be abandoned.

Gullies encroach upon public highways; undermine fills, bridges, and culverts; increase maintenance costs; and make travel unsafe. Livestock grazing near the edge of undermined gully banks is endangered. Gullies occasionally extend through a farmstead, undermining the farm buildings and making it necessary to remove them. The unsightly appearance of a gullied farm reduces its market value. These destructive and unsightly ditches have already caused damage

[1] This bulletin has been prepared in cooperation with C. L. Hamilton and under the general supervision of T. B. Chambers, in charge, Engineering Division. All members of the engineering division, and particularly G. E. Ryerson, have submitted valuable comments and criticisms. The divisions of agronomy, wildlife, and woodland management contributed to parts of the bulletin relating to their respective fields. The earlier gully-control studies of C. E. Ramser have been extensively used as well as material contributed by field engineers of the Soil Conservation Service.

that amounts to millions of dollars—much of it a needless loss, had attention been paid to the things that give gullies their start.

HOW GULLIES START

Before we destroyed the native timber, plowed up the virgin sod, and allowed large herds of sheep and cattle to overgraze the range erosion was not a serious problem. Under nature's cover of vegetation the soil absorbed much of the rainfall and was protected from excessive erosional losses.

In order to make way for the production of crops the land was cleared and plowed. No precautions were taken against loss of soil. On sloping fields the rows were run regardless of the direction of the slope. As a result the amount and velocity of run-off increased until it readily carried away large quantities of soil. At first the soil was removed from the surface in very small rivulets. They gradually became larger, and eventually gullies were formed, which enlarged with each succeeding rain that produced run-off.

Wherever the natural protection of the land is destroyed, the soil is made more vulnerable to erosion. Natural drainageways covered with vegetation once carried the run-off from the land. Stripping these drainageways of their natural cover and cultivating across them (fig. 1, *A*) or subjecting them to other undesirable farming practices is the beginning of many gullies. Steep slopes cleared for cultivation (fig. 1, *C*) will soon be badly gullied.

Poorly placed or poorly protected outlets for farm-drainage systems or improperly designed irrigation channels are points where many gullies begin. Improper location and protection of drains for farm roads or highway systems are an invitation to a gully to take the adjacent land. Trails or ruts over sloping fields (fig. 1, *B*), also contribute to some extent to the formation of gullies. The diversion of run-off into drainageways that are not well enough protected to carry the additional load is one of the surest ways to give gullies a start.

TYPES OF GULLYING

The careless use of land makes it possible for gullies to form. Gullying proceeds by waterfall erosion, channel erosion, erosion by alternate freezing and thawing, or a combination of these three types. Each type of gullying has a characteristic form (fig. 2), which may be modified to a considerable extent by local soil characteristics. If the underlying soil materials are soft and easily incised, deep, straight-walled gullies are formed; if the subsoil consists of plastic, resistant clays, the gullies are relatively shallow with sloping banks.

WATERFALL EROSION

Water falling over the edge of a gully or the bank of a ditch forms deep and very rapidly extending gullies. Their characteristic form is a U-shaped cross section (fig. 2, *A*).

A small vertical overfall usually develops in the lower reaches of a drainageway, and water falling over it undermines the edge of the bank, which caves in, and the waterfall moves upstream. As the overfall advances up the slope its vertical height increases, since it usually leaves a relatively flat slope below. This undermining goes

FIGURE 1.—*A*, A gully is gradually forming in this drainageway. If seeded down and left permanently in grass, the drainageway would carry run-off from the field without gullying. *B*, Sheep traveling up and down the slope made the paths that gradually developed into these gullies. Shifting or rearranging pasture lanes would have prevented this. *C*, Continued cultivation of this hillside will run gullies down the slope. Land as steep as this should be left in permanent cover.

on rapidly, particularly if the surface soil is underlain by sand or easily eroded subsoil.

In this manner gullies often start in the bank of natural watercourses that have been eroded to a great depth. They extend back into the drainage area and grow deeper up the slope, often attaining depths of 50 to 60 feet or more. As they extend backward and cross lateral drainageways or natural depressions, waterfalls are in turn formed in the sides of these depressions, and branch gullies develop. This branching may continue until a network of gullies covers the entire drainage area. Their growth is dependent mainly on soil characteristics, depth of overfall, and the size of the contributing drainage area rather than on the slope of the land. They may extend very rapidly even through almost level land. They frequently grow at a rate of 30 to 50 feet a year, depending on the amount of run-off and the character of the soil. Some of them have been known to advance several hundred feet during a single heavy rain.

In the Pacific Southwest some of the larger gullies formed in this way are known as barrancas; in the Northwest they are called coulees; in the Colorado Basin they are commonly referred to as arroyos or washes.

CHANNEL EROSION

Channel, or ditch, erosion is essentially a scouring away of the soil by concentrated run-off as it flows over unprotected depressions. Gullies formed by channel erosion usually have sloping heads and sides (fig. 2, *B*). In fact, these gullies are often referred to as V-gullies. As the scouring continues the gully becomes longer, deeper, and wider. V-gullies often attain lengths of 1 mile or more and depths and widths of 20 to 40 feet. The extension in length is usually much faster than the widening of the gully because a greater volume of run-off passes over the gully head than over the sides. Usually the gully does not advance beyond the divide of the drainage area, but it may continue to widen and deepen for years. The rate at which the gully deepens is very rapid on the upper part of the area, where the slopes are comparatively steep, and generally it decreases in the lower reaches as the slope decreases. Silting, rather than erosion, may occur if the channel or ditch emerges into a wide, flat-bottom drainageway. The increased volume of water, however, may offset the effect of moderate changes in slope along the lower reaches.

Channel erosion and waterfall erosion are commonly found in the same gully. The extension of the vertical head is usually by waterfall erosion; and the scouring of the sloping bottom and sides by channel erosion extends the depth and width. Gullies frequently start by channel erosion, and as an overfall develops at the head of the gully, the gully continues to develop by waterfall erosion. Gullies formed by channel erosion alone are often a series of closely spaced, parallel, V-shaped gullies in the upper reaches of drainage areas, where slopes are fairly steep and the contributing area small. Channel erosion is usually present in gullies caused by waterfall erosion, except those newly formed.

EROSION BY ALTERNATE FREEZING AND THAWING

Erosion by alternate freezing and thawing is prevalent in parts of the South, where alternate freezing and thawing temperatures are common in the winter and precipitation is generally in the form of rain. Alternate freezing and thawing loosens the soil, which sloughs off and is then carried away by heavy rains. This occurs on all slopes of a gully bank (fig. 2, *C*), and particularly on southern exposures. Gullies formed in this way may extend in all directions, as the direction of growth is not determined by the slopes of a field. Gully erosion by alternate freezing and thawing usually supplements waterfall and channel erosion, particularly in the Southern States. It may continue for years as the only form of erosion in gullies that are near a drainage divide and have little or no contributing drainage area.

In certain localities, during or immediately following prolonged rains, and especially following periods of alternate freezing and thawing, mass movements of soil in the form of slides, earth flows, and slumps take place on steep slopes. A considerable number of gullies have their beginning in these disturbed or galled areas, particularly where the displacement exposes highly erodible subsurface material. These mass movements usually develop conditions that are conducive to gullying.

PREVENTION OF GULLIES

This bulletin deals with the control of gullies that begin as a result of man's abuse of the land. We can prevent these gullies. To prevent the formation of a gully is much better and easier than to control it once it has formed. We can never prevent erosion entirely because natural deterioration will continue as long as there is action by wind, water, or frost. But this natural geologic erosion is generally so gradual and moves at so slow a pace that it does no appreciable harm.

It is the erosion that arises from improper land use and methods of tillage that those who work the land can control. If a farm is free of gullies (fig. 3) it will usually be found that the operator, under a well-developed land use program, uses good farming methods and is constantly on the alert against gullying. Wherever and whenever he finds a danger spot he immediately takes steps to prevent the formation of a gully. It is easy to stop a gully when it has just begun to form, but if the gully is neglected for some time it can usually be checked only with considerable work and at no little expense. It should not be assumed, however, that just because no gullies are visible on a field no erosion is taking place. There may be considerable sheet erosion, which is a forerunner of gullying unless it is checked.

Constant vigilance is necessary to forestall gullying that starts from some practice that may seem of no consequence. The thoughtless driving of a wagon up or down a slope in a field made soft by rain leaves a deep rut that may develop into a large gully unless it is in some way filled in immediately. Filling the rut with straw or manure, or even spading it full, will usually prevent further damage.

Many farms have deep gullies because the feed lots were not properly located or because the stock trails cut deep into sloping pasture lanes (fig. 1, *B*). Such gullies can usually be prevented by shifting fences as need arises and by rotating feed lots if gullies should begin to form. Care should be used in the location and protection of drainageways, of watering places for livestock, and of roads or trails.

Unless caused by such minor things as drain outlets or stock trails, gullying is usually preceded by sheet erosion. A close examination of sheet erosion shows that the soil is not removed in strictly uniform sheets or layers, as is so often supposed, but that numerous small rivulets are formed, which might be classified as miniature gullies. They are so small and close together that this process of soil removal is usually spoken of as sheet erosion. When water from several of these rivulets collects, larger depressions are formed, and they may finally become gullies. It thus is evident that where heavy sheet erosion has been under way for a period of time, gullying is probably imminent. Little sheet erosion occurs on a farm if its steep slopes are covered with trees and shrubs, its flat land and moderate slopes farmed according to approved cropping and tillage practices, and intervening areas reserved for permanent grasses (fig. 4).

The first step in preventing gullies is to plan or replan the farm so as to get the best possible use of the land. This will include the retirement to permanent cover of such areas as are definitely too steep to farm; the utilization of the better agricultural land for cultivated crops; the placing of moderately sloping and eroded areas in meadow or pasture, if such areas cannot be economically tilled. Good land use may require that the general field pattern be considerably changed and only the most suitable areas used for crops. In many instances fences will have to be reset and field roads rerouted to get the best arrangements.

The best known methods of controlling erosion on those slopes that must be tilled are crop rotations, cover crops, strip cropping, and contour cultivation, alone or in combination with terracing where it is required. These practices are discussed in Farmers' Bulletins 1776, Strip Cropping for Soil Conservation; 1758, Cover Crops for Soil Conservation; and 1789, Terracing for Soil and Water Conservation. Insofar as the methods of farming described in these bulletins control erosion they aid directly in preventing the formation of gullies. The use of fertilizers and the conservative grazing of pasture or range also protect the land from erosion.

One of the most important considerations in land use that prevents gullies is the proper disposal of excess run-off water from the fields. Every farm has its own natural drainage pattern, which, in general, it is difficult to change. This pattern includes all water-conveying depressions or channels of either continuous or intermittent flow. Except for minor variations it should be followed in planning fields and, especially, in locating outlets for terraces or diversion ditches. Natural drainageways should be used, wherever possible, and they should be left in sod in order that they may continue to carry water without gullying. The extra ground that may be cropped by farming these drainageways can in no way compensate for the damage that will occur if severe gullying is begun by cultivating the drainageways with the rest of the field.

If the general drainage pattern is not considered or followed in planning for land use, either a very expensive artificial system to dispose of surplus run-off will be required or the water must be concentrated in undesirable locations that will eventually gully. If a natural drainage unit is not complete on one farm, but covers two or more farms, a better water-disposal plan can be worked out if the farms are considered collectively.

If the results of careless cultural practices could have been foreseen or if just a little time had been spent in checking the beginning of a small gully, many of the present large gullies would never have formed. Once the gullies have formed and prevention is too late, it is still possible to stop serious erosion in the gullies and cover most of them with vegetation.

* * * * * * *

If a gully directly menaces a building or a highway structure it is a rather simple matter to determine how much money may be spent to protect this property as its value is usually known or can be readily estimated. Estimating the damage of gullying in a field or the justifiable expenditure for checking it is not so easy, as it is much more difficult to determine the true value of the land and the exact extent of present and ultimate damage.

* * * * * * *

The actual acreage of the land surface destroyed by a gully may be relatively small, but this gully may so dissect a field or threaten adjacent areas as to hamper or endanger farming operations on the entire watershed, either immediately or in the future. And just one badly gullied field on a farm creates an unsightly appearance that reduces the value of the entire farm. The value of a farm on which gullying is active may therefore decrease much more rapidly than the present state of gullying would indicate.

Although severely gullied land has little immediate value some control measures are usually warranted on all such areas if only to protect adjacent lands. But it is well to determine what is the most economical and suitable protection for each gullied area. The cost of controlling a gully and the type of protection should always be considered in relation to the use that can be made of the gullied land as well as the protection to adjacent areas that such control will afford.

* * * * * * *

TREATMENT OF GULLIED DRAINAGE AREAS

Complete gully control includes proper treatment of the drainage area as well as of the gully itself. If the control treatment is applied only to the gully, it is likely that all control efforts are being directed at the results and the cause neglected. Neglecting the drainage area usually makes it much more difficult, if not impractical, to control a gully satisfactorily and often leads to the formation of new gullies faster than the older ones can be checked. For example, if an eroding field is terraced and nothing further is done to control erosion on adjacent fields it is likely that gullies will gradually encroach upon the terraced field and damage it.

If a drainage area is gullied, it will require more complete treatment than if the gullies were not present. For example, regular cropping and tillage practices that conserve the soil may permit the safe production of farm crops on an ungullied drainage area, but if gullies are present it may be necessary to put a permanent cover of vegetation on the area. Or a relatively flat, ungullied drainage area may be cropped without excessive soil and water losses if only crop rotations, contour tillage, and strip cropping are practiced, whereas the presence of a few gullies on that area would make it necessary to use terraces in addition to these other conservation practices if the same crops were to be produced without excessive soil loss and further gullying checked.

The plan of treatment for gully control, in order to be complete, should thus include treatment of the drainage area based on good land use and soil- and water-conserving practices such as strip cropping, contour tillage, crop rotations, and cover crops. These should be combined with terracing or contour furrowing where applicable. Only in this way is it possible to achieve a measure of success in the control of gullies or any assurance that such control will be more than short-lived.

ESTIMATING RUN-OFF

A soil can absorb rainfall up to a certain rate. When this rate is exceeded some of the water begins to run off. That is why heavy and rapid downpours are much more likely to cause harmful erosion than equivalent amounts of rain that fall over longer periods. This fact is so generally understood that special terms have come into use to express this relation. For example, a slow, steady rain is ordinarily called a "ground soaker," and a dashing downpour is commonly referred to as a "gully washer." The more rapid the rate of rainfall, other conditions being the same, the greater the proportion of rain lost as surface run-off.

It is the run-off that loosens the soil and carries it away. Without run-off there would be no gully erosion. Furthermore, the degree of gullying is directly associated with the rate of run-off. A high rate is usually more destructive than a low rate. The rate of run-off is generally a more accurate indication of the probable damage that run-off will cause than the amount of run-off. The rate of run-off is dependent on the size, shape, and slope of the drainage area; the extent, nature, and condition of the soil and cover; the intensity and duration of the rainfall; and the slope and condition of the drainage channels. High rainfall intensities of long duration, poor plant cover, saturated, frozen, or impervious soils, and steep slopes all contribute to high run-off rates.

One must estimate the probable rate and amount of run-off from a particular drainage area before it is possible to design structures to control the run-off or even to know what structures can best be used. Run-off rates are generally of more significance than the total amounts of run-off discharged in computing required discharge capacities for control structures. A possible exception would be where control is obtained through run-off retention or absorption, in which case the amount of run-off may become the deciding factor rather than the maximum rates. Terraces, diversion ditches, drainageways, and spillways must have sufficient capacity to carry the run-off resulting from the maximum rainfall intensities likely to occur during the probable life of the control structure. This period is generally from 5 to 10 years for small check dams, terraces, and diversion ditches and from 20 to 50 years for the more expensive soil-saving dams. The maximum rainfall intensity that is likely to occur during a period of 5 years is usually spoken of as having a frequency of 5 years, and similarly for any other number of years. For example, a dam designed with sufficient spillway capacity to handle the run-off resulting from a rainfall intensity that the records show will probably occur only once in 10 years is said to be designed on a 10-year rainfall intensity-frequency.

* * * * * * *

CONTROLLING RUN-OFF

Applicable methods of handling the run-off to facilitate gully control differ for various types and sizes of gullies and drainage areas. Other important factors that in part determine the best method to be used are soil types and climatic conditions. From the standpoint of economy and practicability, the methods should be considered in the following order:

TABLE 1.—*Run-off from drainage areas*[1] *of 1 to 300 acres, based on a 10-year rainfall intensity-frequency*

Watershed characteristics[2]	Group[3]	Run-off from drainage areas of—																						
		1 acre	2 acres	3 acres	4 acres	5 acres	10 acres	15 acres	20 acres	25 acres	30 acres	35 acres	40 acres	45 acres	50 acres	60 acres	70 acres	80 acres	90 acres	100 acres	150 acres	200 acres	250 acres	300 acres
		Cubic feet per second	*Cubic feet per second*	*Cubic feet per second*	*Cubic feet per second*	*Cubic feet per second*	*Cubic feet per second*	*Cubic feet per second*	*Cubic feet per second*	*Cubic feet per second*	*Cubic feet per second*	*Cubic feet per second*	*Cubic feet per second*	*Cubic feet per second*	*Cubic feet per second*	*Cubic feet per second*	*Cubic feet per second*	*Cubic feet per second*	*Cubic feet per second*	*Cubic feet per second*	*Cubic feet per second*	*Cubic feet per second*	*Cubic feet per second*	*Cubic feet per second*
Rolling timber	1	2	4	5	6	8	16	23	29	35	39	41	49	55	62	73	83	90	97	103	145	185	216	245
	2	2	3	4	6	8	15	21	26	32	35	40	45	50	56	65	74	80	86	91	127	160	186	210
	3	2	3	4	5	7	13	19	23	28	31	35	39	43	48	55	63	68	73	77	105	130	150	168
Hilly timber	1	3	5	6	8	9	18	27	33	40	45	52	57	64	71	85	96	104	114	122	165	213	252	285
	2	3	4	6	8	9	17	25	31	36	41	47	52	58	64	75	85	92	102	108	145	184	217	244
	3	3	4	5	7	8	15	22	27	32	36	41	45	50	55	65	73	79	86	91	120	150	175	195
Rolling pasture	1	5	8	12	15	18	33	47	58	67	77	87	99	110	121	143	164	181	197	208	290	369	433	489
	2	4	8	11	14	17	30	43	53	61	70	79	90	99	109	128	146	160	175	185	251	319	371	419
	3	4	7	10	13	15	27	38	47	54	61	69	78	86	94	110	125	137	148	155	210	260	300	335
Hilly pasture	1	7	13	16	20	24	40	55	67	79	91	102	114	123	142	166	190	210	227	241	334	426	197	572
	2	7	12	15	19	22	37	51	61	72	82	93	104	116	128	148	169	186	202	214	294	369	428	490
	3	6	11	14	17	20	33	45	54	63	72	81	90	103	110	128	145	159	171	180	242	300	345	392
Rolling cultivated	1	8	15	23	29	34	57	77	95	111	128	145	162	182	201	238	271	299	324	344	476	605	720	822
	2	8	14	22	27	32	53	71	87	101	116	132	147	164	181	212	242	265	288	306	419	522	620	703
	3	7	13	20	25	29	47	63	77	89	102	115	128	142	156	183	207	226	244	257	345	425	500	562
Hilly cultivated	1	10	17	27	35	42	70	94	114	134	154	174	196	219	241	285	323	358	389	416	580	727	863	986
	2	10	16	25	33	40	64	86	104	124	139	158	177	198	217	254	288	317	345	369	509	630	742	843
	3	9	15	23	30	36	57	76	92	107	122	138	154	171	187	219	246	271	292	310	420	512	600	675
Terraced rolling cultivated	1	6	9	14	18	21	40	54	68	81	93	107	120	137	155	189	223	264	293	315	465	605	720	831
	2	5	9	13	16	20	36	50	62	73	84	97	109	121	139	168	199	234	260	280	409	522	620	712
	3	5	8	12	15	18	33	44	55	65	74	85	95	107	120	145	170	200	220	235	337	425	500	570

[1] Based on the recommendations of C. E. Ramser.

[2] "Rolling" designates 5- to 10-percent slopes; "hilly," 10- to 30-percent slopes.

[3] Group 1 includes the entire States of Florida and Louisiana, the southern portions of Georgia and Alabama, the southern half of Mississippi, and the southeastern part of Texas. Group 2 includes central Texas, all of Oklahoma except the Panhandle, the eastern half of Kansas, southeastern Nebraska, the southern half of Iowa, western Illinois, all but the southeastern tip of Missouri, all but the northeastern corner of Arkansas, South Carolina, that part of Mississippi, Alabama, and Georgia not included in group 1 (except for the extreme northern parts of these States), the eastern and central portions of North Carolina and Virginia, Delaware, the eastern half of Maryland, New Jersey, and the southern extremes of Connecticut and Rhode Island. Group 3 may be used for all other areas except where local information indicates higher run-off rates.

1. Retention of run-off on the drainage area.
2. Diversion of run-off above the gullied area.
3. Conveyance of the run-off through the gully.

* * * * * * *

RETENTION OF RUN-OFF

Keeping the soil and moisture on a gullied field by means of proper land use and approved cropping and tillage practices lessens considerably the amount of run-off to be carried away through the gully and thereby reduces the amount of treatment needed in the gully itself. If these practices still permit so much run-off to enter the gully that vegetation cannot be established there, the use of subsoiling, contour furrows or ridges, listing, level terraces with closed ends, or earth fills to impound water may also be advisable. These erosion controls are not difficult to construct, and they retain considerable moisture for crops. If they can be applied over the entire drainage area of small gullies or even of medium-sized gullies that have small to medium-sized drainage areas on which the soils are absorptive, the slopes moderate, and rainfall low, little or no control may be necessary in the gully itself.

* * * * * * *

CONTOUR FURROWS AND LISTING

Contour furrows or ridges are primarily small ditches or ridges constructed across the slope with a plow, lister, or terracing machine (fig. 6). Their storage capacity is dependent on their water cross-

FIGURE 6.—There will be little run-away water from this contour-furrowed field to give gullies a start. The stock pond also retains considerable water.

sectional area and their spacing. Various storage capacities can be obtained by regulating the size and spacing of the furrows.

The storage capacity for contour furrows or ridges with various spacings and water cross-sectional areas is given in table 2. If, for example, it is desired to store 2.5 inches of run-off in contour furrows having a water cross section of 1.25 square feet, the furrows should be spaced about 6 feet apart. The storage capacity of the furrows given in table 2 does not include the water that seeps into the soil. The ends of the furrows or ridges are usually turned uphill and the channels blocked at occasional intervals so that, in the event of a break, all the water cannot drain from the furrow and thus encourage gullying. Lister furrows (fig. 7) have been used extensively on cultivated land in the Great Plains to keep rainfall on the fields. On gullied watersheds they can be used to keep water out of the gullies.

* * * * * * *

ABSORPTIVE-TYPE TERRACES

Absorptive-type terraces are constructed larger than contour ridges, and they provide more storage capacity, are spaced farther apart, and can be farmed over if necessary. Their ends can be left partly or completely open, depending on the necessity for some drainage as a factor of safety against overtopping. Where these terraces are to be depended upon to retain all of the excess rainfall, they should be used on only moderate land slopes (generally less than 2 percent) and pervious soil types. It will usually be necessary to provide storage for at least 2 to 4 inches of run-off in areas of low rainfall, and a capacity of as much as 6 to 7 inches may be necessary in areas with high rainfall. The expense and difficulty of providing a storage capacity of 6 to 7 inches usually makes the use of adequate retention measures impractical in areas where rainfall is high. Furthermore, if crops are to be grown on a field containing these terraces, the retention of an excessive amount of water may damage the crops before the water can be absorbed by the soil.

EARTH FILLS

On cultivated areas with absorptive soils, small- or medium-sized gullies with small watersheds can sometimes be reclaimed by placing a series of earth fills across the gullies. Their spacing is dependent on the slope of the gullies. The use of this method is limited to areas where sufficient storage capacity can be provided above the earth fills to retain the major portion of the run-off that is commonly discharged from the drainage area. The earth fills are extended above the ground level, and short diversion ditches or spur terraces are sometimes used to lead overflow away from the ends of the fills in order to prevent damage by erosion. Where a series of earth fills are used and the overflow is diverted from alternate sides, somewhat of a sirup-pan retention is effected. The diverted water is finally either impounded in the gully or absorbed by the soil. The use of this or similar methods requires considerable attention. As the storage capacity of the small reservoirs is gradually reduced by silt deposition, the amount of overflow will increase and may start erosion unless sufficient plant cover has become established to afford protection.

On fields where the measures discussed in this section do not provide sufficient storage capacity to retain the run-off and keep it out of the gullies, drainage-type terraces or diversion ditches may also have to be used to divert part of the run-off.

DIVERSION OF RUN-OFF

Where necessary and practical, run-off should be diverted from a gully head before control measures are attempted within the gully. This principle generally applies to gullies of all sizes except those having so small a drainage area that the run-off is negligible, as for example, a gully with a drainage area of less than an acre. In using either terraces or diversion ditches careful consideration should be given to the disposal of the diverted water. If safe disposal cannot be provided, the water should not be diverted. The disposal of concentrated run-off over unprotected areas may cause gullying.

Terraces are very effective in the control of small gullies on cultivated fields or even medium-sized gullies that are not too deep to be crossed with the terracers (fig. 8). Terraces placed above a gully too deep to be terraced across will divert headwaters from the gully, which may then be treated further if necessary. Terrace construction may be difficult and somewhat expensive on gullied areas, but despite this it is frequently the most satisfactory control measure for terraceable slopes, and particularly where numerous parallel gullies are encountered on slopes that are difficult to vegetate. Figure 9 shows a previously gullied field that has been almost completely reclaimed by terracing. The terraces were constructed across the gullies. This diverted the run-off at points where the terraces crossed the gullies and allowed the gullies gradually to fill with silt.

If the slopes above a gully are too steep to terrace or if the drainage area is pasture or woodland, diversion ditches (fig. 10) may be used to keep run-off out of the gully. Diversion ditches are particularly adapted to areas already covered with trees or grass because ditches below these areas are not so likely to receive silt loads from the drainage area. Diversion ditches are not recommended immediately below cultivated fields not fully protected from sheet erosion unless a permanent filter strip of close-growing vegetation is placed above the ditch to catch the silt carried in the run-off from the fields. This filter strip should have a minimum width of 50 feet and adequate cover to filter out and retain the silt in the run-off. This will reduce silt deposition in the ditch channel and eliminate the need for much subsequent maintenance.

* * * * * * *

CONVEYING RUN-OFF THROUGH GULLIES

If it is not possible to keep water out of gullies by retaining it all on the watershed or diverting it from the gullies, the run-off must be conveyed through the gullies. To do this and to check erosion in those gullies at the same time is possible if vegetation can be established in the gullies or mechanical structures built at critical points to supplement vegetation or to give complete control.

It is emphasized, however, that where a gully is located in a natural drainageway that is to form part of the disposal system for surface run-off, it becomes necessary to convey run-off through the gully. Any erosion control applied in the gully must not reduce the capacity of the drainageway below that required to carry the run-off discharged into it.

It is usually much more difficult to establish adequate vegetation in a gully through which run-off must be conveyed during the period of establishment than in one from which run-off can be diverted. Erodible portions of a gully through which water is conducted must usually be protected by transplanting sod, by the use of specially anchored mulches, or by the use of mechanical structures. The mechanical structures need be only temporary if the plant cover, once established, can provide sufficient protection.

If mechanical measures will ultimately be required for satisfactory control, permanent structures, such as masonry check dams, flumes, or earth dams, supplemented by vegetation, should be provided to convey the run-off over critical portions of the gully. Detailed information on the use of vegetation and mechanical structures in gullies is given in the following sections of this bulletin.

VEGETATION

The methods of preventing gullies that have been recommended, as well as the control measures described on pages 28 to 57, are for the purpose of reducing or controlling run-off so as to make it easier to establish vegetation. The objective of gully control is, in fact, the establishment of an erosion-resistant cover of plants that not only stabilize the gullies but also produce a usable crop, and hence a supplementary income. Just what kind of plants are to be used and how they are planted will, of course, vary according to locality.

In any part of the country, gullied areas are usually the most difficult places on which to try to grow plants. The topsoil is gone; the subsoil is poor or is gone; and the layers finally exposed in the gully may be hardpan or even rock. The revegetation of such places will be difficult. Yet, if the run-off water can be diverted from its old path through the gully, if the soil is given a little preparation, and if the area is temporarily protected from trampling by livestock, a plant cover can be established (fig. 12).

* * * * * * *

NATURAL REVEGETATION

Any gully, no matter how large, and regardless of its condition, will usually be reclothed with vegetation, provided it is properly protected and is in a locality where vegetation will grow. If the water that causes the gully is diverted and livestock, fire, or any other cause of disturbance kept from the gullied area, plants begin to come in. At first they come very slowly because it is hard for them to get a foothold. Later, when the pioneer plants have improved the soil somewhat, other plants appear. The whole natural process may take many years in the drier parts of the country, but where there is more moisture, the process is more rapid.

Plants will always come in naturally on protected areas, but on gullied land there are several things that slow down the final healing of the erosion scar. One is the continued loss of soil caused by freezing, thawing, and washing (fig. 2, *C*). This loss cannot always be stopped, but it can often be reduced by the use of a mulch of boughs, straw, or leaves, which also assists in catching and holding plant seeds. Another thing is the steepness of some gully banks. Until the steep sides cave in and reach a gentler slope (about a 1:1 slope), it is difficult for plants to root themselves. Unless large gullies with steep banks are plowed in, pushed in with a bulldozer, or dynamited, it may take many years for them to become stabilized.

In spite of conditions such as these, hardy, thrifty plants capable of surviving in gullied areas will generally appear naturally. So-called "weeds" will usually come first. They prepare the way for other plants, which always follow them in a year or two. Given sufficient time, this natural process will eventually reclothe the gully with the predominant vegetation of the region, whether that be trees, brush, or grass. Frequently this opportunity to obtain a cheap protective covering is overlooked and unnecessary expenditures are made for structures or plantings. Many previously active gullies have been completely stabilized by a natural growth of vegetation that sprang up after the land was abandoned and livestock excluded (fig. 13). Natural revegetation, however, may be a lengthy process. Where natural growth does not appear to be able to cope with existing erosion or where certain plant species of economic value are desired, it may be necessary to consider ways and means of establishing vegetation artificially.

PLANTING AND SEEDING FOR REVEGETATION

CHOICE OF PLANTS

So far as erosion control alone is concerned, it makes little difference whether trees, shrubs, vines, or grasses are used in a gullied area. Any of these, if well-established, provides good protection for soil. Consequently, the kind of vegetation to use is best chosen on a basis of what the planted area will be used for when it is stabilized.

A stabilized gully can be used as a woodland area, as a wildlife habitat, as a drainageway for water, or as pasture land or hay land. It should not be cultivated, however, nor should it be burned, brushed, or used in a way that will promote erosion. If it is to be used as a drainageway, grass is ordinarily the primary cover (fig. 14). Grass sod can carry more water safely and at higher speeds than can woody plants. A sodded drainageway can be crossed with farm machinery, whereas one with a woody cover cannot. Grassed drainageways produce good yields of hay in some sections. If they are utilized as pasture, grazing should be regulated. It should not be forgotten, however, that good grass growth demands a well-prepared seedbed and reasonably fertile soil, a condition rarely present in gullies.

In certain regions trees and shrubs are easier to establish in gullied areas than grasses, but they should not be used in drainageways unless the amount of water flowing in the channel is relatively small (fig. 15). Trees on a gullied area will produce a crop of fence posts or rough timber for general use about the farm. Shrubs are used particularly to attract wildlife, especially insect-eating birds, if the gully is near cropland, although many trees and grasses are also valuable to wildlife.

Where severely eroded and gullied areas must be retired from crop land or pasture land, woody plants are usually used to help stabilize them, and the areas are then ordinarily reclassed as woodland and wildlife areas.

* * * * * * *

SHRUB CHECKS

In small or medium-sized gullies with small drainage areas it is frequently possible to construct checks consisting of shrubs placed across the flow line of the gully (fig. 17). The shrubs are placed 4 to 5 inches apart in shallow trenches and are sometimes protected by rows of stakes. The stakes are placed about 1 foot down the channel from the shrubs so that the plants will benefit from silt collected by the stakes. The shrub checks reduce water velocities in the gully channel and induce silting, which gives other vegetation a chance to become established. Shrub checks should be closely spaced if they are to be effective. They should be used only in gullies that have a mild grade.

* * * * * * *

SODDING

If an immediate grass cover is required and suitable sod is available, it may be necessary to transplant sod. Sodding is generally too costly for extensive use over large gullied areas. Sod is needed, however, on critical sections at the gully head or at points along the bank or bottom where protection against waterfall erosion is necessary. It is also frequently used in connection with permanent structures. Grass covers can usually be established through sodding on areas exposed to run-off where it would be impractical to secure a cover by seeding. Where the amount of run-off is not too large, and good sod is available, it can be used as a substitute for the more costly masonry and concrete materials. Sod flumes, sod check dams, and sod spillways have all functioned satisfactorily when properly constructed and applied.

* * * * * * *

SODDED EARTH FILLS

Low, sodded earth fills (fig. 20) can often be advantageously used as a substitute for ordinary brush or wire check dams in stabilizing gully channels. Where suitable sod is available they can be constructed at less expense; and they have an additional advantage because of the fact that by their use vegetation is immediately established at intervals along the gully. They have been successfully used in small- to medium-sized gullies with drainage areas of less than 25 acres. The earth fills should be located at strategic points or at regular intervals along the gully, as is necessitated by existing conditions. Frequently these fills are spaced so that the top of each will be as high as the base of the next one above. Side slopes steeper than 3:1 on the upstream side and 4:1 on the downstream face should seldom be used. The fill should be well tamped, and heights in excess of 18 inches should be avoided because of the overfalls created. An average height of 10 or 12 inches is commonly used. The top of the fill should be low in the center and should gradually curve upward to meet the gully sides and provide the necessary spillway capacity. The fill should be solid-sodded on the top and on the downstream face, and it should be carried up the gully sides to a height of 6 to 12 inches above maximum water crest expected over the structure.

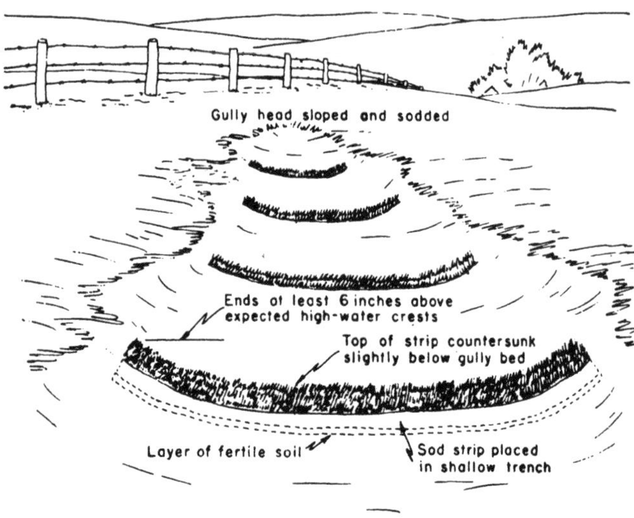

FIGURE 19.— A series of sod-strip checks in a small gully. These checks cannot be used in gullies with steep grades.

Reprinted from *U.S.D.A. Bull. 512* (1971)

UNITED STATES DEPARTMENT OF AGRICULTURE

BULLETIN No. 512

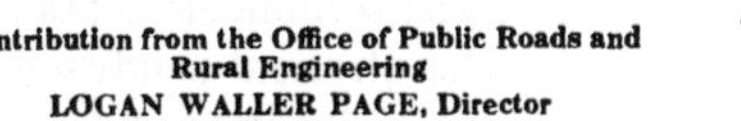

Contribution from the Office of Public Roads and Rural Engineering
LOGAN WALLER PAGE, Director

Washington, D. C. PROFESSIONAL PAPER April 5, 1917

PREVENTION OF THE EROSION OF FARM LANDS BY TERRACING.

By C. E. Ramser, *Drainage Engineer.*

16

CONTENTS.

INTRODUCTION.

The existence of vast areas of so-called worn-out hill lands throughout the United States may be attributed chiefly to soil erosion, due to the natural agencies of wind, frost, and rain. In most localities wind and frost, owing to their comparatively slow processes, play but a minor part in the depletion of the soil and the ultimate destruction of good farm lands. It is the failure of the soil to absorb the rain water which falls upon it that presents by far the most serious aspect of the problem. It is estimated[1] that the Potomac River each year carries off in solution about 400 pounds of solid

[1] Bulletin 17, North Carolina Geological and Economic Survey, p. 21.

Note.—This bulletin treats of terracing as a means of preventing erosion of hillside land. It describes the different types of terraces and points out the applicability of each to the various kinds of soil and topography. It discusses the principles of terrace design. While the investigations upon which the recommendations are based were made in the Southern States, the information is applicable generally to any State in the humid section.

71775°—Bull. 512—17——1

matter per acre of land drained, containing plant food sufficient to produce a crop. Unless this loss be replaced by natural agencies or by the application of fertilizer, it is obvious that the land soon will deteriorate greatly in productiveness and eventually be abandoned.

In addition to the loss of the soluble elements of the soil, a noticeable impairment occurs in the physical condition of the soil. When the moving water washes the soil particles from the surface of the hillside and deposits them on the land below, the heavier particles, or the sandy constituents of the soil, are deposited first, and the finer, or clay, parts last. Since neither pure sand nor pure clay possesses the productive characteristics observed in a soil composed of the proper intermixture of sand and clay particles, it is apparent that the effect of this sorting process is to diminish greatly the fertility or productive power of the soil. Hence, not only the eroded land suffers but also the land at a lower level upon which the eroded material is deposited. Portions of the flood plains of small streams often are covered with a layer of sand, the fertility of the land so covered being practically destroyed, since it is a most difficult task again to build up a productive soil over such areas. Drainage channels, also, constructed at considerable cost, often become filled with soil washed from the hill lands. (See Pl. I, fig. 1.) As a result the adjoining bottom land reverts to swamp and becomes unprofitable for cultivation.

FORMS OF EROSION.[1]

Erosion due to moving water occurs in two forms—sheet washing and gullying. Small areas are practically ruined by gullying (Pl. I, fig. 2), while sheet washing (Pl. II, fig. 1) diminishes the productive power of large areas.

Gullying generally is the most dreaded of the two types on account of its more apparent destructive effects. Where the ravages of erosion proceed unchecked, deep gullies invariably develop in the field. Their appearance causes not only absolute loss of land and inconvenience in cultivating, but a marked lowering in the water table, with a possible accompanying inability of the soil to retain the proper moisture content for the production of crops and to withstand periods of drought.

The injury due to sheet washing, which occurs throughout the United States, generally is underestimated and is regarded by many farmers as of no particular consequence. It is this type of erosion that slowly carries away the very fertility of the soil without apprising the farmer—except through slightly diminished crop yields each year—that the application of remedial measures is imperative in order to save his farm. To the very slowness of its action can be

[1] For a more extended discussion of the translocation of soils, see U. S. Dept. Agr. Bul. 180, by R. O. E. Davis.

ascribed the difficulty often encountered in convincing the landowner that destructive erosion is taking place on his farm.

In some sections of the United States, particularly in the South, erosion is assisted materially by the alternate freezing and thawing of saturated soil. (Pl. II, fig. 2.) The freezing process upheaves a thin layer of the soil near the surface. As this layer of loosened soil thaws, it settles, with a tendency to move slightly down the slope. It is very common for heavy rains to occur directly after the thawing period and wash away the loosened soil from the surface of the field. Probably no other combination of natural conditions could operate more effectually to rob a field of its most fertile soil in the same period of time.

METHODS OF PREVENTING EROSION.

Erosion is due chiefly to the free movement of water over the surface of the land, which carries off particles of soil. If all rain water were absorbed by the ground upon which it falls, soil erosion would be reduced to a minimum. It is obvious, therefore, that in order to prevent or reduce erosive action the soil must receive treatment that is conducive to the admission and the storage of large quantities of rain water; and methods must be employed to reduce the velocity, and thereby the transporting power, of the run-off water.

Since the storage capacity of a soil depends upon its porosity, any treatment which results in an increased porosity of the soil will reduce erosion materially. This porous condition usually is obtained directly by deep plowing and by a thorough incorporation of organic matter in the soil. Methods of subsurface drainage which lower the ground water level improve the porous structure of the soil and increase its ability to absorb surface water. The treatment of cover, such as seeding land to pasture, growing timber, and planting cover crops in the winter, tends to check and diminish erosion greatly. Other methods which retard the flow of the water and conduct the excessive run-off from the field with a reduced amount of erosion, are contour plowing, hillside ditching, and terracing.

It is the purpose of this paper to deal primarily with the prevention of erosion by means of terracing; but since all of the methods of prevention enumerated above tend to mitigate the destructive effects of erosion, some of them should be used invariably in connection with terrace systems. The manner in which each contributes to the prevention of erosive action will be described briefly.

DEEP TILLAGE AND APPLICATION OF HUMUS.

By deep plowing the absorptive power and reservoir capacity of a soil is increased greatly. It is said[1] that 10 inches of loose, plowed

[1] Soil Report N. 3, Illinois Agricultural Experiment Station, p. 16.

soil will absorb 2 inches of rainfall. The incorporation of organic matter or humus in a soil adds materially to its moisture-holding capacity. This is best accomplished by plowing under deeply, manure, stubble, stalks, and various cover crops. This organic matter, in a decomposed state, is capable of absorbing considerable water and forms a richer and deeper top soil.

USE OF COVER CROPS.

Vegetation or cover crops will protect the soil in four ways: (1) by holding rain water on the surface for a time, thus giving the soil a better opportunity to absorb the water; (2) by keeping the soil open through the growth of the roots, which form passages for the water to reach the subsoil; (3) by holding the soil particles together through the binding power of the roots; and (4) by reducing the movement of soil particles through diminishing the velocity of surface water. Cover crops usually are grown during the winter or when the land is not being used for other crops. Their importance as a means of protecting land from erosion at such times can not be emphasized too strongly. Vetch, clover, cowpeas, wheat, and rye are used commonly for this purpose. It can be said generally that good farming and the use of cover crops go hand in hand.

PRACTICE OF LEVEL CULTURE.

Contour plowing and the following in general of practically level lines in farm operations tend to check the surface flow down a slope and to retain the water where it falls. In cultivating crops each row is banked up and a shallow depression which holds the surface water is left between the rows. Thus the absorption by the soil of this impounded water is facilitated and the rapid run-off down the slope, with its destructive eroding power, often is entirely eliminated in case of ordinary rains. Contouring contributes also in a considerable degree to the conservation of moisture on hill lands. The very apparent benefits of this practice merit its universal use on lands subject to erosion.

PASTURING AND FORESTING.

Often it seems impossible to prevent erosion on lands with excessive slopes. No attempt should be made to cultivate such areas but they should be seeded to meadow or pasture and usually retained as such. In well-sodded land the soil is not exposed directly to the erosive action of the water, so that erosion is much less destructive than in cultivated fields.

In many sections of the country timberland on excessively steep slopes has been cleared for cultivation, and in many instances after clearing it was found impossible to control or check the erosion. Such lands should be reverted to timber; otherwise the ravages of erosion will reduce it soon to a state of barrenness. It is known that erosion is least active in forested areas, because of the penetration and binding power of the roots and the accumulation of a thick layer of leaves and organic matter on the soil surface. The soil possesses great coherence and power of resistance to the erosive action of the water and the layer of humus protects the surface and also absorbs considerable water.

UNDERDRAINING.

It can be seen readily that by the underdrainage of land to carry off the excess water from the soil space is created for the reception of more water from the surface. The water falling upon the surface sinks into the soil, percolates through it, and is conducted away by the underdrains to an open drainage channel without running over the surface and causing destructive erosion. Entrapped air, which often prevents the entrance and free movement of water in the soil, finds a means of escape through subdrainage channels. The physical condition of the soil is altered by underdrainage through the aeration and flocculation of the soil particles. A perceptible expansion and a slight upheaval of the soil take place, resulting in an increase in the size of the individual pore spaces. Hence, the rainfall percolates more easily and quickly into the soil and a diminution in the run-off follows. This system of draining is accomplished best by the use of tile drains.

USE OF HILLSIDE DITCHES.

Hillside ditches, as the name implies, are ditches constructed on hillsides to intercept run-off water and carry it at a low velocity to the nearest open drainage channel. Wherever this method of preventing erosion is employed there is likely to be a constant, perceptible draining off of the finer particles of soil, and a continual enlargement of the ditch takes place, the extent depending upon the amount of fall given to the ditch. (See Pl. III, fig. 1.) It is inadvisable, therefore, to resort to this method except when it is necessary to intercept surface water from adjoining higher land on which methods of preventing erosion are not employed. Sometimes hillside ditches are constructed to serve as outlets for systems of graded terraces where natural drainage outlets are not available.

TERRACING.

The greatest benefits from the foregoing methods of prevention come when they are applied in connection with a system of terraces.

Terracing affords the best means of conserving the hillside soils against the washing due to heavy rains.

A field trip was made by the writer through the States of North Carolina, South Carolina, Georgia, Alabama, and Mississippi for the purpose of studying the nature, causes, and effects of erosion, and more particularly the method of preventing erosion by means of terraces. Surveys of terraced fields which afford typical examples of every form of terrace in use were made with a view to deducing from a close study of the field data comprehensive and definite instructions for the design and construction of adequate and efficient systems of terraces. It was found that a great diversity of opinion exists among the landowners as to the best form of terrace and in the rules employed in planning a system of terraces. However, this difference of opinion, in most cases, could be attributed directly to varying conditions of soil and topography or to differences in farming methods.

The subject of the proper methods of terracing was discussed at length with experienced farmers—men who are pioneers in the practice of terracing and who are interested vitally in the preservation of their lands for themselves and their posterity. The deductions and conclusions reached are the result of an endeavor to treat from an engineering standpoint the information obtained from actual observation of field conditions in connection with the data derived from field surveys and the advice and opinions of the best informed and most experienced farmers.

DEFINITION AND CLASSIFICATION OF TERRACES.

As applied to the protection of farm lands, a terrace is any arrangement or disposition of the soil the object of which is to retard the rapid movement of surface water and thereby arrest the process of erosion. According to the earliest practice, terracing consists of building land up in a series of level areas resembling stair steps, the interval between the risers being horizontal and the riser itself being vertical or nearly so. This type of terrace has long been used extensively in Europe and China and is used to a great extent on the steeper lands in the United States. It is known generally as the level bench terrace, but to avoid confusion in the use of the term "level" it will be referred to in this paper as the horizontal bench terrace. Strictly speaking, this is the only true terrace, but the word "terrace" in this country is applied also to ridges of soil thrown up and located in such manner as to prevent the rapid flow of water down a slope. This type of terrace will be referred to in this paper as the ridge terrace to distinguish it from terraces of the bench type. The following classification (fig. 1) of terraces shows the various forms of bench and ridge types.

The bench type of terrace is subdivided into two classes, the horizontal and the sloping, the essential difference between the two being shown clearly by figure 1. Practically all terraces of the bench type are level, which means that they have no fall along the direction of their length to drain off surface water to the edges of the field or to an outlet channel.

The ridge type of terrace is subdivided into two general classes, the graded and the level, depending upon whether it has fall in the direction of the terrace to carry off the surface water. Graded and level-ridge terraces are subdivided further into two classes with respect to breadth of base, namely, the broad-base and the narrow-base forms. The broad-base graded terrace is subdivided again with

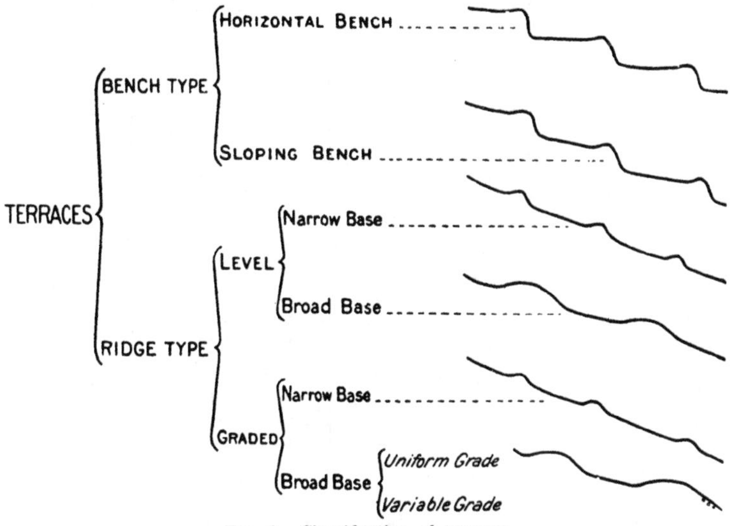

FIG. 1.—Classification of terraces.

respect to grade, the uniform-graded and the variable-graded terraces. Figure 2 shows actual profiles taken on terraced fields and illustrates the various types.

* * * * * * *

COMPARISON OF TERRACE TYPES.

In order to show the relative merits of the bench terrace and the various forms of broad-base ridge terraces, the table below was prepared. Under the column headed "Least amount of erosion" is found the broad-base level-ridge terrace ranking first and the uniform-graded terrace last. Since the primary object of terracing is to reduce erosion, this advantage should have the greatest weight when considering the merits of the different types. The broad-ridge

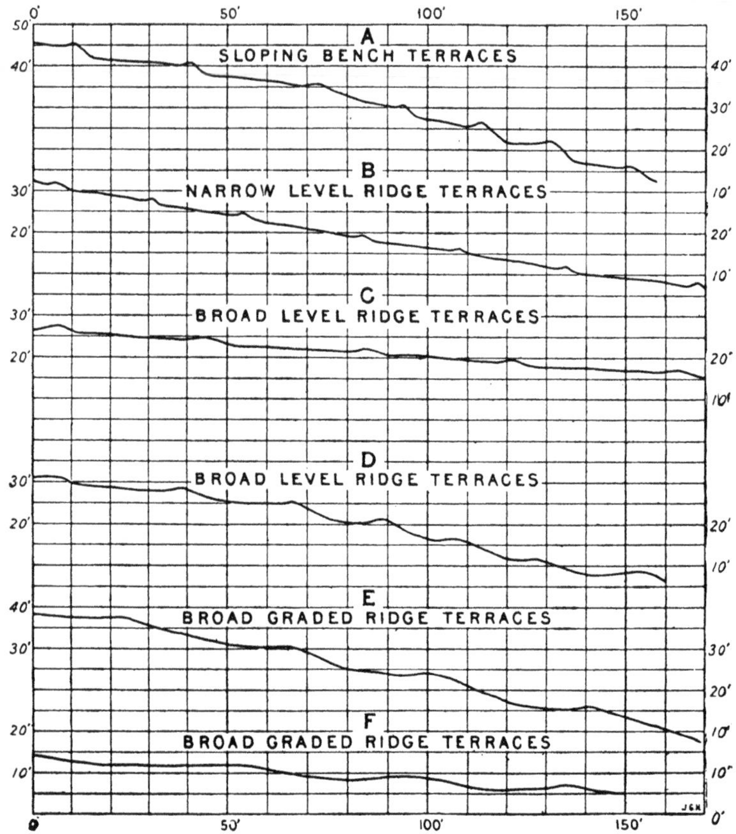

FIG. 2.—Actual profiles of terraced fields, taken across terraces.

terrace is much superior to the bench type of terrace when considering the "Least waste land or weeds." The embankment of broad-ridge terraces can be cultivated successfully and hence no land is lost to cultivation or weeds allowed to grow on the terrace.

Showing how terraces rank with respect to various advantages.

Type of terrace.	Least amount of erosion.	Least waste land or weeds.	Fewest terraces required in field.	Ease of cultivating land.	Best adapted to—			Best land builder.
					Pervious soils.[1]	Impervious soils.	Steep slopes.	
Horizontal and sloping bench	2	2	3	2	2	3	1	2
Broad-base level ridge	1	1	3	1	1	4	4	1
Broad-base uniform-graded ridge	4	1	2	1	4	2	3	4
Broad-base variable-graded ridge	3	1	1	1	3	1	2	3

[1] With reference to least amount of erosion.

Under the column headed "Fewest terraces required in field," the graded terrace ranks ahead of the level type owing to the fact that by giving the terraces considerable fall they may be spaced farther apart than level terraces. However, the greater vertical spacing can be used only at a cost of greater erosion to the field.

Broad-terraced fields are easier to cultivate than the bench type since implements and large machinery can be moved across the broad terraces and if desirable the rows can be run at any angle. With the bench type each bench must be cultivated separately, and difficulty is encountered in getting implements from one bench to another.

From the standpoint of erosion the broad-base level-ridge terrace is best adapted for use on pervious soils; on impervious soils the graded terrace can be used to the best advantage since in the former case most of the water is drained off through the soil and in the latter the water is drained off the field over the surface.

The bench terrace is best adapted for use on steep slopes where it would be practically impossible to build and cultivate a broad-ridge terrace.

The broad-base level-ridge terrace contributes to the building up of land possibly more than does any other form. With this terrace practically no fertile parts of the soil are allowed to escape from the field. The bench terrace also is a good land builder. The greatest objection to the use of the graded terrace is that the water drained off the field usually carries in suspension fertile particles of the soil.

The table below was prepared to assist in the selection of the terrace best adapted to the needs of a particular field. In this connection it is recommended that the design of the terrace system be made from the curves as given in this paper for each type of terrace.

Types of terraces most applicable to land of various slopes.

Kind of terrace.	Average slope of land.	Type of soil.	Grade of terrace.
	Per cent.		
Horizontal and sloping bench	15 to 20	Fairly pervious.	Level.
Broad-base level ridge	3 to 15	...do	Do.
Broad-base graded ridge	3 to 15	Impervious, worn out.	Preferably variable, 0.0 to 0.5 per cent.[1]
Broad-base level ridge with tile drainage	3 to 15	Any type	Level.

[1] Grade will depend upon the length of the terrace, but it is advisable not to exceed a grade of 0.5 per cent if possible.

On the steeper slopes, where the soil erodes easily, clean-cultivated crops, such as cotton and corn, should not be grown. Impervious soils on slopes of 15 per cent or more, and all soils on slopes of more than 20 per cent, are best suited to pasture and timber.

The result that should be attained by a system of terraces and proper farming methods is well expressed in the following quotation taken from a bulletin[1] on Soil Erosion by W J McGee, of the Bureau of Soils, United States Department of Agriculture:

> The primary object is conservation of both solid and fluid parts of the soil through a balanced distribution of the water supply. The ideal distribution is attained when all the rainfall or melting snow is absorbed by the ground or its cover, leaving none to run off over the surface of the field or pasture; in which case the water so absorbed is retained in the soil and subsoil until utilized largely or wholly in the making of useful crops while any excess either remains in the deeper subsoil and rocks as ground water or through seepage feeds the permanent streams.

The above conditions are fulfilled most nearly by the horizontal bench terrace and the broad-base level-ridge terrace, since the movement of the water is reduced to a minimum by both. The graded terrace lacks much in meeting the requirements. The broad-base level-ridge terrace possesses a decided advantage over the horizontal bench terrace with respect to the elimination of weeds and waste land, and over the graded terrace with reference to the movement of the surface water.

In view of the above discussion it is recommended that the broad-base level-ridge terrace be used wherever conditions of soil and topography will permit—that is, where the soil absorbs a portion of the rainfall and the slopes are not too steep. The broad-base level-ridge terrace supplemented by efficient tile drains suitably located would afford the most ideal method for preventing soil erosion on any type of soil. Often the yields obtained and the saving resulting from the absence of soil erosion would justify, in a financial way, the installation of tile.

* * * * * * *

CONSTRUCTION OF TERRACES.

All types of terraces are constructed originally in the same way. The work of construction should begin invariably with the highest terrace in the field and each terrace should be completed before work is started on the one next below. The late fall and early winter is the best time to lay out and build terraces. If one has not time to terrace his whole field well it is better to construct well the first few terraces near the upper side of the field than to terrace the whole field poorly, for a break in a terrace near the upper side of the field is followed by breaks in all below.

The terrace embankment can be built up wholly with an ordinary turning plow. A large sixteen-inch plow with an extra large wing attached to the moldboard for elevating the dirt, is an effective

---

[1] U. S. Dept. of Agr., Bureau of Soils, Bulletin 71, p. 56.

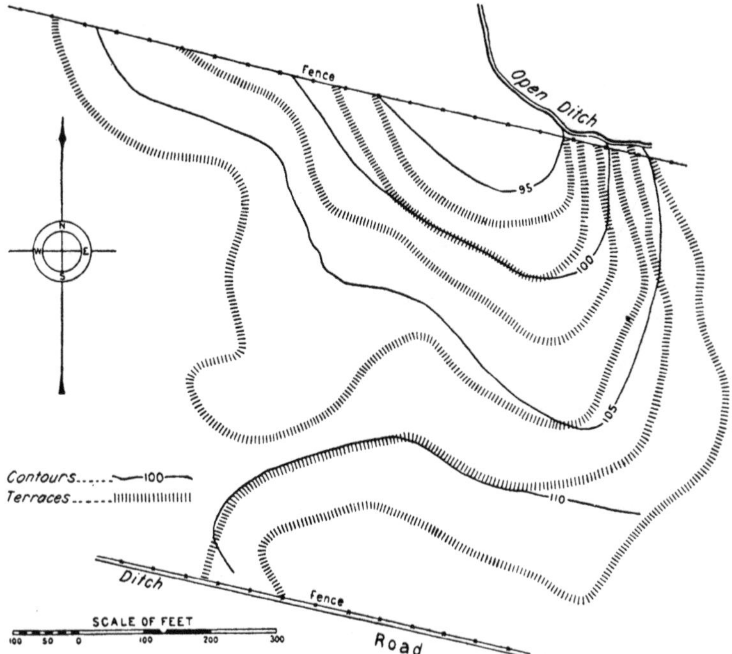

FIG. 16.—Field of graded-ridge terraces.

implement for throwing up a high terrace bank. For broad terraces furrows are thrown toward the center line from each side for a strip 15 to 20 feet in width. Then, commencing at the center again, the strip is plowed in the same manner as before. This procedure is repeated until the terrace has reached the desired height. Many farmers allow the loose earth to be settled by a rain between plowings so that the dirt will turn better. However, it is safer to build the terrace to the desired height at the start for, if a heavy rain, sufficient to overtop the terrace, comes between plowings, much of the original work is undone and considerable damage occurs from erosion. A disk plow can be used successfully to throw up loose dirt, and the ordinary road grader is employed often and is adapted especially to such work.

* * * * * * *

MAINTENANCE AND CULTIVATION OF TERRACES.

A newly built terrace is susceptible to failure until it becomes thoroughly settled. For this reason it is not advisable to cultivate the terrace the first year. It should be sown to some sort of cover crop. Breaks in terraces in the first year tend to discourage a novice in the use of terraces, but unless the embankment is built to an abnormally large size breaks occur often in newly made terraces.

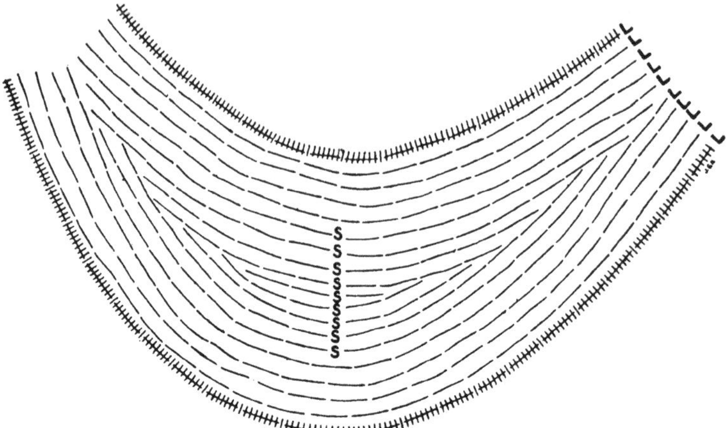

FIG. 18.—Method of laying out rows where distance between terraces varies (L, long rows; S, short rows).

After the terrace has been established permanently, the soil should be thrown toward the center at each plowing of the field, at least once a year. This will increase the breadth and maintain the height of the terrace and the field eventually will assume an appearance of a succession of prominent waves, all of which may be cultivated easily. (See fig. 2–D, E, and F.)

In cultivating a terrace as much of the soil as possible should be thrown toward its center. The best results are obtained where the rows are run parallel with the terraces. At first, usually one row is planted on the top (Pl. VIII, fig. 1), but as the terrace grows broader several rows are planted as shown in Plate VIII, figure 2. These rows invariably produce a greater yield than do those on the land between the terraces. Where large machinery is used, and it is difficult to follow the terrace line, the rows may be run at an angle across the terraces, where the land is not very steep, as in figure 2–F. To do this the terraces must be broad and must be thrown up at least once a year to maintain their height.

Where the rows between two adjacent terraces are to be laid out parallel with the terraces, the same number of rows should be run parallel with each terrace as indicated by the rows marked "L" in figure 18. Owing to the variation in distance between terraces it then will be necessary to fill in with short rows, generally known as "point rows." These rows, marked "S" in figure 18, are run in pairs so as to facilitate the work of cultivation.

RECLAMATION OF GULLIED LANDS.

The best results accomplished in the reclamation of badly gullied and eroded lands were found on the State agricultural experiment farm near Holly Springs, Miss. Plate IX, fig. 1, shows an extreme type of land that was completely reclaimed and made to produce crops. The gullies were partially filled by plowing the soil into them from along the edges, and further filled and levelled off by means of teams and scrapers. As soon as possible a sod of lespedeza or Bermuda grass was started over the levelled-off eroded areas. Most of the land was terraced with broad-base graded-ridge terraces, and terraces, or dams, were constructed across gullies that were too large and deep to be economically reclaimed by filling in. Ponds formed above these dams (see Pl. IX, fig. 2) which served to catch all soil carried into them from above.

Gullying can be effectively checked also by planting trees in the depressions. The native pine and black locust are recommended. Filling in gullies with straw and brush also checks erosion. In one instance a gully was practically reclaimed by dynamiting the bottom to loosen the soil and then stretching wire netting across the gulley below to catch soil particles and vegetation. The use of large pipe through a dam, with a drop drain above, is also a method which can be used effectively.

SUMMARY.

To soil erosion may be attributed the existence of much of the "worn-out" hill lands of the United States. Erosion can be controlled most effectively by the use of terraces. Although terracing is now quite widely practiced in the Piedmont region of the South, in only a few sections are efficient results being obtained. Since the comparatively few well-designed and constructed terrace systems are uniformly successful in preventing soil wash, it follows that the many failures must be ascribed to unsuitable design, faulty construction, or lack of proper maintenance.

The terraces in use in this country are of two general classes, the bench terrace and the ridge terrace, each having variations which are adapted to particular conditions of topography and soil.

The true horizontal-bench terrace is not used widely in the United States, while the sloping-bench terrace is quite common. The disadvantages of the bench terrace are that it can not be crossed by modern farm machinery; the banks can not be cultivated, while each bench must be cultivated as a separate field; weeds and objectionable grasses which grow on the banks tend to sow the entire field. It is best adapted to slopes too steep to permit the use of any form of cultivated terrace, but it can not be recommended for use on slopes exceeding 20 per cent.

The narrow-base level-ridge terrace is used extensively in the Piedmont section of the South. It is cheap to construct and easy to maintain. However, attempts to cultivate this type of terrace have not been successful generally; consequently, as in the case of the bench terrace, considerable land is lost to cultivation, and the growth of weeds and grasses on the embankments tends to seed the entire field as well as sap the strength of the adjacent soil. Outside of these objections, the narrow-base level-ridge terrace, where heavily sodded, renders satisfactory service on pervious soils and slopes not greater than 8 per cent.

The broad-base level-ridge terrace has been developed from attempts to render cultivable the narrow-base form. It has all the advantages of the latter terrace with the added one that no land is lost to cultivation. By the use of this terrace little or no soil is removed from the field. It is best adapted to use on open, pervious soils on slopes not exceeding 15 per cent, but under proper conditions of design, construction, and maintenance can be used on any soil and on slopes somewhat greater than 15 per cent.

As in the case of the level-ridge terrace, the first graded-ridge terraces were small with narrow bases, and they are subject to the same objections that apply to the narrow-base level-ridge type. Moreover, the velocity of flow of the water, due to the grade of the terrace, tends to erode the upper side of the embankment to an extent which a narrow-base terrace can not withstand.

The broad-base graded-ridge terrace (the Mangum terrace) has been adopted in many parts of the country. This terrace, properly constructed, not only can be cultivated but it can be crossed at any angle with large farm machinery. Its broad base and flat embankment slopes render it less liable to damage by the flowing water than is the case with the narrow-base type. The grade may be either uniform or variable, but both practice and theory indicate the variable-graded terrace to be superior to the uniform-graded type.

The graded terrace is adapted particularly for use on impervious and worn-out soils, and on shallow open soils with an impervious foundation—in short, soils that will not absorb much water and that necessitate the removal of most of it over the surface.

By the selection and proper construction of suitable types of terraces erosion can be controlled on slopes up to 20 per cent, or even more. Instances were found where erosion was controlled by the use of terraces on land which had a slope of 30 per cent. However, slopes steeper than 20 per cent usually can be devoted more profitably to grasses or timber than to cultivated crops. Of all types of terraces, the use of the broad-base level-ridge terrace is recommended wherever conditions will permit. This type, supplemented with efficient tile drains, offers the most ideal method of preventing soil erosion on any type of soil.

The success of a terrace system depends largely upon its proper laying off. A good leveling instrument in the hands of a competent and experienced levelman is the best insurance against failure.

Construction always should begin with the highest terrace in the field, and each terrace should be completed before starting the next lower one. The late fall and early winter is the best time to build terraces.

A terrace is susceptible to failure until it has become thoroughly settled. To facilitate settling it is best not to cultivate a terrace the first year, but to sow it to a cover crop. The best results are obtained where crop rows are run parallel with the terraces.

The instructions given herein for the selection and design of terrace systems are based upon the results of surveys, observation, and a study of terraced fields in the best-terraced sections in this country and it is believed that if they are followed carefully a great increase in the efficiency of terrace systems will result and that much better opportunity will be afforded to observe the results with a view to further improving the practice of terracing. At the same time a close study of local conditions—particularly of soil—should be made which no doubt will afford more definite information for improving further the design of a terrace system adapted to a particular locality.

Since the primary purpose of terracing is to hold the soil of the farm in place and thereby both maintain its fertility and render possible an increase of fertility by proper farming methods, all of the benefits, such as greater yields and land values, which result from the preservation and increased fertility of the soil may be attributed directly to the practice of terracing. In short, the terracing of farm lands saves the soils the most substantial and valuable asset of the country.

O

Reprinted from *J. Assoc. Chinese Amer. Engr.*, **8**, 3–13 (1927)

17 Erosion Control in Japan

By W. C. Lowdermilk

"To rule the river is to rule the mountain."

Japanese ancient proverb.

Japan proper would be washed into the sea, or the process of erosion would be so far advanced that agriculture could scarcely be followed, were it not for the rich cover of vegetation which mantles the mountains and slopes. The rock structure of the mainland and islands favors rapid degradation; it comprises chiefly coarse grained granites which readily weather and crumble, and lavas and volcanic ash which are subject to rapid erosion. The land area is strikingly mountainous and picturesque with steep slopes and narrow valleys. The narrow valleys contain a small percentage of level and arable lands. The declivity of the short rivers produces rapid velocities in the waters, which rush back to the sea. Rainfall is heavy and occurs in downpours equalling as much as 300 mm in 12 hours, and totals from 1000 to 2000 mm per annum. The large volumes of stream flow and rapid gradients give to the flowing water great erosive and transporting power. The mantle of vegetation, dense and rich in variety of both shrub and tree restrains the impetuous waters and robs them of their prey of silt, sand, and boulders.

Erosion Caused

Inadvertantly man has here and there in Japan destroyed the mantle of vegetation on the sloping lands, and has thus unleashed forces of erosion. Serious examples have demonstrated to the Japanese the dangers of denuded slopes. Two such examples are given for representative geological formations.

1911 Typhoon

The region about mount Asama, an active volcano, has been from time to time covered with the ejecta and debris of volcanic activity. Layers of andesite lava are intercalated with beds of volcanic ash. Karuizawa, a favorite summer resort, is in this region. The altitudes of 5,000 to 8,000 feet with comparatively short distances to the sea level assigns rapid gradients to the thalwegs of the streams. Water erosion throughout recent geologic periods has been cutting rapidly through the soft and light volcanic debris to from steep sided canyons. The gradients of the valley walls at the headwaters of the streams are sharp, not uncommonly 45°, (100 percent) and even more. The annual rainfall varies between 1000 and 1200 m.m.

But a dense mantle of vegetation covers the surface; reinforces the top soil layers with roots; prevents the rapid accumulation of surface run off of the rain waters; and prevents the active erosion of the surface soils. Previous to 1911 the headwaters of Kirizumi river, which lands are in private ownership, were subject to unregulated cutting of the forest cover (deciduous hardwood type) to supply a thriving charcoal industry. The cutting became general, and utilization close, until 1911 when an unusually heavy typhoon occurred. Rain fell in great down pours. The mantle of vegetation was ruptured; the run-off waters filed their way into the underlying loose volcanic ash. The widening of the cuts and gullies increased the accumulation of run-off waters, and produced acceleration of transporting power. Mud flows of the volcanic ash formed and flowed down the valley unto the plain and began to cover the fertile terraced and productive rice fields. Floods caused great damage in the valley to dwellings and houses of the villages, to roads and to bridges. The mud flows covered up and made of no use valuable rice fields. The loss was accepted as a serioues one.

Forest Cutting

Each succeeding year the mud flows continued and increased in volume. They were covering wider areas each year of the valuable rice paddies. They were enlarging dangerously. After five years it became imperative to do something to check this growing and dangerous menace. The forest Act of 1897 was put into effect, whereby the condition of the forest land was declared dangerous; the area was accordingly declared a protection forest. Cutting was stopped and made subject to the approval of the Forestry Department. The forest engineers were directed to check and control the erosion which had already gained much headway. At the present time (1926) after 10 years of work the head waters of Kirizumi river have been reclothed with the exception of two upper valleys, where erosion is nevertheless under control. The works are not yet complete in these two valleys. Time is required to permit natural forces to take effect following the construction of artificial works. To look upon an achievement of this sort, where 10 years ago great surfaces of loose soil were actively eroding, and eating backward up the high slopes, favoring the concentration of torrential waters; where the rage of wild waters knew no bounds, defying the works of man; filled me with admiration for the faith, for the skill, for the technique and strategy of the forest engineers. They had step by step; with a dam here and a dam there, robbed the torrential water of transporting force; with butress walls strategically placed had prevented undercutting of slopes; with works of soil fixation on slopes where a prostrate alder and grass were called into assistance had reclothed eroding slopes with a preliminary vegetation; by suiting devices to peculiar features of ground and water flow they had gradually diminished the progress of erosion and had replaced it with the processes of plant growth which was healing the gaping wounds in the landscape.

Erosion Checked

It was a costly struggle, requiring over 120,000 Yen. This was a much greater sum than the value of the land on which these works of restoration were done.

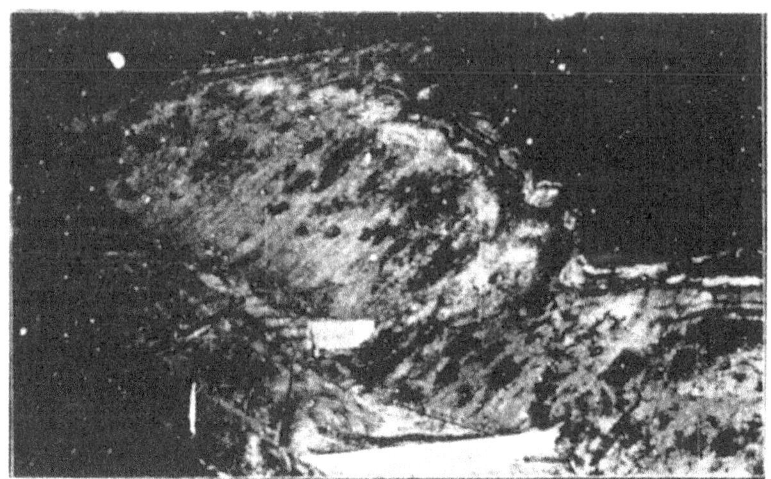

Detail of one of the two head valleys where the chess game of erosion control reaches interesting developments. The check dams have established a base level of erosion. The left slope (looking down the valley) will require a few years to grade down to an angle of repose. Artificial works of soil fixation have been carried out on the right slope.

General view of methods of fixing soil in breaks of the cover of vegetation on slopes. Note the plantation of conifers in a hardwood natural cover above the erosion control works.

View of a slope graded down and the terraces on 5 foot contours where soil is being filled behind sod walls. Shrubs and trees are established on the terraces.

Type of dam used in establishing a base level of erosion in torrential valleys. Watershed above has become completely covered with an excellent stand and growth of alder, oak, and pine. Small resevoirs are maintained in upland valleys, which supply rice fields below with necessary irrigation.

6

It was not so much for restoring the eroding land, as it was for saving the vanishing rice fields from the blight of mud flows, that the expense was justified. Besides the expense of restoration was increasing as the square of the extension of erosion. There was no way of knowing where it would stop. Erosion here was considered a hazard just as forest fire is considered a menace, and must be put out.

The second area typifies the mountains of granite in central Honshu. It lies within Shiga prefecture about the south end of lake Biwar, near Kyoto. The underlying rock is coarse grained orthoclase granite which weathers and crumbles rapidly. The slopes are steep and loose material works rapidly down slope under the influnce of water and gravity. The annual rainfall varies between 700 and 900 m.m.

Hills Denuded

About 400 years ago the great accumulation of Buddhist temples of Mt. Hiari were caused to be burned by a Shogun who was unfriendly to Buddhism. On the accession of another Shogun all the temples were completely rebuilt. The timber was cut from the pine forests on the slopes of Shiga Prefecture. Then followed a time which pemitted fires to burn unrestricted over the granite hills. From 340 to 50 years ago the region was practically barren of forests; so much so that the villagers were scarely able to secure their needed supplies of fuel.

In the 16th year of the restored Meiji, 1871, works of torrent correction were begun in this region under the direction of a Dutch Engineer, Johann Dorehk. First of all, check dams were located strategically in the torrential valleys to rob the impetuous waters of their powers of transportation. Then followed works of fixing the soil on the slopes. The experimentation in the use of various devices and plants was done by the Japanese foresters. Methods were developed which became successful under the skilled application and adaptation of experts. And now one may walk up valleys, past check dams overgrown with vines and bracken over which torrential waters no longer charge, then into the carpet of a thick and absorbent layer of forest litter, through an undergrowth of bracken and moisture and shade loving shrubs, under a full canopy of a pine forest. Now Chamaecyparis, a shade loving (demanding) species, is being planted under the pine forest to complete a forest cover.

Dynamic Geography

The achievement is evident by a contrast with an adjacent valley in which such works were only begun in the past three years. The same rock and soil conditions prevail here that 30 years ago existed in the present reclaimed valleys. The general aspect of the valley is semi-arid. A large portion of the valley, is bare shining hot in the sun; only the more drought resistant plants are growing; the pine trees show evidences of slow growth. Such conditions as these have been changed in the reclaimed valley to those of moist conditions where vegetation has mantled the soil and moisture and shade loving species are thriving. Such work is dynamic geography. It justifies man's control of the forces of erosion under given conditions.

The works of human agency will avail little unless the natural forces such as rainfall, vegetation, and time are enlisted. Thus rainfall must be sufficient to support the necessary vegetation. The unmistakable evidence of this in Japan is the former covered condition of the areas where erosion is at work. Erosion control has its most immediate promise in those regions where unwise treatment by inconsiderate man has set in motion processes which have destroyed a former cover of vegetation. Control is therefore circumscribed by certain definite conditions.

High Cost

The policy of erosion control in Japan is based largely upon saving valuable food producing lands from destruction, rather than upon profit from restored lands. During the year ot 1925 the Tokyo Forestry Board (Tokyo District Office) spent 170,000 Yen on an area of 150 hectares or 375 acres for erosion control. The average cost, including the construction of check dams, was 253 Yen ($225 Gold) per acre. The value of forest land averages about 40 Yen. More than 10 times the marketable value of the land was spent per unit area for the purpose of erosion control. It is therefore impossible to justify this expenditure on the basis of profitable returns from the restored lands. On the other hand, the areas of active erosion under the influence of torrential flows are agents of excavation, transport and deposition of cobble, rock debris, and sterile sands upon rice lands valued at from 240 to 300 Yen per acre. When the increasing imports of rice from alien shores have exceeded 5 per cent for want of rice land enough, including that in Korea and Formosa, any loss in area must be viewed with disquietude and even with anxiety. The preservation of food producing lands becomes a sufficient reason for heavy expenditures in restoring a mantle of vegetation on eroding lands.

Watersheds

This is not the sole benefit from works of torrent correction and erosion control. The experience of Japan is, that a barren or eroding catchment basin of the short rivers yields more sudden and disastrous stages of high water than similar watersheds covered with forest vegetation. The comparative influence of a forested and excessively eroding watershed is not a debatable question in Japan. The Japanese engineers and foresters, unlike those of some western countries, are in agreement on this point. It is not a question of the influence of forests on stream flow; it is rather how shall barren and eroding lands be restored to a cover of forest vegetation. The disasterous experiences in the narrow valleys from angry flood waters issuing from eroding catchment areas have been costly instructors. The maintenance of watersheds in a mantle of forest vegetation has proved to be the most economical policy in the control of flood waters, and necessary to permanency of food production in the valleys.

Japan's Example

Dikes are built in the alluvial plains to protect adjoining fields from inundation and submergence with sands. They are considered essential, even as the mantle of vegetatation is considered necessary. These are supplementary measures not mutually exclusive, nor will one measure alone suffice. Japanese engineers and foresters have progressed beyond a controversial stage to one of cooperation in the control of the natural resources of vegetation, soil and water. Japan furnishes a preeminent example worthy of emulation by western nations in this respect.

The tempering of the destructiveness of flood waters has its its counterpart in the regulation of flow for hydraulic power. The rapid gradient of the numerous rivers issuing from mountains richly supplied with rain assures great potential hydraulic power. This is a very important source of electrical power, which has been used for lighting more widely over Japan than in other countries. Every village and hamlet and most of the farmers' houses are lighted with electricity. Industrial uses of electric power are growing. More than one third of the possible 8,000,000 hydraulic horsepower force is now in use; the development of water power is growing. The permanency and full utility of the resources of hydraulic power is dependent upon a closed mantle of vegetation on the sloping lands.

Forest Laws

These and other objects were the basis of Article 14 of the Forestry Act of Japan (1897) which provides for the setting aside of protection forests where these objectives may be in danger. The purpose is singly or inclusively as follows:

(1) To prevent the denudation of soil;
(2) To prevent sand shifting;
(3) To provide for protection against the devastation of floods, wind and tide;
(4) To prevent avalanches and rolling stones;
(5) To insure a constant source of water supply;
(6) To afford shelters for fish;
(7) To afford landmarks for navigation;
(8) To improve public health and sanitation;
(9) To maintain the scenic beauty of temples, shrines, noted places and historic sites;

Protected Areas

These are the total list of the objectives of protection forests, but the objects numbered (1) and (5) which have to do with erosion control and regulation of water supply assume the first importance in the area devoted to them. The number and areas of the existing protection forests in Japan follow:

Kind of Protection Forests for	No.	Area in acres
1 Protection against soil denudation,	190,840	2,010,857
2 Protection against sand shifting,	10,426	27,465
3 Protection against flood,	15,624	6, 887
4 Protection against wind,	13,650	76,087

5 Protection against tides,	12,093	20,890
6 Protection against avalanches,	4,293	16,850
7 Protection against rolling stones,	418	1,245
8 Conservation of water supply,	73,628	2,187,775
9 Fishery,	23,368	103,097
10 Guiding of navigators,	248	5,475
11 Improving the public health,	153	222
12 Scenery,	8,808	77,487
	353,549	4,534,337

Total Hectares 1,813,734

Of the total area in forests of Japan proper, namely 21,408,965 hectares, the protection forests comprise 1,813,734 or approximately 8.5 per cent. Of these 92.7 per cent of the protection forest area are devoted to the mutually interacting objects of erosion and water control. For torrential run-off involves the two, water and erosion.

Praise for Japan

Thus the regulation of torrents or the restoration of denuding or eroding land has become an important part of the work of the forestry department. Protection forests if intact require only careful management to achieve the object of their establishment. But when the forest manle is broken in steep topography and processes of erosion become active, definite works of erosion control are required. The foresters in Japan have met unique and difficult conditions with these works and deserve the highest commendation from the foresters of the world for their outstanding successes.

Erosion control and torrent regulation are not entirely of recent development. As the population increased in the later middle ages with enlarged demands on the forests, deforestation brought about an increase in torrential development. It became of sufficient importance to require regulatory measures. From the date 1683 AD reforestation was extensively applied and the mountains were maintained generally in a forest cover. "To control the river is to control the mountain," summed up the policy of that period. Little attention apparently was paid to the construction of check dams in the channels of torrents. Following the restoration of the Meiji, about 60 years ago, active and careless cutting took place and produced serious conditions of erosion and torrential development in several areas. The regions of granitic structure suffered most heavily, due to the coarse and easily erodable nature of the rock. Floods became increasingly more disastrous. Systematic works of torrent regulation were begun about 50 years ago.

Check Dams

This time, torrent regulation combined both the methods in practice in Europe with the time honored methods of Japan. The former were carried out first under the plans and supervision of a Dutch Engineer, Johann Dorehk. In addition to the systems of check dams the Japanese foresters worked out improved methods in the fixation and revegetation of eroding soil.

Since the enactment of the Forestry Law in 1897 works of torrent regulation are compulsory in every part of the Empire. The works in the forest areas are generally carried out by forest engineers under the Ministry of Agriculture and Forestry. But the more extensive dam construction, and the works along the larger rivers is generally done under the direction of the Ministry of the Interior. Certain projects of a less important nature which involve little of no revegetation are done by the prefectural offices.

Where needed torrent correction work is required on all private lands. The works are done by the prefectural offices under the supervision of the Ministry of Agriculture and Forestry. The costs are shared, 5/6 by the national and perfectural treasuries and 1/6 by the owner of the land.

The importance of torrent regulation, of erosion control is signified by its general enforcement over the entire Empire of Japan.

Only a brief account of the method of work can be given in this paper, and the general features only will be outlined. It is necessary to treat the erosion control works of Japan proper separately from those of Korea, the reason for which will soon be evident.

Checkmating Erosion

Erosion control as it has been developed in Japan takes on the nature of a great game of chess. The forest engineer after studying and mapping his eroding valley locates and builds one or more check dams. This is his move. He then waits to see what are the responses of the natural forces. The check dam establishes a fixed base level of erosion for the thalweg above it. The slipping and sloughing side slopes may after two to five years come to an angle of repose, whereon vegetation will establish itself naturally. More likely this desired result will not come so quickly. The response of the natural forces in their turn determines the forest engineer's next move, which may be another dam or two, or an increase of the height of the former dam, or the construction of side retaining walls. After another pause for observations the next move is made until erosion is checkmated. The game takes on a lively interest. The operation of natural forces of sedimentation, plant succession and revegetation must be guided and used to best the advantage to keep the costs of this expensive work down to the minimum and to practical proportions.

Fixation

The construction of check dams, or side walls, and of channel works have for their chief purpose the establishment of a base level of cutting or erosion by the stream. This stops the undercutting of slopes by the current of torrential waters. The surface soil is kept from a state of flux and movement. In some instances little more is needed; for the slope material comes to its angle of repose and vegetation naturally takes possession and produces the desired protective, mantles. In other cases it becomes necessary to make use of works to fix the soil on slopes. No attempt is made to fix slopes at a steeper angle than slopes of the same material are held by the natural vegetation in the vicinity. No more is attempted than nature has already done in the region. These works vary in method and application. They are terracing and sod wall terraces, wattle work net works, adapted and suited to the situation by the forest engineer in charge.

A distinctive feature of the Japanese technique is the artificial production of a terrace of soil on steep slopes. The slope is graded down, for this purpose, to the tenable gradient under natural condition. Small horizontal shelves, or terraces are cut in the slope, usually on 5 foot contours. Slabs of grass sod are carried ,to the area and are placed on end (Tsuminae-Ko) or flat and layer on layer (Tsunishiba-Ko) on this terrace. Behind the sod barriers dirt is pulled down from the slope above and tamped in place to a depth of 10 to 15 inches and width of 12 th 24 inches. In this shelf of soil seeds of the hardy Japanese alder, (*Alnus firma*-varieties) are sown. Pine plants (*Pinus densiflora*) are also planted. Fertilizers are sometimes applied to stimulate growth and a quick restoration of a cover.

Gullies of a "bad land" type are graded down to the slope at which they may be held under a cover of vegetation and treated in the same way.

Temporary Vegetation

The underlying objectives are; first, the establishment of a base level of erosion; second, the fixation of the soils on the slopes. The latter objective is obtained either naturally or artificially by means of special works and temporary vegetation consisting of grasses, shrubs, quick growing trees like willows and the hardy pine. After this temporary cover is established more valuable species of trees are introduced to compose the forest cover.

The most common species employed for the temporary cover are:

- Pinus densiflora
- Pinus Thunbergii
- Alnus firma, var. Yasha Winkl firma, var Mullenervis Bgl.
- Robinia pseudoacacia. L.
- Lespedeza bicolor. Turcz var. intermedia Max.
- Alnus japonica S et Z
- Salix japonica Thunb. and other Salicaceae.

Costs of Erosion Control

Within the past 40 years, more than 33 million Yen have been spent on torrent correction dams and works of soil fixation on slopes by the Ministry of Agriculture and Forestry and by the Ministry of the Interior. The amounts spent by the prefectures and by private persons are not recorded but may exceed this amount. The costs of this type of work have been an important item in the budgets of the government.

These expenditures are justified in safeguarding the valuable rice lands, in the regulation of the flow of waters suitable for irrigation of the rice paddies, and in the reduction of eroded material such as silt and debris which would interfer with the control of high stages of run-off waters in the streams, and endanger adjoining agricultural land.

Erosion Control in Korea

A different problem confronts the foresters in Chosen. In Japan proper the work comprises chiefly the repairing of breaks in the cover of vegetation which have occurred on comparatively small areas. The work is chiefly that of repair. The unit area costs are quite high; being applied to small areas they do not reach unmanageable expenditures. On the other hand the southern two thirds of Korea present essentially a barren condition. Serious erosion is far advanced over hundreds of thousands of acres. To apply the measures at the unit area costs applicable in Japan would involve enormous sums beyond practical possibilities. Erosion control is an urgent necessity and has been undertaken energetically as may be observed along the main railway line from Fusan to Antung.

Hardy Shrubs Used

Accordingly the Forest Experiment Station at Seoul has been conducting, under the Directorship of Dr. T. Tozawa, investigations to determine less expensive and effective methods in erosion control. Dr. Tozawa has made use of hardy shrubs as a preliminary vegetative and reclaiming cover. Few attempts are made in these experiments to grade the slopes down. Little dams of sod are made in the small gullies which collect behind them a depth of soil sufficient to support shrub and tree growth.. The desired results will be more delayed but the cost is being cut from 10 to 20 percent of the costs prevalent in Japan proper.

Special attention is directed to these experiments of Korea; for the methods will apply to widely eroded regions, such as north China, parts of western America and other parts of the world.

Summary

Erosion control in Japan has been intensively developed in the last 40 years, combining European and locally developed methods. In particular the Japanese forest experts have advanced to great proficiency in erosion control unexcelled, if equaled, by those of any other nation. The high costs of erosion control are justified by preventing the destruction of valuable rice lands in the valleys from mud flows and debris of torrential erosion; and in the regulated water supply for irrigation of rice fields and for water power. The methods which are now being developed in Korea, make a greater use of vegetation, will be some what slower but will be done at an expense more commenstrate with the wide expense of actively and seriously eroding lands of this part of the Japanese Empire. Erosion control is considered a necessity by the Japanese people regardless of its cost. It is treated as a fire. It must be stopped or controlled. It pays in the long run. It pays in the national economy of the Japanese Empire.

Reprinted from *J. Assn. Chinese Amer. Engr.*, **17**, 22–38 (1936)

18

Soil Erosion and River Regulation With Special Reference to the Yellow River

By Sig. Eliassen

Life Member of the Association

General aspects of the problem

FOR THE PAST seventeen years I have been engaged in the study and regulation of the North China rivers, especially the Hopei river system and the Yellow River. I started out with the firm belief that engineering works alone would suffice to give permanent relief to the evils which the floods of these rivers brought in their wake. But it did not take many years before I was forcefully brought to realize that engineering works alone could only give temporary relief. The silt load which the rivers of north and northwest China carry during the summer flood periods is usually so great that it entirely nullifies the engineers' efforts at finding a permanent or even semipermanent solution to the many problems which the silt flow carries with it. These problems differ with different rivers, but they are all alike in that they do not yield in a permanent way to river engineering treatment. Such treatment merely temporizes with the problem. After some years it comes back, usually in an aggravated form.

The Hai Ho problem at Tientsin is one form, the Yellow River problem is another, which the general silt problem can take. In Tientsin the question is mainly how to exclude the silt, most of it coming from the Yung Ting Ho, from entering the navigable Hai Ho and thus prevent that river from being closed to all ocean-going traffic. But the question has a larger aspect in that the city of Tientsin itself is in danger of becoming invaded by the silt masses. If the situation is allowed to drift it is a question of less than 100 years before this menace will be very serious indeed.

As regards the Yellow River the silt masses prevent the engineers from regulating effectually the river so that the flood flow can be carried to sea without dike breaches and their accompanying terrible inundations.

The main difficulty is that the river engineers are fighting an exceedingly rapid geological land-transformation process. Nowhere else is such a gigantic movement of land masses taking place. The Yellow River transports on the average from the highlands of Shansi, Shensi and Kansu and to a lesser extent from Honan, 2,500 millions of tons of soil every year. This is sufficient to raise a flat area 1,000 sq.km. (about 400 sq.mi.) in extent, 1.5 m. or, say, 5 ft. every year. Comparative figures show that the Yellow River alone transports away from its watershed as much material every year as all the rivers of the United States carry away together. During a real, serious flood the viscous Yellow River silt flow is such that it will not follow the diked channel. As the slope flattens the velocity slows up, silt becomes deposited and the channel area diminished. But water and silt continue to come in from above, pile up and go over the dikes, no matter how high they are built. This was forcibly demonstrated during the flood of 1933, when over a long distance the north dike was completely buried in silt. We cannot close our eyes to this fact, and we also cannot close our eyes to the fact that we are tackling this problem from the wrong end. Instead of tackling the cause we are tackling the effect. The good old saying is, "Prevention is better than cure."

So far it has merely been mentioned that the rivers carry a great amount of silt coming from the highlands of Shansi, Shensi, Kansu and Honan. How is it brought into the rivers? It can be stated with a fair degree of certainty that the silt is mainly brought into the rivers from surface erosion of the loess plateau lands during severe rainstorms and from gully erosion as the plateau runoff rushes down the loess canyon escarpments into the tributaries and the main rivers. The erosion of river-banks and shoals is also involved, but to a much lesser extent. This is evident from the fact that the tributaries are already loaded with silt to their maximum capacity during a freshet period. As such tributaries flow mostly in channels of rocky or gravelly composition which erode but little, there can be only one deduction and that is that the silt must have come from the loess plateau lands usually lying several hundred feet above the tributaries. Numerous gullies cut deeply into the loess plateau carry the runoff into the tributaries. After the water and the silt have reached the tributary and main river channels there is a rearrangement of the silt deposits already formed in these channels, some being washed away and others being formed. This is more or less a balanced effect, at least in the steeper upstream reaches. But during the whole low-water season when there is no surface erosion, these channel deposits are slowly being transported downstream. This is evident from the fact that the tributaries are then clear, but the main Yellow River is about 0.5 per cent silt-laden.

On the whole, the silt percentage during the flood season diminishes towards downstream, showing the tendency towards deposition especially after the river enters the great alluvial plain bordering the coast where the slope flattens out and the silt tends to become deposited in the channel all along, gradually raising it. During the more severe freshets the silting of the Yellow River channel becomes so sudden and so severe that the flow is sent over the dikes, as already described, and enormous tracts of land outside the dikes are becoming raised 3, 4 or 5 m. Hundreds of villages, with many of their inhabitants, animals and all that accompany what we usually call civilization, become buried, some, perhaps, to be excavated after thousands of years and referred to as "ancient relics."

It may be argued that it is a gain to China that the seacoast is being built out and more land gradually added for cultivation to sustain the growing population, and also that the silted-up land becomes more fertile after being flooded. This may be so, but if we stop to reflect on the situation it will at once become evident that the surface erosion covers many tens of thousands of square kilometers of area in the upstream highlands. The fertile and well-prepared top soil taken away from this large area is deposited in a very much smaller area in the plain and along the seacoast, as the soil here is built up vertically. Contrary to common belief, the top soil is the most valuable soil for cultivation as it has been ploughed up, manured and carefully prepared. It is this valuable soil which is transported away and piled up in a small area. Such soil transfer is therefore a huge economic loss to China as a whole. Moreover, that part of the soil which is being deposited in the ocean has had most of the chemicals which originally were in it washed away. This is the reason why near the coast there is a belt of absolutely unproductive land that can be cultivated first only after silt which contains the necessary plant food has been deposited on top. All the soil, therefore, which is deposited at the coast under water and to above the high-tide line is a dead loss and it takes a vertical deposition of at least one meter more before the soil will even yield enough food for sustaining simple weeds requiring a minimum of plant food.

Furthermore, the thousands of gullies which are so rapidly being eroded in the northwestern provinces are usually too steep to be cultivated. A definite destruction of inland farm areas is here going on which is far more serious than people realize. Together surface erosion which impoverishes the surface soils and gully erosion which every year destroys large areas of land are responsible for much of the social unrest in the Northwest and are considerable factors in intensifying the frequently occurring famines. There are large areas in eastern Kansu today where practically nothing will grow beyond scattered wild date bushes as the soil has been so impoverished by surface erosion as to be entirely unfit for agriculture. These areas which are practically uninhabited are a fine operating field for bandits who waylay travelers who have to traverse this region.

The farmers in the Northwest are heroically trying to stem the progress of this erosion. They have terraced the land and in many places tried to check the back cutting of the gullies by providing definite places where the accumulated surface runoff can pass downhill with the least damage. But there is no one to guide them in this effort, and it remains individual and spasmodic. The farmer in the Northwest is fighting a losing game. Famine and social unrest often force him to flee from his holdings and when he comes back after a couple of years he finds his anti-erosion works demolished as they have not been looked after. His terraced fields have developed "sink holes" (erosion holes peculiar to loessic soils) utterly beyond his individual efforts to check. After some more years have passed these holes have connected and his farm land has developed into a gully. I have seen many such cases during my three years' stay on famine relief, irrigation and road work in the

Northwest. I have watched a gully 3 m. deep and 6 m. wide cut back more than 10 m. and deepen from 3 to 4 m. during a single rainstorm; I have seen silt flow rushing down other gullies from the pleatau land above, completely burying sections of a motor car road in silt. If statistics were to be collected of farm lands destroyed by gullying or top-soil stripping, it would be a tale without end.

From an agricultural point of view alone it is highly necessary for the government to step in to see what can be accomplished by organized and concentrated efforts to remedy this truly intolerable and exasperating condition. When one further realizes that surface and gully erosion is the sole cause of all the terrible misery which the Yellow River brings to the densely populated plain then the soil erosion problem becomes one of first-class national importance. There can only be one conclusion if a lasting remedy is to be sought:—soil-erosion control.

The present situation then is this, that instead of keeping the silt of the upstream areas from reaching the river, futile attempts are made at conveying the silt flow to sea. But the river, when it has one of its really silty spasms, refuses to be lead to sea. It spreads over the plain in spite of all human efforts to keep it in a channel. The only logical alternative would be to try to have a controlled silt settling in the plain. This would be easy enough if the plain were uninhabited and the silt could be deposited on it at will. But the plain is one of the most densely populated areas in the whole of China. Over 100 million people live on its alluvial fan land. Ancient land ownership, large towns, large and small villages, customs and habits and all the things which combine to form a very ancient civilization have grown up along the Yellow River. Scarcely a year goes by without a considerable part of this area being visited by disaster from the river. Yet, because the area is so fertile and the roots of this civilization have grown so deep, it would be out of the question to attempt a controlled silt-deposition plan, as it would bring with it social and political problems of a nature which could not be solved. This may readily be understood from the following.

Assuming that such control were to be undertaken it would mean first the depopulating of an area of not less than 1,000 sq.km. with an estimated population of at least 200,000 people living in 500 to 700 villages, some of which might be large and very influential. It may be further assumed that this population could be moved to the Northwest to some rich irrigable area and settled there. The next step would then be to surround the depopulated area with a high strong dike and provide it with a good drainage channel and flow-control works. Villages would have to be built in readiness for another lot of people to move in as soon as the area had been silted up. This would be construction works costing several tens of millions of dollars. On the average the area would lie from 5 to 6 m. lower than the main river course. The silt flow into it in one flood season, allowing only the objectionably high silt flow to enter and settle, would be about 500 million cubic meters of silt. In one year the area would thus be raised 0.5 m. and in ten years' time the area would have reached such a height that no more silt flow could be let in.

The second step would then be to select another area and move the people into the first silted up area and let them settle there. Dikes and control works would be built and during the next ten years the second area would silt up to full height, houses would be built, and the third settling area selected, dikes and control works again built and the population of this area would move into the second silted area, and so on. It is readily seen that it would be a constant migration every eight or ten years of 200,000 people, breaking up their homes and all the attachments which go with this. More than one province would be involved each time a new area was prepared. The plan seems wholly infeasible and would also be extremely costly. One may answer that the river does the silting process in any case. Would it not be much better to have it done under control? But it is just here that the human element comes in. A natural catastrophy is accepted without much questioning. A chance is always taken that it hits the other fellow first. And it is this gambling chance that is the obstacle to any premeditated silt-control plan. Let the will of heaven decide!

Then why not let the people remain in the area, but take care of them while the flooding and silting in the restricted area lasts during a few months of the year. But this also fails for almost the same reason, and besides the flooding of a restricted area would be so deep that the whole population would have to move out for at least four months of the year or they would be drowned. It would be impossible to raise crops and there would be an outcry of cruelty if force were to be resorted to which no government would take the risk to face. This year we are witnessing a refusal to breach an inner Yellow River dike in order to attempt to lead the flow back into the river and thus save tremendous areas from being flooded. Had it been done promptly a few days after the first dike breach had occurred it is quite possible that the whole river flow would have been drawn back to its old channel. But six or eight villages and some thousands of *mou* of farmland,—and let it be specially remarked that it was lying inside the river dikes, would be affected. The protests were so loud that the provincial government involved could not see its way to try it.

We thus find that to attempt to undertake the silt problem from a deposition-control point of view is for social reasons an impossible task and would be a highly expensive task even if it could be done. It has already been mentioned that from a land-gaining point of view it is also a highly uneconomical proposition in that the good surface soil, well fertilized and prepared, is being stripped off from a very large area in the upstream regions and either carried out to sea or deposited in deep layers on the plain within a very small area as compared with the area from which it was stripped. Thousands of years of river regulation practice has also proven that to keep the flow to a channel is also not possible for any length of time. It will always be accompanied by terrible flood disasters occurring on the average every three or four years.

It all goes to prove that the problem is undertaken from the wrong end. Now let us see what the prospects are as regards success if the cause of the trouble were to be tackled instead.

Possibilities of erosion control

Most writers on the subject of erosion control in northwest China are pessimistic about the result. Some of them have formed their ideas as the result of traveling rapidly over long distances through the loess country, generally during spring, when everything has had its most dreary

FIG. 1—THIS TYPICAL NORTHWEST CHINA landscape is barren, tattered and torn.

FIG. 2—IN THIS VIEW from the King Ho watershed the edge of the tableland and the gullies across the valley are to be noted. In the foreground the "mature" stage of the valley which is well-cultivated and semistable may be seen.

FIG. 3—THIS IS ONE of the tributary valleys to the King Ho. Vegetation will grow on the valley sides if only left alone.

FIG. 4—THIS SPLENDID FOREST COVER is on the upper course of the Wei Ho. This forest is now being cut down for construction timber.

appearance. Day after day they have been met with the same disheartening aspect,—a most horribly gullied terrain, showing vividly the destruction caused by torrential rainfall; terraces in the wildest disorder; no vegetation, except the dust-covered, seemingly half-wilted winter wheat and a few scattered trees around the villages; rivers flowing in wide loess gorges several hundred feet deep, their steep loess sides being gullied in the extreme; and then the frequently recurring dust storms doing their share in the erosion destruction,—the whole forming an overwhelming impression of the relentless and destructive forces at work (Figs. 1 and 2). How can there be any hope of remedying such a situation? How can erosion-resisting vegetation be made to grow in such desolate, dry country? So they throw up their hands in despair and say: "Too late! Too dry a climate! Too large a problem for humans to tackle!"

Had these writers seen the same country a few months afterwards in late July and August, perhaps their conclusions would not have been quite so pessimistic. Provided the rainfall had been normal, they would then have seen crops growing abundantly on the terraces, a fair bush growth in the gullies, grass in some abundance and a predominantly green color tone over the whole area where previously there was only one hue, dusty yellow. But then, had they again visited the area upon the approach of winter they would also have seen the population with their cutting knives tearing out from the gullies and slopes every bit of grass and bush growth which during the summer had magically sprouted in non-cultivated places. Fuel is scarce, and it must be collected somehow! How are these farmers to know the damage they do with their cutting knives? As a finality their cattle and sheep finish completely what there is left of the summer's vegetative cover. A few seeds, however, generally manage to survive the ordeal and are ready to sprout after the rains have started the following year. When the natural conditions already are not too favourable there is no wonder that China's Northwest presents such a bleak aspect and that there is little chance for any real erosion-resisting plant cover to grow either on the slopes or in the gullies. There are also summers when the rainfall is so scanty that nothing will grow except some scattered drought-resisting weeds; but in most years the rainfall is sufficient for quite a fair vegetation to develop. Provided it could only be left alone, there is not the slightest doubt in my mind that an erosion-resisting plant cover could be made to grow up without much artificial help almost anywhere it would be required (Figs. 3 and 4). I have seen a wilted and barren countryside in Shensi, noted for its drought famine condition become covered with a healthy grass and bush growth as if by magic after a good early summer's rain. The second year it became still better as the population, remaining after severe famine years, was not sufficient to cut it all down, although they went over the area with their cutting knives to their best ability the whole winter. It was then their main occupation to collect fuel and sell it in the nearest villages. This goes on all over the Northwest, and it seems certain that the fuel question must be solved if there ever is going to be made any successful headway with erosion-resisting vegetation.

Gully erosion

Those who have had experience with erosion control claim that surface erosion is far more serious than gully erosion. In certain places in China the reverse may turn out to be true. In many places the country is so terribly cut up that there are scarcely any decent-sized areas of land left. The gullies cut across from one drainage area to another. Of course the gullies are the most spectacular form of erosion (see frontispiece) and one is apt, under the overpowering impression they make, to forget the importance of sheet or surface erosion, also caused by excessive rainfall, which strips the soil of its most valuable top cover. The best evidence of this destruction may be seen in all the abandoned terraces, usually lying on the higher levels. But the gullies present problems which to an engineer's mind at least have a particular fascination.

I have spent three years in the Northwest on irrigation and road construction in some of the worst erosion-afflicted regions. I have pondered much over the problems which would have to be faced in an erosion-control campaign. It has almost become second nature to me when traveling to stop before gullies and speculate on the best and cheapest methods to adopt to make them erosion-proof and I have not yet come across a gully condition which I think could not be handled successfully. I am firmly convinced that 70 to 80 per cent of the gullies will yield readily to treatment. Of course it is perfectly clear that it will be impossible to reclaim and fill in the thousands of gullies throughout the Northwest. The best that can be hoped for is to stabilize the present condition; save what there is left of the soil in northwest China. Some of the gullies have already reached stages of "semigeological maturity," with a new civilization grown up in their bottom lands, but fighting a desperate battle against the secondary erosion cycle which has started, and threatening to chase it off. (See foreground in Fig. 2.) But the erosion is slower in these older gullies and they may be left alone for the time being. It is the newer and rapidly developing gully formations which must be curbed first, as these are the most destructive both to the land and to the rivers (Figs. 5 and 6).

It is evident that an appraisal must first be made of the whole gully situation before anything can be undertaken in a practical way. This is quite a task in itself as it will require first a good land survey and next a trained staff who will understand how to classify the gullies by inspection with regard to their erosion importance. Then the work to be done to make them erosion-proof will require both engineering ingenuity and agricultural judgment,—the cooperation of engineers and agriculturists. Their work must be done with the local materials which are at hand. Experimental work on various gully conditions must be made to arrive at practical methods. It is self-evident that the cooperation of the farmers must be secured. Another task therefore will be to make the farmers "erosion-minded," to use an American term. The farmers must be educated to see that it is in their own interests that the fight against erosion is undertaken and that they will benefit immensely by it. They must be made to see the whole situation from a new social-economic viewpoint, a northwest China newborn by their own efforts.

Surface erosion

It was briefly mentioned that surface erosion is generally considered a far less spectacular, but a far more serious form of erosion than gully erosion. From an agricultural point of view this may be correct. But from a river-silt standpoint it is possible that gully erosion may be just as serious. It would be extremely valuable to get field data on this.

The North China farmer has in his own way already paid much attention to the surface-erosion question as witnessed by the innumerable terraces which cover the whole of northern China where the land is steep sloping. But he did not do enough and he failed to follow up the terrace work with supplementary work which could have made his anti-erosion campaign a success. He was also careless in the layout and building of this terraces and did not pay sufficient attention to their leveling and proper draining. Due to the multitude of small landholders he can perhaps be excused for not having been able to attend to the drainage problem the way it ought to have been done. When heavy rains occur the water runs from one holder's terrace down on his neighbor's terrace and from there onto the third man's terrace and so on. Perhaps ten or more landholders' terraces may be involved. To work out a practical, non-erosive drainage system when so many individuals are mixed up in it, is not so easy although otherwise the problem may be simple enough. So when there is no one to lead, the whole terrace system is allowed to take care of itself, only being patched up here and there where it has been much damaged after heavy rains. Although the farmers care very much about the matter of drainage they can do very little to improve it. "It is the will of heaven," and they let it go at that.

But they understand well enough that both the gullying and the top-soil stripping mean a terrible loss to them. Just ask some farmer as he is carrying dearly bought manure up to his little terrace which during the summer was stripped of its good top-soil cover. He will answer: "Mei yu fa tze." ("There is nothing to do about it.") Lack of knowledge, lack of leadership, lack of cooperation have brought this fatalism. And so the merry destruction goes on. The best land in the Northwest now lies in the bottom of the loess river gorges on both banks of the rivers which flow there. But river floods and river scour also make the whole land situation here very unstable. The river shifts the land about, takes it away on one side and gives it back on the other. Hundreds of *mou* of land may in one summer freshet season be carried away and built up a little downstream. This river-bank and riverbed erosion thus plays a certain rôle in the silt problem; but during the freshet season this form of erosion does not increase the intensity of the Yellow River silt load since it alternates with deposition all along the course. During the low-water season however, when the tributaries carry almost no flow it is the bank erosion alone which contributes silt to the main Yellow River. This silt load is small, averaging only 0.5 per cent against 20, 30 to 40 per cent at times during the freshet season.

To combat surface erosion it will unquestionably be necessary to get the farmers to straighten out their terraces and assist them in providing proper non-eroding drainage systems. There is no question, but that it can be done if tackled systematically and energetically. The terraces originally were man-made. Surely this must have been a far greater task than that which will be neccssary at present which in most places will amount only to some leveling out.

There will also, unquestionably, have to be introduced radical methods in the system of farming. Soil studies will determine the kind of vegetation to have in the various localities. Ploughing along contours on sloping, non-terraced ground will have to be adopted in order to hold the rainfall so that it does not tend to carry off the top soil or develop gullies. Strip cultivation will have to be introduced whereby strips of erosion-resisting plants alternating with parallel rows of cereals will provide additional safety against surface wash of the soil and gullying. Special consideration will also have to be directed towards the areas near the foot of the mountains where the rainfall comes rushing off the mountain side or down the gullies, tearing away the loess cover. But here there are good materials to work with to make drift and check barriers and there is no doubt that this face of the problem can be successfully tackled. It is being handled successfully in America. Why not in China? Administration, organization, and the necessary enthusiasm and "pep" are all that will be required and in time the transformation will have been done which will best be reflected in the gradually diminishing silt content of the Yellow River.

So far no mention has been made of reforestation, this popular "cure all" for China's bad river conditions. That the development of forests will play its part in the anti-erosion campaign is self-evident especially as a means for providing the necessary fuel and construction timber to a region so poor in such materials. But it is difficult to see how forests alone can be developed sufficiently to curb erosion. Other things must be done. One must consider the present population. It cannot merely be driven off in order to plant forests everywhere. This is realized fully in America where now an erosion-control campaign is in full swing. Here scientific terrace building, and changed methods of farming have been adopted to cope with the surface-erosion problems. Gullies are being plugged with check dams wherever possible and proper non-eroding drainage facilities are being provided to stop additional land from being ruined and washed away. The farmers are behind this fight almost to a man. Only in certain favoured locations is reforesting being undertaken and this matter is being left to agriculturists to decide.

Scientific soil handling is the order of the day in America and the leaders are college men, the vice-director of the soil-erosion service being Mr. W. C. Lowermilk, well-known here in China for his vigorous writings on China's soil condition, and his erosion experiments in Shansi in 1925.

Extent of area to be treated

The despair of most people when discussing erosion-control plans is the enormously large area to be tackled. A conservative estimate places the most important part of the Yellow River drainage area which needs immediate treatment at between 120,000 and 130,000 sq.km., or from 195 million to 212 million *mou* of land. The part of the Yellow River drainage area which gradually would have to be placed under soil-erosion control is probably between 200,000 and 250,000 sq.km.

The area figures quoted are not exactly guesswork. It has been known for the past five years that the Yellow River derives the major part of the intense silt load it carries during freshets from an area which lies down-stream from T'o K'e T'o in Suiyuan, and upstream from Shanchow in West Honan. The drainage area of the river between these two places is about 310,000 sq.km. of which the Wei Ho, which enters the Yellow River at T'ung Kuan, has a drainage area of 145,000 sq.km. There is thus nearly an equal division between the Wei Ho area and that along

FIG. 5—THIS GULLY shows a "runner" which is rapidly extending towards the mountain. There should be no very great difficulty in putting this condition under control.

FIG. 6—THIS TERRACED HILL COUNTRY is not yet too badly gullied. Forestry on the higher slopes, a good bush and grass cover in the gullies, and a straightening out of the terraces will soon stabilize the erosion which evidently is developing rapidly.

the north to south course of the Yellow River between Shansi and Shensi provinces. The two areas also seem to be about equal in importance as silt contributors.

Further localization of these silt-contributing areas can be made. With regard to the Wei Ho, it is known that all the tributaries draining the country to the north of the Wei Ho are heavy silt carriers and that the tributaries, running in from the south, draining the Chin Ling mountain range, are more or less clean-running streams, even during freshets. Of the northern tributaries of the Wei Ho the worst offenders are listed thus:

Tributary	Drainage Area in sq.km.
Ching Ho	50,000
Lo Ho	25,000
Hu Lu Ho	13,000
San Tu Ho	2,000
	90,000

From these areas at least 30 per cent may be deducted as being mountain areas devoid of loess deposits, thus leaving for the Wei Ho an area of about 60,000 sq.km. to be attended to as a first step in the anti-erosion campaign.

For the Yellow River area the most serious silt contributor seems to be the Wu Ting Ho, with a drainage area of 23,000 sq.km., entering from the west or Shensi side. It is also a heavy flood carrier and the amount of silt it brings into the Yellow River is thus serious. Another heavy silt carrier entering from the Shansi side is the San Chuan Ho with a drainage area, however, of only 5,000 sq.km. While so far there have been no gagings taken of the Wu Ting Ho there have been some gagings taken this summer of the San Chuan Ho. This river, which probably is typical of the character of the rivers entering from the Shansi side, carries in freshet a real "mud flow," the silt content being over 30 per cent by weight even when the freshets are small. As its extreme flood flow is estimated at more than 6,000 cu.m. per sec. the silt it then brings into the Yellow River is surely very considerable (Figs. 7 and 8).

The total area of the Shansi streams entering the Yellow River, excluding the Fen Ho, is about 35,000 sq.km. This area is very mountainous and the remaining loess deposits on it perhaps not very large. It is the steepness of this region, however, which makes the loess erosion so intense during heavy rainfall, and it will most likely prove to be a difficult area to handle; but lacking information about it, it is premature to forecast methods of anti-erosion measures. It is thought that less than one-half of this area will require "first consideration" due to its mountainous character. The streams have unquestionably the general physical character of the northwest rivers in that the loess deposits are concentrated in relatively small "basins" on each separate drainage area, and that only these "basins" will need treatment. The area will probably not exceed 20,000 sq.km.

The streams entering from the west or Shensi side, however, are much more flat sloping and their drainage areas are known to have extensive deposits of loess. Exactly how serious a silt carrier each individual stream is, is not known. The total drainage area here is about 60,000 sq.km. of which 45,000 sq.km. may be assigned as of first erosion importance. But it is significant for the whole area on both sides of the Yellow River that measurements taken at the Pao T'ou hydrometric stations in Suiyuan showed this year's flood season an average silt load of 0.8 per cent and a maximum of 1.2 per cent; at the Wu Pao station situated about midway between T'o K'e T'o and T'ung Kuan the maximum silt loan was about 11 per cent and the average about 1.5 to 2.0 per cent for the freshet season as compared with a maximum of over 30 per cent and 3 to 4 per cent average at Lung Men at the downstream end of the area. This shows the progressive effect of the silt influx from both sides of the river.

With regard to the Fen Ho which drains an area of 40,000 sq.km. of central Shansi and enters the Yellow River 120 km. upstream from Tung Kuan, it has been found that this river is heavily silt inflicted during freshets in its upstream parts near Tai Yuan Fu; but that much of the silt becomes deposited along the banks on the 400-km. course towards the Yellow River. Near the place where it flows into the Yellow River the silt percentage has decreased considerably and as the flow even during heavy freshets cannot exceed 2,000 cu.m. per sec. the amount of silt transported to the Yellow River will not be very significant. Under these conditions it has for the present a dilution effect rather than any tendency to increase the silt-load intensity of the Yellow River and for this reason the Fen Ho area may be classified as of second degree importance. The same can be said for the area from Fen Ho down to Shanchow and also below Shanchow.

If we assign, then, 65,000 sq.km. for the whole eastern and western—Shansi and Shensi—drainage area needing "first treatment" it will most likely be found to be a conservative estimate. Together with 60,000 sq. km. for the Wei Ho area it will make a total area of 125,000 sq.km. where anti-erosion measures should first be started. Compare this with the whole effective Yellow River drainage area of 750,000 sq.km. and we are beginning to get a better perspective of the magnitude of the problem.

Soil-erosion control effect on the flood flow of the Yellow River

Since silt occupies space in water it is evident that a diminution of the silt volume will automatically reduce the flood-flow volume. If the silt percentage during floods can be reduced from, say 40 per cent (which several times has been observed at Shanchow in West Honan), it will mean a reduction in discharge of about 18 per cent. Thus, if soil-erosion measures had been effective in 1933 to the extent mentioned, the maximum discharge would have been reduced from 23,000 cu.m. per sec. to about 19,000 cu.m. per sec. It is also probable that a further reduction would have taken place due to the retarding runoff effect inherent in erosion-control projects. If a system of detention basins is also built it would be quite within the realm of possibility to have the maximum floods reduced to well below 10,000 cu.m. per sec. The character of the Yellow River floods would be changed from the present violently flashy condition to a moderately protracted flood wave which could easily be passed through the diked section in the plain and as the silt content has been greatly reduced as well there would be no trouble in the downstream, flatter parts in Shantung.

With the water greatly desilted it would be feasible to have a definitely regulated and relatively narrow river channel where modern navigation would be possible. Irrigation and hydro-electric projects would follow as a matter of course. This is no exaggerated picture of the condition which would result after soil-erosion control had become an established fact.

The soil-erosion service in the United States is only a few years old, but in this short time it has accomplished much. It is now well-advanced with an erosion-control program covering an area of 30 million acres, or about 120,000 sq.km., and expects to have this area fully under control fifteen years from now, if not before. Some of the land is as badly cut up as any in China. This service is thus actively engaged in controlling the erosion on an area comparable to that which would have to be undertaken as a "first treatment measure" in the case of the Yellow River. So why talk about the enormity of the Yellow River erosion problem? Similar areas are being handled in other parts of the world. Merely talking about it and looking at it bring us nowhere here in China. Let us get to grips with it and really size up the problem. It may prove easier than we think.

Preliminary studies which have been made of flood control by means of detention basins seem to indicate that a fair amount of success is possible with this method. During freshets there is likely to be both flow and silt control. But immediately the freshets are over the silt will become flushed out and the constant rise of the riverbed will go on the same as at present. Since dike breaches will practically be eliminated the rapid extension of the delta at the mouth will be further intensified since all the silt will reach the mouth instead of only a part as is the case at present due to the constantly occurring dike breaches. A very difficult situation will be set up near the mouth which rapidly will extend upstream ending in the dislocation of the river unless constant and costly works near the river mouth be made to spread the silt deposition as much as possible. This can all be avoided if soil erosion is checked. One can look at the Yellow River problem any way one likes, discuss any method of control. One will always come back to the same question: how can the silt be eliminated? So why not begin to tackle this side of the problem? And it concerns not only the Yellow River. It is the life line for the whole of northwest China. We engineers should not only welcome, we should invite the cooperation of the agriculturists and foresters in this big problem. It is our plain duty.

Reprinted from *Soil Conservation,* **3,** 93–96 (1937)

PROTECTING BOTTOMLANDS FROM EROSIONAL DEBRIS: A CASE HISTORY

By Carl B. Brown [1]

19

IN many parts of the United States cultivation of fertile bottomlands and gentle alluvial slopes is being menaced by erosional debris swept down in storm run-off from adjacent denuded mountains and steeply sloping uplands. The damage resulting when debris-laden torrents gush out upon flood plains, with sudden dissipation of velocity and deposition of debris load, is usually twofold: (1) The natural and artificial drainageways on the alluvial slopes and bottomlands are clogged with debris, and subsequent flooding of fields by storm run-off occurs; ground-water levels rise, and land becomes waterlogged. (2) Assorted and usually coarser erosional debris derived from subsoils and spread over the bottomland impairs its fertility and often destroys its productive value.

Damage of this type is usually local, restricted at first to the immediate debouchure of tributaries on a flood plain or alluvial slope, but spreading progressively as the process continues. It should be distinguished from similar widespread sanding and aggradation produced during floods by alluvial streams on their own flood plains with sediment often transported from far distant headwater areas. For the latter phenomenon, no immediately effective cure is yet in sight; extensive soil conservation, changes in land use, and direct flood-plain protection will be required. Protection against local damage from erosional debris is, however, frequently obtainable in a short span of time and may be economically justified. It is, indeed, often a problem of immediate concern because alluvial lands are nearly everywhere notably productive and are rarely subject to the hazards of soil erosion common to most upland areas. Developing optimum methods for this protection is proper work for the Soil Conservation Service.

Elaborate and costly means are sometimes justified on thickly-populated alluvial lands. Debris basins formed by concrete dams have been constructed by the Los Angeles County Flood Control District on the alluvial slopes at the base of the San Gabriel Mountains for protection of suburbs of Los Angeles. But for the protection of ordinary agricultural lands, other and more economical methods must be sought. The first hope, of course, is adequately to protect by soil-conservation measures the watersheds from whence the erosional debris comes. This at once brings forth pertinent economic questions involving costs justified for the protection of both upland and alluvial soils.

THIS article describes one solution of this problem, and lays a partial basis for appraising its merits. It is based on a survey made by the Section of Sedimentation Studies, Division of Research, in June 1937.[2]

Doniphan County, Kans., the State's extreme northeastern county, is bordered on its north and east sides by the Missouri River. Intermittent areas of flood plain along the river are intensively farmed because of their fertile alluvial soils.

One of these areas, comprising 4,300 acres, is in Burr Oak Township, approximately 8 miles northeast of Wathena. The area consists for the most part of river-deposited alluvial soils fringed along the river by sand bars and debris, mapped as River Wash. On the south side the area is bordered by a steep and relatively straight bluff line through which a number of small tributary streams have cut steep and narrow ravines leading from the rolling uplands.

All the soils of this alluvial area are members of the Sarpy series,[3] of which Sarpy clay is most commcn, with lesser amounts of Sarpy siltv clay loam and Sarpy very fine sand loam. The Sarpy silt loam which occurs principally along the bluff line represents an admixture of silt and coarser material from the uplands with clay of the river alluvium.

OF the several tributaries heading in the uplands and flowing onto the alluvial flood plain, Chase Creek is the largest. It has a watershed of 1,011 acres. Old residents report that 40 years ago both this small watershed and the bottomlands were well forested, and that water emerging on the flood plain was usually clear or that its occasional small-silt load was deposited in thin layers over the wooded bottom. Between the time of clearing the watershed and 1920, Chase Creek had built an alluvial fan sloping in all directions from its mouth for an average distance of 1,500 feet. The apex of the fan was 6 feet higher than the outer fringe giving a slope of approximately 22 feet to the mile. Much of the material composing this fan is said to

[1] Associate Geologist, Section of Sedimentation Studies, Division of Research, Soil Conservation Service, Washington, D. C.

[2] Acknowledgment is due to E. M. Flaxman, junior geologist, who, assisted by E. H. Moser, Jr., made a topographic survey of these debris basins and collected additional information on their history.

[3] Knobel, E. W., Davis, R. H., and Higbee, H. W. Soil Survey of Doniphan County, Kans., U. S. Dept. Agr. Bur. Chem. and Soils Series 1927, No. 25.

17389—37——2

have been brought down as the result of a cloudburst in June 1917.

The damage resulting from the storm of June 1917 through flooding of growing crops and spreading erosional debris over them, led farmers in the afflicted portion of the flood plain to organize the Doniphan County Drainage District in 1919, taking in 1,900 acres of bottomland. An engineer was employed to survey the area and make recommendations for control of floodwaters and erosional debris and maintenance of drainage. He recommended the construction of a debris basin or diked-off area adjacent to the bluff line to impound both sediment and floodwater until the latter could drain off quietly into a ditch and be carried directly east 2½ miles to natural sloughs draining into the river.

IN 1920 a contract was entered into between the district and C. F. King for rental of 26.6 acres of his land lying just east of the mouth of Chase Creek (sec. 24, R. 21 E., T. 2 S.). This is indicated as the "old debris basin" in the accompanying map. This land was converted into a basin by an earth dike averaging 6 feet in height. The dike on the north side constituted the inner side of the drainage ditch and a second parallel dike constituted the other side. The ditch was dug about 3 feet below ground surface, the excavated material being used in the construction of the dikes. The ditch, which was 10 feet wide, thus had an effective depth of 9 feet. Beyond the desilting basin, no dikes other than spoil banks were constructed along the ditch. The ditch slopes 4.21 feet per mile to the slough where it terminates. No outlet from the basin was provided and water generally escaped by percolation and evaporation, except occasionally when it overflowed the dikes. Whatever damage occurred from overflow was periodically repaired. The flow of Chase Creek was diverted into the basin by a road embankment built across its original channel.

BY 1930 the original basin had completely filled with sediment and the channel was diverted into a similar basin of 73 acres on the west side of the mouth of Chase Creek. Between the spring of 1920 and the spring of 1930, the original basin was filled to an average depth of 5 feet with sediment derived from the 1,011-acre watershed. The depth varied from 8.2 feet near the entrance of the creek to 2.8 feet in the northeast corner. Within this decade, the sediment deposited amounts up to 135.6 acre-feet, equivalent to a removal of 0.13 foot of topsoil over the entire watershed. To this must be added an unknown quantity of sediment that was carried down the drainage ditch, particularly after the basin was nearly full. The ditch was cleared out for most of its length in 1930, a fact which shows that even the spill-over sediment was too coarse or too abundant to be carried by the waterflow on a grade of more than 4 feet to the mile. The indicated rate of erosion, based on sedimentation in the basin alone, would cause complete removal of the 6-inch topsoil layer (average depth for Knox silt loam) in 35 years or less.

THE contract between the district and Mr. King provided for rental of 26 acres of land at the rate of $25 per acre per year, but allowed him to continue such use of the land as he might choose, subject to flood hazard. Mr. King planted the land almost every year and crop production was extraordinarily high when no floods occurred. His production of corn ran 65 to 80 bushels to the acre when no damage occurred, a record that has been maintained since waterflow was diverted to the new basin. He reports that one year he raised 103 bushels to the acre. These figures compare with an average production of about 40 bushels per acre on adjacent alluvial land. They indicate in a most striking way the plant food values lost from the Chase Creek watershed and laid down on this field. Taking into account crop losses by flooding, Mr. King averaged 25 bushels of corn per acre in the debris basin during the 10-year period. King also derives excellent yields from tomatoes, watermelons, strawberries, and beans on the old silting basin. The more sandy soils in the upper part of the basin are used principally for truck crops and the finer-textured soils in the lower part for corn. There have been no appreciable differences in crop yields over a period of years.

THE new desilting area is not surrounded by a well-defined dike. Instead the water is directed along the edge of the bluff by a low dike about 200 feet in front of it. This abruptly checks the incoming current and causes deposition of coarser materials, temporarily protecting the remainder of the basin. A road to the west, north, and east of this basin is 2 to 3 feet higher than the enclosed land, and the drain on the inside connects with the old drainage ditch and serves as an outlet for surplus water.

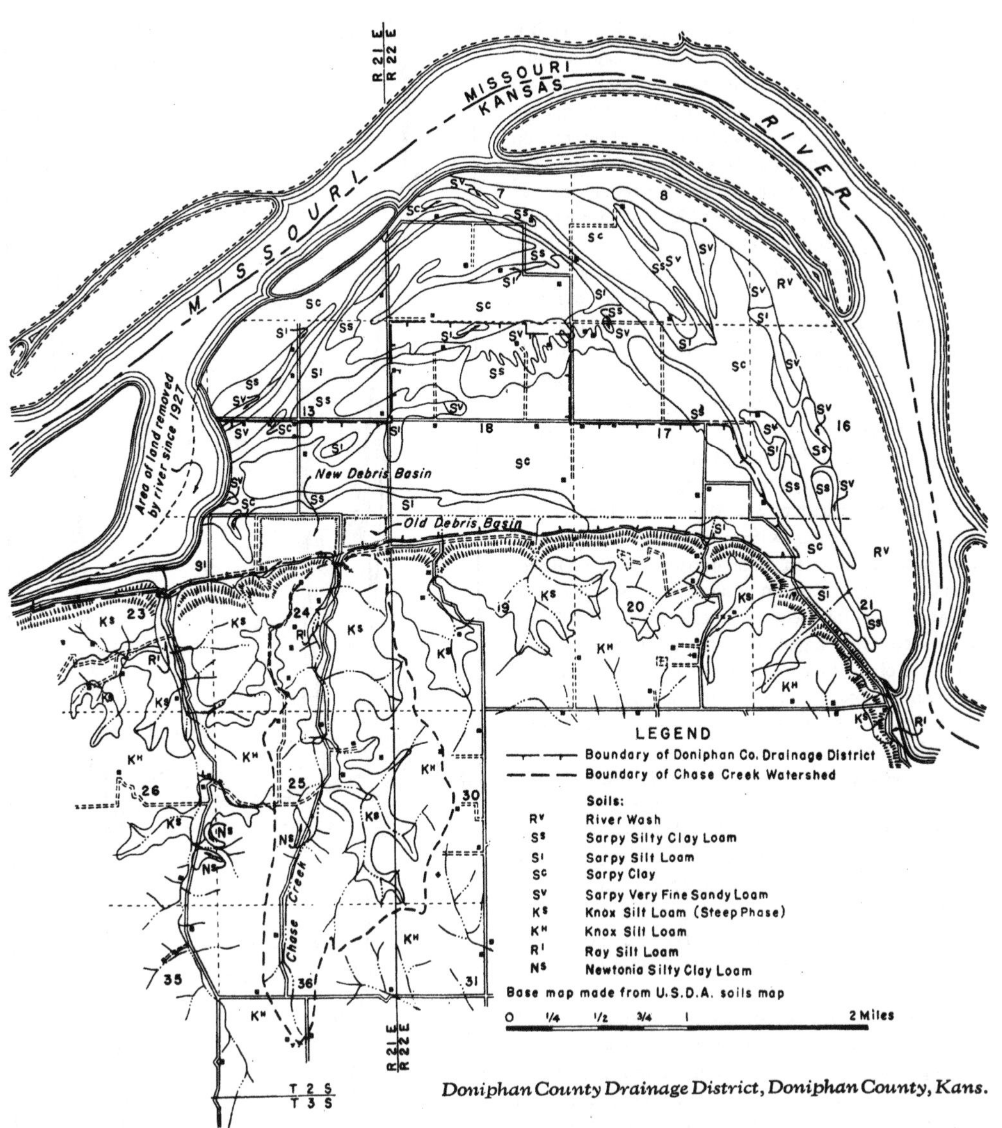

Doniphan County Drainage District, Doniphan County, Kans.

On June 7, 1937, a cloudburst caused a flood which tore out a portion of the low dike where it is buffeted by the direct current from the mouth of the creek. Boulders derived from bed rock were strewn over 1 to 2 acres below the mouth of the creek, and a deposit of medium to coarse sand extended beyond. Most of the sediment, however, was deposited on the western half of the basin. The silt deposits were thick enough to make necessary replanting of the corn crop.

The watershed of Chase Creek is entirely covered by Knox silt loam and its steep phase. More than 90 percent is in cultivation and orchards, woodlands being restricted to the narrow valley of the creek and the steep slopes along the Missouri River bottomlands. The principal crops are raspberries, beans, and other garden truck, with corn, wheat, and rye on the flatter uplands. Apples are the principal orchard crop.

ON the basis of information supplied by King, it is computed that the district has paid a rental of $12,835 for the debris basins from 1920 to 1937. If we exclude all other costs of the district such as the small annual maintenance and interest on the $30,000 of 30-year bonds issued to finance the improvements we find that the average rental cost for debris basins alone has been about 35 cents per acre per year for land protected. That this amount as an insurance against crop loss was fully justified can scarcely be doubted. According to King, it is anticipated that after the bonds have been paid off the annual drainage district special tax will not exceed $1 per acre. It appears from the facts at hand, without the benefits of an economic survey, that the drainage district has been a success with respect to costs versus benefits.

Looking at the other side of the picture, however, the total rental payments to date for the debris basins are equivalent to $12.70 per acre of watershed area. This compares with a cost of $8.51 per acre treated with soil conservation measures on the Muskogee, Okla., demonstration project, or an average cost of $6.09 per acre, for the gross acreage of the 37,000-acre watershed. It is higher than the total costs per acre on seven of the nine demonstration projects put on a maintenance basis at the end of the fiscal year 1937. Therefore, it appears that with the same or a smaller expenditure during the 17-year period for treating the Chase Creek watershed not only would the same protection have been afforded for the bottomlands but at least 415,000 tons of valuable soil (24.5 tons loss per acre per year) would still be in place instead of the debris basins. Soil conservation appears in this case as probably the most favorable alternative even if the debris basins had been constructed as insurance for the initial period of 2 or 3 years until proper land treatment was successfully concluded.

It is believed that two significant points are brought out by this study: (1) Treatment of a watershed of sloping upland solely to protect valley agricultural land to which it is tributary can often be justified from an economic standpoint. (2) Debris basins alone frequently offer a successful method of protecting valley agricultural lands, and adaptions of this practice may prove useful in operation work, under some circumstances, as a relatively cheap and expedient method of protecting bottomlands until soil-conservation treatment of tributary watersheds can be completed; or in very rare cases, use of debris basins may be the only or ultimately the cheapest method to employ.

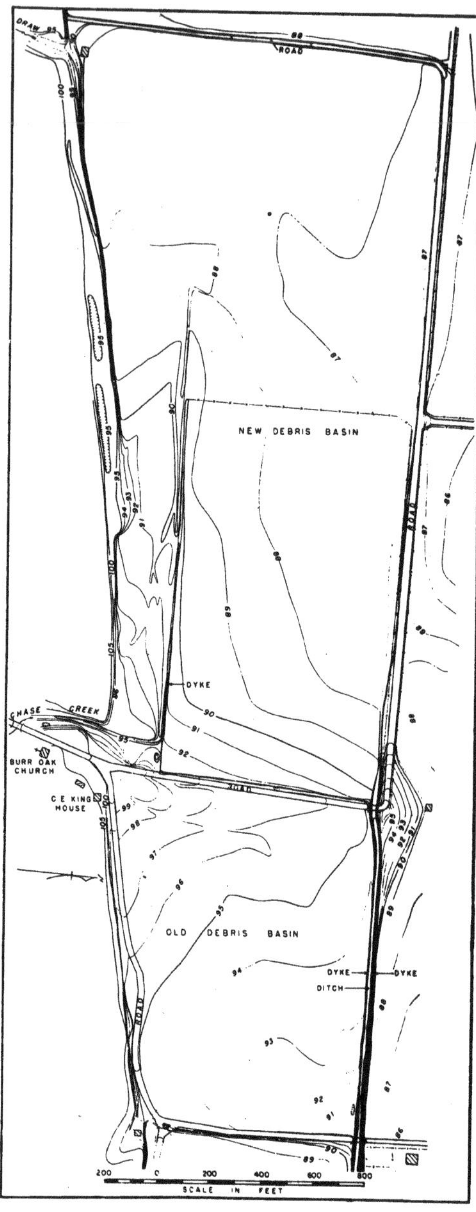

Topographic map of debris basins, Doniphan County Drainage District, Doniphan County, Kans.

Reprinted from *U.S.D.A. Farmers' Bull., 1825* (1939)

SAND-DUNE RECLAMATION IN THE SOUTHERN GREAT PLAINS

By CHARLES J. WHITFIELD, *soil conservationist, Division of Research*, and JOHN A. PERRIN, *junior soil conservationist, Division of Conservation Operations, Soil Conservation Service*

20

CONTENTS

INTRODUCTION

DURING the past 5 years the effects of wind action on soils in the Great Plains have been brought forcibly to the attention of the people of this country by the yearly occurrence of numerous dust storms. At Dalhart, Dallam County, Tex., 61 dust storms were reported in 1935, 45 in 1936, 60 in 1937, and 71 in 1938. Other sections of the Great Plains have experienced similar conditions to a greater or less degree. Another striking manifestation of wind action in certain areas is the formation of sand dunes.

The wind was able to build up these dunes only after surface cover had been destroyed by cultivation, by heavy grazing, by heat and drought, by abrasive action of moving soil particles, by covering of the plants, by in-drifting of soils, or by a combination of these factors. They are a recent development on the Great Plains. Ten years ago there were no active dunes of this origin on many of these lands. These dunes have no plant cover (fig. 1) and should not be confused with the long-established dunes that are also the result of wind action but are now stabilized by vegetation. Nor should they be confused with the "blow-out" type of dune that develops near wells, roads, and cattle trails.

The dunes are valueless in their present condition. Moreover, they are a constant menace to surrounding fertile farm lands, pastures, and buildings because the sands are continually shifting. Many of these immense piles of sand are now found on areas of native prairie sod that were never cultivated. In Dallam County these dunes developed on land that was formerly fine sandy loam and loamy sand—soils that produce good crops but that tend to drift and blow during the spring winds unless protected by adequate cover.

Scattered sand-dune areas of recent origin occur throughout the Great Plains in Texas, New Mexico, Colorado, Kansas, Nebraska, Wyoming, South Dakota, and North Dakota. They are most extensive between Curry County, N. Mex., and Seward County, Kans. At least 12 dune sites are to be found in Dallam County, Tex., alone.

Little or nothing seems to have been done about sand dunes in the southern Great Plains until early in 1936 when the Soil Conservation Service established a study area to determine the best means of controlling these dunes and reclaiming the dune land.

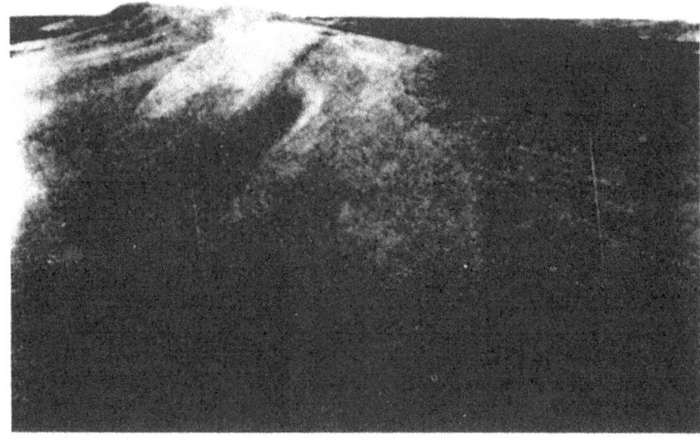

FIGURE 1.—A typical sand dune before being treated. This dune is approximately 22 feet high.

THE STUDY AREA

Two thousand acres of sand-dune land in Dallam County, Tex., 8 miles north of Dalhart, were used for studying and devising methods for stabilizing sand dunes and making them eventually useful for grazing and cultivation. The methods used on this study area are described here to serve as suggestions for treatment of other sand-dune areas.

The sand dunes were of various dimensions. One 470-acre field on the study area that was surveyed in 1936 had approximately 57 sand dunes, ranging from 1 to 9 feet in height and averaging 161 feet in length and 113 feet in width (fig. 2). The subsoil around and between the dunes was hard and eroded to a depth of 10 to 12 inches. In 1930 this field was dominated by native vegetation, with blue grama, side-oats grama, and bluestems as the principal grasses and sand sage as the outstanding shrub. In 1931 it was cultivated for the first time and planted to sorghums in 40-inch rows. It was planted to row crops again in 1932 and in 1933. Because of drought and crop failure only one crop was harvested during this 3-year period. The land lay idle from 1933 until the experimental work was started in the fall of 1936. The front cover shows an aerial view of the area in March 1937. The pure-white part is the actual dune at the time the picture was taken. The dull gray is the hard, eroded land, that part of the original dune site from which the sand has blown away since the area was listed. The striped part is the listed hard area with sand blown into the furrows. Figure 3 shows hegari that was growing on the area in the fall of 1937. In April 1938 nearly all these dunes had disappeared, and their materials had been redistributed into either plant-litter or lister furrows.

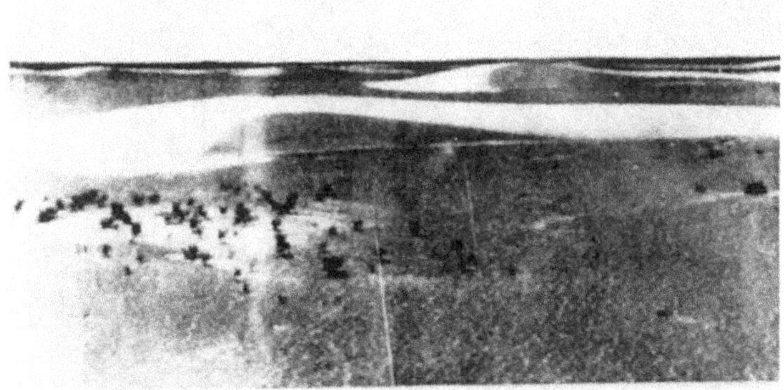

FIGURE 2.—Some of the 57 sand dunes developed on this field during 1934 and 1935. In the foreground is hard eroded land.

FIGURE 3.—Hegari was drilled for a cover crop on part of the field shown in figure 2. This photograph was taken 14 months later.

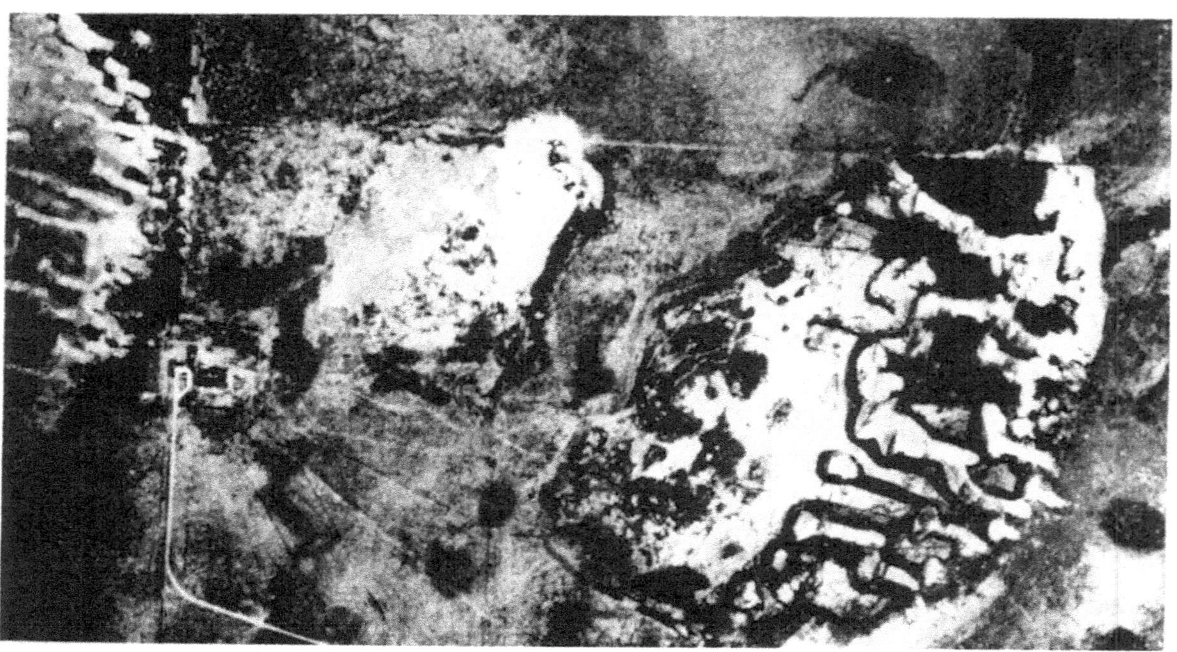

FIGURE 4.—Aerial photograph of the sand-dune study area. On the extreme left is the untreated check area, with dunes and hard eroded land. On the right is the dune site that has been treated. Listing around the dunes and wind channels across the dunes are shown.

In another field, which is some 400 acres in extent and is in the study area, the sand dunes are on land that has never been cultivated (fig. 4). They have developed as a result of wind erosion on 80 acres of land southwest of the present dune site. This land was cultivated from 1907 to 1914 and was then abandoned and used for grazing. Evidences of dune formation were first noticed about 1926. Since 1929 striking changes have occurred; dunes have developed from low mounds to heights of 26 feet. These dunes are large piles or mounds of sand or sandy materials. They range from 50 to 880 yards in length and are usually about 30 yards wide. Their height apparently depends on their age and the direction and velocity of the wind. The highest dune measured was 26 feet in elevation. Owing to the prevailing wind direction, which is generally from the southwest, the axes of most of the dunes is east and west, but a few are slightly southeast and northwest. The dunes are irregularly spaced; the distance between them ranges from 50 to 400 yards. They are generally compact beneath and moist a few inches below the surface.

The subsoil on which the dunes rest, as well as the soil between and around them, is very compact, hard, and fairly level. It consists of the subsoils of uncultivated land that has been eroded by wind to various depths, in some places to as much as 4 feet. The prevailing winds, from the southwest, have left large areas of hard land to the west and south of the dunes and accumulations of sand from 6 inches to 2 feet deep to the north and east.

A border exists between the range grassland and the eroded hard lands and dune lands. It is called the critical area because a hummocked condition and piles of loose sand devoid of vegetation make it susceptible of dune development and because the wind sweeps unhindered across the hard lands, picking up materials as it goes.

STABILIZING THE SAND-DUNE AREAS

Effective treatments of these badly damaged areas must be based on attempts to stabilize them permanently. The best means of accomplishing stabilization seems to be through the reestablishment of a plant cover. Vegetation tends to prevent soil movement (1) by reducing the wind velocity at the surface and (2) by binding the soil with the roots of the plants. In their present state, several factors are preventing these sand-dune areas from becoming stabilized through natural revegetation. In the first place the mechanical effects of the wind, owing to the dust and sand it carries, tend to prevent vegetation from starting and also to cut off the already-existing plants. As a result of obstructing the sweep of the wind by roughing the soil and planting strips of row crops along the border, dense stands of weeds developed over the hard eroded lands on an experimental plot. On the check plot, where no strips were planted, there was very little weed growth. In addition to the mechanical effects of the wind, the physical characteristics of the dunes themselves —height, shape, and continuous sand movement—make it difficult for plants to get started and to maintain themselves.

On the study area the following steps were taken to reclaim the sand-dune sites: (1) Controlling the critical area, (2) leveling the sand dunes so they could be planted, (3) deep listing between and around the dunes to catch the sand and to build on the hard eroded subsoils, and (4) planting to prevent additional soil movement.

CONTROLLING THE CRITICAL AREA

The critical area, that strip of hummocked sandy land lying between the sand-dune area proper and the native pasture, was the first piece of land treated. A tractor, a No. 2 terracer or a road grader, and a railroad iron were used to level the hummocks. The area was deep-listed and planted to sorghums during the planting season. The resultant crop held the sandy material, preventing it from being blown across the hard land and onto the dunes.

LEVELING THE SAND DUNES

Reclamation of wind-eroded land by machinery is sometimes too expensive to be justifiable unless the menace to adjoining areas is serious. The development of methods by which the land could again be made valuable for agriculture or grazing was the principal problem confronting the Soil Conservation Service. The most practical method of lowering the dunes enough so that they could be effectively planted involved the utilization of wind to redistribute the materials that this same force had piled into dunes up to 26 feet in height.

Investigations have shown that far more material is accumulated than lost by untreated sand dunes. In fact sand dunes after they get above 8 to 10 feet in height have a tendency under ordinary conditions to keep building higher. One of the main problems that faced the investigators was to get the sand to move from the dunes out onto the hard eroded land. This was especially true if the leeward slope was steep. In order to decrease the height of the dunes to a point where they could be effectively planted, four means were employed to aid the movement of the sand, namely, wind intensifiers, drag poles, one-way disks, and tractors and blades.

WIND INTENSIFIERS

Three types of wind intensifiers were used—the signboard, the sandbag, and the wind channel. The signboard was constructed by nail-

FIGURE 5.—Gunny sacks placed on the crest of dunes act as wind deflectors and lower as the wind removes the sand from under them.

ing galvanized iron or boards between posts at different heights above the sand level on the crest of the dune (fig. 8). Gunny sacks were filled with sand and placed at different spacings on top of the dune (fig. 5). These wind intensifiers were placed on the sharp edges of the dunes, which are always on the leeward side. Channels generally 3 feet wide and 4 feet deep were dug across the dune. Some of these channels can be seen in figure 4.

These methods proved very efficient in moving sand. Gaps 4, 5, and 6 feet deep were dug out, and the sand was carried out beyond the crest. The sandbags proved most effective since they lowered as the wind removed the sand from around them. During one storm, by the use of sandbags, a dune was lowered about 2 feet. The crest of the dune flattened down, and the dune itself moved forward 6 to 10 feet. During another storm the crest was lowered 1 foot. The only part of the dune affected in this manner was that on which the sandbags had been placed.

The use of wind intensifiers in lowering dunes is practical, however, only if manual labor is available.

DRAG POLE

A second method of destroying the steep slope and crest to the leeward side of the dunes, and thus allowing the wind to carry the sand out beyond, is the use of the drag pole (fig. 6). The drag pole is an 8- by 8-inch timber. A 20-foot pole should be sufficient for the largest dune. One or two horses are hitched to one end of the pole, and two or three horses to the other or top end. It is dragged along the sharp edge of the dune at right angles to the crest. Attaching a disk at the top aids in breaking down the steep slope.

The drag pole can be used more advantageously than the wind intensifiers because it requires less hand labor and a greater area can

FIGURE 6.—A drag pole being used to break the crest of a dune.

be covered in a short period of time. By use of the drag pole, as by the use of intensifiers, the wind is prevented from forming eddies, and as a result it carries huge quantities of sand out beyond the dune. After the crest or the steep leeward slope is destroyed, the dune has a flatter, more oval shape and does not present an obstruction for the wind. Therefore, the wind velocities are not reduced, and the load of sand is not deposited on the dune but moves across and is deposited in the lister furrows, cover crop, or any other obstruction that has been prepared or grown. The entire dune seems to move forward after the crest has been broken down. One dune on which a drag pole was used was lowered 15 feet in 6 months and broadened out along its entire length an average of 24 feet.

ONE-WAY MULTIPLE DISK

The sandy material making up a dune often becomes compact, owing to trampling of stock, to rain, or to the weight of the sand itself. The best way found to loosen this sand so that it can be moved by wind action is by the use of a one-way disk plow. Disk harrowing the dunes with teams is economical as well as effective.

TRACTOR AND BLADE

Another method employed to spread the sand is the use of a tractor and blade to flatten and level the dunes. Quick results are then obtained by wind action. In 6 months one 20-foot dune was lowered to 5 feet by this method.

The procedure with the blade is to make 1 to 3 turns over the highest points of the sand dune. This gives the wind a chance to move the loose sand. From 6 to 12 treatments are usually necessary in order to affect the height of the dunes materially.

DEEP LISTING THE HARD ERODED LAND

One method employed to prevent the sand moving from one dune from accumulating on another is listing the hard, eroded land around and between the dunes. The furrows not only catch the material blown off the dunes (fig. 7) but also prevent more sand from accumulating on the dunes. The great importance of deep listing cannot be over-emphasized. It has been observed that shallow-listed land has continued to blow, whereas deep-listed areas have not blown, owing to the greater amount of claylike cloddy materials brought to the surface. Relisting has been done over much of the area. The soil, even that which was badly eroded, has been mixed with sand from the dunes and with other wind-blown material to such an extent that crops have been produced on most of it, which otherwise would not have grown any vegetation.

As a result of listing the hard eroded lands and cutting off some of the mechanical action of the wind by treatment of the critical area, vegetation—mainly Russian-thistle—has developed on many formerly barren areas. Although this cover is weedy and only temporary, it is

FIGURE 7.—Deep furrows catch the sand that moves off the dunes, and the land is put in a condition to plant.

sufficient to prevent blowing and even to catch and hold the moving sand that blows from the dunes.

PLANTING TO PREVENT MORE SOIL MOVEMENT

The development of a plant cover, whether a cultivated crop or native vegetation, is necessary for complete stabilization. A dense cover will prevent the soil blowing that is the source of dunes. The fine sandy soils composing the dunes, if properly managed and supplied with sufficient moisture, are capable of producing the abundant vegetation that is essential to permanent stabilization. These lands are especially adapted to the production of grain sorghums that will produce a cover crop adequate to protect the soil against erosion during the windy season. On the other hand, these areas are very susceptible to erosion and unless properly handled during good and poor years, the soil will blow, as has been shown. Because of this, it may prove beneficial to return these lands to their original grass cover. In any event, the planting of sorghums or the development of a weed cover is essential, not only to protect the soil from drifting but also to put it in shape so that grasses can be successfully reseeded.

CROP PRODUCTION

The successful farming of these sandy soils depends on conducting tillage and cropping practices in such a manner that the soils will be most resistant to soil movement. Deep listing, drilling, wet cultivation, and the leaving of sufficient crop residues on the ground are important control measures.

Deep listing is much more desirable on sandy soils than shallow listing. Turning up a cloddy, rough surface by deep plowing leaves the soil better protected for long periods and more capable of withstanding the cutting action of the drifting sand. As the furrow produced by deep listing will hold much more eroded material, the effects of deep listing will last considerably longer. Determining whether the crops are to be listed or drilled is of great importance. Although listing is more effective than drilling in conserving moisture, the cover crop produced is not so efficient in protecting the soil from drifting. During the years when the soil moisture and other conditions are favorable at planting time, crops should be close-drilled instead of planted with a lister. During periods of unfavorable conditions however, a cover crop is more likely to be obtained by planting with a lister. If crops are listed in deep sandy soil, listing should be done during or immediately following a rainy period. A crust then forms on the soil surface, preventing soil movement to a considerable extent and thereby keeping crop seedlings from being covered by moving sand. Wet cultivation, for this reason, is very desirable at any time but especially when winter or spring listing is done.

One of the most important factors governing the resistance of sandy soils to blowing is the amount of crop residue left on the ground after harvest. Leaving the entire stalk on the ground after heading seems to be more desirable than leaving only a stubble, provided an average crop has been secured. Measurements indicate that fields with stubble are losing some soil materials, whereas those with stalks are holding the soils. If, however, the stand produced was poor and the soil is in condition to blow, the stalks may cause hummocking—in which case stubble would be better. Leaving the whole stalk protects the lister ridges and prevents them from being eroded. A close-drilled stubble is more effective than a listed one in holding stalks that are mowed and allowed to fall to the ground. The stalks have a greater tendency to blow away on a listed field.

On those fields where it is necessary to remove forage it seems to be much more desirable to cut and remove the feed rather than to graze the crop. If the forage is cut, a stubble at least 10-inches high should be left.

Of the six crops studied—Sudan grass, broomcorn, kafir, hegari, millet, and black amber cane—the four most desirable for cover were broomcorn, Sudan grass, kafir, and hegari. Broomcorn made a better growth under a great variety of soil conditions than any other species with the possible exception of Sudan grass. It is also drought-resistant and effectively protects the soil from drifting.

The planting time for sandy areas depends somewhat on the rainfall. It is advisable to plant the first part of June in order to prevent the plants from being in a critical stage of growth during the dry period in July. If large acreages are to be planted, however, it may be desirable to start operations about May 20, or as soon as favorable moisture conditions prevail. According to results secured at the Dry Land Field Station, near Dalhart, Tex., kafir, milo, and hegari should be planted in the order named.

Rye has been used as a winter cover with varying success on sandy lands. Wherever the stands have been good, soil movement has been controlled throughout the blowing season.

In the spring of 1936 work was started on the denuded sand-dune area shown in figure 8. From the hard eroded land in the foreground

FIGURE 8.—Severely eroded land in the project area. On the crest of the dune is a signboard type of intensifier, made of boards and galvanized iron.

the soil had been removed to a depth of 2 feet. The roughened surface indicates the cutting action of the wind. Eroded materials had piled up as much as 26 feet in the dunes in the background. By the fall of 1937 the excellent cover shown in figure 9 was growing on this

FIGURE 9.—Sudan and kafir growing about 18 months later on the area in figure 8.

area, although only 12.5 inches of rain, about two-thirds the normal amount, had fallen during the entire year.

REVEGETATION

It may prove beneficial to stabilize these sand-dune areas permanently by returning them to a native-grass cover. Two factors have been found to be important in the revegetation of such areas: (1) The development of an adequate cover in order to control soil shifting, and (2) the use of native species adapted to these conditions.

Results indicate that an adequate cover completely protects the soil by preventing shifting, accumulations, or removals. One of the best covers consists of native-weed species, such as Russian-thistle, and annual grasses. Grain sorghums may be efficient in controlling soil drifting if enough stalks and foliage are left to cover both the furrows and the ridges adequately (fig. 10).

FIGURE 10.—The stalks and foliage of the hegari that is being mowed will adequately protect the soil from blowing.

The development of this cover is essential primarily because there must be sufficient litter or trash to prevent the shifting of soil materials during the blowing season. Furthermore, results on the sand-dune project and elsewhere show that the movement of soil particles tends to destroy grass seedlings by removing soil from around the roots, by covering the plants, and by the cutting action of the drifting sand.

Plants that are adapted to the existing conditions should be used for revegetation. On sand-dune areas effective sand-binding species should be used. A number of these, such as sand bluestem, sand reed-grass, Indian grass, side-oats grama, and sand dropseed, occur throughout the southern Great Plains. The most effective method yet found of developing a stand is spreading mature grass hay of these species over fields already protected by a cover. In June 1936, in order to seed a dune, mature sand reedgrass was spread over it. The dune covered 1.1 acres and was more than 16 feet high. Figure 11

FIGURE 11.—A stand of seedlings that developed after mature sand reedgrass was spread in a blow-out.

shows a stand of seedlings in a blow-out in this dune in June 1937. By July 1937 a growth of sand reedgrass and Russian-thistle completely covered the area.

RECOMMENDED LAND USE

Proper land use is essential if these areas are to be kept under control and cease to be a menace to surrounding land. It is much easier to prevent sand dunes from developing than it is to control them after they have developed. The better sandy areas, if farmed so as to prevent soil drifts, are entirely capable of producing good crops of grain sorghums and might well be used for this purpose. The more critical sites should be returned to grass. After they have been completely stabilized with a good grass cover, they can be used for controlled grazing.

Landscape Management

III

This part of the volume considers the large array of interlocking concepts that constitute man's appraisal, use, and evaluation of the nonurban environment as a natural resource. Whereas Part II was largely restricted to conservation ideas and methods regarding soil and its maintenance, Part III includes the much larger arena of the earth's surface, the materials and the waters, as prized elements in man's quest for material growth or spiritual well-being. Thus this part seeks a multipurpose goal in which some of the components are not compatible. Another difference is the scale of action. Part II was devoted to analysis of individual methods that could aid in preventing soil erosion. Part III deals with large-scale integrated management systems that cover areas of hundreds of square miles. In Part II in the United States, the Department of Agriculture was most deeply involved in the practices that were discussed; here a variety of governmental agencies are operating: Bureau of Reclamation, Department of Agriculture (Forest Service), National Park Service (U.S. Department of Interior), Corps of Engineers (U.S. Army), Tennessee Valley Authority (TVA), and so on. This part is divided into two sections: the first considers tangible assets, plans, and construction necessary to harmonize the landscape; and the second analyzes those elements that comprise the more intangible aspects of terrain utilization as an aesthetic and human resource. Since man's use of the landscape comprises so many different forms, it will be necessary to analyze these changes; how to equate them with appropriate management practices and legislative and policy procedures in a large variety of competing directions will provide some focus for the editor's comments.

Resource Planning and Rehabilitation of Damaged Environments

This section concerns the different types of geomorphic planning that are necessary in nonurban regions and provides an historical setting for evolution of laws, legislation, and creation of governmental agencies influential in these changed assessments during the twentieth century. The larger spectrum of conservation in this total framework includes ideas associated with reclamation, rehabilitation, preservation, wilderness, and aesthetics.

Volume I contains an international appraisal of man's destruction of the soil and terrain before 1900, and Part II in this volume continues the story into the twentieth century. Udall (1963) provides an overview in his book of how the "raid on resources" in the United States during the 1800s set the stage for the conservation movements and land-management programs of the twentieth century. He discusses (1) the "Great Giveaway" of land at prices of 10–12½ cents/acre; (2) the granting of areas to railways between 1850 and 1871 larger than France, England, Scotland, and Wales combined; (3) the fires caused by loggers that consumed 25,000,000 acres of timberland annually; and (4) results that occurred from overmining, overlogging, overgrazing, and overfarming.

Just as debates have arisen concerning channelization as a conservation measure, there is also much disagreement concerning the large-scale programs that are being recommended for eradication of phreatophytes in the West. In the 17 southwestern states, phreatophytes annually consume at least 25 million acre-feet of water.

Phreatophyte eradication is a somewhat controversial issue because it pits those who wish to conserve water against those who wish to conserve wildlife. Proponents point out that phreatophytes consume much groundwater, thus depleting water contributed from upstream areas for the downstream regions. Their destruction offers a method to reduce evapotranspiration losses, increasing available water. This control is achieved by replacing the woody plants with grasses, which have a lower consumptive use. However, the benefit of added grass as forage must be compared with the possible loss due to ruining wildlife habitat. Culler (1970) described the Gila River Phreatophyte Project and showed that destruction of phreatophytes and their removal produced significant reduction in evapotranspiration in areas that were cleared.

Water is a vital resource to man in a great variety of other ways (Vol. I, p. 7–30). The use of rivers for navigation and commerce was early recognized as being fundamental in the economy of the United States. The agency largely responsible for river maintenance is the *Corps of Engineers* (see also Happ, 1950). The Corps actually has roots before the Revolutionary War, when there were Army engineers, but the present Corps dates to 1802, when it was officially created by congressional legislation. The first civil works responsibilities were given to the Corps in 1824, when they were allotted $75,000 for removal of navigation impediments in the Ohio and Mississippi rivers. With this rather modest beginning, the programs of the Corps have now expanded to a point where it is the largest single construction organization and biggest developer of water control structures in the world. For example, in May 1967 the Office of Chief Engineers reported that the Corps was custodian of 3828 active civil works projects totaling $30 billion, of which 312 were then under construction. By 1967, they had refined 19,000 miles of rivers and waterways with dams, reservoirs, levees, deeper channels, and other alterations. The Corps had also built more than 800 projects whose primary purpose was flood control. One unusual standing authorization permits the Corps to spend up to $50,000 at a location within a single year for stream bank protection, and up to $100,000 a year for clearing, cleaning, and straightening a stream channel. Volume II discusses shoreline modification programs of the Corps and includes excerpts from their publications. The reservoirs that the Corps-built dams have imponded contain more shoreline than all ocean beaches of the U.S. mainland and the waters covered 4 million acres in 1967.

The Enlarged Conservation Framework

Dasmann (1959) was one of the first to provide a new conceptual definition of "conservation," which he relates to an environmental context, and provides the direction that "conservationists" have taken during the past decade.

> . . . [conservation is] the rational use of the environment to provide a high quality of living for mankind. It involves the planning for and control of man's use of his environment, with a consideration of the long-range future of the human race and with a view to providing environments suitable for the satisfaction of the widest possible range of human aspirations. It involves, therefore, the preservation or creation of diversity in environments, maintaining the greatest variety of living creatures, in order to provide suitable habitat for the varied types of people that still exist and thus to enable the human diversity to survive (p. 6).
>
> The approach to conservation, therefore, that holds the most hope for the future, is an ecological approach, that takes into account the ecology of man. Each natural area, be it defined as a watershed, a vegetation and soil community, or in other terms, represents an ecosystem, a physical and biotic environment, with certain potentialities and certain problems. The objective of management should be to develop that area, or to protect it, in order to provide the greatest yield in improved quality of living for mankind (p. 87).

In the early 1900s Gifford Pinchot was instrumental in bringing some new management practices into the national forests in his position as chief of the U.S. Forest Service; but his brand of conservation became known as a "utilitarian" type and provided a reversal in philosophy to the Forest Reserve Act of 1891. This legislation enabled the President to ". . . set apart and reserve . . . in any part of the public lands wholly or in part covered with timber or undergrowth, whether of commercial value or not, as public reservations." The first three presidents thereafter used these powers to reserve 27 million acres, and Theodore Roosevelt added another 148 million acres. Pinchot's brand of forest management put the government in the timbering business, where today there are 187 million acres of national forests that yield $300 million in revenues, mostly from timbering. Timbering has steadily increased from 5.6 billion board-feet in 1950 to 13.74 billion in 1971. By law, the forests are now multipurpose and are supposed to fulfill several functions: recreation, wildlife preservation, timber production, grazing, and mining.

As early as 1900, Representative John Lacey of Iowa introduced a bill ". . . to establish and administer national parks . . .", but it and subsequent proposals received massive opposition, so that it wasn't until 1916 that the National Parks Act was passed, which established a National Parks Service. By this time there were 12 national parks and a number of national monuments that had been instituted under the Act for the Preservation of American Antiquities in 1906. By 1933, all of these, including other monuments that had formerly been under the jurisdiction of the War Department or the Forest Service, were transferred to the supervision of the National Park Service. The system of national parks now includes more than 30, covering more than 13,000,000 acres. Their utilization by man ranges from education and recreation to aesthetic purposes.

Early settlers in the West were aware of the importance of water and realized that to be successful, crops would have to be irrigated (see Powell, Vol. I). Gardens along the Arkansas River in Colorado were irrigated as early as 1832 by a system of open ditches, and by 1848 the Mormons in Utah were employing irrigation methods on a rather large scale. Finally, after much debate and many changes in wording, President Theodore Roosevelt signed the Reclamation Act on June 3, 1902. This legislation established a Reclamation Service, which was later to become the Bureau of Reclamation in the Department of Interior in 1923. During the early years, the great majority of the effort was in the development of irrigation projects, since the first objective was to change the dry lands into productive farmlands through the use of imported irrigation waters. Today only about one-third of the Bureau's projects are directed toward irrigation; the remainder occurs in benefits they establish in navigation, flood control, fish and wildlife enhancement, power production, recreation, and water supplies for industries and municipalities. The Bureau makes contracts with private companies to build such projects as hydroelectric power plants, dams, canals, and pumping stations and to plan and construct valley projects to store and control waterflow. James (1917, p. 387–388) provides a summary of accomplishments by the Reclamation Service at the end of its first 14 years:

> It has built the highest dam in the world—the Arrowrock, for the storage of the water of Boise River in Southern Idaho. It has built the Elephant Butte dam on the Rio Grande in New Mexico, whose reservoir, when full, will store the greatest quantity of irrigation water. Up to the end of December, 1916, it has dug 9,805 miles of canals and 1,139 miles of ditches and drains, and excavated 91 tunnels with an aggregate length of nearly 26 miles. Dams of masonry, earth, crib, and rock fill have been erected with a total volume of 13,038,109 cubic yards. The available reservoir capacity at this time is approximately 9,007,160 acre feet, or sufficient to cover the states of New Jersey and Delaware with water to a depth of 16 inches. The Service has built 5,594 bridges with a total length of over 23 miles. Its culverts number 6,624 and are 43 miles long. There are now in operation 385 miles of pipe line and 95 miles of flumes. The Service has built 879 miles of wagon road, much of it in what was before inaccessible mountain regions, 83 miles of railroad, 2,804 miles of telephone lines, 438 miles of power-transmission lines, and 1,156 buildings, such as power-houses, pumping-station, offices, residences, barns, and storehouses.
>
> The projects now under way or completed embrace approximately 3,140,976 acres of irrigable land, divided in about 61,310 farms of from

10 to 160 acres each. During the year 1916, water was available from government ditches for 1,734,482 acres on 36,255 farms, and the government was under contract to supply water to 1,313,191 acres. The excavations of rock and earth amount to 146,034,177 cubic yards. The Service has used 2,786,619 barrels of cement, of which it has itself manufactured 1,575,757 barrels of cement and sand cement. The power developed amounts to approximately 47,311 horse-power.

The net investment of the Service to date is approximately $116,000,000. In 1916, crops valued at $38,000,000 were harvested, the gross average yield per acre being $38.25.

By 1967 (Laycock, 1970, p. 79) the Bureau of Reclamation had impounded behind 252 dams water that could cover New York State to a depth of 4 ft. They erected 48 power plants with a combined capacity of 6.9 million kw which distribute electricity along 15,000 miles of high-voltage lines. The Bureau built 344 canal systems with a combined length of 6781 miles.

The United States has had a rapid growth in the development of other reclamation programs and conservation projects of large scale during the twentieth century. One of the excellent early models was developed in Ohio as a result of disastrous floods in 1913. The Ohio legislature passed in the same year the Ohio Conservancy Act, which created the Muskingum Conservancy District, established for the protection of a drainage area extending into 18 counties. The District was authorized to acquire land, and in the process it constructed 16,000 acres of reservoirs designed to contain rainfall that could exceed by 36 percent the rains that caused the severe 1913 flooding. Probably the most famous single water project is the Tennessee Valley Authority (TVA), described in the following terms by Nicholson (1970, p. 179–180):

In terms of water control, the largest, most spectacular and world-famous of conservation projects has been the Tennessee Valley Authority. Although actually created in 1933 under President Franklin Roosevelt, it had its origins in steps taken by his cousin Theodore in establishing the Inland Waterways Commission in 1907 . . . which established the principle of unified river basin management, irrespective of State boundaries, and combining flood control, power development, irrigation, drainage and purification the Commission let loose in the world a potent and constructive new idea which was widely welcomed as enabling conservation principles to be effectively applied in practice to vulnerable regions which had suffered.

It is hardly necessary to describe in more detail this world-renowned conservation project, the greatest and most comprehensive yet achieved anywhere . . . Elsewhere the TVA has been imitated in, for example, the Damodar Valley Project in India, but several attempts to follow it have been castrated, like the original plan for the North of Scotland Hydro-Electricity Board, or nipped in the bud by hostile influences, official as well as private.

The Tennessee Valley, an area of 40,910 square miles, has valuable resources of coal, iron, manganese, copper, marble, limestone, zinc, sand, and gravel. The TVA was created by an act of Congress in 1933 as a federal corporation to develop the resources of the region. This was an outgrowth of 15 years of debate concerning the disposition of a government nitrate plant and Wilson Dam at Muscle Shoals, Alabama. These projects were under the authority of the War Department and had been built because of the National Defense Act of 1916, but had not been completed in time for use during World War I. Although the power and resources of the area were to be important for a wartime economy, the program also called for the construction of dams to control flooding, the deepening of rivers for shipping, the planting of new forests, the conservation of timber already present, and the creation of electric power and mineral resources. Since 1933, the TVA has built 20 dams, which provide one of the most highly regulated rivers systems in the world. Fontana Dam is the highest, 480 ft, and Kentucky Dam is the largest, 206 ft high, ½ mile long, and impounds a 184-mile man-made lake. In addition, the reservoirs provide 14,600,000 acre-feet of storage at the beginning of the flood season. The Tennessee River now provides a 627-mile route for boats with 9-ft drafts, which is connected to the inland waterway system. By 1955 the TVA had reforested 240,000 acres, and the power plants were generating 53 billion kWh/year.

In describing this region today, Frincke (1964, p. 95) reports:

> Only those who saw or lived in the Tennessee Valley before 1933 can fully appreciate the changes that have been brought about in the valley's landscape . . . a series of clear, blue lakes, that extend from the ancient slopes of the Great Smoky Mountains . . . to the hot, rolling midlands of western Tennessee and Kentucky. No longer is this great river system a silt-laden, brown, unpredictably fast-moving stream.

Government action agencies which construct devices that change the terrain and its water regimes, such as the Corps of Engineers, Bureau of Reclamation, and TVA, have their detractors. The opponents range through a broad spectrum from those who feel all engineering works are highly questionable to those who feel certain policies need to be redefined or modified.

> The Engineers, however, are not only straining to fill in the Grand Canyon and to dam the last wild stretch of the Missouri, to wall off the rich estuaries of Long Island and to cover the Great Swamp with asphalt. They are in every section of every state, ripping, tearing, building, changing. Theirs is a rape from which America can never, never recover (Marine, 1969, p. 19).

Laycock's *The Diligent Destroyers* (1970) is totally dedicated against engineers and their works and policies.

> Many of the projects on which the Corps of Engineers works in any given year are included in the annual Rivers and Harbors Bill, informally referred to as "The Pork Barrel."

Typically, the Rivers and Harbors Act of 1965 authorized fifty-three new projects in twenty-three states . . . Significantly [The Act] frequently amends former . . . acts upward. In 1965 the act amended the 1960 act on several projects. It struck out one $2 million item and inserted "in lieu thereof $10,000,000." Another allowance of $200,000 became $500,000, one for $3 million was changed to $10 million, while another project computed to cost $400,000 was increased 25 percent.

When the calculations were completed (on the Tennessee–Tombigbee Waterway project), the benefit–cost ratio came out to 1.24 to 1. To reach this "profit-making" conclusion, the engineers had tossed in all the benefit ingredients they could justify from among those allowed by Congress, including several million dollars for recreation, fish and wildlife "enhancement," and wage payments to those employed to work on the canal. The Chief of Engineers sent the report along to the Secretary of the Army with his opinion . . . [the plan was] economically justifiable.

The Secretary of the Army Stanley Resor disagreed. The report went sailing through the Appropriations Committee. The Corps of Engineers was ordered to get on with this expenditure of public funds, and the Secretary's adverse opinion was not even taken to the floor of Congress for consideration (p. 27–29).

It is frequent practice for the Bureau of Reclamation and the Army Corps of Engineers, when figuring the value of a proposed project to ignore any losses that might result. Some of the losses are incalculable because they destroy such natural wonders as wilderness areas of river valleys. There is a glaring example in the celebrated case of the dams which the Bureau of Reclamation proposed for the Grand Canyon. Economists noted that the Bureau chose to ignore the evaporation losses of 100,000 acre-feet of water annually from the proposed reservoirs. The benefit–cost figures allowed this water a value of zero, but in the thirsty West no such quantity of impounded water has a true value of zero. To replace this water, as they planned to do eventually, with water brought in from the Columbia River, would cost a minimum of $70 per acre-foot. As the Sierra Club noted, ". . . this subsidy amounts to an additional $7 million per year." The Bureau of Reclamation had chosen to overlook this loss (p. 122–123).

When discussing a new dam and reservoir proposed by the TVA on the Little Tennessee River, "Was the reservoir needed for recreation? Within a fifty-mile radius, there were twenty-two other dams and reservoirs. So many dams had been constructed throughout the southeast by TVA and the Corps of Engineers that according to the Tennessee Game and Fish Commission . . . "impoundments no longer provide the attraction they once did to non-residents." Why then did TVA insist on the $41,000,000 dam, flooding 14,400 acres? The answer lay in the fact that TVA was moving into the real estate business. The major purpose in creating the reservoir on the Little Tennessee would be to sell land to the industries needing cheap water transportation. When it was pointed out that the new dam would flood good industrial sites already available in the river bottomland, TVA chose not to hear (p. 196).

An economic survey and analysis of 147 water projects constructed with federal funds in 10 southern states by Haveman (1965) indicated that the benefit–cost ratio for 63 of them could not be substantiated on the basis of accepted economic methods. These projects cost more than $1 billion and, according to Haveman, constituted misallocation of national resources and economic waste.

Leopold and Maddock (1954) provide insight into another problem that is created during engineering management of river systems:

> . . . much of the damage from the 1951 flood in the Kansas River basin . . . may have been actually due to development in the flood plain resulting from the present flood control project (p. 153).
>
> In the short period, the Corps of Engineers have been engaged in major flood control activities, it has spent, largely between 1928 and 1952, $3,600,000,000. It is incongruous that the Corps made essentially no provision to restrict flood-plain development and the consequent increase in damageable values (p. 247).

In addition to the federal legislation and acts already cited, or those that will be discussed in detail later, several additional ones need mention at this time which also have some influence on the manner in which terrain and its resources are managed.

1. The Weeks Act of 1911 extended the principle of protection for navigable streams into their associated forests and watershed area.

2. The Taylor Grazing Act of 1934 crystallized into policy the various and disparate strands of land classification, local usage, and adaption into conservation interests and plans for the future.

3. The Multiple Use–Sustained Yield Act of 1960 specifies two economic goods and three services that the U.S. Forest Service in its management of forests must make available to the nation: (1) timber from the trees, (2) forage from grazing lands, (3) recreation, (4) wildlife management, and (5) protection of rivers and consideration for their constant flow.

4. The Outdoor Recreation Act of 1963 (Fitch and Shanklin, 1970) declared: "The Congress finds and declares it to be desirable that all American people of present and future generations be assured adequate outdoor recreation resources." Thus was enunciated for the first time that recreation is a human resource and needs planning, classification, and evaluation just as any other resource.

5. The Land and Water Conservation Fund Act became effective on January 1, 1965. It authorized the Secretary of the Interior to grant matching funds to the states for the planning, acquisition, and development of outdoor recreation lands and facilities.

6. The Wild and Scenic Rivers Act of 1968 specifies ". . . that certain rivers of the Nation which . . . possess outstandingly remarkable scenic, cultural or other similar values, shall be preserved in free-flowing condition, and that they and their immediate environments shall be protected for the benefit and enjoyment of present and future generations."

7. The Environmental Policy Act of 1969 may become one of the farthest reaching and encompassing legislative mandates in history (Coates, 1971, p. 224). Its declaration of purpose states:

> To declare a national policy which will encourage productive harmony between man and his environment: to promote efforts which will prevent or eliminate damage to the environment biosphere and stimulate the health and welfare of man; to enrich the understanding of the ecological systems and natural resources important to the Nation; and to establish a Council on Environmental Quality.

It provides for the inception of an Environmental Protection Agency and a Council on Environmental Quality. However, one feature that at the time received little notice but which is now becoming the *most powerful* element of the act concerns its requirement that prior to construction of federal projects, an *environmental impact statement* must be filed by the agency which will contain the following information:

(i) the environmental impact of the proposed action,
(ii) any adverse environmental effects which cannot be avoided should the proposal be implemented,
(iii) alternatives to the proposed action,
(iv) the relationship between local short-term uses of man's environment and the maintenance and enhancement of long-term productivity, and
(v) any irreversible and irretrievable commitments of resources which would be involved in the proposed action should it be implemented.

This act provides a lever by which conservation–preservation groups can scrutinize closely all proposed federal construction in nonurban regions. Many federal agencies have compiled their own regulations, which will act as guidelines for their planners, such as the Highway Environment Reference Book (1970) by the U.S. Department of Transportation. The Tocks Island Dam Project on the Delaware River is typical of those schemes that are receiving great opposition from conservationists because of inadequacies in justification and rationale of the impact statements, and it has been currently delayed. Even Governor William T. Cahill of New Jersey has announced that he ". . . cannot support the Tocks Island Dam project as it is now proposed because of the adverse environmental and financial impact on New Jersey" (Hanson, 1972, p. 2). Another project that has been seriously challenged by many groups is the trans-Alaska pipeline project. The final environmental impact statement for this project is contained in a six-volume report that weighs 25 pounds and costs $30. The outstanding contribution for management of earth resources that emerges from this necessity for compiling impact statements is that it forces agencies to plan for the future and to present sound scientific data upon which predictions can be made and decisions justified.

The federal government is not alone in developing a strong conscience on environmental affairs. Many of the states are also moving forcefully in this direction, and New York State has been a leader. In 1965 voters approved the Pure Waters Bond Act, which appropriated $1 billion (the largest single measure of its kind ever enacted) for the abatement of pollution in rivers and lakes of the state. Again on November 7, 1972, the voters overwhelmingly approved by a 2:1 margin the $1.15 Environmental Quality Bond Act, which sets the stage for a full-scale clean-up of the State's land, air, and water. The bond enactment authorizes the spending of $650 million for sewage treatment plants, $150 million to combat air pollution, $175 million to acquire valuable lands in the public interest for such purposes as recreation, and $175 million for solid waste improvement. Thus it speaks to a broad spectrum of environmental needs and management procedures, and takes a strong step toward the rehabilitation of damaged environments.

Of course, many U.S. laws, environmental planning methods, and reclamation projects (see also McComas, 1972) could be repeated with examples from other parts of the world. In England and Wales, for instance, there are more than 150,000 acres of derelict land; 36,000 acres are in urban areas and 114,000 acres of industrial dereliction are in less populous areas. There is an average increase in dereliction at the rate of 3500 acres/year (12,000 acres are laid waste by working for mineral development, but 8500 acres per year are being reclaimed). Redemption of the lands includes reforestation, land contouring for agricultural use, and beautification in terms of other types of landscaping and development of ponds, lakes, and parks, (Anonymous, no date; Ministry of Housing and Local Government, 1963). As Stamp (1964, p. 202) points out:

> . . . the arboreal landscape of Britain is undergoing at the present time a revolution. Large areas, swept bare of woodlands for centuries, are being clothed once again. Where once babbling brooks threaded their silver way through noble forest glades there have long been peat bogs and wet cotton grass or Molinia moorlands of little use to man either economically or aesthetically. These lands are being reclaimed with the help of trees not native to this country. . . ."

Shoor et al. (1964, p. 72–73) show the dimensions of land reclamation and rehabilitation in Israel:

> For 2,000 years, until the turn of this century, natural forests, once cultivated areas, lay in complete neglect overcome by wind and soil erosion. There were hills bare of top soil, burnt and covered with boulders, coastlines invaded by sand dunes, and treeless plains. Springs overflowed, resulting in numerous swamps, and bogs infested by malarial mosquitos formed vast deserted tracts of land in the inner region.

They then describe some of the reclamation projects such as the Hulen Valley Reclamation. "This project has provided some thousands of dunans of farm land, and exploited water, and has freed the area of malarial scourge . . . The Hulen can well be considered an outstanding achievement in reclamation schemes." This was done by deepening and widening the bed of the Jordan River and digging main and subsidiary drainage canals to draw off surplus marshy waters.

Many of the United Nations publications deal with conservation and reclamation in different parts of the world. Typical of these is the UNESCO (1961) report, which deals with reclamation of arid areas and the problems of salinity. For example, Alexander Kovda (of the U.S.S.R.) describes how saline soils may be leached during reclamation of saline soils in the arid zone; I. A. Antipove-Karataev and P. A. Kerzum analyze reclamation methods for exploiting saline and swampy soils in Tadzhikistan, and M. Ayazi discusses problems and procedures being used in Iran in the article "Drainage and Reclamation Problems in the Garmsar Area." Additional source materials and information on land management and reclamation occurs in Volumes I and II. For example, see also Burrows, Marsh, and Powell (Vol. I).

Editor's Comments on Papers 21 Through 24

The articles chosen for this section illustrate both governmental and private investigations and suggestions for landscape management. The first selection deals with the study of geomorphic phenomena—hazards—and how they influence planning in rugged forested terrain. The last three articles are concerned with the restoration and rehabilitation of terrains damaged by mining operations.

R. G. Bailey is a hydrologist with the Forest Service. His study presents both an inventory of anticipated hazardous areas and a set of criteria that can be used for appropriate planning purposes, as in road construction and timber harvest. Although only part of the much longer article (131 p.) could be used here, it is an integral unit and can stand by itself. As Bailey points out in the preface, logging operations in the region had largely been concentrated on the lower slopes by methods of clear-cutting:

> However, as the supply of timber from the more stable lands diminishes, the ever-increasing demand for the timber resources has turned the attention of the timber industry to the more rugged terrain with increased stability hazards. This terrain is thus on the threshold of more intensive use and development.

The study area covered 1.13 million acres and a principal aim was to ". . . provide . . . a framework for managing the forest environment of northwest Wyoming on a more realistic basis than heretofore."

The next three articles form a group that are related by virtue of man's changing the terrain because of mining activities and what steps can and should be taken to reclaim the devastated landscape. The first article, by the U.S. Department of Agriculture is a general overview and treats surface-mined areas, whereas the Beatty and Zube articles each deal with reclamation of a particular site in a certain geomorphic setting. All these articles could also have been included in Part I, since they deal with areas of man-induced destruction of lands and waterways/areas, but the main theme now is on a constructive tone and concerned with their restoration. Before something is to be changed, however, it is necessary to point out that such change is necessary, worthwhile, and beneficial. The greatest single destroyer of landscape in the United States today is the strip mining by the coal industry (see Part I). Reclamation schemes as pointed out in the USDA article; they should consist of at least several phases: careful preplanning of the operation, proper mining procedures, water control and terrain stabilization during mining, and reclamation after mining. This is relevant to all forms of surface mining. Elements that need analysis during the pre-mining stage include determination of characteristics of overburden, such as toxicity of materials and response to water, whether easily erodable. Evaluation of surface and groundwater conditions is necessary so that proper drainageways and water control structures can be developed. During mining, appropriate diversion ditches for water control with sediment ponds must be carefully designed. The method for removal and placement of overburden is critical to ensure minimal leaching of toxic materials into watercourses. The top soil and other reusable materials should be carefully handled to minimize costs when used as cover and for revegetation during the after-mining phase. Actually, as much as possible there should be simultaneous reclamation during the mining itself (Bauer, 1965), so as to limit to the minimum amount of time exposure of deleterious materials to weathering and leaching. In reclamation, whenever possible the site should be regraded to approach the original landscape configuration. Revegetation with grasses and/or legumes should be planted in all possible areas and, when appropriate, trees planted in combination with mulch to aid in inhibiting erosion during the growth period. The final step in the process might be called the *rehabilitation* of the surface-mined land—that is, its design for reuse in some other capacity. Such use might include

1. Cropland. This would occur with level or gently rolling topography with burial of all rocks and sufficient fine materials with nutrients that would support agricultural development.
2. Pastureland. Grading should not exceed 15 degrees to permit farm machinery to plant and develop forage crops. Soil conditions must also be favorable for grasses and legumes.
3. Rangeland. This requires less treatment and grading.
4. Wildlife use. Specific game food and cover plantings are necessary and should be maintained.
5. Recreational use. After grading, potable water supplies and sewage facilities could be developed. Ponds and lakes, even reservoirs where appropriate, could be constructed to enhance the multiple-usage concept and the aesthetics of the locale. Golf courses, boating, swimming, hiking, and so on, might all be part of the overall plan.
6. Industrial or home site use. In suburban settings, the rehabilitated lands can serve for sites that would aid in maintaining a substantial tax base for municipalities.

The articles by R. A. Beatty and E. Zube show other dimensions to the problems associated with reclamation and rehabilitation from a landscape designer's point of view—the creation of something that will not only be practical but visually aesthetic.

The Beatty article might be considered a possible happy ending to the Gilbert article (Vol. I), which showed the vast amount of man-made destruction created by hydraulic dredging in the gold fields of California. The Zube article is interesting since it suggests a *man-created landscape* to replace the *man-created badlands,* in which the newly planned terrain is not necessarily modeled after the original landscape configuration. These articles thus lead into the last portion of the volume, which deals with the matter of aesthetic planning and evaluation of the landscape configuration.

Evaluation of the Land-Water Ecosystem

> There is nothing more practical in the end than the preservation of beauty (Theodore Roosevelt).
>
> In wilderness is the preservation of the world (Henry Thoreau).

The preservation of nature in wilderness form, landscape aesthetics, and evaluation methods for assessing priorities of landscape management have been hotly debated issues throughout the twentieth century. N. Roosevelt (1970, p. 1–2) summarizes one of the new conservation views of the landscape in the following manner:

> Had anyone then (before 1900) suggested that scenery was among the most valuable of our natural resources, only a few dreamers would have understood. Today we know better. Although conservation has many objectives, these call for protection of areas of exceptional natural beauty from destruction or severe mutilation by logging, stripping, draining, damming, and flooding, or by ramming freeways through them with mechanically and unimaginatively cleared rights-of-way. After decades of confused thinking, we see at last that scenery is limited in extent, irreplaceable and invaluable.
>
> Where the original theory of conservation of natural resources was based on ultimate use and consumption, today's concept of conservation centers on use without consumption or destruction—use through enabling an increasing number of visitors to enjoy the natural beauties of areas that have been set aside, and, in pleasant surroundings, to indulge in forms of recreation as varied as motoring for pleasure, boating, swimming, hiking, skiing, fishing, and camping.

However, as early as 1906, Shaler presented some of the first twentieth-century appraisals on this theme in chapters in his book under such headings as "The Future of Power," "The Beauty of the Earth," and "The Future of Nature upon the Earth." Here he offers a few of the arguments that have been repeatedly used by those who would use or manage the terrain in a more utilitarian fashion:

> There are those who feel that an intensely humanized earth, so arranged to afford a living to the largest possible number of men, will lack much of the charm that it has now; that it will become so far artificial that

> its primitive nature will be utterly lost. A careful examination of the conditions will show us that while the order of beauty is doubtless to be greatly changed by the hand of man, there is reason to believe that the alterations will enhance its aesthetic value, making its features far more contributive to spiritual enlargement than they were in their primal wilderness state (p. 172).

> The absolute wilderness, however, noble in its aspect, has aesthetic interest for relatively few persons. Even to them, it lacks the charm of the fields that bear the impress of the hand of man . . . Hence it is that only the more expanded souls can rejoice in the untrodden deserts, the pathless woods, or the mountains that have no trace of culture (p. 181).

> Turning again to the question of preserving the beautiful aspects of the earth, we note that the streams and lakes are the most likely of all natural features to suffer from the action of man. In some ways, this damage is unavoidable. The need of power, the most immediate and far-reaching of all man's necessities, is certain to destroy our waterfalls and rapids and to reduce all our rivers to a series of lake-like pools (p. 188).

Although a few positive steps had been taken in the United States before 1900, and several of the seeds had been planted for preservation of natural settings, wilderness, and aesthetics, it was not until the twentieth century that programs became inplemented on a large scale. However, with programs and land management have come a host of problems and controversy concerning priorities in usage. There has now arisen the need for appropriate evaluation methods along with the requirement for impact statements.

One of the very early statements, considered a forerunner of the national park concept, was given by George Catlin, celebrated painter, upon his return from a trip up the Missouri River in 1832:

> . . . (the Great Plains) might in future be seen preserved in their pristine beauty and wildness, in a magnificant park . . . A nation's park containing man and beast, in all the wild and freshness of their nature's beauty (Wildland Research Center, 1962, p. 18).

Although the establishment of the first national park did not occur until 1872, when Yellowstone was created with more than 2 million acres as ". . . a public park or pleasuring-ground for the benefit and enjoyment of the people," two other events helped pave the way. One of these was the development of Hot Springs, Arkansas, in 1832 as a national reservation and public health resort, and the other was the deeding of Yosemite Valley and the Mariposa Grove of Big Trees to the State of California by the federal government in 1864 ". . . upon the express conditions that the premises be held for public use, resort, and recreation, shall be inalienable for all time." Neither plans for Yosemite or Yellowstone originally anticipated the concept of wilderness values that are in current vogue, and the first large battle on this issue that resulted in positive action for such a public park occurred in New

York. The first years after establishment of Yellowstone National Park were stormy and its retention was threatened from several quarters. Commercial interests wanted to develop the area, prospectors wanted to explore the region for mineral claims, and a railroad right-of-way was proposed. By 1886 the Department of Interior had won out and the new philosophy of the park emerged as enunciated by the Secretary Lucius Q. C. Lamar for ". . . the preservation of wilderness of forests, geysers, mountains . . . with a view to holding for the benefit of those who shall come after us something of the original 'wild west' that shall stand while the rest of the world moves, affording the student of nature and the pleasure tourist a restful contrast to . . . busy and progressive scenes."

The definition and concept of wilderness as currently used embrace fundamental ideas, which are illustrated by the articles reproduced in this section. It is defined in the following manner by the National Park Service (1957):

> A wilderness is an area whose predominant character is the result of the interplay of natural processes, and large enough and so situated as to be unaffected, except in minor ways, but what takes place in the non-wilderness around it.
>
> We accept man using the wilderness for recreation; and we accept a trail, a simple campsite, or even a drift fence or fire lookout, so long as the predominant character of wilderness remains. We also accept certain management and protection practices, so long as they are confined to minimizing the effect of man, and do not interfere with the normal interplay of natural processes. Wilderness also needs to be regarded as a quality—defined in terms of personal experience, feelings, or benefits.

Some of these factors include "A scene or vista of unusual natural interest or beauty . . ." and "An area secluded or removed from the sight, sounds, and odors of mechanization." Thus a wilderness terrain is both an ecological condition and a state of mind. The first recognition of this in the establishment of such a land reserve was in the Adirondack Mountains of New York. In the 1880s, mining and lumbering interests were allegedly stripping and defacing the area, but the real trigger that aroused the outside political and commercial interests was the decline of water levels in the Erie Canal and Hudson River. Control of the watershed areas for the rivers was contended as being necessary to conserve commercial prosperity in down-stream regions. Thus on May 15, 1885, a New York State bill was enacted to establish a "Forest Preserve" of 715,000 acres that was to permanently remain ". . . as wild forest lands." In 1892 the area was expanded to include more than 3 million acres and was relabeled the "Adirondack State Park"; by 1894, at a constitutional convention, Article 7, Section 7, was inserted, which guaranteed permanent preservation for the Adirondack wilderness.

The concept of wilderness on the federal level for forest reserves, as in conservation, became a contest between the *preservationists* as championed by John Muir and the *utilitarian–reasonable use* group headed by Gifford Pinchot. These two men were good friends in the early 1890s, but congressional passage of the Forest Management Act on June 4, 1897, left no doubt that the reserves were not destined to be wilderness. The estrangement came to a head in the summer of 1897 after Pinchot had approved sheep grazing in forest lands, and when the two met in a hotel lobby, Muir told him, ". . . I don't want anything more to do with you. When you were in the Cascades last summer, you yourself stated that the sheep did a great deal of harm" (Nash, 1967, p. 138). Thereafter, Muir considered the forests as lost for preservation of wilderness and put his remaining energies to defend and promote the national parks. He was instrumental in California's returning Yosemite Valley to the federal government and establishing of the Grand Canyon area as a national monument.

Aldo Leopold was influential in helping to establish the first federal wilderness area. He had been a forester with the Forest Service from 1909 to 1918 but resigned when his ideas were in conflict with their policies. However, when the programs of energetic Stephen Mather as head of the newly formed National Park Service in 1916 attracted much interest, Leopold sensed a more receptive climate in Washington, and rejoined the Forest Service in 1919. Along with Arthur H. Carhart, they helped develop the wilderness concept, and Leopold's 1921 article in the *Journal of Forestry* and his advocacy of each western state returning 500,000 acres of wilderness culminated with the setting aside on June 3, 1924, of 574,000 acres of the Gila National Forest to be known as the Gila Wilderness Area. This breakthrough in formulation of governmental policy and recognition of the need for preservation of nature led to a major series of events. For example, in 1927 a 4500-acre ponderosa pine forest in northern Arizona was withdrawn from timber production, the first natural area to be set aside and so designated for scientific study. By 1940 the Forest Service had designated 31 such areas.

With the adoption on July 12, 1929, of what became referred to as the "L-20 Regulations," the Forest Service became the first public agency to issue orders for the establishment of areas that embodied the wilderness concept. As amended in August 7, 1930, these regulations state:

> The Chief of the Forest Service shall determine, define, and permanently record . . . a series of areas to be known as primitive areas, and within which . . . will be maintained primitive conditions of environment, transportation, habitation, and subsistence, with a view to conserving the value of such areas for purposes of public education, inspiration, and recreation.

The L-20 Regulations were superseded on September 19, 1939, with new Regulations U-1, U-2, and U-3 (a), which redefined the terminology and made distinctions between "wild" and "wilderness" on the basis of size, and differentiated them as having less human interference than the primitive areas. The crowning achievement in the legislative process occurred in 1964 with the passage by Congress of the Wilderness Act. It states in part:

The Wilderness Act of 1964

Be it enacted by the Senate and House of Representatives of the United States of America in Congress assembled,

Short Title

Section 1. This Act may be cited as the "Wilderness Act."

Wilderness System Established Statement of Policy

Sec. 2. (a) In order to assure that an increasing population, accompanied by expanding settlement and growing mechanization, does not occupy and modify all areas within the United States and its possessions, leaving no lands designated for preservation and protection in their natural condition, it is hereby declared to be the policy of the Congress to secure for the American people of present and future generations the benefits of an enduring resource of wilderness. For this purpose there is hereby established a National Wilderness Preservation System to be composed of federally owned areas designated by Congress as "wilderness areas," and these shall be administered for the use and enjoyment of the American people in such manner as will leave them unimpaired for future use and enjoyment as wilderness, and so as to provide for the protection of these areas, the preservation of their wildnerness character, and for the gathering and dissemination of information regarding their use and enjoyment as wilderness; and no Federal lands shall be designated as "wilderness areas" except as provided for in this Act or by a subsequent Act.

(b) The inclusion of an area in the National Wilderness Preservation System notwithstanding, the area shall continue to be managed by the Department and agency having jurisdiction thereover immediately before its inclusion in the National Wilderness Preservation System unless otherwise provided by Act of Congress. No appropriation shall be available for the payment of expenses or salaries for the administration of the National Wilderness Preservation System as a separate unit nor shall any appropriations be available for additional personnel stated as being required solely for the purpose of managing or administering areas solely because they are included within the National Wilderness Preservation System.

Definition of Wilderness

(c) A wilderness in contrast with those areas where man and his own works dominate the landscape, is hereby recognized as an area where the earth and its community of life are untrammeled by man, where man himself is a visitor who does not remain. An area of wilderness is further defined to mean in this Act an area of undeveloped Federal land retaining its primeval character and influence, without permanent improvements or human habitation, which is protected and managed so as to preserve its natural conditions and which (1) generally appears to have been affected primarily by the forces of nature, with the imprint of man's work substantially unnoticeable; (2) has outstanding opportunities for solitude or a primitive and unconfined type of recreation; (3) has at least five thousand acres of land or is of sufficient size as to make practicable its preservation and use in an unimpaired condition; and (4) may also contain ecological, geological, or other features of scientific, educational, scenic, or historical value.

National Wilderness Preservation System—Extent of System

Sec. 3. (a) All areas within the national forests classified at least 30 days before the effective date of this Act by the Secretary of Agriculture or the Chief of the Forest Service as "wilderness," "wild," or "canoe" are hereby designated as wilderness areas. The Secretary of Agriculture shall—

(1) Within one year after the effective date of this Act, file a map and legal description of each wilderness area with the Interior and Insular Affairs Committees of the United States Senate and the House of Representatives, and such descriptions shall have the same force and effect as if included in this Act: *Provided, however,* That correction of clerical and typographical errors in such legal descriptions and maps may be made.

(2) Maintain, available to the public, records pertaining to said wilderness areas, including maps and legal descriptions, copies of regulations governing them, copies of public notices of, and reports submitted to Congress regarding pending additions, eliminations, or modifications. Maps, legal descriptions, and regulations pertaining to wilderness areas within their respective jurisdictions also shall be available to the public in the offices of regional foresters, national forest supervisors, and forest rangers.

Classification. (b) The Secretary of Agriculture shall, within ten years after the enactment of this Act, review, as to its suitability or nonsuitability for preservation as wilderness, each area in the national forests classified on the effective date of this Act by the Secretary of Agriculture or the Chief of the Forest Service as "primitive" and report his findings to the President.

Presidential recommendation to Congress. The President shall advise the United States Senate and House of Representatives of his recommendations with respect to the designation as "wilderness" or other reclassification of each area on which review has been completed, together with maps and a definition of boundaries. Such advice shall be given with respect to not less than one-third of all the areas now classified as "primitive" within three years after the enactment of this Act, not less than two-thirds within seven years after the enactment of this Act, and the remaining areas within ten years after the enactment of this Act.

When conservation–preservation policies of other countries are considered, England and Japan show contrasting histories of development. Japan has a somewhat parallel growth with the United States, whereas in England such national policies are more delayed. As Nicholson (1970, p. 159–161) pointed out:

> Already in 1908, it could be said that conservation in America had arrived; the same could not be said for Britian before 1963 at earliest—a timelag

> of over half a century. It was not until 1963 that all bodies interested in care and use of the countryside were convened with agreement to "oversee and contribute to the maintenance of the amenities of the countryside as a whole."

The Nature Conservancy was started in England in 1949 and

> Within fifteen years of its creation, the Nature Conservancy had acquired more than a hundred National Nature Reserves, had designated some 2,000 statutory Sites of Special Scientific Interest, and had built up a multidisciplinary staff of well over a hundred scientists . . . .

Sato (1964, p. 52–54) described the growth of the park system in Japan in the following manner:

> Let us look over our policies concerning the preservation of rural landscape. The first policy is found in the law of 1873 about parks. This was the Minister's order to establish public parks throughout the country, after the model of the Western countries. With this law, public parks were born and many beautiful spots and resorts have developed, and been preserved as public parks. In 1907, the Parliament passed a law providing the "forest reserves." According to this law, the Minister of Agriculture may designate beautiful forests or woodlands as "landscape forest reserves," and the trees in this area cannot be cut down without the Governor's permission. It was a protection of beautiful sceneries from capricious destruction.
>
> Next, in 1919 we had the Law of Preservation of Historic Sites and of National Monuments. By this law, the famous beautiful sights, as well as precious plants, animals, historic sites and old famous gardens, are protected under the control of the Important Cultural Properties Protection Commission. At present, the landscapes protected by this Commission are of small scope.
>
> The most important act was passed in 1931; that is the National Park Law. It designated twelve national parks up to 1937. After the war, in 1956, this law was amended and changed its title to "Natural Park Law." Up to this time, nineteen national parks and twenty quasi-national parks have been designated. The main purpose of this law is to protect the excellent landscapes and to promote their adequate utilization.
>
> In telling how they are chosen, it depends on six factors, one of which is natural landscape. The national park shall be of great scope, competent to rank as outstanding scenery of the world and simultaneously representative of Japanese landscape.
>
> We greatly regret that the dams of hydro electric power, large gas tanks and factories are constructed without considering our natural property of beautiful landscape . . . The damages of landscape are increasing

> every year, in spite of many means of protection and of the eager activities of the Nature Protection Association, Parks and Reservation Association, National Park Association, Urban Beauty Association, and so forth.
>
> This is obviously because of lack of appreciation of the natural beauty on part of the public. For the low living standard, temporary need may hinder the people from considering the value of the vicinity landscape. By such temporary need, our traditional disposition of love of nature would be easily surrendered.

Case Histories

1. The earliest and most notable battles in which the issue was clearly drawn between wilderness preservation and its utilization was the *Hetch Hetchy case* in California. The Hetch Hetchy Valley is a spectacular high-walled segment of the Tuolumne River on the western slopes of the Sierra Nevada Mountains. When Yosemite National Park was created in 1890, the Hetch Hetchy part and its surrounding area were designated a wilderness preserve. In the early 1900s, San Francisco started action that requested use of the valley as a reservoir site. Although the request was at first denied by the Secretary of the Interior, it was approved on May 11, 1908, apparently as an aftermath of the 1906 earthquake, when fire had devastated the city. Thereafter, a five-year struggle developed that encompassed administrations of Theodore Roosevelt and Woodrow Wilson. Two of the chief antagonists were Gifford Pinchot and John Muir.

> As to my attitude regarding the proposed use of Hetch Hetchy by the city of San Francisco . . . I am fully persuaded that . . . the injury . . . by substituting a lake for the present swampy floor of the valley . . . is altogether unimportant compared with the benefits to be derived from its use as a reservoir.
>
> . . . the fundamental principle of the whole conservation policy is that of use, to take every part of the land and its resources and put it to that use in which it will serve the most people.
> (G. Pinchot as quoted in Nash, 1967, p. 160, 171.)

On the other side, Muir had taken the position of the importance of aesthetics and the essential character of wilderness:

> . . . for everybody needs beauty as well as bread, places to play in and pray in where Nature may heal and cheer and give strength to body and soul alike.
>
> These temple destroyers, devotees of ravaging commercialism, seem to have a perfect contempt of Nature, and instead of lifting their eyes to the God of the Mountains, lift them to the Almighty Dollar.
> (Muir as quoted in Nash, 1967, p. 165, 160.)

In review of the public debate that was enjoined on this issue, it is a tribute to preservationists that they were able to stall implementation of the project for five years. Even Roosevelt, who earlier had allowed approval of the reservoir stated in his eighth annual message on December 8, 1908, that Yellowstone and Yosemite ". . . should be kept as a great national playground. In both, all wild things should be protected and the scenery kept wholly unmarred." Hundreds of newspapers throughout the country carried articles and wrote editorials about the Hetch Hetchy controversy and many leading magazines—such as *Colliers, Independent, Nation,* and *Outlook*—rallied to the cause of the preservationists. Key opposition was aroused by such groups as the Sierra Club and the Appalachian Mountain Club. In order to prosecute the case, the wilderness enthusiasts were obligated by law to combine their case under a single organization, which became the California Branch of the Society for the Preservation of National Parks. Although their arguments were that the dam and reservoir would desecrate the landscape, proponents of the project shrewdly pointed out that instead of spoiling the beauty, the man-created lake would enhance it and allow for walks and roads to the region, so that it could be used for public recreation in much the same manner as already developed European mountain–lake resorts. Gradually such staunch advocates of natural areas as Congressman William Kent of California became advocates for the reservoir. (Kent had given some of his property to the federal government and requested that it honor John Muir; this was done by presidential proclamation on January 9, 1908, designating the land the Muir Woods National Monument.) On September 3, 1913, the House passed the Hetch Hetchy bill 183 to 43, with 203 not voting. The bill was also passed by the Senate on December 6, with 43 for, 25 against, and 29 not voting. On December 19, Wilson approved it and in his statement declared that ". . . the bill was opposed by so many public-spirited men . . . that I have naturally sought to scrutinize it very closely. I take the liberty of thinking that their fears and objections were not well founded" (U.S. Congressional Record, 1913, p. 1189).

Although the preservationists lost this battle, they did not necessarily lose the war, because the debate had generated a large following, had educated the general public on the virtues of nature preserves, and was largely instrumental in stimulating sufficient national strength so that a wilderness enthusiast, Stephen T. Mather, could become head of the new agency formed by the National Park Service Act of 1916.

2. A second classic area with an unusual evolutionary concept and struggle is the *Minnesota border region* between the United States and Canada. Here Quetico Provincial Park, Canada, was established in April 1909 and was followed a few weeks later by President Roosevelt's creation of the Superior National Forest. This area has constantly been besieged by commercial interests, which coveted its fur, timber, minerals, hydroelectric power potential, and resort possibilities. In the early 1920s, roads were proposed into the heart of what is now the *Boundary Waters Canoe Area.* After a five-year fight, conservationists, canoemen, and the U.S. Forest Service saw the creation of the first Superior Primitive Area. The next threat came from a proposal to create a giant hydropower complex with seven dams that would flood the area to depths of 80 ft. An organization known as Quetico-Superior Council was formed to fight this proposal. For nine years the struggle continued, until in 1934 the International Joint Commission denied the power development application, and stated:

> The boundary waters referred to . . . are of matchless scenic beauty and of inestimable value. The Commission takes the position that nothing should mar the beauty of this last great wilderness (Olson, 1970, p. 30).

A totally new, unique, threat developed after World War II when real estate developers realized that the region had hundreds of landing areas suitable for float planes. Since the region still had thousands of acres of private land, these areas now became ideal potential summer homesites and fly-in resorts. This triggered a land development boom and Ely, Minnesota, became a major seaplane base.

> So swift was this development and so intense the traffic that the U.S. Forest Service and the President's Quetico-Superior Committee, backed by all major conservation groups, urged the establishment of an air-space reservation over the area. President Truman signed the Executive Order in 1949, the first airspace reservation ever established for the protection of wilderness. There were violations immediately, but the District Court, the Circuit Court of Appeals, and finally the U.S. Supreme Court upheld the order (Olson, 1970, p. 31).

The Ontario Department of Lands and Forests followed suit and established its own airspace reservation, and went further in the elimination of private structures, and banned mining from all provincial parks. Although in the United States the Multiple Use Act of 1960 makes clear that economic value alone is not to be used as an overriding judgment for use, and the Wilderness Act of 1964 gave official sanction that the area be preserved as wilderness, a new threat of mining is now present. Since the old mining law of 1872 permits anyone holding mineral rights to prospect, a group claiming surface mineral rights to 30,000 acres is testing this law. Thus this region of 1,063,000 acres, formerly known as the Superior Primitive Area, then the Roadless Area of the Superior National Forest, and now the Boundary Waters Canoe Area, has been placed in jeopardy once again.

3. A third example is *Mineral King,* a beautiful, secluded valley in a tract of national forest in California. It is nearly surrounded by Sequoia National Park and includes the magnificent Kern River back country of the High Sierra with its approaches to Mount Whitney. Mineral King got its name because it was formerly an intensive mining area. The existing claims have expired, and the area has returned to near wilderness. Although many conservationists assumed that this area would be added on the park, the U.S. Forest Service in 1949 issued a prospectus seeking private capital to develop the area for winter sports. This unusually picturesque area contains five major streams, 22 lakes, and an open alpine character with mountains that rise more than 4000 ft above the valley floor. The meadows and slopes contain a great variety of flowers mixed with cottonwood and aspen groves with adjoining glacially polished bedrock exposures.

The Disney Corporation won the bidding in 1969 for development and have planned a $35 million year-round resort. Plans call for more than 1 million visitors annually, which would provide four times the density as that in overcrowded Yosemite Valley. The facilities would include a huge complex of hotels, restaurants, parking areas, and more than 22 ski lifts, extending 4 miles into the mountains. Additional facilities would be necessary, such as a highway across the Sequoia National Park, and water storage, sewage treatment, and power generation. Conservationists wonder how much traffic the natural setting can stand before it is destroyed, since the thin soils are fragile and provide a serious challenge to the biota even under natural conditions.

> Not only has the Forest Service assumed that ecological problems could be overcome, it has also assumed that there were no real ecological limits on what it might do in the valley. The concepts of any natural carrying capacity for the basin is absent from its planning.
>
> What is worse, historically the Service has assumed that the bigger the project, the better. What was conceived of as a project costing a couple of hundred thousand dollars in the 1940s, jumped to a $3 million minimum project in February 1965, and to a $35 million project in late 1965 (McClosky and Hill, 1971, p. 174).

The Sierra Club invoked the Administrative Procedure Act, which accords judicial review to a person suffering a wrong because of an action by a federal agency, and sued the secretaries of Agriculture and the Interior. The Federal District Court in San Francisco granted a temporary injunction, but the Ninth Circuit Court of Appeals ruled that the club had no legal standing to sue. On April 1972, the U.S. Supreme Court set aside the suit by the Sierra Club to block the Disney Corporation from building the $35 million resort. In a 4–3 decision, the court ruled against the Sierra Club on a single procedural issue. While reaffirming recent decisions that injury to "aesthetic, conservational and recreational" values are as much a basis for suing as injury by economic loss, the Court held the individuals seeking judicial redress still must show they have suffered personal loss. Thus an organization cannot claim such damage or be a surrogate for the entire public. Justice Harry Blackmun suggested ". . . an imaginative expansion of our (Court) traditional concepts of standing in order to enable an organization such as the Sierra Club, possessed, as it is, of pertinent, bona fide and well-recognized attributes and purposes in the area of environment, to litigate environmental issues" (Coates, 1972, p. 115).

4. The controversy surrounding a proposed jetport in south Florida may be typical of battles that will be waged in the last quarter of the twentieth century. *Everglades National Park* was established in 1947 in 2035 square miles of the southern tip of Florida. Conservationists are convinced that it and adjacent regions to the north will become ecological disasters if the proposed plans for development nearby are consummated.

> The future health of Everglades National Park will be seriously affected by any major new international jetport built in south Florida, especially if it is located near the park. In 1968, the Dade County Port Authority purchased 39 square miles of land in the Big Cypress Swamp for a new jetport; the southern boundary of the property lay only 7 miles north of Everglades National Park. The National Audubon Society and other

> conservation organizations challenged the Port Authority at a series of hearings, and secured dramatic and favorable coverage of the jetport controversy in the national news magazines and on television. As a result of the efforts of these conservationists, the Department of Interior, the National Academy of Sciences, and a private group headed by former Interior Secretary Udall, investigated the jetport during the summer of 1969. All three produced reports that expressed alarm over the likely impact of the jetport on the National Park (Harte and Socolow, 1971, p. 192–193).

The assessments by Luna Leopold and the U.S. Department of Interior are typical of these reports:

> Development of the proposed jetport and its attendant facilities will lead to land drainage and development for agriculture, industry, housing, transportation, and services in the Big Cypress swamp that will destroy inexorably the south Florida ecosystem and thus the Everglades National Park ... [this] will take place through the medium of water control, through land drainage and changed rates of discharge. It will come about through decrease in quality of water both by eutrophication and by the introduction of pollutants, such as pesticides (Leopold, 1969, p. 11, 13).

> The south Florida problem is merely one example of an issue which sooner or later must be faced by the Nation as a whole. How are the diffused general costs to society to be balanced against the local, more direct and usually monetary, benefits to a small portion of the society? Concurrently, the society must ask itself whether the primary measure of progress will indefinitely be the degree of expansion of development, such as housing, trade, and urbanization, even at the expense of a varied and, at least in part, a natural landscape.

> ... the benefit to society from the maintenance of an ecosystem is of a different order than that due to the preservation of a few species.

> ... the training airport is intolerable ... because the collateral effects of its use will lead inexorably to urbanization and drainage which would destroy the ecosystem.... Elimination of the training airport will inhibit land speculation and allow time for formation of public awareness of environmental degradation which is the prerequisite for effective and practical action in the field of planning and land-use control.
> (U.S. Department of Interior, 1969, p. 150–153.)

At the present time the Port Authority has been restrained from continuing development of this particular site.

Case histories from other countries could add to such illustrations: for example, Stoddart (1968) and the Russian view on what constitutes the ecologic viewpoint (Sochava, 1971).

Reprinted from *Landslide Hazards Related to Land Use Planning in Teton National Forest, Northwest Wyoming* (1971)

Chapter V 21

Forest Land Use Implications

R. G. BAILEY

Effects on Sedimentation

Many streams descending from the mountain ranges in Teton National Forest have aggraded their lower reaches and have built alluvial fans flanking the ranges. The bulk of these alluvial deposits is doubtless an admixture of both glacial outwash and material derived from accelerated fluvial erosion and loss of stream volume accompanying post-glacial changes in climate. Stream channels which flow over these deposits meander and braid widely. The condition of these streams indicates a sediment supply too large to be transported with the available discharge. As a result, deposition of sediment load in the channel forces an alteration of course on practically an annual basis.

Many of these choked streams can be traced upstream to massive landslides which have effectively narrowed, or blocked for a time, the channel (fig. 62). At these points, sediment is introduced into the stream by erosion of the landslide debris alone or by diversion of the flow which increases erosional attack upon the opposite valley-side slope. At numerous locations, active movement is taking place as stream erosion removes material from the channel allowing landslide material covering the slide scar to creep and slide down into the channel. The effects on channel morphology of this process are dramatically illustrated where a relatively clear, stable stream passes through an area of active landslide encroachment to emerge highly charged with sediment and with a characteristic braided pattern.

Removal of debris produced by landsliding slopes probably produces a significant part of the total load carried by streams in the Teton region. As just one example, outflow from the lower Gros Ventre slide dam cut a channel nearly 300 feet wide and about 100 feet deep beneath the crest of the dam which in the process supplied over 3 million cubic yards of sediment to the river. Other direct evidence is lacking as to absolute yearly amounts. However, on similar slide-prone hillsides in the coast ranges of northern California, Kojan (1967) found that soil creep and landslides account for as much as 70 percent of the sediment reaching major streams. The practical importance of this relationship is clear because disturbance resulting from land use may aggravate these mass movements, thus increasing the delivery ratio. For example, in 1961, landslides from roads in the Cascade Range of Oregon produced average sediment concentrations of about 16 times expected from the pre-road relationship (Fredriksen, 1963). This is further emphasized by the fact that much of the landslide debris has yet to be removed and that millions of cubic yards of potential sediment are temporarily stored immediately adjacent to an effective transport mechanism. Lateral erosion by streams is keeping many of the landslides in an unstable condition and this situation will continue in the future.

As an example of the sediment production of unstable slopes, Brown and Ritter (in press) report that runoff from the Eel River basin in northern California stripped away 320 million tons of soil and rock from a 3,100-square mile area between 1957 and 1967. These documented erosion rates may be the highest recorded in the continental United States. Apparently, most of the sediment was derived from landslides and earthflows which are widespread throughout the basin. Many of the slides were probably produced or accelerated by logging and road construction.

Not included in the above estimate, of course, are the often considerable and long-term erosion and sedimentation caused by landslides. Kojan (in press) has described this relationship in northern California as follows:

> Landslides frequently cause aggradation upstream of their point of incursion into the stream channel, raising the channel out of its bedrock gorge, leading to lateral corrasion and spreading slope instability. Sediment derived by streams from the toes of active or dormant landslides may adversely affect fish habitat or lead to a permanent loss of reservoir storage capacity downstream.

Large bouldery deposits in channels within the study area illustrate that large volumes of water have periodically moved down these channels. Some of these flows probably are related to landslides which have blocked the steep valleys, the breaching of which yielded abrupt surges similar to that resulting from the lower Gros Ventre slide-dam failure. The effect of such catastrophic floods may be the triggering of periods of increased mass movement by oversteepening of slope profiles and removing slide debris from the channel. Tricart (1961) noted that a large flood in the French Alps activated a period of intensified soil creep and slumping.

Effects of Forest Management Practices on Slope Stability

In steep topography, the soil and rock mantle are in precarious balance with the physical factors of the environment, and any material change in the environment results in instability. It is only reasonable to expect that activities of man which upset this balance will also increase instability. Two of the principal methods by which man usually increases instability in mountainous terrain are by road construction and logging activities. Therefore, the practical problem faced by land managers is the decision to accept the consequences of disturbance of unstable slopes or to control the effects of these activities in order to minimize the occurrence of mass movement. Control may be done by application of direct methods of slope stabilization or avoidance of areas of known or expected instability.

It should be realized that not all landslides can be prevented. Probably, the exceptionally large landslides are primarily controlled by the rate of geologic erosion and are largely uninfluenced by activities of man, but smaller earth movements initiated by man's activities are, in turn, at least partially preventable or controllable by man. Included in this concept is the decline of the view of a landslide as an "act of God" and the growing realization that many landslides are initiated by "acts of man." Increasingly, the land manager is faced not only with the increased possibility of landslides, but also with greater economic losses and greater chances for injuries and deaths when landslides occur. In addition, he must necessarily be aware of the effect of management activities in unstable areas on the sustained yield of high quality water because of the increasing need for improved water management. Furthermore, he must consider the effect of management on landslide activity which tends to reduce the soil mantle. The importance of soil loss is apparent when its very slow recovery rate is considered.

One objective of this study was to provide an indication of where geologically unstable conditions occur. A second objective was to outline relative hazards which present definite built-in limitations to management practices in an area where improper land treatment can trigger a sequence of events requiring decades, or longer, to control.

Although numerous studies have been made to determine the effects of timber cutting and roadbuilding on hydrologic conditions leading to surface runoff and soil erosion (Packer, 1966), only a few research studies have been made to determine the effects of these practices on deep-seated slope failure. A review of the literature on this subject, however, reveals the overwhelming concensus which maintains that **forests do in fact play an important soil protective role and that clearcutting can promote not only erosion, but also landsliding** (Gray, 1969). Literally hundreds of commercial logging operations indicate that this conclusion is well founded. Too often, however, failure to recognize the limits of disturbance allowable beforehand in areas of inherent instability has led to seriously deteriorated water quality. **These findings highlight certain hydrologic principles and slope stability considerations that should be applicable in northwestern Wyoming.**

In the sections which follow, the disturbance allowable in the various unstable mapping units will be discussed. Included will be a consideration of where actual and potential hazards exist. A few suggestions for management based on field reconnaissance, photo interpretation, and a knowledge of the land-forming processes are given. These suggestions are greatly simplified and do not constitute a management prescription since there are many other factors, such as economics, road design and location, esthetics, silvicultural practices, etc., which affect the success of any management practice. An excellent discussion of guidelines for use in planning land management activities in geologically unstable areas is presented by the U. S. Forest Service (1966), Arnold (1963), Federal Water Pollution Control Administration (1970) and Kojan (in press).

The principles used in evaluating and preparing recommendations for each of the following areas are basic in approach to other areas in the mountainous West, having similar problems. All of the landslide treatments which improve the stability of an active or potential landslide mass do so either by reducing the activating forces which tend to induce movement, or by increasing the shearing resistance or other forces that resist the movement. It is apparent, therefore, that any treatment which accomplishes either of these two effects will be of some benefit in preventing or minimizing landslide movement. For any particular landslide, however, not all types of treatment will be equally effective or economical. The selection of the best method of treatment is an engineering problem, requiring the evaluation of many factors which are beyond the scope of this study. A discussion of various preventative treatments is presented by Root (1958).

Slump, Earthflow, and Landslide Deposits

Active slides are characterized by an excess amount of surface and subsurface moisture. This moisture is derived from precipitation which drains into the mass from the slump basin or depression areas on the hummocky surface (fig. 12). Lack of well defined surface drainage allows much of the surface water accumulated in the depression areas to infiltrate. Cracks in the surface resulting from recurrent movement also allow ingress of water. The resultant effect is the flow of subsurface moisture through the slide mass. When the soil and earth material becomes saturated, the natural internal friction and cohesive forces are reduced to the point that gravitational forces frequently cause movement. Excess moisture, coupled with incompetent bedrock, steep slopes, and fine-textured cohesive soils, greatly increases the probability that an increase in moisture may contribute to accelerated movement of the mass.

Under certain conditions, the presence of loose, deep soil (characteristic of disturbed areas) with high moisture content is conducive to forest growth. Such conditions constitute a much better growth medium than a shallow, well-drained soil over a very competent bedrock. Accordingly, many of these flows are covered by a dense growth of high quality timber. In other words, some of the best timber often occurs on highly unstable slopes. It seems reasonable that, because of timber values and the relatively gentle slopes in these areas, the desire to log these areas can be expected.

It is generally accepted that the forest growth has two stabilizing functions: (1) drying out of the surface layers, and (2) their consolidation by a network of roots. Landslides with deep-lying slide planes cannot be detained by vegetation, although in this case, too, it can partly lower the infiltration of surface water into the slope and thus contribute indirectly to the stabilization of the slide (Zaruba and Mencl, 1969, p. 147).

The hydrologic effect of forest cover on slope stability derives from the fact that trees transpire through their leaves and this, in turn, depletes soil moisture. Soil moisture depletion produces negative pore-water pressure which, as seen earlier, is conducive to slope stability. A forest can also intercept moisture either in the crowns of trees or in the ground litter. The implication of this process is summarized by Gray (1969, p. 14-15):

> Interception and transpiration tend to either mitigate or delay the onset of waterlogged conditions in a slope. All other things being equal, a forested slope might not reach critical saturation quickly nor exhibit such high pore water pressures (piezometric levels) after intense storms as a denuded (clear-cut) slope. This is somewhat equivalent to saying that forested slopes may be able to tolerate storms of greater intensity and duration than cutover slopes; i.e., before a critical failure condition is reached.

Bethlahmy's (1962) findings from a clearcut plot and adjacent forested plot in the Cascade Range of Oregon clearly support the view expressed above. These trends are schematically illustrated in figure 63. This behavior is by no means peculiar to Douglas fir forest in the Cascade Range. Bethlahmy cites several studies by other investigators which show similar trends in areas of widely differing climatic and soil conditions.

Clearcutting also has its own special influence on slope hydrology by affecting snow accumulation and melt. Maximum accumulation of snow is greater in the cut than in the uncut forest. Anderson (1956) reports it, for example, as 12 inches greater as a long term average in the central Sierra Nevada of California. Clearcuts that create large open areas not only store more snow than does a forest, but they have much more rapid melt. Anderson estimated a long term average of 16 inches of snow water left in the forest at 7,600 feet elevation on June 9 when all snow had melted in the large open areas. Clearcuts, therefore, would develop maximum saturation and higher pore-water pressures more quickly than uncut forest.

The importance of changes in the ground water regime in initiating and sustaining movement of unstable slopes is well known. Most studies, however, have emphasized changes due to seepage from artificial sources such as dams, irrigation systems, and septic tanks. Few studies have been made to determine the effects of timber cutting in producing movement by increasing the amount of available moisture in the slope. Of the work that has been done on the effects of logging, most have stressed the influential role of root systems as a strength factor on shallow-to-bedrock soils.

In recent years, most quantitative research has been dircted toward evaluating the effects of timber cutting on streamflow characteristics. Most of these have been concerned with some aspect of water field, with mass slope stability considerations almost completely lacking. Increased streamflow following clearcutting of mature forest has been well documented in numerous studies throughout the United States. Experiments of all types within the temperate zone of the world suggest increases in streamflow up to about 10 inches per year as a result of clearcutting forested watersheds (Hewlett and Hibbert, 1961). The effect of clearing is most pronounced at extremely low levels of discharge (i.e., base flow) and becomes negligible at high discharges. Consequently, the increases derive from reduction in transpiration and increased snow accumulation, and not from overland flow. Accordingly, increases in streamflow also reflect increases in ground water of at least the same order of magnitude. Presumably, such an increase could also be affected by an increase in precipitation. Schumm (1965, p. 786), based on a review of the literature on the subject, states the belief that a 10-inch increase in precipitation from that of the present was typical during Pleistocene glaciation. Others suggest that the increase may have been somewhat less than 10 inches. Discounting the effects of evaporation which would have been less during the Pleistocene because of cooler temperatures, it seems possible that removal of timber may temporarily produce roughly the same moisture regime in the slope as during the Pleistocene. Since many of the slump-earthflows were initiated during this moister period and probably ceased to move as the climate became dryer, it is reasonable to

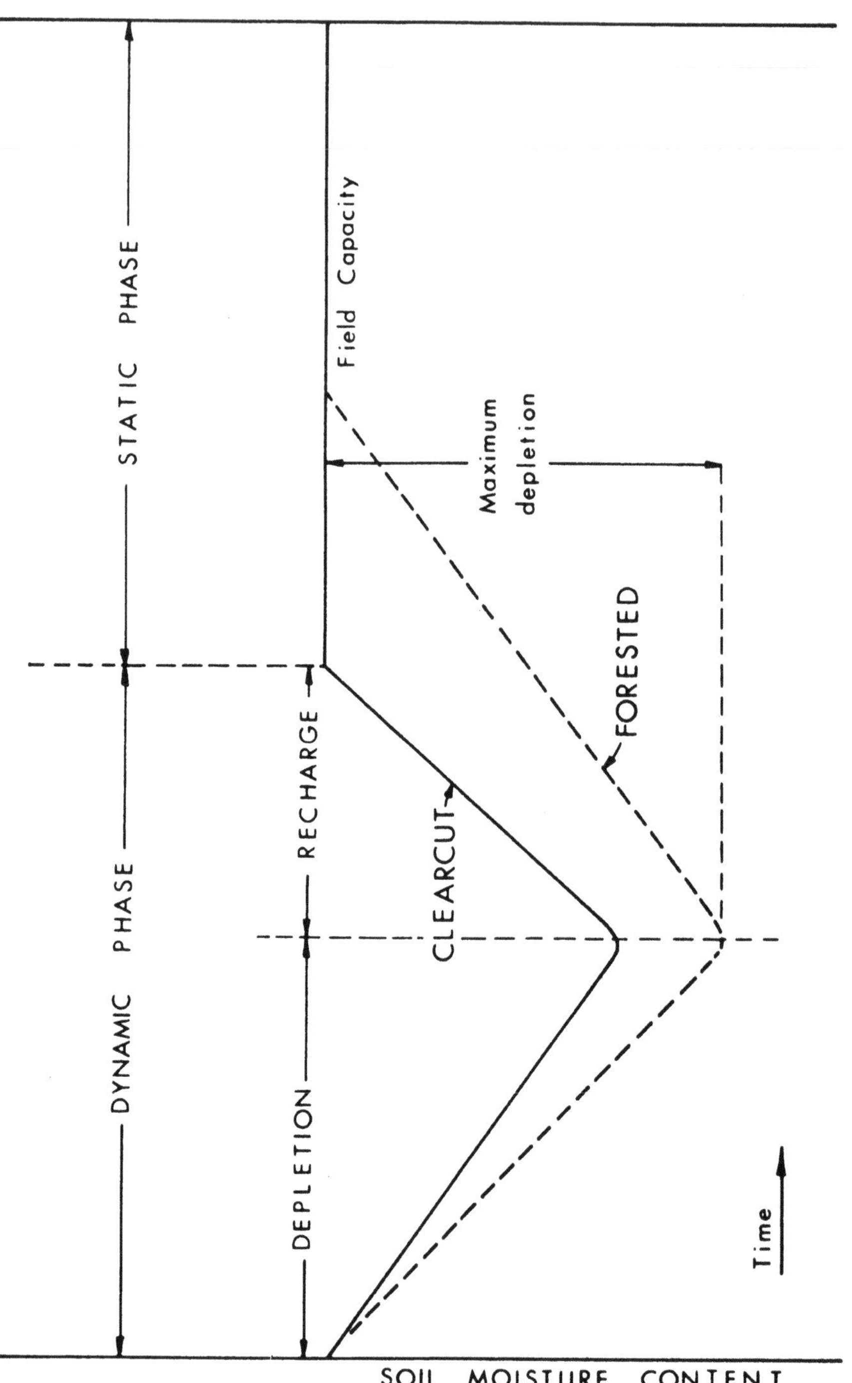

Figure 63.—Idealized representation of soil moisture cycle in forested and clearcut areas (adapted from Gray, 1969, fig. 4).

assume that increased moisture of comparable quantity may cause acceleration of movement.

The few instances where these effects were studied indicate that extensive clearcutting should be avoided, because it disturbs the stability of the slope as a result of the change in the surface and ground water system (U.S. Forest Service, 1966; Zaruba and Mencl, 1969; Gray, 1969). The roots of trees maintain the stability of slopes by mechanical effects and contribute to the drying of slopes by absorbing part of the ground water. The deforestation of slopes impairs the water regime in the surface layers. Reporting on the influence of vegetation cover in stalibilizing unstable slopes in southwestern Colorado, Varnes (1949, p. 12) states:

> Although the long cold winters leave only a short growing season, the abundant precipitation helps perpetuate a heavy cover of vegetation and aids in the reestablishment of cover over barren areas created by new roads cut or by recent landslides. The valley is blanketed with a thick growth of aspen and scattered evergreens. An undergrowth of ferns, grasses, and low brush forms a continuous retaining mat over all but the cliffed slopes. The abundant vegetation cover has apparently been effective in preventing earth movements under normal conditions, in spite of the adverse factors of weak component material and oversteepened slopes. Earth movements on the South Fork of the San Miguel River generally occurred where the vegetation cover or the slopes have been disturbed by man or where there was abnormal supply of water.

In the Fishlake National Forest in Utah, defoliation of aspen has been performed experimentally to increase water yield (Max Robinson, oral communication, 1969). Following defoliation, the landslide deposits on which the aspen was growing became reactivated and are currently moving downslope.

Most of the problems involving stability of active slides and flows are associated with the construction of unbraced cuts for roadways. Experience has shown that excavation of even a very shallow cut with standard slopes (1-1/2:1) may cause the soil to move toward the cut, and the movement may spread to a distance from the cut equal to many times the depth. The literature on this subject is extensive. For an introduction, see Wagner (1944), Varnes (1949), and U.S. Forest Service (1966). The foregoing descriptions have pointed out locations where roadcuts have triggered accelerated movement making road stabilization virtually impossible (figs. 64 and 65). Road cutbanks and fill slopes, because of the excessive moisture and high clay content of the soil, often produce numerous failures by removing lateral support and intersecting the subsurface flow of moisture.

Recommendations

When a landslide stops moving, equilibrium between gravity and the slide material's shearing resistance is established. Removal of the toe or increase in ground water may upset the equilibrium, thus permitting repeated movement of the slide.

1. **Active slump, earthflow, and landslide deposits.** — Accordingly, logging, with its accompanying roads and skid trails, should be avoided on active landslide deposits because an increase in subsurface moisture can be expected to accelerate the movement. Cuts in landslide material are always troublesome because the material came to rest as soon as its factor of safety with respect to further movement became equal to unity. If these locations cannot be avoided, construction should be preceded by radical and permanent drainage. These areas are identified on the landslide hazard map (pl. 1) as Qsa.

2. **Inactive (?) slump, earthflow, and landslide deposits.** — Some inactive landslides, after thorough onsite investigation, may be logged without causing excessive slope movement. However, road and cutting units should be laid out so as to not modify the existing hydrology of the slide. Many old slides have developed defined drainage channels that often carry water throughout the year. If roads are constructed through these areas, it is imperative they do not restrict the natural flow of water and that all water be returned to the natural channel below the road. Road alignment should closely follow

Figure 64.—Disturbance caused by road construction along U. S. 187-189 in Hoback Basin probably caused accelerated movement of this pre-existing slide.

Figure 65.—Aerial view of a road failure in an active slide area. Direction of movement is toward left of photo. Location is about 100 feet east of Togwotee Pass on U. S. 26-287.

the surface topography and cuts and fills, which often impede the natural flow of subsurface moisture, should be kept at a minimum. Switchbacks should be avoided if possible, due to the extreme hazard of cut and fill slope failures. If switchbacks are used, various measures such as internal and surface drainage and minimizing the size of cuts and fills should be applied to reduce the potential for massive earth movements. Roadbase stabilization and surfacing needs are high because the soil of many landslide areas has poor trafficability and becomes very sticky and slick when wet (fig. 66). These areas are identified on the landslide hazard map (pl. 1) as Qsi.

3. **Active frost rubble, talus, and rock glaciers.**—Most fresh talus and associated deposits are so unstable that slope undercutting, or dropping of a large rock fragment from the cliff above, will easily set off sliding of the surface particles. When saturated by summer cloudbursts, enormous rock and mudflows ensue. These areas should be left undisturbed. They are identified on the landslide hazard map (pl. 1) as Qfa.

Oversteepened Lands

Oversteepened slopes (greater than 30°) commonly are characterized by numerous rock outcrops, shallow soils, avalanche paths, and colluvial and talus deposits. Where concentrated movement of soil and rock during both wet and dry periods has occurred, erosion chutes are found. The dominating forms of movement are rockfalls and debris avalanches, although mudflow also plays a significant role. Roads constructed in these areas tend to have high cutbanks, long fill slopes, and poor alignments. Because of the deeply incised drainages on the steep terrain, stabilization of cutbanks and fill slopes by mechanical or vegetative methods seldom is achieved. Recurrent and frequent slides produce high maintenance needs. Stream runoff peaks tend to be high and are charged with debris derived from slides in the natural channels or above cutbanks. As a result, roadside ditches and culverts often become plugged causing failure of the road structure.

Illustrative of this are landslide frequencies on oversteepened slopes at the H. J. Andrews Experimental Forest in western Oregon. Dyrness (1967) found that 72 percent of the mass soil movement events were associated with roads, 17 percent in clearcut logged areas, and 11 percent in undisturbed areas. Although 72 percent of the mass movements occurred in connection with roads, only 1.8 percent of the total area of the forest was in road right-of-way. The most common single type of mass movement on the Experimental Forest was the road fill failure. Backslope failures were most common in areas where mass movements had occurred repeatedly in the past. Backslope failures were also instrumental in fill slope failures where slump material had blocked the interior drainage ditch, thus saturating the fill slope. In other cases, culverts carrying perennial or intermittent streams failed either because of inadequate size or, more probably, because they became blocked by debris. Dyrness (1967) concluded that disturbance may cause some small and, by itself, insignificant change which is nonetheless sufficient to upset the tenuous equilibrium and trigger mass soil movement.

Other problems related to development of oversteepened lands derive from the nature of large areas of surficial deposits which characterize these slopes. Along the base of slopes and in valley bottoms are accumulations of coarse, unconsolidated debris which has momentarily come to rest after movement by gravity. None of this debris is stationary. If it is mixed with ice or saturated with water, the whole mass may slowly flow downslope as a rock glacier (fig. 48). Talus slopes, a by-product of rapid frost disintegration of exposed cliff faces, are formed by rockfall (fig. 47). Glaciated valleys in places not dominated by talus slope development are usually plastered by unconsolidated glacial deposits (fig. 67). This material commonly is precariously perched along the steep valley walls. Potential shear zones exist at the discontinuity between the morainal till and the underlying bedrock; and, when disturbed, large debris avalanches are likely to develop. In local areas, remnants of glacial outwash deposits and old stream gravels form terraces well above the present stream levels. Road

Figure 66.—View or road traversing a slide area in Ditch Creek, showing ruts caused by insufficient bearing strength of the materials in these areas, especially when wet. Ruts are 12 inches deep and 6 to 8 inches wide.

Figure 67.—Deep glacial trough of Flat Creek canyon in Gros Ventre Range, air oblique view south. Cirques are visible at head of drainage; former direction of ice flow toward observer. Valley walls are veneered by unstable talus and glacial till. Timber along right wall has partially stabilized an old talus slope.

construction through these deposits has resulted in considerable cutbank sloughing with continual road maintenance problems (fig. 68).

Although it is possible to anticipate the performance of these materials in a general way by taking full advantage of the existence of precedents, many problems are encountered when trying to predict the performance of a specific slope. This is due, in part, to physical properties of the materials which are likely to change over short distances in every direction on account of erratic variations in the degree of weathering. Therefore, the consequences of undercutting existing slopes, the effects of seepage toward such slopes, or the degree of stability of slopes to be produced by excavation can only be expressed in general terms. An example of this difficulty is provided by a recent slide on the H. J. Andrews Experimental Forest. Although there was ample evidence of old slides on the timbered slopes of this area, Fredriksen (1963, p. 4) concluded:

> In spite of precautions taken in road location and construction, it appears that the presence of the road and culvert triggered this mass soil movement. Even a "full benched" road inevitably upsets the balance of forces within the soil mantle. A decision to build a road in an area of unstable topography constitutes a calculated risk no matter how well the road is designed and constructed to minimize damage.

Another example of the harmful effects of roads on mass slope stability is provided by recent flood damage in the Pacific Northwest. Rothacker and Glazebrook (1968, p. 10) reported that roads were involved in over 60 percent of all major storm damage reports. Mass soil movements were listed as the primary cause of damage. In many cases, roads not only were damaged by massive movements of soil onto road surfaces or road failure itself, but they also were the primary cause of the mass soil movement. Hillside road construction often undercut naturally unstable slopes or deposited excessive soil material on slopes already at the "angle of repose." Most severe damage occurred where unstable soil and rock formations were not avoided or where special construction practices were not employed in unstable areas which could not have been avoided. Water that eroded road surfaces and stream water that undercut roads by streambank erosion or channel changes were other important factors causing road damage. In many cases, roads had been constructed in or were crowding the natural flood channel of the stream. Culvert failure, another common cause of road damage, was most often related to plugging with debris. In most cases, the hydraulic capacity of the culvert was sufficient to carry the volume of water as long as it remained unplugged.

Slope gradient alone characterizes these areas as primed for mass movements of all types. The influence of slope is clearly indicated by Dyrness (1967) who found that only about a sixth of the mass movement events occurred on slopes with a gradient of less than 24°. Apparently, road related slides are also positively correlated with slope gradient as revealed by the frequency of slides in Zena Creek watershed on the Payette National Forest in central Idaho. Gardner et al. (1969) reported that 96 percent of road failures occurred on slopes of 30° or greater. They also found that all but two of these failures were on fill slopes and that 73 percent occurred in a watercourse or draw. They concluded that all of the mass slope failures in the Zena Creek area were attributable to or aggravated by the following factors: (1) location of the road on too steep terrain; that is, slopes of 35° or greater, (2) inadequate control of surface and subsurface water flows; (3) too much emphasis on alignment, thereby not conforming sufficiently to the topography; (4) fill slopes that were too steep; that is 35° or greater; (5) cut and fill construction on sections where a full bench road without fill was necessary for stability; (6) failure to remove all of the logs and debris from the fill; and (7) failure to compact fill slopes to densities consistent with design standards and material properties.

Where numerous slides and falls are found on oversteepened slopes in the absence of well-developed cohesive soil, they indicate problems of larger magnitude. Areas between slides commonly support extensive forest

Figure 68.—Debris slide bordering U. S. 26-89, 2.5 miles southwest of Hoback Junction. Light material is unconsolidated river gravel overlying dipping strata of the Phosphoria formation (Permian). Gravel terrace resulted from damming of Snake River by a landslide.

cover, indicating that they have remained stable for a number of years despite the action of forces tending to induce sliding. Indications are that slide resistance is imparted by tree rooting through the soil and into cracks in the underlying bedrock (Swanston, 1967a). Destruction of this rooting system would greatly increase the susceptibility of these slopes to slide.

Although timber harvesting by itself generally does not increase the occurrence of mass movement to the extent that roads do (Dyrness, 1967), there is evidence from other areas which suggest that under certain conditions logging triggers sliding. Bishop and Stevens (1964) indicated that clearcutting of old-growth forests apparently accelerated debris avalanche and flow occurrence in southeastern Alaska (fig. 69). These mass movements developed on steep slopes in a glacial till soil, apparently triggered by high pore pressures developed during periods of intense rainfall. They reported the number and acreage of slides increased by a factor of 4.5 in the 10 years since logging. Loss of strength reflecting death and gradual deterioration of root systems after clearcutting was concluded to be an important factor in accelerating mass movement. It is estimated that approximately 31 percent of a mile-square clearcut in Maybeso Valley in southeastern Alaska has been affected by these forms of erosion since 1959 (Harris, 1967).

Similar findings in southern California indicate the important role of dense deep-rooted vegetation in binding thin rocky soil in place on slopes whose gradients equal or exceed the angle of repose. Rice et al. (1969) found that the conversion of brush to grass, often initiated after wildfires have partially removed the brush, frequently is followed by a substantial increase in slope instability. Following heavy rainstorms, four times more debris slides occurred on areas converted to grass than on deep-rooted brush areas. On some watersheds up to 10 percent of the area was stripped of soil and vegetal cover. They estimated that conversion from brush to grass resulted in about a sevenfold increase in the volume of debris slide erosion since 1965.

Another study in timbered areas of the Wasatch Mountains of Utah also points out the effectiveness of vegetation for retarding slides of the "surface-sliding type." Croft and Adams (1950, p. 2-3) concluded that before modern-day land use "landslides . . . were . . . rare, if not entirely absent, in this drainage basin since the present mineral mantle developed on its slopes." They attribute the recent occurrences of landslides largely to loss of mechanical support by root systems of trees and plants, chiefly by timber cutting and burning and, to some extent, by excessive livestock grazing. In line with these observations they concluded further that "young second-growth timber . . . falls far short of providing the root support that its much larger and older predecessors did."

Another possibility for loss of shear strength is that the soil is locally compacted under the load of trees, thereby increasing shear strength. Logging — and local unloading — may cause "decompaction" by shrinking and swelling with soil moisture variations causing lowered shear strength (Bishop and Stevens, 1964, p. 14).

Logging debris or slash may also create a problem if it is permitted to remain in the stream channel or if it reaches the stream channel as part of a landslide. A great increase in erosive power occurs when this debris is brought into suspension by earthflows in the channel (Bishop and Stevens, 1964). The effect is similar to that produced by debris avalanching, except that it may be amplified by the increase in available debris.

Recommendations

Evidence from the research reviewed above, as well as from many commercial timber-harvest operations, suggests some generalizations about management planning activity on these kinds of landscapes.

1. **Active debris avalanches and mudflows.** —Probably the most practical and direct management policy, at present, is avoidance of areas of maximum slide susceptibility. The lower limit of internal friction angles for soils commonly found on these slopes is about 35°. Slopes with gradients equal to or greater than this angle are highly susceptible to sliding, particularly if they are severely disturbed.

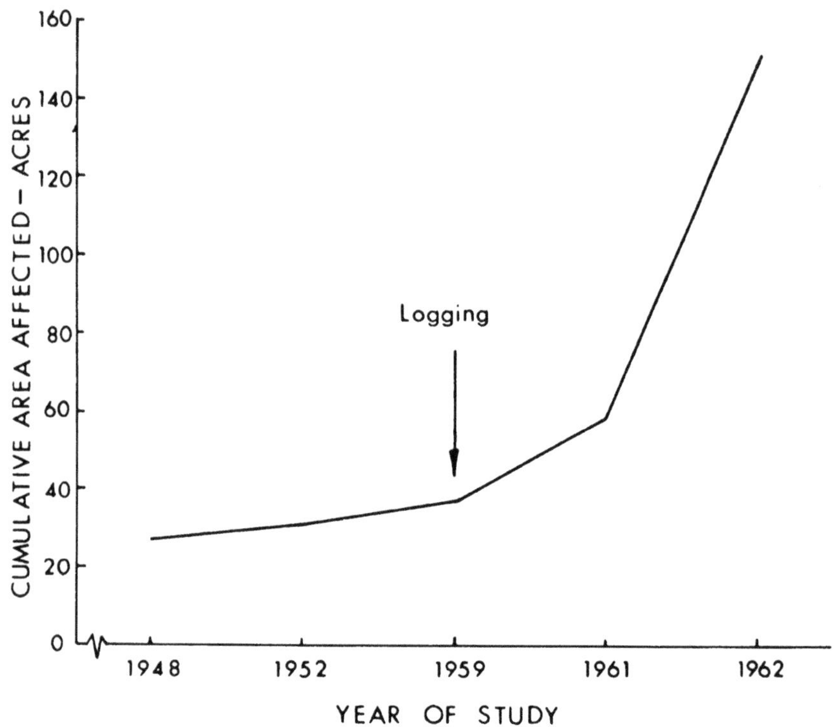

Figure 69.—Acreage of slides before and after logging in Maybeso Creek, Alaska (adapted from Gray, 1969, fig. 2).

With this information, general areas of maximum instability can then be identified. On aerial photos, these areas can be delineated by the occurrence and concentration of recent and old debris avalanche and flow scars. Slopes characterized by such features indicate that mass-wasting processes are proceeding at a rate more rapid than vegetation can heal the slide scars. Under these conditions, the slope may be classified as active. Disturbance caused by logging can only accelerate the process of active removal by sliding. Therefore, the total vegetative cover should be maintained. These active areas are identified on the landslide hazard map (pl. 1) as Qca.

2. **Inactive (?) debris avalanches and mudflows.**—On oversteepened slopes which have achieved some degree of stability, as evidenced by a relatively low frequency of recent movement, the potential energy available for movement, due to the slope gradient, is still very great. In many instances, the basal sapping mechanism has been removed. Meanders have migrated downstream, for example, or when the stream after shifting its channel, commences vertical downcutting, so that the old flood plain becomes a terrace. Under these conditions, with no removal of mass-wasting material from the base, slope angle decreases until the climatic and botanical conditions permit a plant cover to stabilize the slope. These slopes exist under a state of "conditional stability." Disturbances caused by road construction and timber harvest may be of sufficient magnitude to initiate sustained movement similar to that witnessed on active slopes. These areas of imminent slide hazard, lying above a critical angle of stability, should be given special management considerations. Careful evaluation should be made before project action is taken. These areas are identified on the landslide hazard map (pl. 1) as Qci.

Since roads are so often an important factor in causing mass movements in these areas, the problem is to determine means of minimizing their effect. Perhaps the most obvious means of minimizing their effect is to reduce road mileage to an absolute minimum. In steep, mountainous terrain, this may be done by the use of skyline and possibly, balloon logging methods. In many areas, it is possible that improvement in road location may appreciably reduce the frequency of mass soil movements. Unstable soils and landforms should be identified, and the route selected should avoid these areas wherever possible. Road location through these areas should ascend the more stable sections, irrespective of grade, to reach the bench landscape areas. In addition, improvements in road design, construction, and maintenance may contribute substantially to increased mantle stability. Modification of waste handling to avoid sidecasting on steep slopes and provision for adequate road drainage are two of the more important means to minimize mass movement hazards where avoidance of unstable areas is not feasible.

Potential Landslide Areas

Once landslides are located, it serves as a warning that the general area has been unstable in the past, and that new disturbance may start new movements (fig. 70). However, many slides are too small to be readily detected in small-scale photography or, where visible, to be practically mapped to scale. In addition, while many studies have been made on mass movement phenomena, it is generally not possible to effectively predict the exact location of future mass movements or what precise conditions are necessary to reactivate old movements. Recognition of existing landslides, although important, is not sufficient. A geologically ancient landslide may now be quite stable, so far as being affected by proposed construction. On the other hand, the excavation or loading involved in the construction may induce land movement even where there is no evidence of previous landslides. This is due, in part, as Sowers and Sowers (1951, p. 28) point out:

> In most cases, a number of causes [for landslides and flows] exist simultaneously, and so attempting to decide which one finally produced failure is not only difficult but also incorrect. The final factor is nothing more than a trigger that set in motion an earth mass that was already on the verge of failure.

At best, determination of areas with grossly analagous conditions to the movements in-

Figure 70.—Old debris slides on southfacing slope adjacent to Hoback River near Camp Davis; U. S. 187-189 in foreground. Three separate slides with concave scars and convex depositional elements can be identified. Minor slope failures such as these indicate distressed conditions and usually presage problems of much greater magnitude when the slope is disturbed.

volved, seems feasible at present. Presumably, such areas have a relatively high potential for movement under the proper conditions of moisture and stress.

On the basis of reconnaissance information, areas were identified in which there are special conditions conducive to mass movement. These vulnerable areas, in general, are:

1. In fine-textured or incompetent materials.
2. In areas having slopes beyond the common angle of stability of the material in that area.
3. In rocks whose bedding planes dip steeply with the slope.
4. In areas characterized by accelerated stream channel downcutting (basal sapping) and,
5. In areas where thick layers of competent rock overlie a plastic substratum.

Identification of these areas was made by reference to their observed performance, geologic maps, and to topographic expression which gave a recognizable homogenity to the unit (fig. 71).

Each of these units, together with the mass movement features within them, poses relatively distinct problems to the land manager, particularly from the standpoint of mass movement susceptibility. Since the geologic and physiographic conditions which produced massive slides and flows in the past are still present, mass movment phenomena can be expected to continue in vulnerable areas. Generally, it is more feasible to say where landslides are likely to occur than to predict when they will occur. Nevertheless, many potential slope failures lack only one ingredient — a triggering force. Such a force can be supplied artificially by timber removal and/or road construction (fig. 72). As with actual slides and flows, the potential for accelerated movement greatly increases with moisture content; so does the potential for slope failure increase with moisture content in the adjacent vulnerable areas. Disturbance caused by logging may be all that is necessary to increase moisture and trigger mass movements.

Recommendations

These areas are where excessive subsurface moisture; steep slopes; clay-textured, plastic soil; easily weathered bedrock; and rapid downcutting have produced a high frequency of slope failures.

1. **Areas built up of weak rocks and/or undercut slopes liable to sliding.**—In any area of inherently low stability, especially where slides are known to be prevalent, the development of roads and timber harvest should be preceded by thorough investigation. Included should be detailed soils-hydrologic surveys and engineering studies to provide information on the possible effects of the proposed disturbance. In this case, the limits of the potential incipient landslide are necessarily unknown, in contrast to slide analysis in which an active slide of definite extent already exists. These areas are identified on the landslide hazard map (pl. 1) as Qsp.

Man is not capable, nor is money available, to study in detail to guarantee the stability of all the slopes along most proposed disturbances. As a general rule, the amount of investigation that is warranted is a function of the landslide susceptibility of the surrounding country. Studies of existing landslides in the region are thus warranted in order to determine their geologic setting and their causes. If similar conditions are present along the proposed disturbance, it can be assumed that slides may occur there, too.

After a general knowledge of the environment has determined the area to be susceptible to landsliding, an onsite investigation is in order. Within these susceptible areas, special attention should be given to the slopes, changes in slopes, and their relationship to the different materials involved. Cracks and other evidences of motion, as well as all sources of water, should be noted. The structure of the underlying bedrock, as well as the depth of overburden should be determined carefully.

Evidence of soil creep and of "stretching" of the ground surface, should also be sought. Stretching is here distinguished from soil creep because it indicates comparatively deep-

Figure 71.—Air oblique view west, showing the Gros Ventre River valley and lower Gros Ventre slide. Slopes on the south side of the valley have a high mass movement potential as evidenced by past history of massive slope failures. Rocks of similar lithology dip steeply toward the valley where the river has downcut into them leaving them without support.

Figure 72.—Timber access road cutting through an area of high mass movement potential near upper Hoback Canyon. Failure of cutbank was apparently caused by removing lateral support from a slope with excessive subsurface moisture and plastic soil.

seated movement whereas soil creep is of surficial origin. The phenomena of stretching is most commonly observed in non-cohesive materials that do not form or retain minor cracks readily. The best evidence of stretching consists of small cracks that surround or touch some rigid body, such as a root or boulder, in otherwise homogeneous material. These cracks form because the tensional forces tend to concentrate at or near the rigid bodies.

Vegetative patterns should not be overlooked. Locally, a small difference in soil moisture condition is often detected by a corresponding change of vegetation. A detailed study of such local changes is very helpful in landslide investigations. For instance, wet vegetation, represented by dark spots or "tails" is a clue to seepage in slopes. Aspen trees are commonly good indications of wet ground conditions.

Slopes which are covered by dense vegetation may conceal numerous small slumps commonly referred to as "cat steps." The head region may remain greener than surrounding areas because of the swampy conditions. Occasionally, these slump basins are occupied by a small pond and aquatic vegetation. The foot region (which is a zone of uplift) is marked by seeps, springs, and marshy conditions. Tilted trees commonly are found on the toe. The most common road construction problem through these areas derives from undercutting the slump toe on the downfacing slope and thus intercepting the water table. Roads constructed across the relatively flat slump basin often pond water unless adequately drained.

Recognition of potential landslide conditions where no apparent landslide movements have yet taken place, may be facilitated by exploratory boring, coupled with shear measurements in the laboratory. Such a combined field and laboratory investigation, backed by at least general knowledge of the underlying rocks, should reveal the soil profile and ground water conditions along a proposed route even where surface features alone do not provide sufficient warning. It must be remembered, however, that there are some rather severe limitations on the applicability of shear measurements to landslide problems. In addition, the theory of the stability of slopes can be used to absolute advantage only in those rare instances when a cut is to be made in a

Table 7.—Area of ranger districts by landslide type, Teton National Forest.[1]

Landslide Type	Acreage in Type[3]	Ranger District[2]			
		1	2	3	4
Stable	483,754 (42.6)	102,790 (41.1)	100,799 (51.4)	202,820 (51.4)	77,345 (26.3)
Qsi	91,420 (8.1)	6,710 (2.7)	18,258 (9.3)	30,832 (7.8)	35,620 (12.1)
Qci	40,435 (3.6)	18,457 (7.4)	3,518 (1.8)	2,928 (0.7)	15,532 (5.3)
Qsp	335,356 (29.6)	54,279 (21.7)	54,003 (27.5)	87,752 (22.3)	139,322 (47.3)
Qsa	65,752 (5.8)	3,413 (1.4)	14,739 (7.5)	40,983 (10.4)	6,617 (2.2)
Qca	72,628 (6.4)	39,588 (15.8)	4,835 (2.5)	12,350 (3.2)	15,855 (5.4)
Qfa	45,339 (3.9)	24,622 (9.9)	0 (0.0)	16,622 (4.1)	4,095 (1.4)
TOTAL	1,134,684	249,859	196,152	394,287	294,386

Source: Compiled by writer.

[1]*Determined from the stability map by acreage grid.*

[2]*District symbols: 1—Jackson; 2—Buffalo; 3—Gros Ventre; 4—Hoback.*

[3]*Parenthesis denotes percent of total area.*

fairly homogeneous mass of soft or medium clay (Terzaghi and Peck, 1967).

Summary of the Teton Land Stability Situation

The area of each ranger district by landslide type is given in table 7. For land use planning purposes, land of each landslide type causing problems was further grouped into the following broad land stability classes (table 8):

Those areas which were not classified are those which are stable and suitable for land use development with existing technology.

Table 8.—Qualitative summary of unstable land.

Landslide Type	Degree of Stabilization	Stability Class	Land Use Limitations
Qsa, Qca, Qfa	Active (Rapid movement)	Highly Unstable	**Unusable;** unsuitable for logging, road construction, or other soil disturbance.
Qsi, Qci, Qsp	Inactive (Slow movement)	Slopes of Questionable Stability	**Potentially usable;** May be suitable for land use development with improved technology and with more detailed land data to describe limits of use.

The 1,134,684 acres of land in the Teton National Forest includes 650,930 acres of unstable land. Of the unstable, 183,719 is highly unstable. Severe limitations make these unsuitable for logging, road construction, or other soil disturbance and therefore have been classed as unusable. Nonetheless, many areas have good capability for grazing, wildlife, and for extensive recreational use. The remaining 467,211 is classed as potentially usable, that is, it probably could be used with improved technology and with more detailed land data to describe limits of use (table 9). Careful management or restricted use are in order in these areas.

Table 9.—Teton National Forest land area by broad land stability class.

Land stability class	Area in acres	% of total area
Stable land	483,754	42.6
Unstable		
Potentially usable (Qsi, Qci, Qsp)	467,211	41.2
Unusable (Qsa, Qca, Qfa)	183,719	16.2
TOTAL	1,134,684	100.00

Source: Compiled by writer.

Needs and Priorities in Further Research

The information reviewed here points out several shortcomings of previous research concerned with the effects of timber harvest on slope stability. The most serious shortcoming, mentioned earlier, is that only a few studies have had the assessment of the effect changes in ground water regime have on slide susceptibility. Consequently, there is little quantitative data to aid in the prediction of how much disturbance can be allowed without affecting the stability of the slope. Research should be directed toward increasing our knowledge of the hydrology of mountain slopes, the factors involved in the inherent stability of the slope, and how forest management practices alter the relationship.

In addition to the search for new and better methods of alleviating slides, research should also be increasingly directed toward developing criteria for the recognition of potentially unstable areas. An understanding of the geology is needed in planning what can be done to stop such landslides before they start, as well as in the planning stages of any major development project, so that areas in which conditions favor earth movement can be avoided, if at all possible. If construction in such areas cannot be avoided, a complete understanding of the geologic and hydrologic conditions may help in controlling the activating causes before slides occur or become serious.

For adequate planning, there is a need for information on the probability of the occurrence of slides of a given size and some idea of the lands affected. Since most slides are associated with periods of excessive moisture, there is need for forecasting the frequency of slide-producing climatic events. Basic data on temperature, precipitation, soil moisture, streamflow, and snowmelt will be required in developing a forecast system for high elevation regions.

Before information of this kind can be used in developing management plans, there must be a better understanding of the processes of landscape development. Underlying the successful application of methods to prevent and control landslides, there must be a better understanding of fundamental processes which operate on steep mountain slopes.

There is need within the study area to understand the basic physical properties of the soils or rocks that are involved in landslides to provide a basis for prediction of their behavior under conditions which change the local stress conditions or ground water conditions. Much more research on the basic properties of different materials as related to different environments is needed before more precise performance predictions can be made.

Current knowledge of both the role and the occurrence of subsurface water in relation to landslides is highly inadequate. The influence of hydrostatic pressures on slope stability is only known in general terms, as are the quantitative effects of water on the mechanics of failure. Further research on methods for building roads and timber harvest which will not impair the subsurface hydrology of unstable areas is obviously desirable. Drainage remains of prime importance in improving stability, but little is known as to how much disruption in drainage a slope can tolerate before failure occurs.

Literature Cited

Albee, H. F., 1968, Geologic map of the Munger Mountain quadrangle, Teton and Lincoln Counties, Wyoming: U.S. Geol. Survey Map GQ-705, Scale 1:24,000.

Alden, W. C., 1928, Landslide and flood at Gros Ventre, Wyoming: Am. Inst. Mining Metall. Eng. Trans., v. 76, p. 347-360.

Anderson, H. W., 1956, Forest-cover effects on snowpack accumulation and melt, Central Sierra Snow Laboratory: Amer. Geophys. Union Trans., v. 37, p. 307-312.

Anderson, H. W., Coleman, G. B., and Zinke, P. J., 1959, Summer slides and winter scour — dry-wet erosion in southern California mountains: Pacific SW Forest & Range Expt. Sta. Tech. Paper 36, 12 p.

Antevs, E. V., 1948, Climatic changes and pre-white man, *in* A symposium on the Great Basin, with emphasis on glacial and post-glacial times: Univ. Utah Bull., v. 38, p. 168-191.

Arnold, J. F., 1963, Road location to retain maximum stability, *in* Forest watershed management symposium: Ore. St. Univ. p. 215-224.

Atwood, W. W., 1909, Glaciation of the Uinta and Wasatch Mountains: U.S. Geol. Survey Prof. Paper 61, 96 p.

.... 1915, Eocene glacial deposits in southwestern Colorado: U.S. Geol. Survey Prof. Paper 95-B, p. 13-26.

.... and Mather, K. F., 1932, Physiography and Quaternary geology of the San Juan Mountains, Colorado: U.S. Geol. Survey Prof. Paper 166, 176 p.

Baker, R. F., and Chieruzzi, R., 1959, Regional concept of landslide occurrence: Highway Research Board Bull. 216, p. 1-16.

Beaty, C. B., 1956, Landslides and slope exposure: Jour. Geology, v. 64, p. 70-74.

Behre, C. H. Jr., 1933, Talus behavior above timber in the Rocky Mountains: Jour. Geol., v. 41, p. 622-635.

Behrendt, J. C., Tibbetts, B. L., Bonini, W. E., and Lavin, P. M., 1968, A geophysical study in Grand Teton National Park and vicinity Teton County, Wyoming: U.S. Geol. Survey Prof. Paper 516-E, p. E1-E23.

Bethlahmy, N., 1962, First year effects of timber removal on soil moisture: Int. Assoc. Sci. Hydrol. Bull., v. 7, p. 34-38.

Bishop, D. M. and Stevens, M. E., 1964, Landslides on logged areas in southeast Alaska: U.S. Forest Serv. Res. Paper NOR-1, 18 p.

Blackwelder, E., 1912, The Gros Ventre slide, an active earth-flow: Geol. Soc. America Bull., v. 23, p. 487-492.

.... 1915, Post-Cretaceous history of the mountains of central western Wyoming: Jour. Geology, v. 23, p. 97-117, 193-217, 307-340.

.... 1926, Earthquakes in Jackson Hole, Wyoming: Seismol. Soc. America Bull., v. 16, p. 196.

.... 1935, Talus slopes in Basin Range Province (abs.): Geol. Soc. America, Proc. for 1934, p. 317.

.... 1942. The process of mountain sculpture by rolling debris: Jour. Geomorphology, v. 5, p. 325-328.

Braddock, W. A. and Eicher, D. L., 1962, Block-glide landslides in the Dakota Group of the Front Range foothills, Colorado: Geol. Soc. America Bull., v. 73, p. 317-324.

Brown, W. M. III, and Ritter, J. R., Sediment transport and turbidity in the Eel River basin, California: U.S. Geol. Survey Water-Supply Paper 1986, (in press).

Bryan, Kirk, 1934, Geomorphic processes at high altitudes. Geog. Rev., v. 24, p. 655-656.

Capps, S. R., 1910, Rock glaciers in Alaska: Jour. Geology, v. 18, p. 359-375.

Carter, J. R. and Green, A. R., 1963, Floods in Wyoming, magnitude and frequency: U.S. Geol. Survey Circ. 478, 27 p.

Crandall, D. R. and Varnes, D. J., 1961, Movement of the Slumgullion earthflow near Lake City, Colorado: U.S. Geol. Surv. Prof. Paper 424-B, p. B136-B139.

Crockett, D. H., 1966, A geologic-soil stability report of the proposed Snake River Canyon-Alpine Impoundment area, Teton, Bridger, and Targhee National Forests: In-Service Rept., U.S. Forest Serv., Intermountain Region, Ogden, Utah, 9 p.

Croft, A. R. and Adams, J. A. Jr., 1950, Landslides and sedimentation in the North Fork of Ogden River, May 1949: Intermountain Forest & Range Expt. Sta. Res. Paper No. 21, 4 p.

.... and Ellison, L., 1960, Watershed and range conditions on Big Game Ridge and vicinity, Teton National Forest, Wyoming: U. S. Dept. of Agri., Forest Service, 37 p.

.... 1967, Rainstorm debris floods: Univ. Arizona, Agri. Expt. Sta. Rept. 248, Tucson, Ariz., 36 p.

Curry, R. R., 1966, Observation of alpine mudflows in the Tenmile Range, central Colorado: Geol. Soc. America Bull., v. 77, p. 771-776.

Dickinson, R. G., 1965, Landslide origin of the type Cerro Till, southwestern Colorado: U.S. Geol. Survey Prof. Paper 525-C, p. C147-C151.

Dishaw, H. E., 1967, Massive landslides: Photogramm. Eng., v. 33, p. 603-608.

Dorr, J. A. Jr., 1956, Post-Cretaceous geologic history of the Hoback Basin area, central western Wyoming, *in* Wyoming Geol. Assoc.

Guidebook 11th Ann. Field Conf. 1956: p. 99-108.

Dyrness, C. T., 1967, Mass soil movements in the H. J. Andrews Experimental Forest: Pacific NW Forest & Range Expt. Sta. Res. Paper PNW-12, 42 p.

Eden, W. J., 1967, Buried soil profile under apron of an earth flow: Geol. Soc. America Bull., v. 78, p. 1183-1184.

Edmond, R. W., 1951, Structural geology and physiography of the northern end of the Teton Range, Wyoming: Augustana Library Pub. 23, Rock Island, Ill. 82 p.

Emerson, F. B., 1925, 180 ft. dam formed by landslide in Gros Ventre Canyon: Eng. News-Rec., v. 95, p. 467-468.

Farmer, E.H. and Fletcher, J. E., Precipitation characteristics of summer storms at high-elevation stations in Utah: Water Resources Research (in press).

Federal Water Pollution Control Administration, 1970, Industrial waste guide on logging practices: Northwest Region, Portland, Ore., 40 p.

Fenneman, N. M., 1931, Physiography of western United States: McGraw-Hill, New York, 534 p.

Flaccus, E., 1959, Revegetation of landslides in the White Mountains of New Hampshire: Ecology, v. 40, p. 692-703.

Flint, R. F., 1957, Glacial and Pleistocene geology: John Wiley & Sons, Inc., New York, 553 p.

.... and Denny, C. S., 1958, Quaternary geology of Boulder Mountain, Aquarius Plateau, Utah: U.S. Geol. Survey Bull. 1061-D, p. 103-164.

Foster, H. L., 1947, Paleozoic and Mesozoic stratigraphy of northern Gros Ventre Mountains and Mount Leidy Highlands, Teton County, Wyoming: Am. Assoc. Petroleum Geologists Bull., v. 31, p. 1537-1593.

Fredriksen, R. L., 1963, A case history of a mud and rock slide on an experimental watershed: Pacific NW Forest & Range Expt. Sta. Res. Note PNW-1, 4 p.

.... 1965, Christmas storm damage on the H. J. Andrews Experimental Forest: Pacific NW Forest & Range Expt. Sta. Res. Note PNW-29, 11 p.

Fryxell, F. M., 1930, Glacial features of Jackson Hole, Wyoming: Augustana Lib. Pub. 13, Rock Island, Ill., 129 p.

.... 1933, Earthquake shocks in Jackson Hole, Wyoming: Seismol. Soc. America Bull., v. 23, p. 167-168.

.... and Horberg, L., 1943, Alpine mudflows in Grand Teton National Park, Wyoming: Geol. Soc. America Bull., v. 54, p. 457-472.

Gale, B. T., 1940, Communication, Further Earthquake shocks in Jackson Hole, Wyoming: Seismol. Soc. America Bull., v. 30, p. 85.

Gardner, J., 1969, Observations of surficial talus movement: Zeitschr. Geomorphologie, Band 13, p. 317-323.

Gardner, R. B., 1967, Major environmental factors that affect the location, design, and construction of stabilized forest roads, *in* Loggers handbook: Pacific Logging Congress, Portland, Ore., v. 27, p. 1-5.

.... Gonsior, M. J. and Martin, G. L., 1969, Zena Creek Road and logging system investigation: In-Service Rept., Intermountain Forest & Range Expt. Sta.

Gray, D. H., 1969, Effects of forest clear-cutting on the stability of natural slopes: Univ. of Michigan, Ann Arbor, 67 p.

Griggs, R. F., 1938, Timberlines in the northern Rocky Mountains: Ecology, v. 19, p. 548-564.

Hack, J. T. and Goodlett, J. C., 1960, Geomorphology and forest ecology of a mountain region in the central Appalachians: U.S. Geol. Survey Prof. Paper 347, 66 p.

Hadley, J. B., 1964, Landslides and related phenomena accompanying the Hebgen Lake earthquake of August 17, 1959, *in* the Hebgen Lake, Montana, earthquake of August 17, 1959: U.S. Geol. Survey Prof. Paper 435, p. 107-138.

Hall, W. B., 1960, Mass-gravity movements in the Madison and Gallatin ranges, southwestern Montana, *in* West Yellowstone — earthquake area: Billings Geol. Soc. Guidebook 11th Ann. Field Conf. 1960: p. 200-206.

Hansen, W. R., 1961, Landslides along the Uinta fault east of Flaming Gorge, Utah: U.S. Geol. Survey Prof. Paper 424-B, p. B306-B307.

Hardy, R. M., 1963, The Peace River Highway Bridge — a failure in soft shales, *in* Stability of rock slopes: Highway Research Board Pub. 1114, p. 29-39.

Harris, A. S., 1967, Natural reforestation on a mile-square clearcutting in southeast Alaska: Pacific NW Forest & Range Expt. Sta. Res. Paper PNW-52, 16 p.

Hayden, E. W., 1956, The Gros Ventre slide (1925) and the Kelly Flood (1927), *in* Wyoming Geol. Assoc. Guidebook 11th Ann. Field Conf. 1956: p. 20-22.

Hewlett, J. D., and Hibbert, A. R., 1961, Increases in water yield after several types of forest cutting: Int. Assoc. Sci. Hydrol. Bull., v. 6, p. 5-17.

Highway Research Board, 1958, Landslides and engineering practice: Eckel, E. B., ed., Washington, D.C., Highway Research Board Spec. Pub. 29, 232 p.

Hoffman, G. J., Curry, R. B., and Schwab, G. O., 1964, Annotated bibliography on slope stability of strip mine spoil banks: Ohio Agric. Expt. Sta. Res. Circ. 130, 92 p.

Holmes, G. W. and Moss, J. H., 1955, Pleistocene geology of the southwestern Wind River Mountains, Wyoming: Geol. Soc. America Bull., v. 66, p. 629-654.

Horberg, L. 1938, The structural geology and physiography of the Teton Pass area, Wyoming: Augustana Library Pub. 16, Rock Island, Ill. 85 p.

.... Edmund, R. W., and F. M. Fryxell, 1955, Geomorphic and structural relations of Tertiary volcanics in the northern Teton Range, Wyoming: Jour. Geology v. 63, p. 501-511.

Hough, B. K., 1957, Basic soils engineering: Ronald Press Co., 513 p.

Howe, E., 1909, Landslides in the San Juan Mountains, Colorado, including a consideration of their causes and their classification: U.S. Geol. Survey Prof. Paper 67, 58 p.

Ives, R. L., 1941, Vegetative indicators of solifluction: Jour. Geomorphology, v. 4, p. 128-132.

Johnson, R. B., 1967, Rock streams on Mount Mestas, Sangre De Cristo Mountains, southern Colorado: U.S. Geol. Survey Prof. Paper 575-D, p. D217-D220.

Jones, F. O., Embody, D. R., and Peterson, W. L., 1961, Landslides along the Columbia River valley, northeastern Washington: U.S. Geol. Survey Prof. Paper 367, 98 p.

Keefer, W. R. and Love, J. D., 1956, Landslides along the Gros Ventre River, Teton County, Wyoming, *in* Wyoming Geol. Assoc. Guidebook 11th Ann. Field Conf. 1956: p. 24-28.

.............., 1957, Geology of the Du Noir area, Fremont County, Wyoming: U.S. Geol. Survey Prof. Paper 294-E, p. 155-221.

.... 1963, Karst topography in the Gros Ventre Mountains, northwestern Wyoming: U.S. Geol. Survey Prof. Paper 475-B, p. B129-B130.

Kesseli, J. E., 1941, Rock streams in the Sierra Nevada, California: Geog. Rev., v. 31, p. 203-227.

Kojan, E., 1967, Mechanics and rates of natural soil creep: Proc. 5th Ann. Eng. Geol. and Soils Engineers Symposium, Pocatello, Idaho, April 1967, p. 233-253.

.............., Road location and design criteria for use in landslide terrain: Pacific SW Forest & Range Expt. Sta. Res. Paper (in press).

Ladd, G. E., 1935, Landslides, subsidences and rockfalls: Am. Rwy. Eng. Assoc. Proc., v. 36, p. 1091-1162.

LaMarche, V. C., Jr., 1968, Rates of slope degradation as determined from botanical evidence, White Mountains, California: U.S. Geol. Survey Prof. Paper 352-I, p. 341-377.

Langenheim, J. H., 1956, Plant succession on a subalpine earthflow in Colorado: Ecology, v. 37, p. 301-317.

Lawrence, D. B., and Lawrence, E. G., 1958a, Bridge of the gods legend, its origin, history and dating: Mazama, v. 40, p. 33-41.

.... and ...., 1958b, Historic landslides of the Gros Ventre valley, Wyoming: Mazama, v. 13, p. 42-52.

Leopold, L. B., Wolman, M. G., and Miller, J. P., 1964, Fluvial processes in geomorphology: W. H. Freeman & Co., San Francisco, 522 p.

Liang, T. A., 1952, Landslides — an aerial photographic study. Ph.D. Thesis, Cornell Univ., 274 p.

.... and Belcher, D. J., 1958, Airphoto interpretation, *in* Landslides and engineering practice: Washington, D.C., Highway Research Board, Spec. Pub. 29, Chap. 5, p. 69-92.

Love, J.D., Keefer, W. R., Duncan, D. C., Bergquist, H. R., and Hose, R. K., 1951, Geologic map of the Spread Creek — Gros Ventre River area, Teton County, Wyoming: U.S. Geol. Survey Oil and Gas Inv. Map OM-118.

.... Weitz, J. L., and Hose, R. K., 1955, Geologic map of Wyoming: U. S. Geol. Survey, Scale 1:500,000.

.............., 1956a, Cretaceous and Tertiary stratigraphy of the Jackson Hole area, northwestern Wyoming, *in* Wyoming Geol. Assoc. Guidebook 11th Ann. Field Conf. 1956: p. 76-94.

.............., 1956b, Summary of geologic history of Teton County, Wyoming, during late Cretaceous, Tertiary, and Quaternary times, *in* Wyoming Geol. Assoc. Guidebook 11th Ann. Field Conf. 1956: p. 140-150.

.............., 1956c, Geologic map of Teton County, Wyoming, *in* Wyoming Geol. Assoc. Guidebook 11th Ann. Field Conf. 1956: Map in pocket.

.... and Montagne, J., 1956a, Pleistocene and recent tilting of Jackson Hole, Teton County, Wyoming, *in* Wyoming Geol. Assoc. Guidebook 11th Ann. Field Conf. 1956: p. 169-178.

.... and .... 1956b, Road log: Gros Ventre slide, stratigraphy and structure of Mt. Leidy Highlands, *in* Wyoming Geol. Assoc. Guidebook 11th Ann. Field Conf. 1956: p. 219-226.

.... 1968, Stratigraphy and structure — a summary, *in* A geophysical study in Grand Teton National Park and vicinity, Teton County, Wyoming: U.S. Geol. Survey Prof. Paper 516-E, p. E3-E12.

McConnell, R. G., and Brock, R. W., 1904, Report on the great landslide at Frank, Alberta Territory (Canada): Canada Dept. Interior Ann. Report, 1903, pt. 8, p. 1-17.

Mather, K. F., and Wengerd, S. A., 1965, Pleistocene age of the "Eocene" Ridgway Till, Colorado: Geol. Soc. America Bull. v. 76, p. 1401-1408.

Matthes, F. E., 1930, Geologic history of the Yosemite Valley: U.S. Geol. Survey Prof. Paper 160, 137 p.

.............., 1938, Avalanche sculpture in the Sierra Nevada of California: Int. Assoc. Hydrology Bull. 23, Riga, p. 631-637.

.... 1939, Report of the committee on glaciers: Am. Geophys. Union Trans., 20th Ann. Mtg., pt. 4, p. 518-523.

..............., 1940, Report of the committee on glaciers: Am. Geophys. Union Trans., 21st Ann. Mtg., pt. 1, p. 396-405.

Melton, M. A., 1957, An analysis of the relations among elements of climate, surface properties, and geomorphology: Tech. Rept. 11, Proj. NR 389-042, Office of Naval Research, Geography Branch, 102 p.

..............., 1960, Intravalley variation in slope angles related to microclimate and erosional environment: Geol. Soc. America Bull., v. 71, p. 133-144.

Moyle, R. W., and Olson, E. P., 1967, Gros Ventre Road #30015, Teton National Forest: In-Service Rept. U. S. Forest Service, Intermountain Region, Ogden, Utah, 16 p.

Mudge, M. R., 1965, Rockfall-avalanche and rockslide - avalanche deposits at Sawtooth Ridge, Montana: Geol. Soc. America Bull., v. 76, p. 1003-1014.

Naismith, H., 1964, Landslides and Pleistocene deposits in the Meikle River valley of northern Alberta: Canadian Geotechnical Jour., v. 1, p. 155-166.

Orme, A. R., and Bailey, R. G., 1970, The effect of vegetation conversion and flood discharge on stream channel geometry: The case of southern California watersheds: Assoc. Am. Geog. Proc., v. 2, p. 101-106.

Oswald, E. T., 1966, A synecological study of the forested moraines of the valley floor of Grand Teton National Park, Wyoming: Ph.D. Thesis, Montana State Univ. Bozeman, 101 p.

Pacific Northwest River Basins Commission, 1969a, Climatological handbook, Columbia Basin States, temperature: v. I, pt. A, Vancouver, Wash., p. 5.

..............., 1969b, Climatological handbook, Columbia Basin States, precipitation: v. II, Vancouver, Wash., p. 5.

Packer, P. E., 1966, Forest treatment effects on water quality, *in* International symposium on forest hydrology: Pergamon Press, New York, p. 687-699.

Packer, R. W., 1964, Stability of slopes in an area of glacial deposition: Canadian Geographer v. 8, p. 147-151.

Peltier, L. C., 1950, The geographic cycle in periglacial regions as it is related to climatic geomorphology: Assoc. Am. Geog. Annals, v. 40, p. 214-236.

Peterson, R. M., 1954, Studies of Bearpaw Shale at a damsite in Saskatchewan: Am. Soc. Civil Eng. Proc., v. 80, Separate No. 476, p. 1-28.

Pierce, W. G., 1968, The Carter Mountain landslide area, northwest Wyoming: U.S. Geol. Survey Prof. Paper 600-D, p. D235-D241.

Potter, N. Jr., 1969, Tree-ring dating of snow avalanche tracks and the geomorphic activity of avalanches, northern Absaroka Mountains, Wyoming: Geol. Soc. America Spec. Paper 123, (INQUA Volume) p. 141-165.

Prostka, H. J., 1967, Effect of landslides on the course of Whitetail Creek, Jefferson County, Montana: U.S. Geol. Survey Prof. Paper 575-B, p. B80-B82.

Rahn, P. H., 1969, The relationship between natural forested slopes and angles of repose for sand and gravel: Geol. Soc. America Bull., v. 80, p. 2123-2128.

Rapp, A., 1959, Avalanche boulder tongues in Lappland: Geografiska Annaler, v. 41, p. 34-48.

..............., 1960, Recent development of mountain slopes in Karkevagge and surroundings, Northern Scandinavia: Geografiska Annaler, v. 42, p. 65-200.

Reiche, P., 1937, The Toreva-block — A distinctive landslide type: Jour. Geology, v. 45, p. 538-548.

Retzer, J. L., 1954, Glacial advances and soil development, Grand Mesa, Colorado: Am. Jour. Sci., v. 252, p. 26-37.

..............., 1965, Significance of stream systems and topography in managing mountain lands, *in* Forest-Soil Relationships in North America: Oregon State Univ. Press, Corvallis, Ore., p. 399-411.

Rice, R. M., Corbett, E. S., and Bailey, R. G., 1969, Soil slips related to vegetation, topography, and soil in southern California: Water Resources Research, v. 5, p. 647-659.

Richmond, G. M., 1948, Modification of Blackwelder's sequence of glaciation in the Wind River Mountains, Wyoming (abs.) Geol. Soc. America Bull., v. 59, p. 1400-1401.

..............., 1957, Three pre-Wisconsin glacial stages in the Rocky Mountain region: Geol. Soc. America Bull., v. 68, p. 239-262.

..............., 1962, Quaternary stratigraphy of the La Sal Mountains, Utah: U.S. Geol. Survey Prof. Paper 324, 135 p.

..............., 1964a, Glacial geology of the West Yellowstone basin and adjacent parts of Yellowstone National Park, *in* The Hebgen Lake, Montana, earthquake of August 17, 1959: U.S. Geol. Survey Prof. Paper 435, p. 223-236.

..............., 1964b, Glaciation of Little Cottonwood and Bells Canyons, Wasatch Mountains, Utah: U.S. Geol. Survey Prof. Paper 454-D, 41 p.

..............., 1965, Glaciation of the Rocky Mountains, *in* Wright, H. E. Jr., and Frey, D. G., eds., The Quaternary of the United States — A review volume for the VII Congress of the International Assoc. for Quaternary Research: Princeton, N. J., Princeton Univ. Press, p. 217-230.

Ritchie, A. M., 1958, Recognition and identification of landslides, *in* Landslides and engineering practice: Washington, D.C., Highway Research Board Spec. Pub. 29, Chap. 4, p. 48-68.

Rohrer, W. L., 1968, Geologic map of the Fish Lake quadrangle, Fremont County, Wyoming: U.S. Geol. Survey Map GQ-724, Scale 1:24,-000.

.... 1969, Preliminary geologic map of the Sheridan Pass quadrangle, Fremont and Teton Counties, Wyoming: U.S. Geol. Survey, Scale 1:24,000.

Root, A. W., 1958, Prevention of landslides, *in* Landslides and engineering practice: Washington, D.C., Highway Research Board Spec. Pub. 29, Chap. 7, p. 113-149.

Ross, C. P., and Nelson, W. H., 1964, Regional seismicity and brief history of Montana earthquakes, *in* The Hebgen Lake, Montana, earthquake of August 17, 1959: U.S. Geol. Survey Prof. Paper 435, p. 25-30.

Rothacher, J. S., and Glazebrook, T. B., 1968, Flood damage in the National Forests of Region 6: Pacific NW Forest & Range Expt. Sta., 20 p.

Rouse, J. T., 1934, The physiography and glacial geology of the Valley region, Park County, Wyoming: Jour. Geology, v. 42, p. 738-752.

Schroeder, M. L., 1969, Geologic map of the Teton Pass quadrangle, Teton County, Wyoming: U.S. Geol. Survey Map GQ-793, Scale 1:24,000.

Schumm, S. A., 1965, Quaternary paleohydrology, *in* Wright, H. E. Jr., and Frey, D. G., The Quaternary of the United States—Review volume for the VII Congress of the International Assoc. for Quaternary Research: Princeton, N. J., Princeton Univ. Press, p. 783-794.

Scott, K. M., and Gravlee, G. C., Jr., 1968, Flood surge on the Rubicon River, California — Hydrology, hydraulics, and boulder transport: U.S. Geol. Survey Prof. Paper 422-M, p. M1-M40.

Sharp, R. P., 1942, Mudflow levees: Jour. Geomorphology, v. 5, p. 222-227.

Sharpe, C. F. S., 1938, Landslides and related phenomena: Columbia Univ. Press, New York, 137 p.

Sharpe, C. F. S., and Dosch, E. F., Relation of soil creep to earthflow in the Appalachian Plateaus: Jour. Geomorphology, v. 5, p. 312-324.

Shroder, J. F., 1967, Landslides of Utah: Ph.D. Thesis, Univ. Utah, 337 p.

Smith, H. T. U., 1936, Periglacial landslide topography of Canjilon Divide, Rio Arriba County, New Mexico: Jour. Geol., v. 44, p. 836-860.

..............., 1949a, Periglacial features in the driftless area of southern Wisconsin, Jour. Geology, v. 57, p. 196-215.

..............., 1949b, Physical effects of Pleistocene climatic changes in nonglaciated areas: Eolian phenomena, frost action, and stream terracing: Geol. Soc. America Bull., v. 60, p. 1485-1516.

Smith, J. F., Jr., Huff, L. C., Hinrichs, E. N., and Luedke, R. G., 1963, Geology of the Capitol Reef area, Wayne and Garfield Counties, Utah: U.S. Geol. Survey Prof. Paper 363, 102 p.

Soil Survey Staff, 1951, Soil survey manual: U.S. Dept. Agr. Handbook No. 18, 503 p.

Sowers, G. B., and Sowers, G. F., 1951, Introductory soil mechanics and foundations: The MacMillan Co., New York, 228 p.

Stout, M. L., 1969, Radiocarbon dating of landslides in southern California and engineering geology implications: Geol. Soc. America Spec. Paper 123 (INQUA Volume), p. 167-179.

Strahler, A. N., 1940, Landslides of the Vermilion and Echo Cliffs, northern Arizona: Jour. Geomorphology, v. 3, p. 285-301.

Swanston, D. N., 1967a, Debris avalanching in thin soils derived from bedrock: Pacific NW Forest & Range Expt. Sta. Res. Note PNW-64, 7 p.

.... 1967b, Soil-water piezometry in a southeast Alaska landslide area: Pacific NW Forest & Range Expt. Sta. Res. Note PNW-68, 17 p.

.... 1969, Mass wasting in coastal Alaska: Pacific NW Forest & Range Expt. Sta. Res. Paper PNW-83, 15 p.

Swenson, F. A., 1949, Geology of the northwest flank of the Gros Ventre Mountains, Wyoming: Augustana Library Pub. 21, Rock Island, Ill., 75 p.

Terzaghi, K., 1950, Mechanism of landslides, *in* Application of geology to engineering practice: Geol. Soc. America (Berkey Volume) Chap. 4, p. 83-123.

.... and Peck, R. B., 1967 Soil mechanics in engineering practice: John Wiley & Sons, Inc., New York, 2nd edition, 566 p.

Thomas, C. A., Broom, H. C., and Cummans, J. E., 1963, Magnitude and frequency of floods in the United States, pt. 13, Snake River Basin: U.S. Geol. Survey Water-Supply Paper 1688, 250 p.

Thorarinsson, S., Einarsson, T., and Kjartansson, G., 1959, On the geology and geomorphology of Iceland: Geografiska Annaler, v. 41, p. 135-169.

Thorp, J., 1931, The effects of vegetation and climate upon soil profiles in northern and northwest Wyoming: Jour. Soil Sci., v. 32, p. 283-301.

Tricart, J., et collaborateurs, 1961, Mecanismes normaux et phenomenes catastrophiques dans l' evolution des versants du bassin du Guil (Htes-Alpes, France): Zeitschr. Geomorphologie, v. 5, p. 277-301.

U.S. Forest Service, 1966, Typical landscape characteristics and associated soil and water management problems on the Mt. Baker National Forest: Pacific Northwest Region, Portland, Ore., 89 p.

U.S. Geological Survey, 1956, Compilation of records of surface waters of the United States through September 1950, pt. 13, Snake River Basin: U.S. Geol. Survey Water-Supply Paper 1317, 566 p.

..............., 1963, Compilation of records of surface waters of the United States, October 1950 to September 1960, pt. 13, Snake River Basin: U.S. Geol. Survey Water-Supply Paper 1737, 282 p.

U.S. Soil Conservation Service, 1967, Summary of snow survey measurements, Wyoming: U.S. Dept. Agr. Soil Conserv. Service, 152 p.

Van Burkalow, A., 1945, Angle of repose and angle of sliding friction: An experimental study: Geol. Soc. America Bull., v. 56, p. 669-708.

Vanderwilt, J. W., 1934, A recent rockslide near Durango, in La Plata County, Colorado: Jour. Geology, v. 42, p. 163-173.

Varnes, D. J., 1950, Relation of landslides to sedimentary features, *in* Applied sedimentation, John Wiley & Sons, New York, p. 229-246.

.... 1958, Landslide types and processes, *in* Landslides and engineering practice: Washington, D.C., Highway Research Board Spec. Pub. 29, Chap. 3, p. 20-47.

Varnes, H. D., 1949, Landslide problems of southwestern Colorado: U.S. Geol. Survey Circ. 31, 13 p.

Wagner, W. T., 1944, A landslide area in the Little Salmon River Canyon, Idaho: Econ. Geology, v. 39, p. 349-358.

Waldrop, H. A., and Hyden, H. J., 1962, Landslides near Gardiner, Montana: U.S. Geol. Survey Prof. Paper 450-E, p. E11-E14.

Ward, W. H., 1945, The stability of natural slopes: Geog. Jour. v. 105, p. 170-197.

Washburn, A. L., and Goldthwait, R. P., 1958, Slushflows: Geol. Soc. America Bull., v. 69, p. 1657.

Watson, R. A., and Wright, H. E. Jr., 1963, Landslides on the east flank of the Chuska Mountains, northwestern New Mexico: Am. Jour. Sci., v. 261, p. 525-548.

Weed, W. H., 1893, The glaciation of the Yellowstone Valley north of the Park: U.S. Geol. Survey Bull. 104, 41 p.

Witkind, I. J., 1969, Geology of the Tepee Creek quadrangle, Montana-Wyoming: U.S. Geol. Survey Prof. Paper 609, 101 p.

Yeend, W. E., 1969, Quaternary geology of the Grand and Battlement Mesas area, Colorado: U. S. Geol. Survey Prof. Paper 617, 50 p.

Zaruba, Q., and Mencl, V., 1969, Landslides and their control: American Elsevier Publishing Co., Inc., New York, 205 p.

Reprinted from *U.S.D.A. Miscellaneous Publication 1082* (1968)

22

RESTORING SURFACE-MINED LAND

By the U.S. Department of Agriculture

Introduction

A power shovel as big as an office building bites into the earth, piling up row on row of rock and soil to get at a vein of coal . . . .

An auger with a 7-foot bit bores into a hillside, and coal works its way out like wood shavings . . . .

A floating barge dips its big chain-bucket into a streambed for a load of sand and gravel . . . .

An ore-laden train snakes its way out of a giant open pit . . . .

Through these and other operations man carries on the big activity of surface mining. He get many minerals, fuels, and building materials tha help our Nation grow and that provide jobs i rural America.

In the process, the land is changed—laid bare rearranged into parallel ridges, or scooped out lik a soupbowl. Properly treated and managed, it ca be returned to safe and productive use, even be come a greater asset to the community than it wa before mining. Left alone, it may produce onl

With today's large excavating equipment we can not only surface mine faster and cheaper but also reshape th landscape and rehabilitate the site easier.

TABLE 1.—*Land disturbed by strip and surface mining in the United States, by commodity, Jan. 1, 1965*[1]

[In thousands of acres]

Mineral	Strip mining			Into hillside	Quarry-open pit below ground level	Total	Dredge, hydraulic, and other methods	Grand total[2]
	Contour	Area	Total					
Coal[3]	665	637	1,302					1,302
Sand and gravel	38	258	296	82	371	453	74	823
Stone	6	8	14	100	127	227		241
Gold		8	8	1	3	4	191	203
Clay	10	26	36	22	44	66	7	109
Phosphate	28	49	77	13	93	106		183
Iron	7	31	38	30	96	126		164
All other	11	12	23	59	81	140		163
Total	765	1,029	1,794	307	815	1,122	272	3,188

[1] Acreage by method of mining estimated from random sampling survey.

[2] Compiled from data supplied by U.S. Department of the Interior; from Soil Conservation Service, U.S. Department of Agriculture; and from estimates prepared by the field study group.

[3] Includes anthracite, bituminous, and lignite.

stream-fouling sediment and acid and ugliness.

For many years the U.S. Department of Agriculture (USDA) has been helping private-land owners restore their surface-mined land as part of their regular programs of wise land use and conservation treatment. USDA also has done restoration work and research studies on the public land it administers. Its experience and skills range all the way from preplanning mining to prevent offsite damage to development of a mined area for highly intensive uses.

Through studies and experience and through participation in the 2-year National Surface Mine Study under Public Law 89–4, USDA has gathered a great deal of information about surface-mined-land conservation progress and needs. In this report highlights of the data are given as well as ideas for future action, suggested by research and experience, that can speed restoration of the surface-mined land that is intermingled with farm, ranch, forest, and other land in rural and suburban America.

SURFACE-MINED LAND—BY STATES.—An estimated 3.2 million acres of land—some in every State—had been disturbed by surface mining by January 1, 1965 (tables 1, 2).

DISTANCE FROM POPULATION CENTERS.—Surface-mined-land conservation is a *rural* opportunity. More than four-fifths of the mined land surveyed is at least a mile from communities with a population of more than 200. More than half are more than 4 miles from town. And 40 percent of the mined land cannot now be seen from any U.S. highway or passenger railroad. Most areas were close enough to communities, though, for a family to reach for an afternoon recreation outing. No urban growth was evident around two-thirds of them, which suggests that these areas are likely to continue in agricultural and related uses.

OWNERSHIP.—Ownership of the land and its minerals holds the key to use and conservation of these resources. Since most surface-mined land is privately owned, opportunity for improvement lies largely in local assistance programs of mutual interest and value to landowners and their neighbors—the kind of program already being carried on by the Nation's 3,000 soil and water conservation districts and by State forestry agencies with USDA help. Increased assistance through these going programs could do the job. And since the mining industry owns more than half of the surface-mined land, it has a challenge to restore its property to a useful state and to prevent offsite damages.

A survey of 693 surface-mine sites[1] in 1966 showed that many were scattered small acreages best treated as part of the total conservation management of the farm and other areas with which they are intermingled. Nearly 80 percent of the sites were in forest, farm, or grassland or reverting to forest at the time of survey. These same uses were being made of land adjacent to 86 percent of the sites. Less than 2 percent of the acreage had

[1] Sites were selected at random from mined land throughout the Nation to represent the surface-mining situation. Of the total, 180 sites were mined for coal; 149 for sand and gravel; 100 stone; 49 clay; 49 iron; 48 gold; 40 phosphate; and 78 for eight other commodities.

TABLE 2.—*Condition of surface-mined land, by State, Jan. 1, 1965*

[In thousands of acres]

State	Land needing treatment [1]	Land not needing treatment [1]	Total land disturbed [2]
Alabama	83.0	50.9	133.9
Alaska	6.9	4.2	11.1
Arizona	4.7	27.7	32.4
Arkansas	16.6	5.8	22.4
California	107.9	66.1	174.0
Colorado	40.2	14.8	55.0
Connecticut	10.1	6.2	16.3
Delaware	3.5	2.2	5.7
Florida	143.5	45.3	188.8
Georgia	13.5	8.2	21.7
Hawaii [3]			
Idaho	30.7	10.3	41.0
Illinois	88.7	54.4	143.1
Indiana	27.6	97.7	125.3
Iowa	35.5	8.9	44.4
Kansas	50.0	9.5	59.5
Kentucky	79.2	48.5	127.7
Louisiana	17.2	13.6	30.8
Maine	21.6	13.2	34.8
Maryland	18.1	7.1	25.2
Massachusetts	25.0	15.3	40.3
Michigan	26.6	10.3	36.9
Minnesota	71.5	43.9	115.4
Mississippi	23.7	5.9	29.6
Missouri	43.7	15.4	59.1
Montana	19.6	7.3	26.9
Nebraska	16.8	12.1	28.9
Nevada	20.4	12.5	32.9
New Hampshire	5.1	3.2	8.3
New Jersey	21.0	12.8	33.8
New Mexico	2.0	4.5	6.5
New York	50.2	7.5	57.7
North Carolina	22.8	14.0	36.8
North Dakota	22.9	14.0	36.9
Ohio	171.6	105.1	276.7
Oklahoma	22.2	5.2	27.4
Oregon	5.8	3.6	9.4
Pennsylvania	229.5	140.7	370.2
Rhode Island	2.2	1.4	3.6
South Carolina	19.3	13.4	32.7
South Dakota	25.3	8.9	34.2
Tennessee	62.5	38.4	100.9
Texas	136.4	29.9	166.3
Utah	3.4	2.1	5.5
Vermont	4.2	2.5	6.7
Virginia	37.7	23.1	60.8
Washington	5.5	3.3	8.8
West Virginia	111.4	84.1	195.5
Wisconsin	27.4	8.2	35.6
Wyoming	6.4	4.0	10.4
Total	[4] 2,040.6	1,147.2	3,187.8

[1] Compiled from data supplied by Soil Conservation Service, U.S. Department of Agriculture.
[2] Compiled from data supplied by U.S. Department of the Interior; from Soil Conservation Service; and from study-group estimates.
[3] Less than 100 acres.
[4] Does not include 108,000 acres of National Forest land needing treatment.

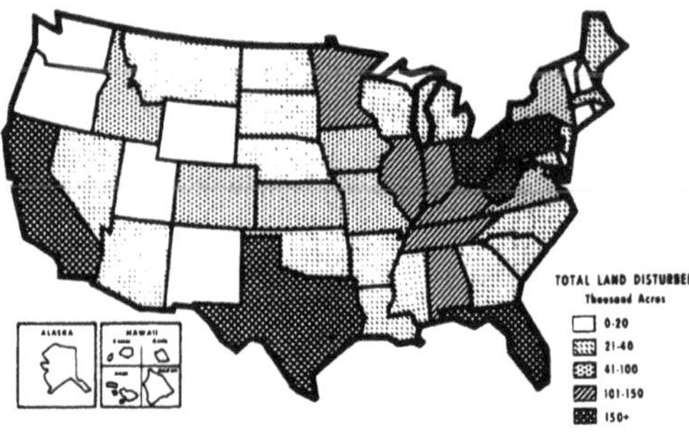

Surface-mined land—by States.

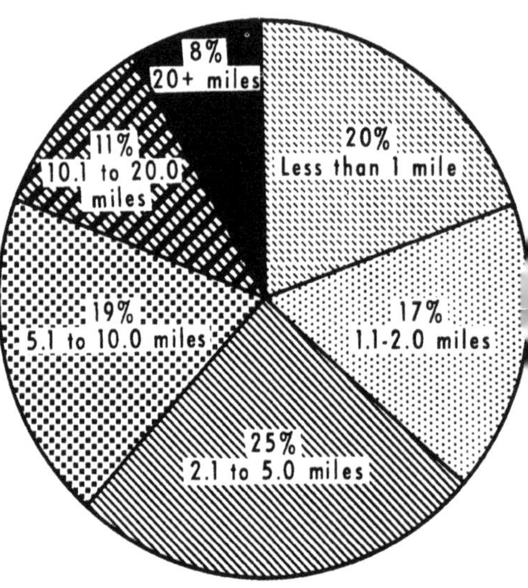

Proximity of surface-mined areas to population centers.

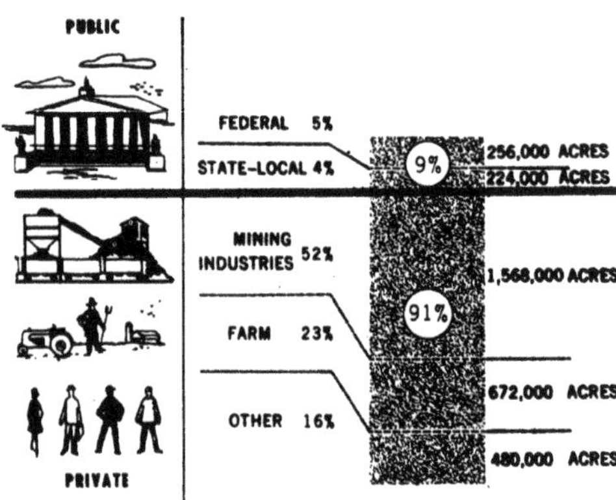

Ownership of surface-mined land.

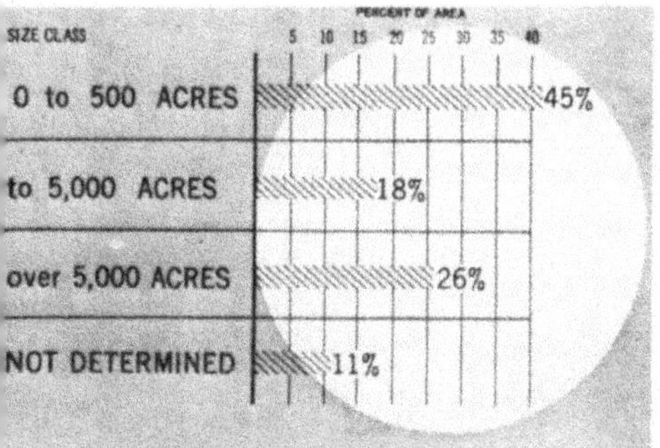

Small ownerships predominate.

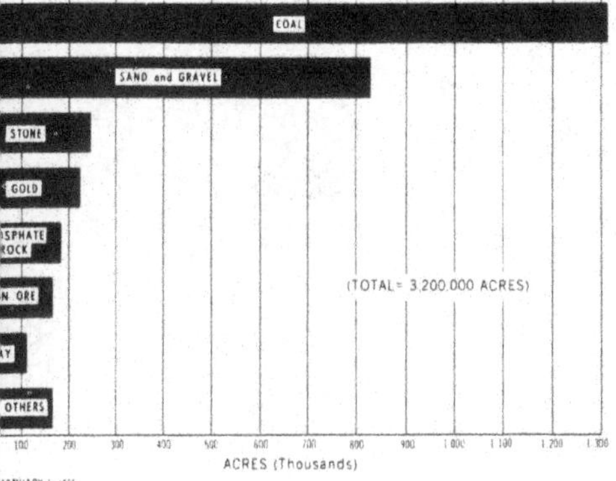

Surface-mined land—by commodities.

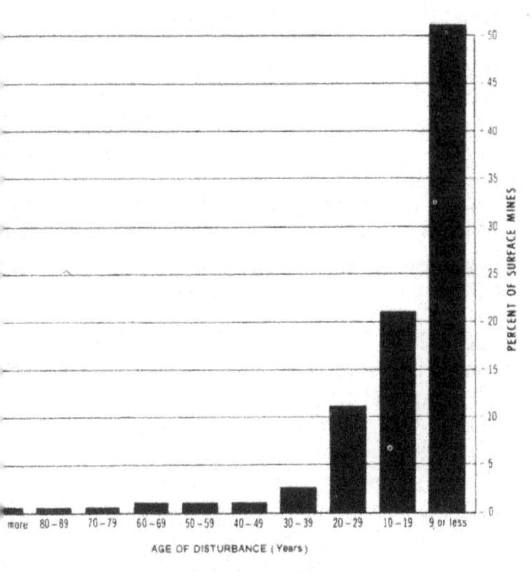

Age of surface mines.

been set aside solely as outdoor recreation or wildlife areas; usually these are compatible with other uses of the land.

SURFACE-MINED LAND—BY COMMODITIES.—More than 50 minerals are produced by surface mining in the United States. About 95 percent of the acreage disturbed by 1965 was for seven commodities: Coal, about 40 percent; sand and gravel 25 percent; stone, gold, clay, phosphate, and iron 30 percent. On two-thirds of the areas surveyed, the mineral deposit being mined was over 9 feet thick. This means great value from an acre but difficulty in reshaping the land to its original contours. Grading enough to satisfy intended land use is more practical. Some thin deposits might better have been left unmined where restoration costs would be proportionately high.

AGE OF SURFACE MINES.—Of the 693 sites sampled in 1966, 10 were mined more than a century ago. But most spoil banks and other disturbances are less than 10 years old, indicating a rapid rise in surface-mining activity. The acreage mined has more than doubled in the last 20 years.

DURATION OF SURFACE MINING.—More than half of the sites sampled were quarries or pits that had been operated for more than 10 years. Only a third of the sites had been operated for less than 5 years. Most were active long enough to have a significant economic impact on the community, and usually other surface-mining operations began later within the same watershed or drainage area.

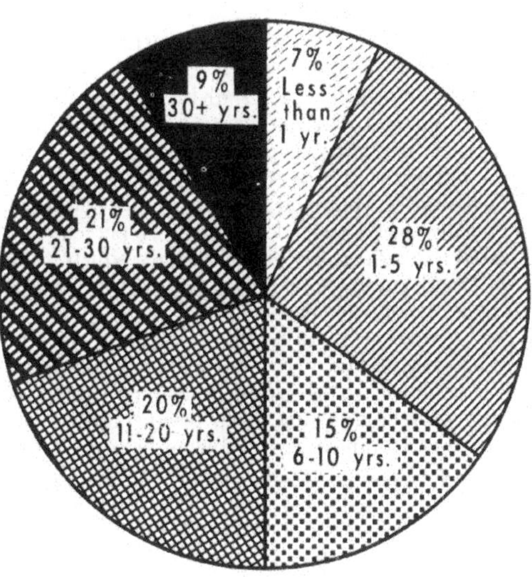

Duration of surface mining in years.

Where storm-water control is inadequate, surface-mine spoils and the community may be damaged.

Characteristics and Physical Condition

Of the 3.2 million acres disturbed by surface mining, about a third needs no further treatment to prevent sediment or other damage to adjacent land and water. About 46 percent of these 1.1 million acres that need no treatment was stabilized by nature over a period of years; 51 percent was treated through efforts of the mining industry and individual landowners; and the rest was treated by government at some level.

On the other two-thirds, newness of the disturbed area, distance from natural seed sources, or other problems make establishment of protective plants slow or difficult. Steep or unstable slopes, acidity, or stoniness are problems in some areas. These are susceptible—in varying degrees—to erosion and may contribute sediment and other pollutants to streams that drain them.

Spoil banks

In surface-mining operations the layers of soil and rock above the mineral deposit are shovele out and piled up in "spoil" banks. These banks ar a mixture of soil, subsoil, and unweathered roc that is far from resembling a soil formed in natur Their characteristics vary greatly among mine and even within the same mine. Prediction of sit suitability thus is best done with the help of prc fessional soil scientists, agronomists, foresters, an other specialists.

TEXTURE.—Spoil texture influences the amoun of moisture available for plant growth. In genera spoil composed largely of sand has good aeratio but is apt to be droughty. Clay banks compac easily and crust over during dry periods. Loam and silty shales usually have enough fine materia to hold moisture. On about 80 percent of th surface-mined land, spoil texture is adequate fo growing adapted grasses and legumes for quic erosion control and to supplement tree or shru

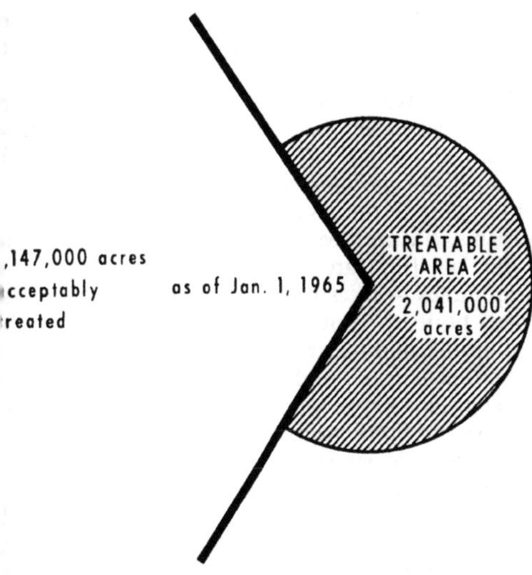

Area acceptably treated.

lantings. Rock content on about three-fourths of he banks, however, restricts the type of equipment hat can be used in revegetation. On about one-ourth of the banks the spoil is suitable for farm rops.

CIDITY.—Acid problems are associated largely vith coal mining. They are caused when minerals eft exposed to air and water react to form toxic r corrosive substances.

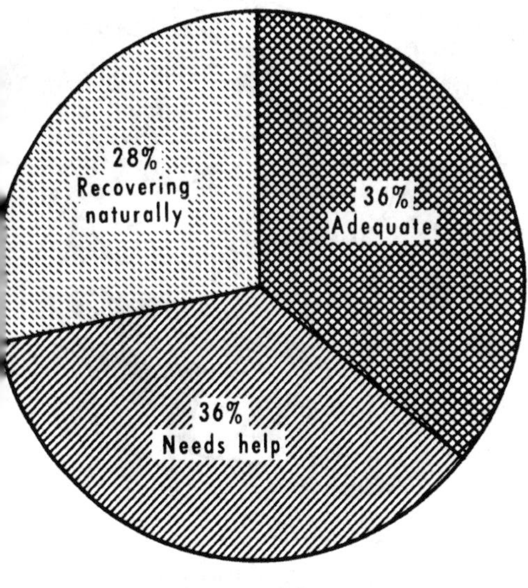

Cover conditions.

By itself, acidity does not directly influence plant growth. But it affects the availability of soil nutrients—dissolved minerals—and the number of soil micro-organisms. Strongly acid soils may, however, dissolve enough elements to injure or destroy plants that absorb them. More than half of the sites have acid soils; 20 percent are acid enough to be a limiting factor in establishing plant cover; only 1 percent is so acid that plants will not grow. Acidity usually is reduced through weathering and leaching of the acid-forming materials.

SLOPES.—More than 2 million acres (about 75 percent) have been mined on areas with original slopes of less than 20 percent—in the small watershed projects with which USDA has been working most of the mined areas have slopes of less than 10 percent. Only about 8 percent of the mined areas were on hillsides with slopes of more than 40 percent.

Four-fifths of the affected areas were on side slopes, ridgetops, or isolated knobs from which storm-water flows need to be guided into defined stream channels—with grass waterways or chutes, for example. The other one-fifth were on valley floors close to rivers and subject to local flooding.

Climate

About four-fifths of the surface-mined land is in areas where rainfall and temperatures are adequate for plant growth. With adequate spoil conditions and proper preparation, plant establishment and growth should be possible. On the other one-fifth, plants grow slowly because of too little or too much moisture, high temperatures, or unfavorable evapotranspiration ratio. Here special treatments and plants are needed to offset poor ecological conditions.

Erosion

About 2 million acres have evidence of sheet erosion. Some erosion is inevitable on fresh spoil banks, as it is on any bare soil. How severe it is depends on steepness and length of slope, extent of freezing and thawing, amount and intensity of precipitation, and how water is concentrated on the spoil. Thus, the quicker a plant cover is established to protect against erosion the better.

Forty percent or 1.2 million acres have eroded enough to form rills and small gullies. On 12 percent or 400,000 acres, gullies more than a foot deep have formed; these seem to be associated with long slopes created by grading.

Sheet erosion is not a serious problem in either area stripping or dredging since most of the soil movement is between spoil banks and little leaves the mine area. Sheet erosion is more serious in contour stripping.

Erosion danger is greatly increased at the point where storm water drains from a surface mine because of the concentrated force of water.

SLIDES.—On about 3,600 miles of slopes left by contour and area stripping (called outslopes), massive slides are a problem—especially where the subsoil is unstable. Slides may enter streams and even block channels. Their stabilization or removal would be costly and would involve geology, soils, engineering, hydrology, and forestry skills. Slides of this size occur on about 10 percent of the total mileage of outslopes.

ACCESS ROADS.—Mining haul roads are responsible for much erosion, especially in mountain areas. About 1,650 miles of these roads have eroded so badly they need major repairs. Another 3,300 miles are moderately eroded. Access roads for most mines surveyed were under 7 miles in length, and many were of half a mile or less. Many would best be revegetated rather than kept as roads. The rest need careful management after hauling stops.

Plant cover

For newly mined land, the great need is to establish plant cover as quickly as possible. Adequate plant cover reduces erosion and siltation in almost all cases, but it takes time. There is no "instant cover." Examination of sites capable of supporting vegetation showed that 36 percent had plant cover of 40 percent or more. About 28 percent of the sites had less than 40 percent cover at the time but, in the judgment of the survey team, would develop adequate protective cover naturally in time. The other 36 percent of the sites will require seeding, planting, fertilizing, and other attention to develop adequate protective cover.

It was estimated that three-fourths of the vegetation had occurred naturally on ground with more than 10 percent plant cover, and one-fourth through the efforts of man. Variations in vegetation appear to be associated with climatic conditions, spoil characteristics, nearness to natural seed sources, and age of the spoil banks. Half of the banks are less than 10 years old.

Water quality and streamflow

Surface mining in some areas is a source of water pollution, mainly sediment and to a less extent acid. Of the sites surveyed, 56 percent showed no pollution; 23 percent showed some intermittent pollution; and 21 percent produced considerable pollution. The survey team estimated that about a third of the surface-mined land needing conser vation treatment, or about 665,000 acres, need some action to reduce offsite water pollution.

Of the streams receiving direct runoff from sur face-mined sites, 31 percent of those examine contained noticeable amounts of mineral precipi tates. Water discoloration, suggesting chemical o physical pollution, was noted in 37 percent of th streams. Natural seepage from unworked coal an other pyritic material—from both surface an deep mines—causes limited local pollution. Acces roads built of pyritic waste material also may b sources of acid water.

Sediment is a problem where inadequate plan cover permits erosion and water is allowed to ru off the site from roads, terrace outlets, outslopes or slides. It is particularly severe in areas of high intensity storms and steep slopes.

Sediment generally was not present in smal streams more than 2 miles from the mine area. Bu of 14,000 miles of stream channels affected by sur face mining, half have had their water-carrying capacity reduced; along 4,500 miles capacity wa moderately reduced, and along 2,500 miles capacit had been affected only slightly.

Self-contained mining sites—quarries, dredge areas, and some area-stripped sites—do not hav enough runoff to warrant costly storm-water con trols. Contour-stripped areas can be used to man age runoff in much the same way as broad-base

Stream capacity reduced by sediment from surface-min ing operations.

423-412 O - 71 - 2

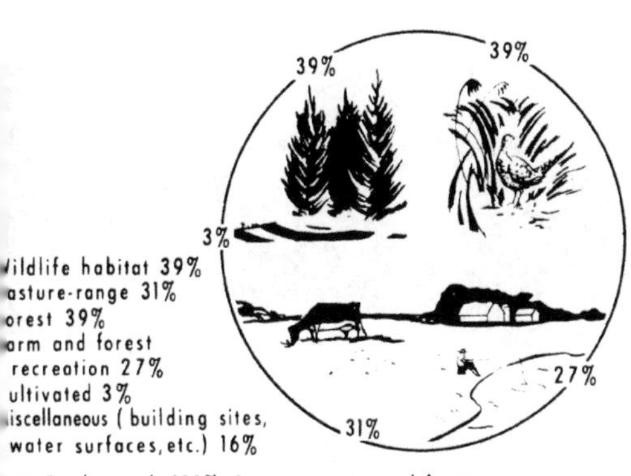

ote: Total exceeds 100%. Some areas are used for more than one purpose.

Potential uses of surface-mined land.

erraces. But on 98 percent of the surface-mined and studied in Appalachia—where most contour tripping is done—storm-water runoff control was ot adequate to prevent erosion, sediment, or looding.

On these areas, vegetative and mechanical meas- ires or a combination are needed. An example is he need for grading within some surface-mine pits o control storm runoff. About 75 percent of the ites need some grading, and only 45 percent have eceived any. Grading too much or on the wrong oil material, though, may make matters worse; pecial care and technical assistance are needed. In ome areas of the West, minor reshaping of some anks is adding to the beauty of the landscape.

Ponds

Many surface-mined areas have ponds or depres- ions, especially where area stripping has been one. Forty-two percent of the ponds are smaller han an acre, 40 percent are 1 to 10 acres, and 18 ercent are larger than 10 acres. Two-thirds are nore than 5 feet deep.

Acidity is a problem in some ponds—one-fifth f those studied had a pH rating of less than 4.5. The other four-fifths are less acid and include he larger and deeper ponds that have greater po- ential use. Some are being used even for municipal vater supplies.

Animal life was present in four-fifths of the onds, but scarce in the acid ponds.

Effect on wildlife

Disturbing land and water for mining naturally isrupts wildlife habitat. State fish and game commissions reported to U.S. Bureau of Sport Fisheries and Wildlife that nearly 2 million acres of wildlife habitat had been damaged by surface mining—68 percent of it east of the Mississippi River. Most damage resulted from:

- Stream widening, affecting water temperature and depth of spawning beds.
- Lake draining.
- Burying or removing spawning gravels.
- Diverting surface flow.
- Sediment.
- Chemical changes in soil and water quality.
- Removing food, nesting, and escape cover plants.
- Forming high walls that limit animal access or movement (a problem on about one-fourth of the high wall mileage studied).

Where proper restoration measures have been taken, fish and wildlife habitat has improved and often is better than before mining. Since the same kinds of wildlife use the mined site and adjacent land, there is opportunity for managing both areas together for wildlife habitat on private and public property.

Safety

One-third of the mined areas studied had some safety hazard, usually water. On 22 percent of the inactive areas there was evidence of abandoned buildings, equipment, debris, or rubble—some hazardous and nearly all unsightly. Ten percent had one or more deep-mine openings—without shaft sealing. Restoration measures, well planned and carried out, reduce the danger to public safety.

Properly treated and managed surface-mined areas can help meet growing needs for outdoor relaxation.

A 4-year-old shortleaf pine plantation on surface-mined land.

Accomplishments

USDA's participation in surface-mined-land conservation began in the 1930's. The Forest Service then began research on revegetating mined land and keeping acid and sediment out of streams. The Soil Conservation Service at the same time began helping landowners improve their soil and water resources and solve many land use and land treatment problems, among them surface mining.

During one 5-year period, 1960–64, more than 5,000 land owners and operators in 500 local soil and water conservation districts in 31 States applied conservation measures to nearly 128,000 acres of surface-mined land with USDA help (table 3). The survey team noted that the conservation districts considered restoration of mined areas as part of the total conservation job on individual properties or whole watersheds and not a separate or special activity.

During the same 5-year period State foresters, through Federal-State cooperative programs, provided technical help to more than 1,250 ownerships in replanting about 37,000 acres to trees.

Industry

Many mining firms are giving increased attention to the challenge of surface-mined-land con-

ABLE 3.—*Surface-mined-land treatment by cooperators with soil and water conservation districts, 1960–64*

State	Area	Landowners participating
	Acres	*Number*
abama	931	88
izona	1,000	3
kansas	809	66
lifornia	3,246	116
lorado	557	38
rida	6,326	89
orgia	970	252
aho	268	22
nois	6,367	94
diana	848	68
va	122	18
nsas	1,515	73
ntucky	13,784	1,139
uisiana	1,038	127
aryland	403	16
chigan	1,378	130
ssissippi	3,350	229
ssouri	2,546	33
ontana	4	1
braska	1,896	194
w York	824	93
io	23,613	433
lahoma	5,866	232
nnsylvania	13,043	306
th Carolina	8,441	140
th Dakota	2,945	77
nnessee	1,830	42
xas	1,110	26
ginia	10,102	58
st Virginia	11,890	887
sconsin	725	165
Total	127,747	5,255

vation. Reclamation associations formed in a mber of States have their own professional staff foster restoration work. Individual firms and eir associations have restored many mined acres, nducted demonstration projects and experimen- plantings, carried on substantial research work,

perly selected and planted, trees can aid in stabilizing nined area and increase its usefulness.

To speed the restoration job, professional assistance is needed in classifying spoil banks and interpreting the data for various land uses.

and in general promoted effective conservation treatment of surface-mined land.

For example, reclamation associations in the Appalachian region have done reforestation and seeding on 74,000 surface-mined acres. The National Sand and Gravel Association's members rehabilitated 52 percent of the acreage they mined in 1965, compared with only 25 percent just 2 years earlier. Phosphate mining firms in Florida, between 1961 and 1966, voluntarily restored 75 percent of the acreage mined during that period. Where mines are near urban areas, many phosphate miners have made plans *before mining* for later development of the site as residential, commercial, or recreation areas. And surface-mine operators in 22 States have formed the Mined Land Conservation Conference to promote restoration of mined land for useful purposes.

Many other firms and commodity groups have yet to follow these examples and respond to the challenge of surface-mined-land conservation.

Government

Fourteen States have laws requiring restoration work, most enacted fairly recently. Their provisions are compared in table 4, with three exceptions. Georgia and Kansas enacted laws that call for establishment of State boards to license and regulate surface mining and enforce restoration. Montana's law authorizes its Bureau of Mines and Geology to enter into contracts with coal strip-mine operators. Amounts spent for restoration work can be credited against the coal license tax. No bond is required. A restoration-plan map is called for that includes covering exposed seams, grading ridges that are near highways, constructing earthen dams, and planting recommended species for later forest or grazing use of the mined area. Work usually must be completed within 3 years after mining.

In addition, Iowa, North Carolina, and North Dakota have established advisory boards or committees to suggest restoration programs. In Colorado, coal-mining firms and the State's Department of Natural Resources have a voluntary contractual agreement dealing with restoration. This arrangement will be watched with interest.

A few other States have some control over sur face mining through water-pollution-control sta tutes. State funds have been made available t universities and foundations for research an demonstration activities.

Some local governments have used zoning regu lations to control mining and require restoratio of the land.

Beyond assistance to private-land owners, th Federal Government also has made some headwa in restoring surface-mined areas on public lan and is engaged in research work.

Research

USDA is the recognized leader in basic researc on surface-mined-land conservation. Most researc now underway in government and industry is i six categories:

(1) *Revegetation*—developing plant species tha will provide quick cover or permanent growth an comparing various combinations of seedings.

Water is a natural companion to surface-mined areas. Planning for its collection, control, and storage can enhanc the usefulness of the site and reduce offsite damages.

11

TABLE 4.—*State surface-mining licensure,*

State	Minerals covered	License or permit	Bonding requirement
linois	All	Required. $50 for 1st acre, plus $5.50 to $11.50 per additional acre depending upon quantity.	$200 per acre with a minimum of $1,000. $600 per acre with a minimum of $3,000 for gob which cannot be vegetated.
ndiana (new law effective January 1, 1968). Administered under policies of natural resources commission.	Coal, clay, and shale	$50 plus $15 per acre	Minimum $2,000 at the rate of $300 per acre. $225 per acre is released on completion of grading requirements.
owa	Coal, gypsum, clay, stone, sand, gravel, or other ores of mineral solids for sale or processing.	Original license, $50. Renewals annually, $10 per year.	Equal to the estimated cost of rehabilitating the site as required in sec. 17 of the act.
entucky	Coal, clay (except ball clay).	Required. $50 per year, plus $25 per acre. Approved reclamation plan before issuance of permit. Map accompanying request for permit must be certified by a registered professional engineer.	$100 to $500 per acre, with a minimum of $2,000.
laryland	Coal	Required to register with Bureau of Mines, and $100 1st year and $10 renewal each year.	$200 per acre for land affected with a minimum of $1,600. Cash deposit is acceptable.
hio	Coal	Required. $75 per year, plus $15 per acre.	$300 per acre, with a minimum of $2,000. Cash deposit is acceptable.
klahoma	All	Permit. $50 per year flat fee	$50 per acre or assessed value of the land the preceding year, whichever is lesser. Cash or government securities acceptable.
ennsylvania	Anthracite and bituminous coal.	License or permit. $300 per year flat fee. Mining and backfilling plans required before permit issued.	In any case not less than $500 per acre, but regulating agency may bond to $1,000 per acre if conditions warrant. Minimum bond $5,000.
ennessee	All solid materials in natural deposits except limestone, marble, or dimension stone.	Required. Fee of $250 per year plus $25 per acre, not to exceed $750.	Not less than $100 nor more than $200 per acre, as determined by State Commissioner of Conservation.

12

Reclamation required	Refuse	Substitution of sites
Conditioning to make suitable for productive use including forestry, grazing, cropping, wildlife, recreation, and building sites, according to a plan. Ridges to be struck off a minimum of 10 feet for forestry, 18 feet for pasture, and graded to allow use of farm machinery for cropland. Plant species to be used must be approved by the department of conservation. Acid forming material to be covered by 4 feet of water or other material capable of supporting plant life. To be completed within 3 years following the permit year except planting will be delayed where weathering is needed to establish plants.	Slurry to be confined in depressions or by levees and screened with border plantings. Infertile gob to be covered by a minimum of 4 feet of productive material. To be reclaimed within 1 year after active use.	Subject to approval of Department of conservation.
Grade to a rolling topography to reduce erosion and permit best land use. Acid materials in final cut to be covered with water or earth. Establish satisfactory vegetative cover prior to bond release. Submit plan of reclamation prior to mining. Roads constructed to minimize erosion. Damming of final cut to form lakes.	Remove or bury all metal, lumber, and other refuse resulting from operation.	Subject to approval of regulating agency. Substitute site must be an equal area previously mined by the operator requesting such substitute.
Spoil bank when feasible. Avoid acid forming materials on surfaces of spoil. Grade spoil banks to regular slope of not over vertical rise of 1 foot for 3 feet horizontal except when original slope was greater. Control drainage. Cover acid forming materials with 2 feet of earth.	No specific provisions	No specific provisions.
Spoil grading concurrent with operation and reclamation completed within 12 months of permit expiration. Stand percentage complete prior to bond release. Coverage of acid producing material.	Remove or bury all metal, lumber, and other refuse resulting from operation. No depositing refuse or spoil material into public roads, streams, lakes, subterranean waters, or other public property.	Allowed with respect to planting only, subject to approval of division if investigation shows that revegetation of original site may not be successful.
Rules and regulations of Bureau of Mines. Spoil graded to minimize erosion, depressions, and steep slopes. Overburden graded to cover final pit. All openings from underground mining sealed off. Impoundments approved by Department of Water Resources.	No provisions	No provisions.
Grade to rolling topography to reduce erosion and permit logging or grazing. Grading of isolated peaks. Access roads and fire lanes. Construction of earth dams in final cut. Reclamation complete 2 years from completion of stripping. Substitution.	Loose coal, mine refuse, and other debris to be graded so as to reduce the piles of such material and make possible its submergence in water. If not covered by water, material shall be covered with overburden.	Subject to approval of Division of Reclamation. Affected area to be substituted must be equal acreage to the original area affected.
Grading to rolling topography. Cover acid material with at least 2 feet of soil. Fire lanes will be built in forested areas.	No provisions	Subject to approval of Department of Mines and Mining.
Bituminous.—Original slopes 12° or less, backfill to original contour. Original slopes more than 12°, terrace, high wall backfilled to 45° angle. Stream pollution from acid drainage not permitted. Anthracite.—Pits near roads and buildings must be completely backfilled. Peaks and ridges must be rounded off. Backfilling to be completed within 6 months after end of operation. Disturbed areas to be planted to trees, shrubs, or grass within 1 year after completion of mining operation, or forfeit $100 per acre.	No specific provisions	No statutory provisions; however, the operator may option not to plant and to pay $100 per acre to State instead.
Coal.—Covering of exposed coal, drainage, and water control. Grade to preserve existent roads and provide favorable conditions for revegetation. Minimum crest of spoils 20 feet wide. (If restored for normal cultivation, operator relieved of further rehabilitation.) Other materials.—Minimum crest of spoils 15 feet wide. Plant trees, shrubs, grasses, and so forth up to $25 per acre, one time. Operator may pay full estimated cost of revegetation and be relieved of further responsibility. (If restored to permit normal cultivation, operator relieved of further rehabilitation.)	Remove or cover all metal, lumber, and other refuse except vegetation.	No provisions for substitution Act appears ambiguous: One section states operators obligation not discharged until revegetation meets Commissioner's standards; another that maximum planning expenses of $25 per acre shall be required.

13

State	Minerals covered	License or permit	Bonding requirement
irginia	Coal only	Required. $150 initial fee. Permit approval based on approved reclamation plan and initial bonding of $75 per acre, based upon number of acres of land the operator estimated will be disturbed by strip mining during next ensuing year, with minimum of $2,500, before issuance of permit. In approving plans of reclamation and issuing rules and regulations, soil and water conservation district supervisors may be asked to advise, assist, and provide local facilities.	Within 30 days following anniversary date of issuance of permit, operator shall post additional bond in amount of $50 per acre for each additional acre of land estimated by him to be disturbed during the next year following anniversary date of permit.
Vest Virginia	Coal, clay, manganese, iron ore.	Prospecting permit: $150 per acre for area disturbed during prospecting. Permit to surface mine: $100 initial fee, $50 annual renewal. Special reclamation fee: $30 per acre for land disturbed.	Minimum $3,000. $100 to $500 per acre disturbed. Director of Department of Natural Resources will set rate per acre.

(2) *Chemistry of overburden and spoils*—iden-ifying soil and rock mixtures, soil and water characteristics, and effects of fertilization and weathering.

(3) *Hydrology*—studying water and drainage effects, sedimentation, and ground-water movement and storage.

(4) *Earth movement and placement*—finding new or adapted equipment and methods for mining and more economical restoration.

(5) *Haul roads*—designing better and safer access roads as well as better hauling equipment.

(6) *Land use potentials*—making guidelines for finding the best use for a mined area consistent with the community land use pattern and needs, characteristics of the mined land, and cost-return factors.

There are many areas of study in which more research is needed to improve both surface mining and the reuse of the mined areas:

Comprehensive knowledge of physical and chemical characteristics of spoil materials is needed, as well as interpretations or ratings of surface-mined-areas land use potentials or limitations.

Better methods are needed for lifting, moving, piling, and relocating overburden, especially on sloping land.

More knowledge is needed about the responses of many different plants and about their usefulness for landscaping, screening, protective cover, wildlife habitat, and soil building.

Improved methods of preparing surface and subsurface water storage are needed to make effective onsite water use and prevent pollution and excess runoff.

Potential and challenge

Properly planned, treated, and developed to blend with adjacent land use patterns, most surface-mined areas have great potential (table 5). Thirty-two percent of the areas surveyed provide an outstanding view of mountains, valleys, or lakes. Haul roads can open up many areas to visitors for the first time. Ponds can give an area greater economic value than it had before mining. And most areas can be kept in private ownership.

With today's growing land use demands, particularly farm and forest recreation, these opportunities deserve attention. The challenge to USDA is to assist in developing resource uses in surface-mined areas that will be compatible with one another and with uses of adjacent land.

A similar challenge is to make sure that the optimum benefit—both to the landowner and the community—is derived from each dollar spent in mining and land restoration. Some shallow deposits would better be left unmined where restoration costs would be prohibitive. Some mined sites would best be treated to prevent offsite damage but not developed. In some areas, mined land can be treated and managed for intensive use.

bond, and reclamation requirements—Continued

Reclamation required	Refuse	Substitution of sites
Regrade the area in a manner to be established by rules and regulations which provide the following: 1. All surface deposits of removal overburden will assure a surface of gently rolling topography. 2. Preservation of existing access truck roads and needed roads for recreation or forest fire protection. 3. Plant vegetation upon parts of areas where revegetation is practicable. 4. No water impoundments can be built by operator for wildlife, recreation, or water supplies without prior approval.	Remove metal, lumber, and other debris resulting from mining operations. Grade loose coal, refuse, and other debris on bottom of last cut so as to reduce the piles of such material in accordance with good conservation practices.	Subject to approval of Department of Conservation and Economic Development
Cover the face of coal. Bury all toxic material, roof coal, pyritic shale, and material determined to be acid producing. Seal off any breakthrough of acid water caused by the operator. Impound, drain, or treat all runoff water. Plant species adapted to site as prescribed in a planting plan, within 1 year after mining is finished unless the planting is deferred by the Director of the Department of Natural Resources. Operator may satisfy the requirement for reclamation by contracting with local soil conservation district or a private contractor.	Remove or bury all metal, lumber, equipment, and other refuse resulting from the operation.	None.

TABLE 5.—*Potential multiple and alternative uses of surface-mined areas in several States* [1]

[In percent]

State	Cropland	Pasture land	Rangeland	Woodland	Wildlife habitat	Ponds and reservoirs	Farm and forest recreation	Residential, institutional, industrial	Other unspecified
Arizona	0. 5	0. 5	19. 7	0. 2	17. 6	2. 4	9. 6	6. 4	49. 8
Arkansas	. 3	6. 2	14. 7	44. 4	15. 5	10. 0	6. 8	1. 1	1. 3
California	. 5	8. 6	26. 9	17. 8	62. 5	5. 2	46. 2	10. 4	7. 3
Florida	2. 8	10. 8	42. 6	48. 4	46. 6	14. 9	53. 2	6. 1	3. 7
Illinois	15. 2	49. 6	. 9	27. 3	31. 8	9. 0	28. 9	7. 3	3. 0
Indiana	2. 4	15. 9	(2)	55. 9	49. 2	12. 7	47. 5	12. 9	6. 0
Kansas	1. 3	14. 7	20. 5	24. 2	32. 1	5. 8	11. 7	3. 0	3. 7
Louisiana	. 2	2. 8	(2)	59. 3	30. 5	21. 7	25. 2	1. 7	3. 3
Michigan	3. 3	7. 0	. 1	34. 4	24. 8	5. 6	12. 3	3. 7	13. 6
Missouri	. 2	38. 1	5. 8	42. 0	39. 9	8. 8	18. 1	2. 2	17. 6
Nebraska	1. 3	13. 0	23. 1	5. 7	37. 3	30. 1	49. 7	8. 9	3. 4
Oklahoma	. 4	56. 8	32. 4	33. 2	50. 6	10. 4	25. 6	14. 1	1. 7
Pennsylvania	10. 0	20. 0	(2)	80. 0	92. 0	2. 0	12. 0	13. 0	5. 0
West Virginia	9. 0	5. 0	(2)	75. 9	10. 0	. 1	34. 0	. 5	. 5
Average	3. 4	17. 8	13. 3	39. 2	38. 6	9. 9	27. 2	6. 5	8. 6

[1] The percentages exceed 100 for individual States and the national average because more than 1 potential use may apply on some areas.

[2] Less than 0.1 percent.

efore this area was stripped, the mining firm planned as part of its operation to reshape and seed the site for this asture—at no cost to the landowner.

Principles for a National Surface-Mined-Land Conservation Effort

The mining industry, conservation districts, and l levels of government should work together to ıt practical principles into surface-mining operions at every site:

REPLANNING.—Make good mine housekeepıg and practical restoration measures an integral ırt of plans for the site—before any mining ;tivity begins. Include a plan for both interim and ıal land use where practicable.

TABILIZATION.—While mining is going on, ke steps to control erosion on the site and on ıul roads, including establishing quick-growing lants. Plant permanent cover to protect the area fter mining, and reseed or replant where previous vegetation has failed.

TORM-WATER CONTROL.—Plan control of ırface runoff on a watershed basis to fit stream pacities and prevent harmful sediment deposits.

/ATER QUALITY.—Place highly toxic spoil aterial only where it can be covered with other verburden or a permanent body of water. Seal off ıger holes and any breakthrough to former underground mines. Control drainage from sites and haul roads to keep toxic substances and sediment out of adjacent streams.

WATER STORAGE.—Create as many lakes as practicable, to aid water control and increase potential use of the mined site. Dams and ponds should be designed properly to guard against failure.

AIR QUALITY.—Help prevent offensive noises and air contamination by controlling use of explosives, fire, and motorized equipment.

NATURAL BEAUTY.—Plan operations so they have a minimum impact on the landscape. Make treatment work practical and pleasing to the eye.

HEALTH AND SAFETY.—Take steps before, during, and after mining to minimize hazards from equipment, structures, and water areas.

Mined land should be devoted to the highest and best possible uses compatible with the use patterns of adjoining land and with the geographic location, topography, and other site characteristics.

Information

Those involved in surface mining and restoration of the areas—and those who use the products—must be kept abreast of social, scientific, and economic developments that affect their efforts.

Education in both the program responsibilities and scientific aspects should be fostered by the Federal Government. Universities and colleges provide formal knowledge in this field; the less formal is supplied by trade schools, correspondence courses, field days and workshops, and on-the-job training.

Lectures, field demonstrations, and onsite guidance in solving mined-land problems—the how-to-do-it—would aid in extending new ideas, new methods, and new techniques.

Field trials or tests should be expanded to follow through on basic research in plants, techniques, and methods and to demonstrate their effectiveness. USDA offices located in nearly every county in the Nation can fill many of these information needs in their everyday dealings with local citizens and groups.

Leadership and assistance

Federal and State agencies should make use of experience gained in activities closely related to surface mining as guides to assistance in surface-mining operations and conservation.

For example, USDA has leadership in developing and interpreting soils information and in helping land operators make effective use of it. This information with interpretations specifically for surface-mined land would have great value both in finding potential sources of surface-mine deposits and in restoring surface-mined land to safe, productive use.

Since the problems and opportunities concerning surface-mined land are largely on private rural property, USDA has a major responsibility to provide Federal leadership and assistance in its restoration.

The 186 million acres of National Forest under USDA jurisdiction are managed for mineral resources as part of overall resource management. Since much National Forest land is intermingled with privately owned land, the use and management of one is coordinated with the other to provide maximum private and public benefits.

USDA works closely with private landowners and with State and local governments. Its assistance on private land is channeled through soil and water conservation districts, State foresters, and State and county extension programs. Each conservation district has a program that fits its local problems and is a central source of help in solving these problems. Most surface-mined land is in a soil and water conservation district.

USDA endorses the type of national mined-land conservation effort outlined in these pages. It is a use of the same principles USDA has followed for years in its cooperative work with private landowners. Accomplishments already made by soil and water conservation district cooperators, the mining industry, and Government show that such a program can do the job.

Conclusions

Proper treatment of surface-mined land is an integral part of the total resource conservation effort on private and public land. To this end, USDA recommends as a four-point course of action:

1. *That Federal agencies demonstrate leadership by restoring their surface-mined land.* Each agency managing public land should develop a plan for completing the job within 10 years. Each agency should establish adequate safeguards to prevent harmful effects from surface mining on its land in the future.

2. *That treatment of old mined areas be accelerated.* The Federal Government should participate with States, counties, municipalities, the mining industry, associations, conservation districts, private individuals, and others in developing long-range, comprehensive restoration programs—designed on a watershed or drainage-area basis. Federal technical and financial aid should be on a long-term contract basis.

3. *That to deal with the problem of future rehabilitation of surface-mined land, Federal agencies extend their knowledge and assistance to States and producers of the 50-odd commodities involved.* Technical information should be disseminated as it is developed. Federal agencies should study existing State statutes on mined-land restoration (table 4) and develop model statutes. The goal should be the blending of knowledge and trust between all levels of industry and government in the interest of mining with a minimum of adverse effects.

4. *That Federal research programs, studies, and field demonstrations be expanded.* Many problems of treating mine spoils have not been solved and many opportunities remain unrealized. Present research efforts are inadequate. The problems examined in this report need specific attention.

Reprinted from *Landscape Architecture,* Jan., 125–128 (1966)

The inert becomes "ert"

23

A new approach for reconstructing California's old gold fields

by Russell A. Beatty

Aerial photo of typical "severe" gold dredged land. Elevation differential between rows of arching piles is approximately thirty feet.
(Photo courtesy of California Department of Water Resources)

Gold

"Gold!" A word, a cry, a mineral, but above all—an era. Ever since that first innocent call went out in 1848 at Coloma, the economy, the history, and the lustre of California were boosted to such great heights that California and gold became synonymous. Until that time California was a far-off land saved for adventurers. The cry of gold brought out ingenuity that men never knew they had. And the men were the Forty-niners, a rough, tough lot who extracted the precious bits of metal from the womb of the Sierras. They picked, dug, panned, washed, and hydraulicked the streams and hills of the Sierran foothills. The scars they left are great, but are forgiven in the romance they provided.

Sandbox exercise can be useful preliminary to re-design of mined-over tailings area in old gold fields. In three photographs (right, and at bottom) author illustrates simple variations of contours to be achieved in re-grading. Larger number of variations are possible when mining is pre-planned with finished landscape as a goal. Center photo below shows variations in gravel and cobbles. The large cobbles can become a strong visual element, or else can be buried as a base for land forms.

Gold mining did not begin and end with the Forty-niners. Gold recovery became a serious business which lasted until 1962. The chief method of recovery in the 1860's was hydraulic mining, which ended abruptly with a court decision that prohibited the dumping of mining debris in the Sacramento-San Joaquin River systems.

Deep, gold-bearing gravels had been discovered along the Feather, Yuba, and American rivers. The only feasible means of extraction was mechanical, and the machine to be used was the floating bucket dredge. Adapted from a New Zealand dredge, the California dredge began operating in 1898 along the Feather River. Increasing in size and capacity, dredges were developed to handle 125,000 cubic yards of gravel a week. Gold was recovered from gravel dug as deep as 100 feet; but the average was 50 feet. During the period 1898 to 1954, bucket dredging yielded a total of $255,000,000 of gold in three principal counties—Yuba, Butte, and Sacramento.

Today thousands of acres of derelict land along the streams and rivers at the eastern edge of the Great Valley remain just as they were left 25 or 50 years ago. Some of the tailings have been removed for aggregates by sand and gravel companies or for use in the construction of the Oroville Dam. A few isolated areas have been leveled, but most of the dredged land is unusable in its present form.

Physical Characteristics of Tailings

Dredge tailings in California are unlike any other form of surface mining spoil. The unique characteristics of climate, topography, and geology combine to give dredged land its own physical properties.

The dredging operation essentially turned over the soil profile. The fine sand and clay silt were sluiced out the rear of the dredge and deposited at the bottom of the pond. Coarse gravels and cobbles were deposited on top of this by the tailing stacker, which piled the material in broad, sweeping arcs. Since no real volume of material was extracted and the fines were partially separated from the gravels, the resulting grade was about 15% higher than the original terrain. Some dredges had multiple stackers which mixed the tailings more uniformly and reduced the "swell."

Even though the standard dredge inverted the original soil profile, a wide variation in the resulting profile exists. Shallow dredging, multiple tailing stackers, and varying original profiles have caused a non-uniform underground "soil" condition. Test holes dug by the State Division of Highways on dredged land reveal a variation in the amount of fine sand and clay material at various depths. Coarse gravels averaging four inches in size usually occur on the top but fine material capable of supporting vegetation can occur within the top ten feet, instead of at the bottom of the tailings.

Other underground characteristics are more constant. Dredge tailings are extremely porous. Water can percolate immediately to the ground-water level. This characteristic prohibits any land use in which waste material can possibly pollute the ground water (sanitary land fill, etc.).

Similarly, because of the roundness of the gravels, dredge tailings are difficult to compact. Therefore, the problems of growing plant materials encountered when coal strip-mine spoil is graded will not occur in dredged land rehabilitation. If compaction is required, a vibrating roller and a sheepsfoot roller with wedged lugs will suffice.

Since the surface of dredged land whether graded or not, consists of large gravels, erosion and ponding are not problems. Nearly 100% of all water received can percolate to the water table. Likewise, since the soil structure and normal capillary action are destroyed, little water is lost by evaporation. The surface cobbles, being poor heat conductors, act as a large-scale mulch. Like heated bricks the surface cobbles retain a great amount of solar heat. The micro-climate is affected and severe frosts are probably inhibited or at least lessened to some degree.

Even though the coarse tailings will repose at an angle of 1:1, the surface is rather unstable. When a force is applied, such as when walking, the cobbles slide like marbles, making unvegetated piles almost impossible to climb easily and safely. This disadvantage can be turned to good advantage when the tailings are graded by machine.

Grading and Earthwork

Once the clean layer of cobbles is moved and the fine material mixed with gravels, tailing piles are relatively stable and easy to move. The Natomas Company reports that simple leveling with a bulldozer costs about 25¢ per cubic yard or about $2500 per acre. Land previously unsalable can then be sold for prices up to $10,000 or more. The use of carryalls to move the material will increase the costs to about 50¢ per cubic yard.

If large scale earthwork is required, a wheel bucket excavator with a portable conveyor belt would accomplish the job more efficiently. At Oroville one of these excavators is used to remove tailings for construction of the Oroville Dam. A 30-foot wheel with eight two-cubic yard buckets literally eats 6,800 tons of tailings an hour to a depth of thirty feet. Since little grading has been done, the costs of completely rehabilitating dredged land are not known. Two facts are certain, however: tailings can be graded easily, and the value of the land will increase proportionately with the effort and imagination used.

Vegetation

On any land which has been destroyed by fire, landslides, or surface mining vegetative cover soon begins to soften the evidence of destruction. On gold dredged land the number, species, and physical condition of volunteer plant materials indicate that moderate to good vegetative growth is possible. On the more severe tailings vegetation occurs primarily in the ravines between tailing piles where finer materials have settled. Most of these indicator plants are trees and shrubs native to the Sierran foothills and the Central Valley of California.

Leveled tailings seem to support a fairly uniform natural vegetation. Fine sand and clay silt are usually mixed sufficiently with the gravels so that seeds are able to germinate during the rainy season and develop into healthy plants. The predominant trees growing on the dredged land near Folsom on the American River are cottonwood (*Populus fremontii*) and digger pine (*Pinus sabiniana*). The tree cover is sparse with volunteer grasses, lupines, and other annual herbaceous plants covering the surface of the leveled or less severe tailings. This savannah appearance suggests a pleasant park-like landscape.

On an area of leveled tailings near Folsom a grove of eucalyptus (*E. globulus*) and a grove of olives (*Olea europaea*) had been planted in 1909. Both plantings still exist and are growing although rather poorly. Due to neglect and a lack of data on their history, they serve as poor examples of exotic plantings.

A more recent attempt indicates that tree planting on dredged land is possible. The State Division of Highways has planted trees at a highway interchange built with dredge tailings near Oroville. One-gallon aleppo pine (*Pinus halepensis*) and five-gallon Chinese pistache (*Pistacia chinensis*) were planted in March 1964. Imported sandy loam was placed in an oversized hole and the trees were planted. A fertilization and irrigation program has resulted in normal established trees with good growth and no signs of weakness. The native trees and shrubs, as well as the few exotic plantings I have observed, seem to indicate that woody plant materials can be grown successfully on graded dredge tailings. Lawns and herbaceous ground covers are impossible unless imported topsoil is provided.

Design Implications

The development potential of this seemingly worthless land is as rich and exciting as the imagination will allow. To a child a sandbox is a private world where the silica particles—dry, glaring, and hard—can be transformed into great castles, battlefields, villages, farms, and a thousand forms limited only by the child's imagination. By adding a little water, a few sticks and stones, and a toy tractor the child can create myriad worlds reflected by his moods and desires, limited in extent by the sides of the sandbox. The inert becomes "ert." The static particles become a plastic mass at the hands of the creative being—the child.

With dredge tailings we have a similar situation but at a greatly magnified scale. The great acreage of dredged land is limited physically only by its extent or by a river and theoretically limited by property lines imposed by man. The material, like the sand in the box, is hard, dry, and loose, but is potentially plastic. The tractor is larger, as are the other impressive forms of earth moving machinery of today. Man still has the imagination, displayed earlier in his sandbox, if he will only use it.

The unique physical properties of dredged land combined with the visual characteristics of the regional landscape can suggest forms of development. In place of leveled tailings and standard urban or even park development, a highly livable yet economically feasible type of development can be created. By taking advantage of the plasticity of tailings, visually exciting land forms are possible. Combined with clustered, high-intensity land use, the cost of land shaping can be more than offset by the economy of condensed development. The result will be a highly efficient landscape with a predominance of open space.

The coarse gravels and cobbles can be used to form the base for land forms. By removing 20 to 30 feet of surface material to form low hills, a fairly fine soil would be uncovered. The land form can be used to support tree growth or for clustered buildings. The fine soil material of the lower excavated land is then capable of growing grass and other plant materials.

For example, if a residential subdivision is to be built, a variation of some of the newer cluster developments is suggested. Clusters of living units with small, low-cost gardens and recreation areas would leave up to 50% of the land open for visual relief. Only a small amount of imported topsoil would be required for cultivated plantings. The gravelly soil could be planted with trees and some woody ground covers but would remain essentially as low-maintenance open space. Similarly a park or large recreation area could be developed on the same principle. Grading and tree planting could comprise the major portion of the development. Small, intensive-use areas can be designed to limit the necessity of importing great quantities of topsoil. Paths and trails can interlace the open space, providing a pleasant natural environment without the extensive use of lawns and exotic shrub masses.

This same principle can apply equally as well to industrial parks. Space and the character of the regional landscape can dominate with only small intensive gardens developed in conjunction with the major building groups.

These are only a few ideas derived from an initial study of dredged land rehabilitation. A much wider spectrum of development forms is possible. The "sandbox spirit" can result in a landscape that will best satisfy the ideals, purposes, and needs of the people of the region.

The man in control of the dredge tailings or any other similar land has a great advantage over the child in the sandbox. At his command is a vast cumulation of human and material resources backed by years of experience and knowledge. The summation of imagination added to the available facilities can result in the creation of a great human landscape. By using only a fraction of this formula the results can be as devastating, if not worse than the existing conditions.

Too often man has arrogantly ignored the full potential of available land and has destroyed what was once a beautiful and livable landscape. He has used but an ounce of imagination or none and the easiest tool—the bulldozer—in a headlong attempt to make money by providing living facilities for the burgeoning populace. Today's urban and suburban landscapes all too clearly demonstrate that the results have been negative and that the facilities are not livable after all. In considering the uses of dredged land we have one great advantage; the land has already been destroyed. What can be recreated depends upon the ability of man to use this formula to its fullest and best extent.

If the most sensitive and best use of gold dredged land is to be achieved, a coordinated research program is needed. Mined land rehabilitation is a complex process involving a whole gamut of environmental sciences concerning soils, water, horticulture, engineering, and others depending on the ultimate land use. The key member of this team should be the landscape architect. No other profession is involved with all the sciences of the environment, particularly those related to the rehabilitation of devastated land. This is landscape architecture in its broadest sense.

Example of strong landforms in this functional earth sculpture at Aerojet General's rocket engine testing plant near Sacramento, Calif., built from gold dredge tailings.

Reprinted from *Landscape Architecture*, Jan., 136–140 (1966)

A New Technology for Taconite Badlands

New mining techniques offer a chance to do more than create a "moonscape" in Wisconsin

by Ervin Zube

24

What's left of "suburbia" around Hibbing, Minnesota, after generations of iron mining. The same moonscape faces scores of other regions unless old procedures are revolutionized to make by-products of mining fit into new landscape.

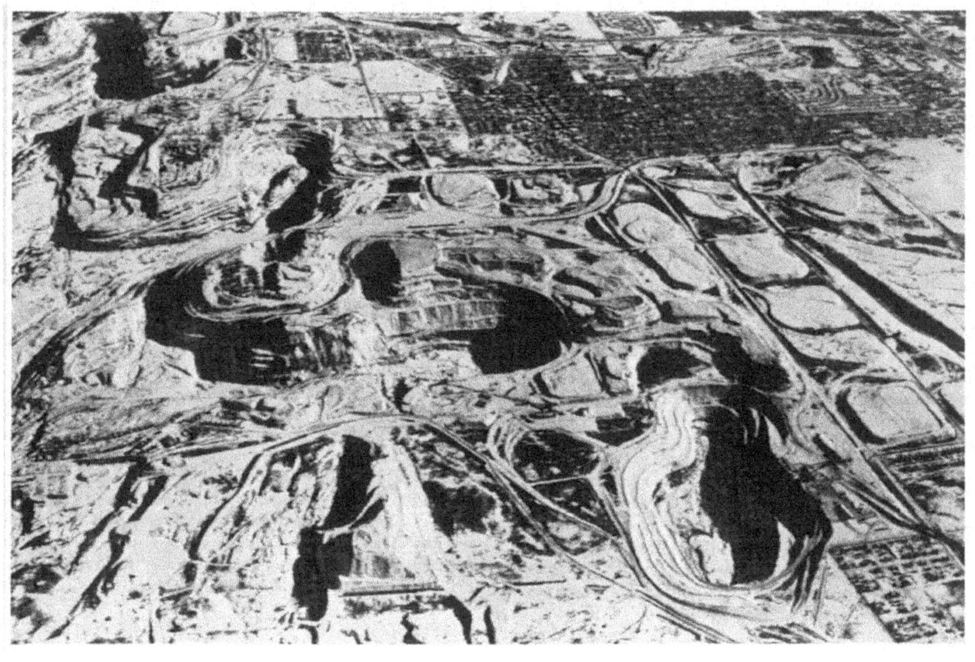

Pits and dumps—man-made forms imposed upon the landscape . . . forms that are the product of and a testimony to the giant technological strides of Twentieth Century America . . . forms generated by a demand for iron ore to feed a steel-oriented economy . . . forms that have developed in chaotic juxtaposition to the cultural patterns associated with human habitation and to the ecological processes of the landscape.

The first signs of an emerging, man-made American landscape of open-pit mines and waste dumps appeared in the late 19th century when the then vast deposits of high-grade iron ore of the Lake Superior region were first tapped. The subsequent development of diesel- and electric-powered earth-moving equipment accelerated the mining activity and provided the means for expanding the scale of operation. Planning had not preceded the mining, and pits and dumps began to encroach on towns and cities. The built-up areas then had to back away from the growing mining operation to provide the space required for this industry—the base of the regional economy. Unplanned land use patterns and land forms proliferated into a discordant environment, frequently described as "man-made badlands."

This method of mining iron ore and its resultant landscape have by-passed the state of Wisconsin. But they are on the horizon. Our purpose here is to learn from past experience and to attempt to set forth ideas and principles that will generate a new landscape as a direct result of open-pit mining. It must be one that has the potential for supporting future economic use and also that will relate physically and visually to the surrounding unmined landscape. We are not interested in "man-made badlands" for northern Wisconsin.

When open pit mining commences in Wisconsin, and it will, because of the great reserves of low-grade ore in the state, it should be on the basis of a firm commitment to a sound conservation ethic. This conservation ethic must take into account not only the wise use of the ore that is extracted, but also the wise use of the waste material that is a by-product of the mining; the land on which the waste material is deposited; the pit that is created; and an understanding of the innumerable interrelationships of man's activities with the natural landscape. In other words, we must conscientiously concern ourselves with that which will be left when the iron ore is gone.

The geologic forces that created the Gogebic Range in Wisconsin and Michigan provided a set of conditions that were initially more favorable to shaft mining, but technological advances and changing economic conditions after World War II contributed to the demise of this shaft mining. At the same time, the giant strides made in improved earth-moving equipment during and immediately after the war worked to the advantage of the open-pit mine. Now, pits three to five miles in length and 400 to 500 feet in depth can be seen on the Mesabi Range, a testimony to man's ability to literally move mountains.

However, the once vast reserves of high-grade direct-shipment ores containing 50 to 55 percent iron are rapidly being depleted. The great future of the Lake Superior Iron Ore Region, including northern Wisconsin, now appears to lie in the vast deposits of low-grade ores commonly called taconites, containing 25 to 30 percent iron.

Taconite mining, like the mining of high-grade ore, is essentially a quarrying operation. The ore body is exposed by stripping off the covering mantle of glacial till or overburden, which can vary in depth from practically nothing to about 100 feet. This is the first waste material. As the ore is removed, areas of non-iron-bearing rock, or lean ore, are encountered. This is the second waste material. The ore is then crushed and separated magnetically. The coarse tailings from the initial separation, the third waste material, consist of particles about one-tenth inch in size. A second and final separation occurs after the ore has been ground to the fineness of face powder and the fine tailings, the fourth waste material, are water borne in suspension to tailings or settling basins. The remaining particles containing 63 to 65 percent iron are then agglomerated into sinter clinkers or nodules of usable size and strength by intense heat (No. 2).

The waste material produced is equal in volume to approximately one and one-half to two times the volume of ore extracted. It is created in part by the conversion of a dense, hard, raw material into granular materials of varying sizes containing a great amount of air space in their aggregate mass. A pit measuring five miles in length, 1000 feet in width at the surface, and five hundred feet in depth, will yield in excess of 750 million cubic yards of waste material.

A program that considers the future as well as the present must consider these by-products of the taconite operation as resources for the creation of a new dynamic landscape rather than as waste materials. Each of the resource materials possesses characteristics that should influence its use in the design and development of the new landscape.

Overburden. Susceptible to erosion. The natural angle of repose of the specific material comprising the overburden should be a determining factor in the design of the land form. It will support vegetation; field investigation of overburden dumps indicates success with tree plantations and also vegetative cover established by natural invasion.

Rock reject. Coarse textured material of a rust red color. Vegetative cover will be very sparse, occurring only in minor pockets where sufficient soil fines settle to support minimal plant growth.

Coarse Tailings. Fine textured material of a light gray color. Vegetative cover is unlikely.

Fine Tailings. Very fine textured material susceptible to wind erosion when dry. Field investigation indicates partial success with seeding of grasses and the establishment of sparse cover by natural invasion. Method of disposal produces a tailings plain (No. 4).

These then are the resources for the creation of new land forms. The forms created and their relationships to the landscape will be determined on the basis of several factors in addition to those listed above. The economics of transportation, the operational requirements of earth-moving equipment, and fragmented land ownership patterns have been traditional determinants. Section lines and individual property lines create arbitrary boundaries to land forms that should relate to the landscape and not the political subdivision thereof. But, even within the bounds of these determinants, one can find a variety of forms ranging from stepped ziggurats through rounded mounds of varying sizes to large plateaus occasionally measuring over one and one-half miles in length (No. 3).

We must think in terms of these man-made forms designed in relationship to the existing topographic pattern—not filling valleys and drainage ways—and relating to the form and scale of the surrounding undisturbed landscape. The land forms can be composed on the basis of the positive qualities of each material into a unique landscape expression that minimizes the problems of dust and erosion by establishment of vegetative cover. They must also relate in scale and form to the pit, the configuration of

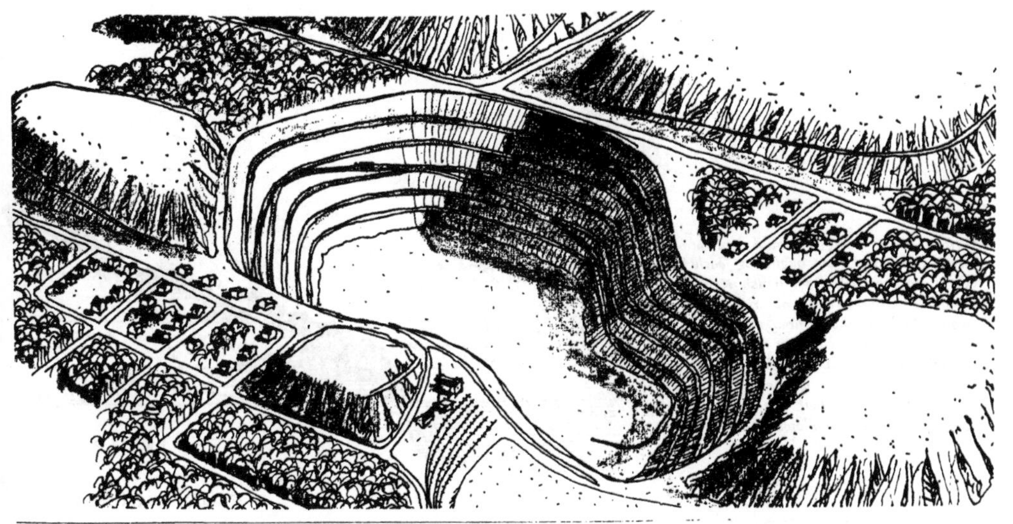

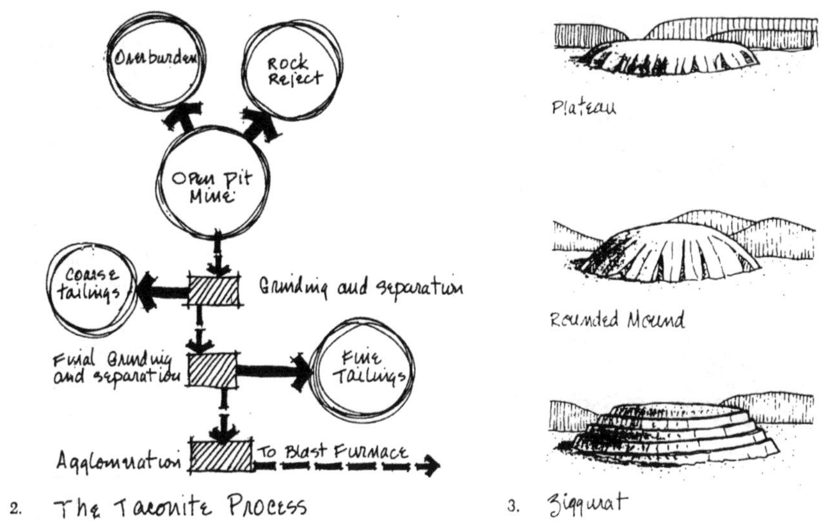

2. The Taconite Process

3. Ziggurat

Raw materials for new landscape come from holes such as this in the traditional ruination called open pit mining (top). Taconite process yields varied raw materials for new landscape forms, however: plateau, rounded mound, ziggurat and others. Sketch at right shows how each material can be used to form landscape most suitable to the material's potential.

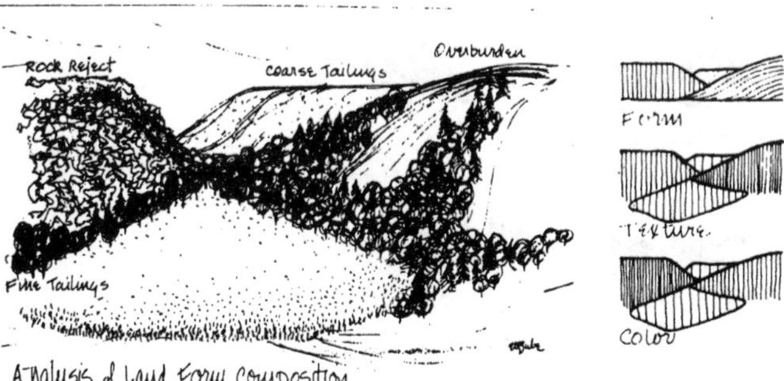

Analysis of Land Form Composition

138

Land form designed for summer cabins

Rounded cut thru over-burden - eases transition and reduces erosion

Existing Overburden

Access to Water surface

Flooded pit

Partial section thru Pit

Strong landscape forms could dominate the scene if planned from beginning, each form suited to materials from which it is made (below). Cross-section at right through water-filled pit shows how over-burden of fertile soil is retained, then re-graded to the water's edge to provide vegetation and natural surroundings for housing developments around the edge. Shelf just under water's edge provides safe shallows for swimming.

5.

Land Form Resources

Overburden

Rock Reject

Coarse Tailings

4. Fine Tailings

which is mainly determined by operational requirements, such as access for equipment and egress for ore; geologic determinants including the dimension and tilt of the ore body; and land ownership boundaries. The overburden, rock reject, and tailings dumps can be utilized as elements of transition in bridging the gap between the obvious man-made qualities of the pit and the natural landscape. This element of transition can be accomplished by forms either in harmony, as the plateaus or rounded mounds, or in contrast, as the stepped ziggurats, with the natural land form. They should also be designed to accommodate some future economic use, be it visual amenity, wildlife habitat, or building sites for recreational uses (No. 5). Much of the Lake Superior Region is benefited by a relatively high water table; thus many of the pits and shafts, when abandoned, are rapidly filled by ground-water seepage. Constant dewatering operations are required during the active lives of most mines. Recent investigations on the Gogebic Range indicate conditions favorable to the creation of major water bodies in the pits after completion of mining. With adequate consideration in the early phase of the mine development the pits could become elements of great economic value as the demand for water-oriented recreation increases and the supply of available water surfaces is exploited. If the water table does not indicate the likelihood of flooding, the pit can still be a strong topographic feature to be util ized for winter recreation. But some consideration must be given early in the development to provide adequate means of access to the water surface in the one case and to provide appropriate slopes with the proper solar orientation in the other (No. 6).

All of the required information for the ultimate refinement of this proposed landscape is not yet available. But broad land-planning and design principles must be adopted before the open pit mining starts. A review of the basic resources and the existing economic base of the region can provide the basis upon which to plan for future uses which might include game habitat areas, forestry, winter and summer recreation, and, if adjacent to urban areas, space for urban growth. This review will also help in defining the existing landscape pattern and provide the framework for the design of the new landscape.

Hopefully, if such a program is adopted and vigorously pursued, continuing research and investigation will provide us with new information on more effective means of establishing vegetative cover on land forms built from the by-products of taconite mining. Continued development of earth-moving equipment can provide greater flexibility and freedom in the design of land forms. And, additional uses can be identified for these lands that can have a profound effect on the emerging landscape.

A unique opportunity exists in Wisconsin, as in other similar areas throughout the world, for the development of a program that will leave a legacy of a landscape designed for future use rather than one exemplary of an obsolete philosophy. History has shown that idealistic, altruistic motives alone will not suffice, but history has also shown that we live to regret the short-sighted management of our basic resources (Nos. 6 & 7).

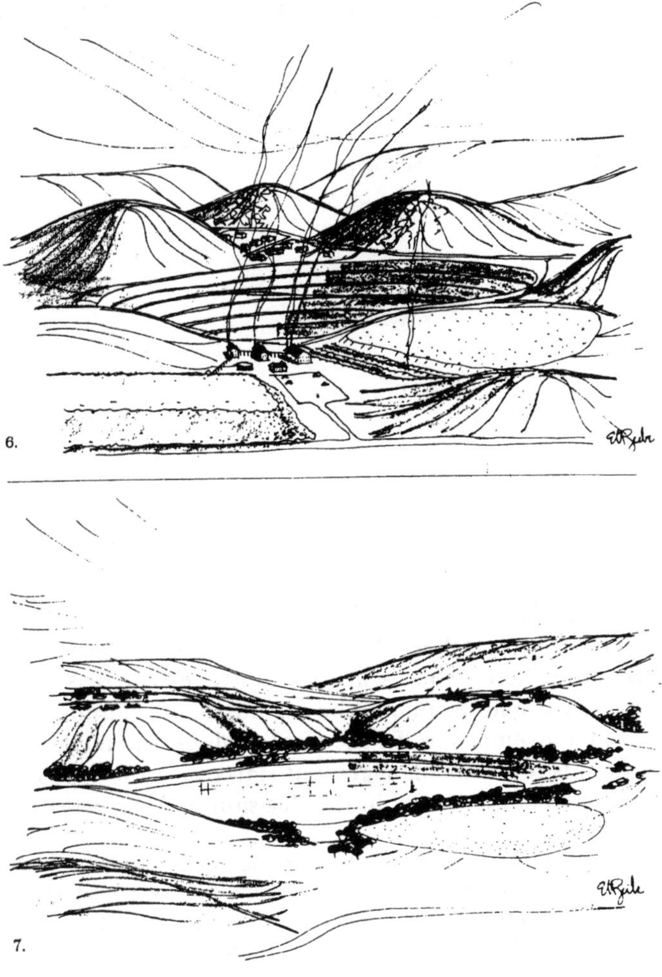

6.

7.

Sketches by the author

Leftovers of careless industrialization at top, and the more ordered scene below. This is the choice which faces Wisconsin and other places where taconite mining could ruin whole regions unless carefully planned.

Editor's Comments on Papers 25 and 26

25 **Craighead and Craighead:** *River Systems: Recreational Classification, Inventory and Evaluation*

26 **Leopold:** *Landscape Esthetics*

The last two articles in this volume and their associative ideas constitute environmental concepts that are much less tangible than those dealing with material resources. Because these subjects are so controversial, it is vital that appropriate evaluation standards and techniques be formulated to minimize controversies that might ensue from a given action. In the final analysis, the methods for utilization and evaluation of lands for their management starts with the type of model or system of priorities that is to be employed in the decision-making process.

> The problem before society is not whether conservation is good or bad, whether to conserve or not to conserve, whether to make use of the environment or not to make use of it, but the problem of alternatives. The decisions to be made are management decisions. How can we best manage the complicated and interrelated earth–people system to the maximum benefit of society now and in the future. In order to weigh alternatives, some method must be developed for assigning values to the various factors that make up an alternative (Flawn, 1970, p. 196).

The articles by F. C. and J. J. Craighead and by L. B. Leopold were elected because they contain some of the earliest rating systems for landscape evaluation that are still held in highest regard. Increased leisure time is one of the facets of modern society that is placing additional stress on the environment (Dower, 1970; Mercer, 1970).

Bryan (1933) wrote one of the early books that deal with the cultural landscape, and one chapter is devoted to "Recreation and Gratification of the Aesthetic Sense." However, no attempt was made to quantify what many would consider subjective opinions. In the last few years, a series of articles has presented different systems for classification and evaluation of landscapes; for example, there are those by Sargent (1967), Shafer et al. (1969), and Morisawa (1971). The most comprehensive approach, by far, is the system devised by Leopold et al. (1971), whereby a 100 × 88 factor matrix is developed. The purpose of this model is to serve as a guide for those in management positions who are called upon to make impact statements in accordance with the Environmental Policy Act of 1969.

> This circular suggests an approach to accomplish . . . the analysis and numerical weighing of probable impacts. This type of analysis does not produce an overall quantitative rating but portrays many value judgments. It can also serve as a guide in preparing the statement called for . . . [in] the Act. A primary purpose is to insure that the impact of the alternative actions is evaluated and considered in project planning (Leopold et al., 1971, p. 1).

The matrix plots the "existing characteristics and conditions of the environment" (including physical processes, biological conditions, and such cultural factors as aesthetics and recreation) against "proposed actions which may cause environmental impact" (such as modification of regime, land transformation and construction, resource extraction, land alteration, etc.).

The ultimate human test of man's actions and changes of the natural environment rests in the hands of the courts and the legal processes that can be brought to bear when the issue is in contention. Coates (1971, p. 229–231) presents a brief review of typical cases where the question concerned how far could man go in creating "aesthetic pollution" of the landscape without being held accountable for his actions.

It is only a short step from man's physical pollution of materials to pollution of his other senses that also affect his well-being. There is increasing recognition of the harmful effects of other distortions in the natural environment, which include noise and aesthetic pollution. Can a price tag be placed on aesthetics when the beauty of the landscape is marred or destroyed? A brief review of the court's attitudes on such matters may prove informative.

In Crance v. State of New York (1955) the Court of Appeals awarded the claimant $10,000 because a road caused an impaired view of Seneca Lake along with a loss of easy access to the lake. When such matters are brought to trial, however, it must be remembered that the skills of the attorneys and the quality of the expert witnesses play an important role in the final decision. In Bostick v. Smoot Sand & Gravel Corp. (1957) a land developer had invested in construction of costly houses along the Potomac River because of the scenic view of the river and had designed each house so that all rooms would have a view. The defendant moved into the area with cranes, power shovels, and barges. In this case, however, the court awarded no damages for aesthetic displeasure and instead distinguished a visual nuisance from noise and odors which can affect physical discomfort (154 F. Supp. 744 D. Md. 1957) in Anonymous, 1970, p. 742).

The case of the Potomac River should be contrasted with the ruling provided by the Appellate Division in New York, Keinz v. State of New York (1957), which helped establish a precedent for the importance of aesthetic factors in the environment:

> Two properties might be physically identical yet their market value could be different because of their surroundings . . . The "view" might be a

> mountainside or a valley as well as a lake. In either event, the view augments the value of the premises and if a portion thereof is taken and the view spoiled the market value of the premises remaining is reduced. It may be a matter of judgment but it is also a matter of dollars and cents and the constitutional policy requires that such reduction in value not be borne by the owner whose property is taken for a public purpose without his consent (161 NY 2d 604, 4th Dept., 1967).

In a slightly different vein the erection of billboards near a major highway was prevented by the court in New York State Thruway Authority v. Ashley Motor Court (1962), which stated:

> . . . billboards can be as destructive of the beauties of the countryside as a plague of locusts. "Beauty may not be queen but she is not an outcast beyond the pale of protection or respect. She may at least shelter herself under the wing of safety, morality or decency" (10 N.Y. 2d 151, 1962).

The New York Court of Appeals in People v. Stover (1963) went slightly farther in stating:

> . . . aesthetics is a valid subject of legislative concern and that reasonable legislation designated to promote an aesthetic end is a valid and permissable exercise of police power (Searles, 1969).

Thus the stage was set for the policy statement that occurs in the Department of Transportation Act passed by the Congress in 1964:

> It is hereby declared to be the national policy that special effort should be made to preserve the natural beauty of the countryside and public park and recreation lands, wildlife and water-fowl refuges and historic sites.

Perhaps the classic case in a locality with a potential for causing aesthetic pollution is Scenic Hudson Preservation Conference v. Federal Power Commission (1966), in which a storage reservoir power project of Consolidated Edison Company was fought for seven years in the courts. Sive (1970, p. 364) stated:

> Comparatively few administrative decisions and environmental litigations reviewing such decisions have involved and will involve areas of scenic beauty to be as unique as that of Storm King Mountain and the Highlands and Gorge of the Hudson. Many, however, pose and will pose serious questions of the relative benefit and value of the preservation of natural beauty and historic shrines against purely economic considerations. Moreover, much current and future environmental litigation is under and will be under statutes which in their policy provisions accord particular significance to considerations of the preservation of natural beauty and historic sites and other aesthetic considerations.

Searles in his review article on environmental law concludes (1969, p. 217),

> . . . the law of aesthetics is becoming more important today than it has been in the past. The courts have come to the realization that the aesthetic factors previously considered outside the context of public use are equally as important as other factors not only in connection with value but also with the public health and welfare.

One of the most recent decisions given by the Court of Claims in New York in this area, and upheld by the Court of Appeals, is Dennison v. State of New York (1969).

> As a result of the appropriation, a new highway has been constructed . . . In place of the beautiful view of forest and mountain [Lake George, N.Y.], which claimants could see from their westerly windows and living areas . . . has been substituted the new highway supported by an embankment approximately 27 feet above grade . . . All of the sylvan beauty afforded by the forest pre-existing the highway and the privacy and quiet it provided are gone . . . In awarding damages . . . the Court of Claims took into consideration "the loss of privacy and seclusion, the loss of view, the traffic noise, lights and odors all as factors causing consequential damage to the remaining property" (Flavin, 1969, p. 411–412).

As we come to the end of this three-volume series, we have come full circle with man's involvement of the land-water ecosystem . . . indeed the total landscape. Man influences the environment, and in turn, is influenced by it (see, for example, Semple, Vol. I). The values of scenery have also left their imprint (Vol. I, p. 31–33). Therefore, it is fitting to close this environmental appraisal with articles written by outstanding and recognized world leaders in ecology and conservation. John and Frank Craighead have been in the forefront of the new ecological movement for more than a decade. Their article is one of the earliest classification systems for rating rivers as recreational resources on the basis of a quantification of parameters. This is an abridged article, since the original contains a series of tables with a rating system for boating, fishing, and hunting resources. The evaluation of parameters varies with the purpose of the proposed activity. Such factors as accessibility, type of barriers, and water quality are important. Density of use is significant because one activity can conflict with another, thus negating some possibilities for multiple use.

The article by Leopold has a common denominator with that by the Craigheads inasmuch as their goal is to categorize landscape in a rigorous and orderly fashion. Luna Leopold is one of the greatest innovators in the field of hydrology and fluvial geomorphology. He has served the federal government in a variety of capacities with the U.S. Geological Survey, including several years as Chief, Water Resources Division, and has served as President of the Geological Society of America. He emphasizes the property of uniqueness, and landscapes are given high ratings when judged to be wild, spectacular, and natural as contrasted to man-created scenes of construction or urbanization.

The properties of wilderness, beauty, and pleasurable aesthetics of the landscape are generally viewed as ends in themselves and necessary ingredients for the spirit of man and his salvation:

> One cannot state the degree to which man is dependent upon some contact with wild nature and natural environments. Although we are becoming increasingly an urban people, this is a recent phenomenon, and there is no evidence by which we can judge what kind of people would be produced after many generations of separation from a more natural world. We are after all a wild species, only recently separated from environments that were mostly wild (Dasman 1959, p. 279).

However, it must not be overlooked that scenery can also be a business. Management of such picturesque regions can be a business and can constitute a profession, although from some points of view it should be termed "a lack of management or interference by man." Publishers have also found there is a lucrative source of income in producing especially handsome books that can transport the "armchair urbanized stay-at-home" into the wondrous world of Nature. The Time–Life series of books on "The American Wilderness" is typical of this trend (already having published *Baja California, Atlantic Beaches, Wild Alaska, The Northeast Coast,* and *The Grand Canyon*). The Sierra Club has also published a handsome set of pictorial essays that show the matchless beauty of nature.

Ian McHarg in his classic book *Design with Nature* (1969, p. 197) provides a fitting finale for this trilogy on environmental geomorphology and landscape conservation:

> The land will be raped and creatures extirpated because the inconsistencies will be so loud; who can plan for the longer term when survival is today? (Instead the ecological model) . . . contains the possibility of an inventory of all systems to determine their relative creativity in the biosphere . . . Ecosystems can be viewed as fit for certain prospective land uses in a hierarchy. It is then possible to identify environments as fit for ecosystems, organisms and land uses. The more intrinsically an environment is fit for any of these, the less work of adaptation is necessary. Such fitting is creative. It is then a maximum benefit/minimum cost solution.
>
> In the quest for survival, success and fulfillment, the ecological view offers an invaluable insight. It shows the way for the man who would be the enzyme of the biosphere—its steward, enhancing the creative fit of man-environment, realizing man's design with nature.

Reprinted from *Naturalist*, **13**, 2–19 (1962)

Water is probably the greatest of all outdoor recreational attractions. River systems and their shore environments are major recreational resources, yet at the present time they are uninventoried and unclassified. Brooks, streams, and rivers, together with their lake impoundments, offer millions of stream miles and millions of acres of water or water-influenced recreation.

25

Recreational Classification, Inventory and Evaluation

FRANK C. CRAIGHEAD, JR., Ph.D.
President, Outdoor Recreation Institute

JOHN J. CRAIGHEAD, Ph.D.
Leader, Montana Cooperative Wildlife Research Unit

Water is probably the greatest of all outdoor recreational attractions. River systems and their shore environments are major recreational resources, yet at the present time they are uninventoried and unclassified. Brooks, streams, and rivers, together with their lake impoundments, offer millions of stream miles and millions of acres of water or water-influenced recreation. The sky-rocketing and often-quoted statistics on fishing, boating, swimming, and reservoir and park or forest visits, are well-known indicators of the growing importance of the nation's water-oriented outdoor recreation activities.

The nation is rapidly developing and utilizing its tremendous outdoor recreational resource, especially that portion of the resource that is water-oriented, without knowing the total scope of the resource, the rate at which it is being developed, the characteristics of the resource and how it can be classified, rated and evaluated, preserved and improved. Public acquisition of key lands is important, but even more essential is the wise and efficient management of both private and public recreational resources already available.

Rivers and their watersheds extend from the high interior mountains to the coastal plains, bisecting major ecological areas. A river channel and its watershed are inseparable ecological entities transcending the boundaries of national parks and forests, private and public land, municipality, county, and state. The ecological unity of the channel and its shore environ-

PHOTOS BY THE AUTHORS

ment must be considered in recreation use, planning, and development. Misuse of the watersheds or drainage areas adversely affects the channel water as well as the many environmental factors that contribute to the quality of recreation activities. The same applies to the water itself and its reciprocating influence on purely shore forms of recreation activity. This becomes readily apparent when the criteria presented in the evaluation forms (p. 14) are studied. Because of this, river system recreational resources must be considered as a complex rather than singly. For example, it is not enough to consider fishing alone. Other activities such as boating, swimming, hunting, picnicking and camping, with their corresponding resources must be simultaneously considered as they represent an appreciable proportion of the overall river system recreation resource.

Recreation use in turn must be coordinated with other water uses such as municipal, industrial, waste, transportation, power production, flood control, and irrigation. From a recreation standpoint, however, the prime and urgent need is to classify, inventory, and evaluate our rivers and streams on a basis of their present condition and use so that resource managers can plan for their optimum use and can compare recreation with other uses. This will provide a foundation of fact for the preservation and management of these undocumented and already deteriorating resources and it is essential for discriminate acquisition of additional recreational land and water in the future. In this paper we have proposed a recreation classification and a systematic method of inventorying and evaluating river system recreational resources. Our purpose is to indicate some necessary steps and to present methodology for getting the job done. The system proposed is only a start and must be explored, altered and perfected in the process of applying it to the task.

The field of outdoor recreation has received great stimulus in the past few years from the establishment of the President's Committee on Outdoor Recreation in June, 1958, from the report of this Committee to President Kennedy, and finally from the creation of the new Outdoor Recreation Bureau under the Department of the Interior.

I. A Recreational Classification

In order to inventory and evaluate water bodies available or potentially available for recreation it is necessary to provide a classification. Rivers and streams can be classified according to size and according to condition and use. They can be evaluated for quality on a basis of specific criteria or characteristics of the resource. The terms excellent, good, fair and poor quality, as used on the evaluation forms, designate these attributes. The term *quality* when used in a general sense refers to excellence.

Size

R. H. Horton[1] proposed what he termed the stream order. A first order stream is one so small it receives no tributary channels. A stream of second order is one which receives only first order tributaries. Where a second order stream meets another of comparable size or order it becomes a third order stream. Leopold and Kinnison[2] proposed 10 orders and classified under these orders most of the streams of the United States. From such a list (Table 1) an estimate can be made of channel lengths or river segments of various sizes available for recreational purposes.

Table 1. Relation of stream order to the number of streams in the United States, their average length and drainage area.

Order	Number	Average length (miles)	Mean drainage area (square miles)	Total length (miles)
1	1,572,000	1	1	1,572,000
2	352,000	2.3	4.75	809,000
3	80,000	5.3	23	424,000
4	18,250	12	109	219,000
5	4,170	28	518	116,000
6	948	64	2,460	61,000
7	206	147	11,700	30,000
8	41	338	55,600	14,000
9	8	777	264,000	6,200
10	1	1,790	1,250,000	1,800
				3,253,000

(1) Definition used is that of Strahler; e.g., order 2 defines the length segment between junction upstream of order 1 channels and junction downstream with an order 2 channel.

The Susquehanna in Pennsylvania and the Brazos in Texas would be streams of eighth order while the Mississippi is considered a tenth order stream.

Under this type of classification a channel of the fourth order averages about 100 feet in width and drains about 100 square miles of land. This might be considered the smallest channel size having substantial recreation value. Table 1 indicates that there are 219,000 miles of order four river channels in the United States. The total mileage of fourth to tenth order channels is 448,000 miles. It is apparent that streams and rivers provide an enormous potenial for water-oriented recreation.

Condition and Use

Not all waterways have the same recreational potential since the kind of recreation depends upon a variety of factors such as the purity of the water, the type of stream or river, uses of the water and watershed, and the character of the shore environment[3]. Some entire river systems have high recreation potential while only portions of others can be utilized. In general, rivers and streams with channels of the fourth order and larger can be grouped into four major classes with regard to the degree to which the main channel and watershed have been manipulated by man. These are:

1. Wild Rivers — Those that are free of impoundments and inaccessible except by trail. These streams and their watersheds are essentially primitive. They furnish a dispersed wilderness-type recreation. Few such rivers exist today.

2. Semi-wild Rivers — Those accessible in places by roads but with watersheds still largely in primitive condition and shorelines undeveloped. They furnish semi-wilderness type recreation for small groups.

3. Semi-harnessed — Developed Rivers — those readily accessible by road, impounded or diverted in their lower stretches with developed shorelines including urban development and characterized by heavy land use on the watersheds. Extensive parts of the upper reaches are still unimpounded and undeveloped.

4. Harnessed — Developed Rivers — Those rivers that are either largely harnessed through impoundments and/or with highly developed watersheds and shorelines. They are characterized by impoundments, structures, artificial channelling, dyking, and varying degrees of pollution. Though developed, they may have extensive stretches of water and shoreline highly valuable or potentially valuable for recreation.

The primary function of *wild rivers* is the protection of watersheds and scientific, educational, and particularly recreational use. Recreation is wilderness-type and unique. Travel by hiking, horseback riding, or boating is necessary to reach and enjoy these waters. In the West they are important if not essential to dude ranching. The water resource can be utilized in many ways further down the major river system.

The recreation function of *semi-wild rivers* is similar to wild rivers. They are, however, more accessible, and recreation, though of unusual quality, is not of the wilderness type. With proper planning there can be more use of these rivers and their scenic watersheds for hunting, fishing, boating, and camping without abusing the resource. Impoundments should be small or perhaps altogether avoided except in cases where the natural watershed is not stable.

The *semi-harnessed — developed river* has potential for dispersed recreation, and for mass recreation including recreation development sites. The realization of the recreation potential depends upon managing the water and watersheds in such a way that the condition of the water, its biota, and the immediate shore environment are not adversely affected for recreation.

The *harnessed — developed rivers* offer tremendous potential opportunities for mass outdoor recreation. These are generally the lower portions of a river system located close to the users. The present recreation use is usually of poor or mediocre quality. These rivers and their shorelines, if managed for their recreational resources could provide large and varied recreational opportunities for expanding urban populations. Pollution, littering, inadequate access, and inadequate public ownership are major drawbacks that must be remedied.

Ecological Sub-classes

These four major classes can be further sub-divided to meet a wide range of ecological conditions. A wild river meandering through a brackish water swamp would have entirely different recreation potential than a white water stream spilling through mountain country. The same applies to the other three classes. In a nationwide inventory it would be necessary to develop sub-classes based on ecological criteria. These in general would reflect the quality of the land and water for recreational purposes. Thus the first step in a national inventory would involve cataloging stream and river channels first by size, second by condition and use, and third by a combination of both. In general the larger channels with the least misuse would possess the greatest water-based recreation potential.

UPPER SNAKE RIVER, WYOMING

A Semi-wild River. Until recently the stretch of river from Moran to Jackson would have been classified as semi-wild. A major highway has increased access and commercial boating introduced mass float trips. Here a unique stretch of river was opened up through ready access to conform to the standards of other rivers found in only too great abundance. If such areas are inventoried and evaluated, our planners may decide to keep them as they are rather than destroy their uniqueness.

The waterways placed in these classes and subclasses must be rated, evaluated and assigned a quality designation. Many factors, both objective and subjective, determine this quality.

II. Inventorying Resources and Evaluating Quality

We can evaluate an entire stream or any desired unit of a stream or watershed using the detailed rating procedures we have developed (see inventory and evaluation forms). Both the inventory and the evaluation are conducted on an acreage basis. The area evaluated can range in size from an entire stream and watershed to a stream and shore section of only a mile or less. The first step is to delimit the area or unit to be evaluated. Where a recreational resource, for example hunting, has been subdivided into big game, small game and waterfowl habitat (see hunting evaluation form), the inventory can be expressed in actual land or water acres as well as in resource acres. Where big game, small game and waterfowl acreages overlap and their totals exceed the actual land acreage, they are referred to as resource acres. This constitutes only part of the task as neither acre-feet of water or resource acres can alone express the quality relation.

Quality Evaluation

An evaluation of quality for any desired area, such as a ten mile stretch of stream and the surrounding shore drainage, can be obtained by rating the area with a set of selected criteria considered essential to, or characteristic of, a specific recreational activity and resource such as fishing, boating, and hunting. Thus, if the hunting resource is to be evaluated, the hunting evaluation form is used. If fishing and boating are significant the proper forms are used to evaluate these resources for the designated area as well. In this system of evaluation, quality is determined by a combination of charactistics and the values used to rate quality are both objective and subjective. Thus some criteria pertain largely to tangible components of the environment, while others relate to intangible values associated with recreational activity in the environment. The result is an evaluation of a particular recreational resource in its broadest sense — a physical and biotic entity as well as a human activity.

Rating Criteria

All recreational activities are an interaction between people and a suitable environment.[1] The term recreational resource refers to this interaction of people and

A BIG GAME HUNTING AREA ADJACENT TO THE BITTERROOT RIVER, MONTANA

If evaluated, this would rate high for big game and small game hunting.

POTOMAC RIVER NEAR WASHINGTON

This picture illustrates a stretch of a Semi-harnessed River. The Potomac is unusual in that it offers relatively wild country close to a large city. Increased pollution over the years has detracted from the recreation value of fishing, swimming, boating and even hunting. Nevertheless this area from Little Falls to Seneca dam is unique for a river of this kind. A controversy is now raging over whether or not to dam it. It may eventually be necessary but those who wish to develop it know little of its unusual recreational values. Its recreational resources as they are now should be carefully evaluated before a decision is made to destroy them. Small game hunting and waterfowl hunting are fair to good and could be improved. The Potomac in this area was once an unusually productive river—smallmouth bass and channel cats. 25 years ago it was a warm-water fisherman's dream. It still can be but fishing has gradually deteriorated with increasing pollution and turbidity.

environment which creates fishing, hunting, boating, and other recreation activities. For this reason we will refer interchangeably to hunting, fishing, and boating as well as to the actual physical environment to mean the recreational resource. A perusal of the three evaluation forms reveal that certain characteristics are common to all the resources being rated and that these characteristics can be designated as rating criteria. These criteria are: Environmental Effect, Populations, Success or Satisfaction, Accessibility, Crowding, Research and Management, Seasons, Conflicts, Size, Habitat, Pollution or Littering, Hazards and Barriers. Each of these criteria has a somewhat different relationship to the particular resource — fishing, boating or hunting, and this is clarified by the descriptive choices offered in the evaluating forms.

In order to provide uniformity, differentiate between degrees of quality, and to permit comparison of rated units of a given resource it is essential that the rating be expressed using a common denominator. This is achieved by assigning each criterion maximum numerical values. For example, the population criterion on the fishing resource form is given a maximum value of 5 and accessibility 3. Thus the evaluator scores the various criteria separately and the sum of these numerical scores is a measure or indication of the quality of the particular recreation resource. The maximum score that could possibly be attained for a fishing resource on this basis is 44.

Quality Designation

This total score is then expressed by a quality designation of Excellent, Good, Fair or Poor. Two different units or areas can be given different combinations of numerical ratings, yet both receive the same quality designation.

The fixed numerical values of the criteria reflect the relative importance of each criterion as parts or characteristics of the resource. The established minimum ratings that designate Excellent, Good, Fair or Poor are only approximate ratings and leeway in their use is reserved to the evaluator.

The total scores for any resource for the same unit of water and shore environment can be averaged to give a recreational resource rating. Ratings of separate units can also be averaged to rate an entire river system.

As stated earlier, both an objective and a subjective approach is deemed essential to proper evaluation but wherever possible the criteria have been made objective. The evaluator must utilize data and information gathered through research, administrative studies, management inventories and through direct observation and investigation. The quality designations, therefore, provide more than a quality rating. They are a record of the condition of the resource, its assets, its limitations, its trend and its changes. The data, if carefully gathered and interpreted will indicate proper recreational use and the management programs needed. Pres-

CROWDING ALONG POTOMAC —

Snagging herring and shad along the Potomac at Chain Bridge. Crowding detracts from fishing pleasure but is bound to increase in such locations adjacent to metropolitan areas as population increases.

A harnessed developed section of river—The water from here to Washington, D. C. is highly polluted. It provides some boating recreation but this could be greatly improved. If improved, use would increase. This is where recreational developments and mass recreation should be encouraged.

ent evaluations when compared with future ones will reveal the effectiveness of past or existing management practices. Thus the evaluation provides the basis and the starting point for effective, comprehensive and enlightened management of our streams and watersheds for multi-purpose recreational use.

III. Discussion of Rating Criteria for Determining Quality

On the evaluation forms a number of choices are provided for the use of the evaluator. In some instances he must make mental comparisons between the area under consideration and others with which he is familiar. For this reason evaluators should preferably have a wide background of experience. On the other hand the method tends to keep to a minimum the need for past experience and technical competence in all fields involved.

Environmental Effect

The general scenic or esthetic tone or mood of an area and its effect on the evaluator is rated. Often this effect is the result of experiencing certain kinds of value or beauty whose physical components are combinations of water, sky, rock and soil formations, vegetation, light patterns and effects of the elements. Wildlife variety and abundance are considered a part of the environmental effect. The opportunity to observe and enjoy wildlife adds pleasure to recreation activities. The inclusion of environmental effect as a rating criterion presupposes that the esthetic effect of the environment or surroundings contribute measurably in providing fishing, hunting and boating satisfaction and thus must be considered in determining the quality of the resource. Natural beauty is a part of the mold that has shaped man down through the ages. It may well be necessary to his continued progress and welfare. It is most certainly an important consideration in outdoor recreation.

Populations

High, moderate and low population densities are used as an index to rate fish and game populations. Wherever possible ratings should be based on censuses, relative abundance counts, hunter kill statistics, creel census data and other indices of abundance. Where population data are available from State or Federal agencies, research bulletins, or management reports, the figures used in the evaluation should be recorded as well as the references. Populations can change relatively quickly and drastically. Specific and accurate figures or estimates are valuable for future comparisons.

Success or Satisfaction

A choice of success or satisfaction is given here because hunter or fisherman success figures will not be available for all areas evaluated. Success ratios should be used when available. They are often expressed as an average number of fish taken per hour, per trip or per acre; as a success ratio of so many ducks per hunter day or a deer hunter success of a certain percent. Hunting, boating or fishing satisfaction when used may have to be estimated or obtained through interview or field observation. Generally speaking, the satisfaction is high when the catch or kill is high but there are times of high fishing and hunting satisfaction when the kill or catch is low.

Accessibility

This critèrion is intended to evaluate the degree to which accessibility is appropriate rather than the relative ease with which the areas may be reached. Recreation quality may be decreased if entire shorelines, watersheds, or water areas are too accessible by road. The suitability of access has been considered in constructing the river system classification but should be evaluated here in greater detail. Adequate access may consist of roads in one area and only trails in another.

Crowding

Crowded conditions are those where too many people tend to decrease the enjoyment for all, where safety of participants becomes a problem and where actual conflict may exist. The practice of good technique is often impaired. Crowding is, of course, a relative condition but the extremes are readily recognized.

Research and Management

This reflects the value of scientific data and its effective application toward the goal of optimum management of the recreation resources. The quality of fishing, boating and hunting activities and resources are becoming increasingly dependent upon research and management. This will become a more important factor as recreation use and pressures increase, thus in general the resource is rated higher if it is wisely managed than if it is not. Specific information relative to this criterion can be obtained through state Fish and Game Departments, the U. S. Fish and Wildlife Service and other public agencies.

Season

Legally established seasons affect recreation activities and resources. Hunting seasons for any given area are not always favorable to hunters nor do they always insure proper management. Often political rather than technical considerations predominate. Some seasons may not provide for an adequate harvest of fish or game. Other seasons may coincide with unpleasant weather conditions. The natural length of the season as in boating is also a factor to be considered. Long seasons should normally rate higher than short ones.

Conflicts

Other recreation activities as well as other land and water uses often conflict with hunting, fishing or boating activities; or adversely affect a particular recreation resource. Such conflicts should be recognized and evaluated. Their effect may become more pronounced in the future.

UPPER SNAKE RIVER—WYOMING

Excessive accessibility must be considered in evaluating boating, fishing and hunting resources. Careful planning is necessary to provide appropriate access for any specific recreational resource.

Size

The units of water and watershed to be evaluated may vary greatly in size. Generally speaking the larger units provide greater opportunities and should receive higher ratings. Small areas are more subject to exploitations and disturbances.

Habitat

Optimum fish and game populations are dependent upon good habitat. The abundance or dearth of habitat requirements — food, cover, water and space can be observed or measured. Factors such as the population itself, plant successional changes, overgrazing, soil erosion and other natural or man-made disturbances will alter the favorableness of habitat for different species. Limiting factors, habitat condition and trend should be noted or measured and then evaluated.

Pollution and Littering

Littering and pollution detract from the quality of the recreation activity. Where industrial or domestic pollution occurs to the extent that fish life cannot survive or fishing and boating is made unpleasant or unsafe, such waters would be rated low. The extent of pollution can be determined through observations, interviews, checking of private and public records, and through tests such as dissolved oxygen content. Public Health authorities often are familiar with pollution conditions, have established pollution criteria for certain streams and operate pollution sampling stations. Data on dissolved oxygen, coliform bacteria count, turbidity, pH, and other conditions can be obtained through this service. The U. S. Geological Survey in cooperation with the States publishes and can provide

MIDDLE FORK OF THE FLATHEAD, MONTANA, TAGGING DOLLY VARDEN

Where information is scientifically and periodically gathered for management purposes and where the water and watershed reflect good land and fish-water management, it will be recorded on the criteria scores (See valuation forms).

This is one of Montana's few remaining wild rivers. The upper reaches are shallow, the water clear. Here lie the spawning beds of the Dolly Varden trout. Vegetation and undisturbed soil control the runoff. There is unlimited choice of campsites and long stretches of excellent boating water.

information on stream flow, chemical character of water, silting and other conditions.

Hazards and Barriers

Hazards, barriers and obstacles may either detract or add to the quality of a particular recreation activity. Rocks and rapids may add zest to canoeing but interfere with motorboating. Log jams may present serious hazards to boats, but provide excellent cover for fish. Extensive windfall may detract from big game hunting but provide necessary sanctuary for the game. Hazards and obstacles should be rated accordingly.

IV. Determining Permissible Use

After the Nation's river system recreational resources have been classified, inventoried and evaluated, then the task remains of determining permissible use or the recreational carrying capacity of the various river system resources evaluated. At the present time there are no carrying capacity figures available for any of our streams or rivers. There is a limit to the outdoor recreation use any given area or resource can provide without destroying or impairing the physical resources or adversely affecting the quality of the benefits or values received. Recreational capacity figures, when obtained, will show the degree of utilization of a resource, provide a basis for comparison of use, and are necessary to compute the recreational resource needed to satisfy a known or projected demand.

V. Comparing Recreational Use With Other Uses

There is a very real and a growing need to express certain recreational values or resources of river systems in dollar terms. This applies particularly to the Semi-harnessed—Developed Rivers and to the Harnessed—Developed Rivers where there are sharply competing claims on the use of the resources. We have not yet developed an adequate method or methods for doing this. Once recreation carrying capacities have been determined it will be feasible to use these figures to assign or compute a dollar value for any particular resource unit. If recreational capacities are expressed in units of time (man-days) and related units of resources are expressed in units of area (acres) then we can express a relation between time and recreational resource acres or the interaction between man and a suitable recreational environment.

The next logical step in assigning values to the recreational resource then would be to assume that the benefits or satisfactions that accrue to the individual during a recreational day are to a large extent inherent in the environment, since he seeks a particular environment for the recreational value it offers. This would mean that the average dollar value of a recreationist's time can be used to measure the dollar value of the resource and on an acreage basis if necessary.

Lowell Sumner[5] applied a straight-time wage rate computation to all Alaska vacationists irrespective of occupation, sex, age, or color to determine the monetary value of a recreation day. The straight wage approach results in a high value per recreation day but it has merit in that many people consciously and willingly forego the money they could make had they not used this time in seeking recreation. Also, because the individual makes this choice he is considering all the values involved — both tangible and intangible.

An important purpose of this paper is to emphasize in outdoor water-oriented recreation the need to recognize, measure, and manage for *quality*. In assigning dollar values we must recognize the dilemma of using man-days or visitor-days as related to acreage in evaluating areas such as a wilderness stream and watershed or a unique or outstanding canyon where actual use statistics may be low but quality of recreation high. Increased economic value tends to go hand in hand with greater intensity of use; yet the *quality* of recreational activities such as fishing, hunting, or canoeing diminish with rising use intensity.

In respect to our Wild and Semi-Wild Rivers, we should probably not attempt to assign a dollar evaluation but let the *quality* speak for itself. There is a great unmeasured value in an uncrowded minority-type recreation that embodies the very principles of our democratic free-enterprise society. It should be emphasized here that minority use is not special interest use. The opportunity is available to all. The present day trend is to minimize this type of recreation, putting emphasis on use statistics as an indicator of successful and proper administration of outdoor recreation resources. The classification, evaluation and rating system we have proposed emphasizes the qualitative aspect of the water resource and its esthetic role, yet properly recognizes that use statistics and total quantity of the resource are also important considerations in a nation where mass recreation is well established and provides stimulation to the national economy, and important financial returns to regional and local economies.

ST. MARY'S RIVER, FLORIDA

Though quite different from the other rivers and watersheds, illustrated, it can be inventoried and evaluated in the same way.

LITERATURE CITED

1. Horton, R. A., 1945. Bulletin of the Geological Society of America, Volume 56—Pp. 275-370.
2. Leopold, L. B., and Kinnison, H. B., 1962. Water for Recreation: Values and Opportunities.
3. Craighead, J. J., and Craighead, F. C., Jr. Water and Outdoor Recreation. Testimony before the Senate Select Committee on National Water Resources, Missoula, Mont.
4. Craighead, F. C., Jr. Outdoor Recreation Research. The Naturalist Magazine. Volume 9, Number 2—Glacier, Land of Shining Mountains, Pp. 23 to 27, 3 illustrations.
5. Sumner, Lowell, 1956. Your Stake in Alaska's Wildlife and Wilderness. Reprinted from the Sierra Club Bulletin, Dec. 17 pp.
6. U. S. Forest Service, 1959. Work plan for the National Recreational Survey in a review of the Outdoor Recreational Resources of the National Forests

Reprinted from *Natural History*, Oct., 36–45 (1969)

Landscape Esthetics

How to quantify the scenics of a river valley

26

by Luna B. Leopold

There are an increasing number of bills before Congress that in one way or another affect the landscape or the environment. Each of these requires seemingly endless numbers of congressional hearings, which are recorded upon endless reams of paper.

And if, for some reason, you happen to read the voluminous testimony surrounding one of these environment-affecting proposals, you will generally find a marked contrast between the volume and kind of information presented by those who are pressing for technical development—building a dam, constructing a highway, installing a nuclear power plant—and the testimony of those who either oppose the development or wish to alter it in some way. The developer usually employs numerical arguments, which tend to show that there is an economic benefit to be obtained by constructing something—whatever that something may be. The argument is usually expressed in terms of a "cost-benefit ratio." It is typically argued, for instance, that the construction cost of a given project will be repaid over a period of time and will yield a profit or a benefit in excess of the development costs by a ratio of, let us say, 1.2 to 1. The argument is further supported with great numbers of charts, graphs, tables, and additional figures.

In marked contrast, those who favor protection of the environment against development are fewer in number, their statements are based on emotion or personal feelings, and they usually lack numerical information, quantitative data, and detailed computations. Perhaps this is the reason why this latter group seems to be continually fighting rearguard actions—losing battle after battle.

The time has come when the argument of the environmentalist might best be presented by (1) separating facts from emotions in relation to the environment, and (2) by providing him with a means of quantifying his arguments: using numbers to talk about the landscape. While to some of us this may be a little like using a computer to describe Shakespeare, it seems that society still has the right to have all aspects of any proposed development presented in a way that is as objective as possible.

One strategy used by environmentalists or conservationists to combat this paucity of statistical barter involves an attempt to describe society's interest in landscape integrity in monetary terms, which make a region's esthetic attributes appear to be similar to the kinds of benefits ascribed to the planned development. From this has arisen the unfortunate and unsound procedure of evaluating recreation in terms of what is called the "visitor day." This argument is based on the idea that the average visitor to a particular place spends one dollar a day (or some other amount) there, which he would not have spent had he not visited the spot, and that the enjoyment derived therefrom is in direct proportion to the amount of time or money spent in the given area. To me, this procedure misses the whole point of recreational activity since we know, by experience, that recreational enjoyment is by no means dependent upon either of these factors. Rather, it appears more sound to develop a way of directly describing the quality of the recreational experience. A first step toward this goal is to objectively describe the landscape itself, which the recreationist visits for enjoyment without regard to expenditure of money or time.

Toward this end, in August, 1968, I began a study designed to produce a method that would quantify the esthetic features of the environment so that the resultant data could be used in many planning and decision-making contexts. Such data could be especially useful when choices must be made among alternative courses of action. They would tend to provide a more prominent consideration of nonmonetary values to society.

The event that gave rise to this report was the application made several years ago to the Federal Power Commission for license to construct a hydropower dam in the Hell's Canyon area of the Snake River in Idaho. Three dams are already operative in the upper parts of Hell's Canyon, but there is still about a hundred miles of undammed river that retains a special character because it is not easily accessible. The river in this section flows through a deep, narrow gorge whose rocks and vegetation give it a particular grandeur. Proposals for damming of the Snake have been vigorously opposed by the country's leading conservation organizations. In the ensuing controversy over this issue, the following alternatives have been proposed: to dam another, nearby river instead of the Snake, to change the size and location of the proposed dam, to abandon the dam proposals altogether, and so on—the usual gamut of moves and countermoves.

The problem posed, then, was to determine objectively if (and in what sense) Hell's Canyon is indeed esthetically unique. The answer to this question would hopefully guide the decision as to whether the river should be dammed in this area. The study thus undertook to evaluate the factors influencing the esthetic appeal of Hell's Canyon. Toward this end, a comparison was made between Hell's Canyon and other river valleys in central Idaho. Then another comparison was made between Hell's Canyon and several well-known valleys in presently established national parks.

Three types of factors appear relevant to landscape esthetics. These groupings of factors and their "subfactors" are listed in table 1. The first group involves the physical features of an area—the presence of mountains and valleys, width of valleys, height and type of mountains. The second group includes those features that have to do with the region's biology, especially—in the case of river valleys such as Hell's Canyon—the vegetation near the stream and on the mountainsides, and the biology within the water itself. A third class encompasses what I have called "human interest factors." These are often more intangible than either the physical or the biological ones, but they are nevertheless influential in determining how the landscape impresses us. For example, if one is at the point on the Delaware River where George Washington is supposed to have thrown the silver dollar, that historical incident, however apocryphal, gives that place a distinct meaning. The phrase "Lincoln slept here" is of a similar sort—there is a human interest associated with the sites where certain phenomena exist or where unusual events have occurred.

Another set of circumstances, also related to the human interest factors, is the presence or absence of vistas or scenic outlooks. The many travelers who pull off to the side of the road at a turnout marked "scenic viewpoint" or "scenic outlook," are an indication that the ability to look from some vantage point across great distances, often to mountains or into far valleys, gives the landscape some special character—whether at a mountaintop or in a river bottom.

The question of access also falls within the realm of human interest. In the listing of criteria I have broken accessibility down into two parts: access to the individual, especially the hiker, and separately, mass use, meaning availability to motorized transport.

Human interest is affected by the general level of urbanization; it can make a piece of landscape more interesting or the opposite. The view from the Berkeley hills, for example, across the bay to San Francisco is made attractive by the skyline of San Francisco itself. In this case, the presence of the city seen from afar seems to make the landscape more interesting. On the other hand, in certain kinds of mountain country the presence of a great many cottages along a road may tend to detract from its inherent character, and have the opposite effect.

Within the list of human-interest criteria, I have included the term "misfits." One often remarks on how certain kinds of architecture fit into a particular landscape; probably one of the things that makes the Swiss mountain landscape so appealing, especially to American visitors, is that the type of architecture and the handling of building location seem to fit especially well into the particular environment. On the other hand, if you were to put a flashing neon sign advertising hamburgers on one of the Swiss chalets, it would be a cultural shock, and I would call it a misfit. So, also, in American landscapes one might be enjoying a drive through a rural countryside and suddenly be confronted with a tremendous dump of car bodies or even an obnoxious roadside billboard. These I consider to be misfits: they are out of character with natural surroundings. One does not mind seeing a large neon sign in the city nearly so much as seeing it in the countryside where it seems out of place.

The other two classes of factors are less complicated than the human-interest group. The physical factors are the easiest to measure in the field. Such factors as river width, river depth, and certain other characteristics require only a recording device or an elementary observation of the river channel. Under biological factors the list includes water color, turbidity, amount of algae, and the kind and extent of water plants, which are often indicative of stream purity or pollution. Under the three categories, a total of 46 criteria were chosen to describe a landscape's esthetic character.

After the factors were chosen, twelve river valleys in central Idaho, including Hell's Canyon, were chosen for evaluation sites that would have some potential for power development. Locations with such potential were selected in order to restrict the sites under discussion to those having something in common with Hell's Canyon. Each site was physically evaluated by standing at the edge of the river, thus providing uni-

Table 1: *In this study, the esthetic qualities of sixteen river valleys are described by means of the 46-factor checklist at right—a first step in determining the relative uniqueness of each site.*

FACTOR NUMBER	DESCRIPTIVE CATEGORIES	EVALUATION NUMBERS 1	2	3	4	5
	PHYSICAL FACTORS					
1	River width (ft.) (at low flow)	<3	3-10	10-30	30-100	>100
2	Depth (ft.) (at low flow)	<.5	.5-1	1-2	2-5	>5
3	Velocity (ft. per sec.) (at low flow)	<.5	.5-1	1-2	3-5	>5
4	Stream depth (ft.)	<1	1-2	2-4	4-8	>8
5	Flow variability	Little variation		Normal	Ephemeral or large variation	
6	River pattern	Torrent	Pool & riffle	Without riffles	Meander	Braided
7	Valley height/width	≦1	2-5	5-10	11-14	≧15
8	Stream bed material	Clay or silt	Sand	Sand & gravel	Gravel	Cobbles or larger
9	Bed slope (ft./ft.)	<.0005	.0005-.001	.001-.005	.005-.01	>.01
10	Drainage area (sq. mi.)	<1	1-10	10-100	100-1000	>1000
11	Stream order	≦2	3	4	5	≧6
12	Erosion of banks	Stable		Slumping		Eroding
13	Sediment deposition in bed	Stable				large-scale deposition
14	Width of valley flat (ft.)	<100	100-300	300-500	500-1000	>1000
	BIOLOGIC & WATER QUALITY FACTORS					
15	Water color	Clear colorless		Green tints		Brown
16	Turbidity (parts per million)	<25	25-150	150-1000	1000-5000	>5000
17	Floating material	None	Vegetation	Foamy	Oily	Variety
18	Water condition (general)	Poor		Good		Excellent
	Algae					
19	Amount	Absent				Infested
20	Type	Green	Blue-green	Diatom	Floating green	None
	Larger plants					
21	Amount	Absent				Infested
22	Kind	None	Unknown rooted	Elodea, duck weed	Water lily	Cattail
23	River fauna	None				Large variety
24	Pollution evidence	None				Evident
	Land flora					
25	Valley	Open	Open w. grass, trees	Brushy	Wooded	Trees and brush
26	Hillside	Open	Open w. grass, trees	Brushy	Wooded	Trees and brush
27	Diversity	Small				Great
28	Condition	Good				Overused
	HUMAN USE & INTEREST FACTORS					
	Trash & litter					
29	Metal (no. per 100 ft. of river)	<2	2-5	5-10	10-50	>50
30	Paper (no. per 100 ft. of river)	<2	2-5	5-10	10-50	>50
31	Other (no. per 100 ft. of river)	<2	2-5	5-10	10-50	>50
32	Material removable	Easily removed				Difficult removal
33	Artificial controls (dams, etc.)	Free and natural				Controlled
	Accessibility					
34	Individual	Wilderness				Urban or paved acce
35	Mass use	Wilderness				Urban or paved acce
36	Local scene	Diverse views and scenes				Closed or without diversity
37	Vistas	Vistas of far places				Closed or no vistas
38	View confinement	Open or no obstructions				Closed by hills, cliffs or trees
39	Land use	Wilderness	Grazed	Lumbering	Forest, mixed recreation	Urbanized
40	Utilities	Scene unobstructed by power lines				Scene obstructed by utilities
41	Degree of change	Original				Materially altered
42	Recovery potential	Natural recovery				Natural recovery unlikely
43	Urbanization	No buildings				Many buildings
44	Special views	None				Unusual interest
45	Historic features	None				Many
46	Misfits	None				Many

KEY: < less than, > greater than, ≦ less than or equal to, / divided by

SITE NO.	LOCATION	TOTAL UNIQUENESS RATIO
1	Wood River, 6 miles above Ketchum	11.07
2	Salmon River, ¼ mile above Stanley	11.00
3	Middle Fork Salmon River at Dagger Falls	11.87
4	South Fork Salmon River, near Warm Lake	13.93
5	Hell's Canyon, below Hell's Canyon Dam	16.09
6	Weiser River at Evergreen Forest Camp on Highway 95	11.17
7	Little Salmon River, 6 miles north of New Meadows	23.10
8	Little Salmon River, 4 miles south of Pollock	13.78
9	Salmon River, 2 miles below Riggins	10.25
10	Salmon River, at Carey Falls, 20 miles above Riggins	14.31
11	French Creek, 1 mile above junction with Salmon River	11.95
12	North Fork Payette River, near Smiths Ferry	10.21

Table 2: *"Total uniqueness value" (see text for full explanation) is an objective measure of how different each site is from other sites studied, without regard to "positive" or "negative" esthetic values. River pollution at Site 7, for example, makes this area relatively unusual and gives it the highest "uniqueness ratio."*

formity in the way the observer looked at the environment. One could just as well have chosen evaluation sites that were part way up the valley sides, but this would have had the disadvantage of putting the observer at varying distances from the river.

At each site the checklist of 46 items was filled out. It can be seen from table 1 that most of the physical factors could actually be evaluated with some common unit of measure. Others, however, had to be estimated in terms of categories—erosion of stream banks, for example. In all cases there were five evaluation categories specified in the checklist. During evaluation, each site was described by assigning to each factor a number from 1 to 5, according to its physical, biological, or human interest characteristics. Where physical measurements were involved, the five categories varied in their span in an unbroken progression. For example, the five categories of river width were: less than 3 feet, 3 to 10 feet, 10 to 30 feet, 30 to 100 feet, and more than 100 feet. In this way, the categorization of a given site with regard to one of the factors could always be fitted into the category quantities.

One of the purposes of the study was to eliminate personal subjectivity in landscape analysis. Accordingly, the "evaluation numbers" for each of the 46 factors in the checklist serve a descriptive function only; evaluation number 5, for example, is not to be interpreted as "superior" to evaluation number 1, or vice versa. If a given site has a river width of more than 100 feet, our analysis does not rank this area above one whose river width is, let's say, less than 3 feet, but merely assigns different evaluation numbers to each of these locations.

The results of such a comparative study depend in part on the sites chosen for comparison. This being the case, another set of comparisons was made between Hell's Canyon and a series of rivers in four national parks of the United States. In this way I could find out not only whether Hell's Canyon is very different or rather like other river sites in Idaho, but also how the region compared esthetically with some of the great beauty spots that the nation has already recognized by giving them national park status.

Any scheme for comparing landscapes must rest on some philosophical framework. The philosophy underlying the scheme I used is the following: *Landscape that is unique either in a positive or negative way is of more significance to society than one that is common.* A place of great scenic beauty is of importance because of its scenic qualities. On the other hand, one could imagine a unique site which is extraordinarily unattractive—a large, neglected, pestilential dump, for example. This also has significance for society, but in the opposite sense.

Having obtained the checklist data from 12 river valley sites in the Idaho region, the next step was to compare the sites factor by factor in order to determine the relative uniqueness of each factor at each site. Let us take river width, for example. As mentioned, with regard to this factor each site would be placed in one of five categories. Most of the sites had river widths falling into category 4—somewhere between 30 and 100 feet. A small number had widths greater than 100 feet, and a few fell in category 3, having a width from 10 to 30 feet.

To discover the relative uniqueness of each site's river width, we then determine how many among the 12 sites had river widths falling within each of the 5 categories. If it happened, for example, that there was only one river more than 100 feet wide, no other sites would share the 5 categorization, and this would determine what I call the "uniqueness ratio" of this one river. The uniqueness ratio for the river in question is equivalent to the reciprocal of the number of sites sharing the category value. The number 1, or unity, divided by the number of sites sharing the value, in this case 1, gives a uniqueness ratio for that particular site of 1.0. If there were two sites that shared category 5 in river width, each would be assigned a uniqueness ratio for river width of unity divided by 2, or ½ (.50). If, for a particular factor, all 12 sites fell in the same category, each of the 12 would be assigned a uniqueness ratio of 1 divided by 12, or .08.

By this method, then, each site had a uniqueness ratio for each of the 46 factors that were measured in the field. Adding uniqueness ratios for all 46 factors for a given site yields a "total uniqueness ratio." In this way, the total uniqueness ratios for the 12 sites may be compared one with the other; the higher the ratio, the more unique the site. This is a way of measuring numerically the relative uniqueness of each of the sites chosen for comparison. The results of this uniqueness ratio technique, when applied to the 12 Idaho river valley sites, are given in table 2 on the page opposite.

There is a technical difficulty involved with the simple addition of the uniqueness ratios for the 46 factors at a given site. By the process of addition, each of the 46 factors is given essentially equal weight in determining the total uniqueness ratio. On further consideration of the list of 46 factors, one may decide that some are far more important than others, and therefore, for certain purposes, selected groups of the factors can be used for other analyses. Nevertheless, the uniqueness ratios for the 46 factors present a general means for quickly comparing a group of sites.

The uniqueness ratio techinque is objective in that it does not distinguish whether a given site is uniquely esthetic or uniquely unesthetic. The valley of the Little Salmon River near New Meadows (Site 7), for instance, was indeed the least attractive, most uninteresting, and unspectacular of the 12 sites surveyed. But because this site was different from the others in being a stream that was sluggish, algae-infested, murky, and slow flowing, it rated a high uniqueness ratio—the highest of all the 12 sites. It was unusual in a negative sense.

As can be seen in the table, the second largest total uniqueness score is Site 5, the Snake River in Hell's Canyon. In interpreting its rank order in total uniqueness score, one can say that Hell's Canyon was different from all others in ways that made it scenically interesting and therefore unique in a positive sense. Surprisingly, then, the two highest uniqueness scores represented the two sites that might be called the two opposite ends of the scale of esthetic interest.

A visual picture of the position of the various sites in the uniqueness scale can be obtained from the graph in figure 1 in which the ratios of human interest and biological uniqueness form, respectively, the ordinate (vertical, or y-axis) and abscissa (horizontal, or x-axis) with the physical uniqueness ratio written in numerical form next to each plotted point. It can be seen that Site 7 stands alone on the graph, primarily because of its uniqueness score on the biological scale—a result of its being the only polluted river among the sites surveyed. It can also be seen that Sites 5, 10, and 8 stand more or less alone because of low values in the biological factors and high values on the human interest scale. The graph, then, is an easy way of seeing that the total uniqueness score of Hell's Canyon, Site 5, and the polluted river, Site 7, are indeed at the opposite ends of a scale of landscape desirability.

Having demonstrated that the total uniqueness score does not involve personal preference or preference bias, one can now proceed to select combinations of factors from the checklist in order to perform additional types of analyses. In this study, checklist factors were chosen for their particular significance with regard to the impression that the

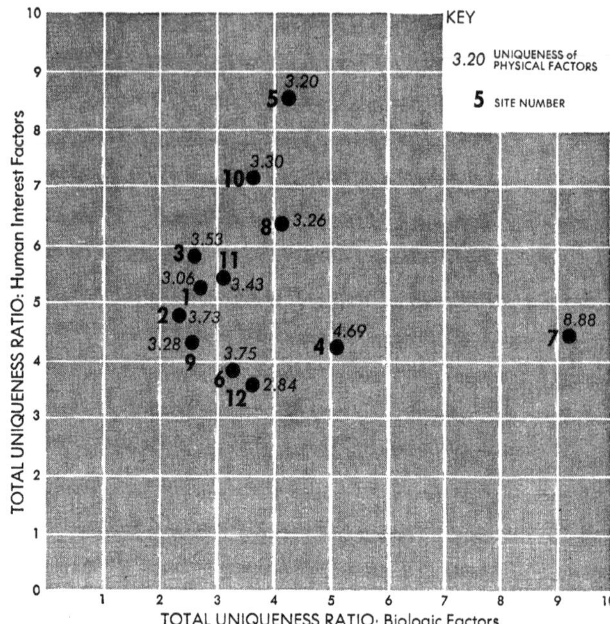

Figure 1: *The graph provides a measure of the relative uniqueness of each of 12 Idaho river valley sites in terms of the three groups of esthetic factors listed in table 1—biologic uniqueness is measured on the horizontal axis, human interest uniqueness on the vertical axis, and the physical uniqueness ratio is listed alongside each site number.*

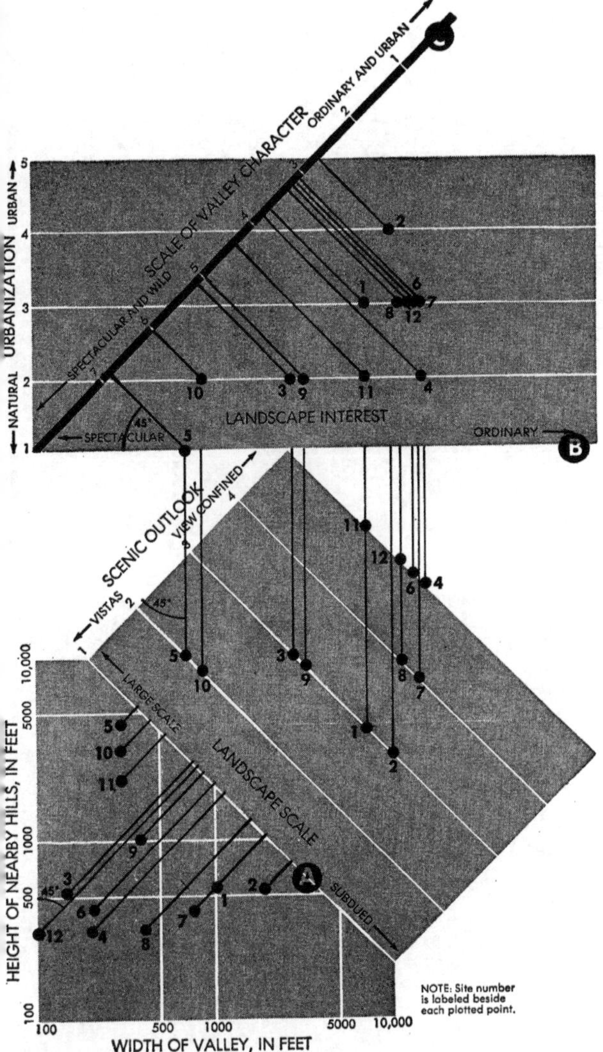

Figure 2: *"Height of nearby hills" plus "width of valley" equals "landscape scale." This value plus "scenic outlook" equals "landscape interest," which, combined with "degree of urbanization" yields "scale of valley character," a measure of the viewer's esthetic impression of the landscape at each site. See text for complete explanation.*

site gives to the viewer. The *selection* of factors does involve personal judgment as to which ones appropriately describe the landscape characteristics; but the selected factors themselves remain independent of this judgment. Factor selection may therefore be thought of as a subdivision of the objective basic data.

In this step factors were selected from the 46-factor checklist for the purpose of evaluating each site in terms of two characteristics—valley character and river character. Valley character I conceived as a *combination* of the scale or grandeur of the landscape, the availability of distant vistas, and the degree of urbanization. Valley character is certainly influenced, in part, by the bigness of landscape features. The spectacular character of many of the grandest scenic views in the world comes in large part from the presence of high peaks in close proximity to the valley floor from which the viewer is seeing the landscape. Specifically, where the valley floor is narrow and the adjacent hills or mountains exceptionally high, the viewer has the concept of a large-scale landscape. Where, in contrast, the valley floor is very wide and the adjacent hills low, the impression is one of flatness—the opposite of grandeur. The Swiss Alps are so spectacular because the valleys are narrow and the mountains are nearby and extremely high. The same can be said of the Teton Valley, in the Jackson Hole area of Wyoming. One, therefore, can make a scale that shows the combination between two of the measured factors, the width of the valley floor and the height of the adjacent hills or mountains, in order to evaluate this grandeur aspect of each site, or what I have here called "landscape scale."

The method of obtaining a value for landscape scale for each of the 12 Idaho river values is dependent upon a diagram, as shown in figure 2. For each of the sites, the height of the mountains was plotted as ordinate and the width of the adjacent valley floor plotted as abscissa on the bottom graph of the diagram. On this graph, Sites 5, 10, and 11 fall in a zone of large-scale landscapes (high mountains, narrow valleys) and other sites fall near the other end of the scale, which I call a "subdued landscape." To simplify the combination of height of hills and width of valley to the single value, landscape scale, the position of each plotted point in the bottom graph in figure 2 is projected orthogonally onto a diagonal line (line A). (The projection of each point is indicated by those lines leading from the points at a 45° angle upward and to the right in the lower graph.) Line A now serves as the line on which the sites are ranked according to their respective landscape scales. Having a value of landscape scale for each site, this joint value can be used for the construction of a new graph, as has been done in the central portion of figure 2, using line A as a horizontal, or x-axis.

At this point, it was reasoned that the impression of a landscape on the viewer is partly determined by this factor just derived, landscape scale, and partly by the "degree of view confinement." Where distant vistas are available in large-scale landscapes, one has the impression of spectacular scenery. In contrast, where the view is confined by heavy cover or by adjacent hills, the result is esthetically ordinary from the scenic point of view. Again this is one of the reasons why the Alps give a strong impression of scenic grandeur; not only is one looking out from a narrow valley to the high mountains immediately adjacent, but one can see up and down the valleys for long distances, viewing ranges of mountains as background vistas. In contrast, when one is in the bottom of a narrow, tortuous gorge, with no distant outlooks, as in certain portions of the Black Canyon of the Gunnison River in Colorado, the impression of the view is distinctly less spectacular.

On the central graph of figure 2, then, scenic outlook (the presence or absence of distant vistas) serves as the y-axis and landscape scale as the x-axis. And the position of each point on the graph is again projected upward at a 45° angle onto line B providing a scale which I have labeled "landscale interest." The position of each site on the landscape interest scale is thus a ranking of the individual sites using a combination of three factors: width of valley floor, height of adjacent mountains, and availability of vistas.

The scale of landscape interest can now be combined with another factor—degree of urbanization—which also affects the viewer's impression of the total landscape. The degree of urbanization is defined as the totality of buildings, houses, roads, utilities, and other earmarks of man-made change present in a given site. As before, using the scale of landscape interest as the x-axis, an ordinate, or y-axis, is constructed representing the degree of urbanization. This pro-

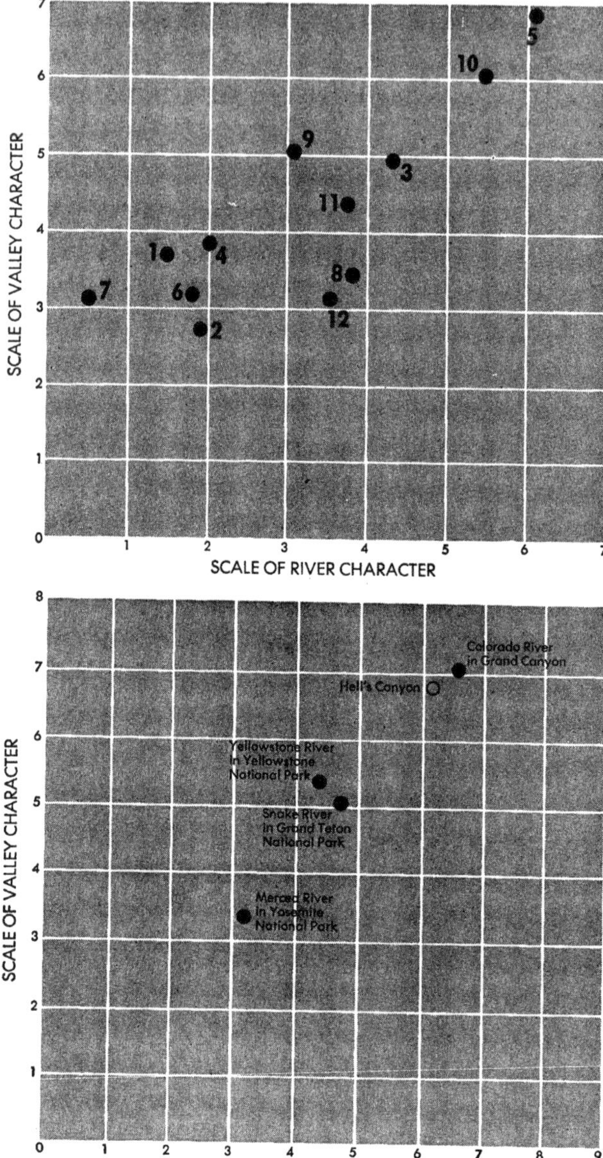

Figures 3 & 4: *The viewer's esthetic impressions of Idaho river valleys and river valleys in national parks are evaluated in terms of "valley character" and "river character." Relative to other locations, numbered sites falling in the upper right-hand portion of the graphs tend to have large rapid-flowing rivers, large-scale landscapes, scenic vistas and little urbanization. On the basis of this assessment, Hell's Canyon ranks second only to the Grand Canyon of the Colorado.*

vides the final or upper graph in figure 2. The sites having a combination of spectacular scenery and minimal urbanization fall in the lower left-hand zone of the upper graph, whereas those having ordinary and urban conditions fall in the upper right. The combination of landscape interest and urbanization, each given equal weight, is represented by another 45° projection onto the diagonal in the upper graph (line C), yielding a scale of what I call valley character. We thus constructed a ranking scale called "scale of valley character" representing a combination of four of the factors from the checklist in table 1: width of valley, height of adjacent mountains, ava.l-ability of vistas, and degree of urbanization.

When the position of each of the 12 sites is considered on this final scale of valley character, we see that Sites 5 and 10, Snake River at Hell's Canyon and the wild reach of Salmon River at Carey Falls, are highly unusual. They are characterized by narrow valley floors, high adjacent mountains, availability of distant vistas, and little or no urbanization. All of the other sites in figure 2, fall to the right of Sites 5 and 10.

After combining the four factors contributing to valley character, we can perform a similar analysis on the rivers themselves, obtaining a measure called "scale of river character." Experience indicates that the grandeur or majesty of a river is dependent upon a combination of its size and apparent speed. Rivers tumbling over a succession of falls tend to be more impressive or esthetically appealing than those that appear sluggish. The latter characteristic is not so much dependent upon true velocity, which is usually poorly estimated by the untrained eye, as it is on the *appearance* of speed judged mostly by waves and surface riffles caused by rapids or falls. We therefore wish to find a combination of factors that gives the viewer the impression of the river's grandeur itself. For this we use river width, river depth, and the presence or absence of rapids, riffles, and falls, chosen from the checklist factors.

Eventually we arrive at a ranking of the sites according to river character; those rivers that are wide, deep, and have rapids, falls, and riffles rank high on this scale. In this instance, Sites 5 (Hell's Canyon) and 10 (Salmon River at Carey Falls) again outrank the other locations. As the last step in our analysis, we use a graph to compare the 12 sites according to both valley character and river character (figure 3), and thereby obtain a final rank reflecting a total of 7 of the factors chosen from among the 46 on the initial checklist.

It can be seen that on this graph Hell's Canyon is indeed unique, for it falls in the farthest position in the upper right-hand part of the graph. Nearest to it again is Site 10, the wild reach of the Salmon River near Carey Falls. Nearly all of the other sites fall in the central cluster in the center of the graph. Site number 7 (Little Salmon near New Meadows), the least interesting of all those surveyed, falls at the opposite end of the scale, appearing in the lower left-hand portion of the figure.

Having evaluated the Idaho sites, a new comparison was made in which Hell's Canyon of the Snake was viewed along with rivers in four areas officially recognized as having great natural scenic beauty. These are the Merced River, in Yosemite National Park in California's Sierra Nevada; the Grand Canyon of the Colorado River; Wyoming's Yellowstone River near Yellowstone Falls in the national park; and the Snake River in the Teton National Park below Jenny Lake, again in Wyoming. How does Hell's Canyon compare with them?

The same kinds of data that were used in the construction of the graphs just described were tabulated for Hell's Canyon in combination with these four national park rivers. A similar set of graphs was derived, the results of which are shown in figure 4. In this comparison, it can be seen that the points representing Hell's Canyon and Grand Canyon fall nearly together in the upper right-hand portion of the graph. They are comparable in the combination of valley character and river character and stand in exceptional positions. By comparison, the Merced River in Yosemite is of a lower order of interest. Truly the Yosemite Valley as a whole is one of the great scenic beauty spots of the world, but the river within the valley is not of a special character. As far as the Snake River in Teton Park is concerned, its position in the center of the graphical comparison can be attributed in part to the higher degree of urbanization present there than in the Grand Canyon or Hell's Canyon, and to the comparatively large width of the valley floor, a characteristic that makes the Snake River at that point less of a grand spectacle than either Hell's Canyon or the Grand Canyon. Relatively great urbanization, along with the absence of scenic vistas, also diminishes the site of the Yellowstone River.

The result of the data collection and analysis indicates that it is possible to set up a list of factors that influence the esthetic nature of a given location. The factors can be considered all together, in this case by the computation of a total uniqueness ratio, or they can be selected and used in various combinations to express certain aspects of a landscape's characteristics. It is hoped that the study will indicate both the need for, and the possibilities of, objective description of landscape. Specifically, this analysis shows that the Hell's Canyon of the Snake River, the site proposed for a hydropower dam, has unique characteristics that give it an exceptional esthetic rating and place it in a category shared by few other landscapes within the United States.

Common scenery? Above average? Unique? The Leopold study seeks to answer these questions by assigning numerical values to "landscape scale," "degree of urbanization," and to other features that contribute to a region's esthetic impression. The study concludes that Hell's Canyon, right, is an area whose superb esthetic qualities are shared by few other landscapes within the United States.

References

Albert, F. A., and Spector, A. H. 1955. A new song on the muddy Chattahoochee, in *Yearbook of Agriculture 1955*, p. 205–210.

Anonymous. No date. Derelict land, Civic Trust, London, 68 p.

Anonymous. 1970. Private remedies for water pollution, Columbia Law Review, v. 70, n. 4, p. 734–756.

Aschmann, H. 1966. The head of the Colorado Delta, in *Geography as Human Ecology: Methodology by Example*, S. R. Eyre and G. R. J. Jones, eds., St. Martin's Press, New York, p. 231–263.

Askew, G. P., Moffatt, D. J., Montgomery, R. F., and Searl, P. F. 1970. Soil landscapes in North Eastern Mato Grosso, Geog. Jour., v. 136, p. 211–227.

Ayres, Q. C. 1936. *Soil Erosion and Its Control*, McGraw-Hill, New York, 365 p.

Bauer, A. M. 1965. Simultaneous excavation and rehabilitation of sand and gravel sites, National Sand and Gravel Assoc. Project 1, University of Illinois, 41 p.

Bennett, H. H. 1939. *Soil Conservation*, McGraw-Hill, New York, 993 p.

Bennett, H. H., and Lowdermilk, W. C. 1938. General aspects of the soil-erosion problem, in *Soils and Men*, Yearbook of Agriculture, U.S. Department of Agriculture, Washington, D.C., p. 581–608.

Bormann, F. H., Likens, G. E., Fisher, D. W., and Pierce, R. S. 1967. Nutrient loss accelerated by clear cutting a forest ecosystem, in *Primary Productivity and Mineral Cycling in Natural Ecosystems*, Ecological Society of America/AAAS, University of Maine Press, Orono, p. 187–196.

Bouillenne, R. 1962. Man, the destroying biotype, Science, v. 135, p. 706–712.

Brice, J. 1971. Measurement of lateral erosion at proposed river crossing sites of the Alaska pipeline: U.S. Department of Interior, Geological Survey, Alaska District, 39 p.

Bridges, E. M. 1965. Soil erosion in the lower Swansea Valley, in *Rates of Erosion and Weathering in the British Isles: A Geomorphological Symposium*, British Geomorphological Research Group–Institute of British Geographers, Bristol, England, p. 7–10.

Brown, C. B. 1950. Effects of soil conservation, in *Applied Sedimentation*, P. D. Trask, ed., Wiley, New York, p. 380–406.

Brown, E. H. 1970. Man shapes the earth. The Geographical Journal, v. 131, pt. 1, p. 74–85.

Brune, G. M. 1948. Rates of sediment production in midwestern United States, U.S. Soil Conservation Service, SCS-TP.65, Washington, D.C.

Bryan, P. W. 1933. *Man's Adaptation of Nature*, Holt, Rinehart and Winston, New York, 386 p.

Casey, H. E. 1972. Salinity problems in arid lands irrigation, a literature review and selected bibliography, Office of Water Resources Research (U.S. Department of Interior), Washington, D.C., 300 p.

Chamberlin, T. C. 1908. Soil wastage, in *Proceedings of a Conference of Governors in the White House,* N. C. Blanchard, ed., U.S. House of Representatives Doc. 1425, 60th Congr., 2nd Sess., p. 75–83.

Chorley, R. J. 1969. *Water, Earth, and Man,* Metheun, London, 588 p.

Coates, D. R., 1971. Legal and environmental case studies in applied Geomorphology, in *Environmental Geomorphology,* D. R. Coates, ed., (Publications in Geomorphology), State University of New York, Binghamton, p. 223–242.

Coates, D. R., ed. 1971. *Environmental Geomorphology* (Publications in Geomorphology), State University of New York, Binghamton, 262 p.

Coates, D. R. 1972. *Environmental Science Workbook* (Publications in Geomorphology), State University of New York, Binghamton, 412 p.

Coates, D. R., ed. 1972. *Environmental Science Workbook* (Publications in Geomorphology), State to 1900, Dowden, Hutchinson & Ross, Stroudsburg, Pa., 485 p.

Coates, D. R., ed. 1973 (in press). *Environmental Geomorphology and Landscape Conservation:* Vol. II, Urban Areas: Dowden, Hutchinson & Ross, Stroudsburg, Pa.

Coppock, J. T., and Coleman, A. M. 1970. Land use and conservation: Geog. Jour., v. 136, p. 190–210.

Culler, R. C. 1970. Water conservation by removal of phreatophytes: Amer. Geophys. Union Trans., v. 51, n. 10, p. 684–689.

Dasmann, R. F. 1959. *Environmental Conservation,* Wiley, New York, 375 p.

De Bell, G., ed. 1970. *The Environmental Handbook,* Ballantine Books, New York, 367 p.

Detwyler, T. R., ed. 1971. *Man's Impact on Environment,* McGraw-Hill, New York, 731 p.

Doehring, D. O. 1968. The effect of fire on geomorphic processes in the San Gabriel Mountains, California, Contributions to Geology, v. 7, n. 1, University of Wyoming, p. 43–65.

Douglas, I. 1967. Natural and man-made erosion in the humid tropics of Australia, Malaysia, and Singapore, in *Symposium on River Morphology,* Assoc. Intern. d'Hydrologie Scientifique, Gentbrugge, Belgium, p. 17–30.

Dower, M. 1970. Leisure—its impact on man and the land: Quart. Jour. Geographical Assoc., v. 55, p. 253–260.

Duley, F. L., and Miller, M. F. 1923. Erosion and surface runoff under different soil conditions: Agric. Expt. Sta. Res. Bull. 63, University of Missouri, 46 p.

Ellis, W. S. 1972. Bangladesh: hope nourishes a new nation, National Geographic, v. 142, n. 3, p. 295–333.

Emerson, J. W. 1971. Channelization: a case study: Science, v. 173, p. 325–326.

Emiliani, C., Harrison, C. G. A., and Swanson, M. 1969. Underground nuclear explosions and the control of earthquakes: Science, v. 165, p. 1255–1256.

Evans, D. M. 1966. Man-made earthquakes in Denver; Geotimes, v. 10, n. 9, p. 11–18.

Fitch, E. M., and Shanklin, J. F. 1970. *The Bureau of Outdoor Recreation,* Praeger, New York, 227 p.

Flavin, J. M. 1969. *New York Reports Court of Appeals,* v. 22, 2nd ser., 1111 p.

Flawn, P. T. 1970. *Environmental Geology:* Harper & Row, New York, 313 p.

Fortier, S. 1928. Silt in the Colorado River and its relation to irrigation: U.S. Dept. Agric. Tech. Bull 67, 94 p.

Frincke, H. C. 1964. A new regional landscape: the Tennessee Valley Authority, in *Shaping Tomorrow's Landscape,* Vol. 2, S. Crowe and Z. Miller, eds., Djambatan, Amsterdam, p. 94–100.

Glymph, L. M., and Storey, H. C. 1967. Sediment—its consequences and control, in *Agriculture and the Quality of Our Environment,* N. C. Brady, ed., Amer. Assoc. Advan. Sci. Publ. 85, p. 205–220.

Golomb, B., and Eder, H. M. 1964. Landforms made by man: Landscape, v. 14, n. 1, p. 4–7.

Golze, A. R. 1950. Problems of irrigation canals, in *Applied Sedimentation,* P. D. Trask, ed., Wiley, New York, p. 364–379.

Grasovsky, A. 1938. A world tour for the study of soil erosion control methods, Imperial Forestry Institute, University of Oxford, Paper 14, 76 p.

Gray, D. H. 1969. Effects of forest clearcutting on the stability of natural slopes, Assoc. Engr. Geologists Bull., v. 7 (1 and 2), p. 45–66.

Halpenny, L. C., et al. 1952. *Ground Water in the Gila River Basin and Adjoining Areas, Arizona—A Summary:* U.S. Geological Survey, Tucson, 224 p.

Hanson, G. D., ed. 1972. Tocks Island Dam Project halted, Conservation News, v. 37, n. 20, p. 2.

Happ, S. C. 1937. Fertile valleys laid waste by upland erosion, Soil Conservation (U.S.D.A.) v. 2, n. 9, p. 194–198.

Happ, S. C. 1950. Stream-channel control, in *Applied Sedimentation,* P. D. Trask, ed., Wiley, New York, p. 319–335.

Harte, J., and Socolow, R. H. 1971. The Everglades: wilderness versus rampant land development in south Florida, in *Patient Earth,* J. Harte and R. H. Socolow, eds., Holt, Rinehart and Winston, New York, p. 181–202.

Harte, J., and Socolow, R. H., eds. 1971. *Patient Earth,* Holt, Rinehart and Winston, New York, 364 p.

Hartman, W. A., and Wooten, H. H. 1935. Georgia land use problems, Georgia Agric. Expt. Sta. Bull. 191.

Haugen, R. M., and Brown, J. 1971. Natural and man-induced disturbances of permafrost terrane, in *Environmental Geomorphology,* D. R. Coates, ed. (Publications in Geomorphology), State University of New York, Binghamton, p. 139–149.

Haveman, R. H. 1965. *Water Resource Investment and the Public Interest,* Vanderbilt University Press, Nashville, 199 p.

Healy, J. H., Rubey, W. W., Griggs, D. P., and Raleigh, C. B. 1968. The Denver earthquakes, Science, v. 161 p. 1301–1310.

Herfindahl, O. C. 1961. What is conservation?: Resources for the Future, Reprint 30, August, Washington, D.C., 12 p.

Hill, A., and McClosky, M. 1971. Mineral King; wilderness versus mass recreation in the Sierra, in *Patient Earth,* J. Harte and R. H. Socolow, eds., Holt, Rinehart and Winston, New York, p. 165–180.

Hill, J. J. 1908. The natural wealth of the land and its conservation, in *Proceedings of a Conference of Governors in the White House,* N. C. Blanchard, ed., U.S. House of Representatives Doc. 1425, 60th Congr., 2nd Sess., p. 63–75.

Hilton, K. J., ed. 1967. *The Lower Swansea Valley Project,* Longmans, Green, London, 329 p.

Ireland, M. A., Sharpe, C. F. S., and Eargle, D. H. 1939. Principles of gully erosion in the Piedmont of South Carolina: Agric. Tech. Bull. 633, 142 p.

Jacks, G. V., and Whyte, R. O. 1939. *Vanishing Lands,* Doubleday, New York, 332 p.

James, G. W. 1917. *Reclaiming the Arid West:* Dodd, Mead, New York, 411 p.

Kennedy, D. J. L. 1961. A study of failures of liners for oil wells associated with the compaction of oil producing strata, unpublished Ph.D. thesis, University of Illinois.

Kotschy, F. 1964. Changes in silvicultural system in the Mondsee district in retrospect, Allg. Forstztg., 75(23/24), p. 271–273.

Laycock, G. 1970. *The Diligent Destroyers:* Doubleday, New York, 223 p.

Lobeck, A. K. 1939. *Geomorphology,* McGraw-Hill, New York, 731 p.

Leopold, A. 1921. The wilderness and its place in forest recreational policy, Jour. Forestry, v. 19, p. 718–721.

Leopold, L. B. 1969. Everglades jetport, summary, National Parks Mag., Nov., p. 11–13.

Leopold, L. B., Clarke, F. E., Hanshaw, B. B., and Balsley, J. R. 1971. A procedure for evaluating environmental impact, U.S. Geol. Survey Circ. 645, 13 p.

Leopold, L. B., and Maddock, T. 1954. *The Flood Controversy,* Ronald Press, New York, 278 p.

Leopold, L. B., Wolman, M. G., and Miller, J. P. 1964. *Fluvial Processes in Geomorphology:* W. H. Freeman, San Francisco, 522 p.
Linton, R. M. 1970. *Terracide:* Little, Brown, Boston, 376 p.
Linzon, S. N. 1971. Economic effects of sulfur dioxide on forest growth: Jour. Air Pollution Control Assoc., v, 21, n. 2, p. 81–86.
Lofgren, B. E. 1969. Land subsidence due to the application of water, in *Reviews in Engineering Geology,* Vol. II, Geol. Soc. Amer., p. 271–303.
Marine, G. 1969. *America the Raped:* Discuss Books, New York, 331 p.
Marx, W. 1967. *The Frail Ocean,* Ballantine Books, New York, 274 p.
McCaskill, M. 1966. Man and landscape in North Westland, New Zealand, in *Geography as Human Ecology: Methodology by Example,* S. R. Eyre and G. R. J. Jones, ed., St. Martin's Press, New York, p. 264–290.
McComas, M. R. 1972. Geology and land reclamation: Ohio J. Sci., v. 72 (2), p. 65–75.
McHarg, I. L. 1969. *Design with Nature,* Natural History Press, New York, 197 p.
McKenzie, G. D., and Utgard, R. O. 1972. *Man and His Physical Environment,* Burgess, Minneapolis, 338 p.
Mercer, D. C. 1970. The geography of leisure—a contemporary growth-point: Quart. Jour. Geographical Assoc., v. 55, p. 261–273.
Miller, J. N. 1970. Rape on the Oklawaha, Readers Digest, June, p. 54–60.
Miller, J. N., and Simmons, R. 1970. Crises on our rivers, Readers Digest, Dec., p. 78–83.
Ministry of Housing & Local Government. 1963. New life for dead lands, H. M. Stationery Office, London, 30 p.
Minor, H. E. 1925. Goose Creek oil field, Harris County, Texas, Amer. Assoc. Petroleum Geologists, v. 9, n. 2, p. 286–297.
Mitchell, J. G., and Stallings, C. L., eds. 1970. *Ecotactics: The Sierra Club Handbook for Environment Activists,* Pocket Books, New York, 288 p.
Morisawa, M. 1971. Evaluating riverscapes, in *Environmental Geomorphology,* D. R. Coates, ed. (Publications in Geomorphology), State University of New York, Binghamton, p. 91–106.
Morisawa, M., ed. 1972. *Quantitative Geomorphology: Some Aspects and Applications* (Publications in Geomorphology), State University of New York, Binghamton, 315 p.
NAS-NRC. 1961. Principles of resource conservation policy: with some applications to soil and water resources, National Academy of Science–National Research Council Publ. 885, Washington, D.C., 50 p.
Nash, R. 1967. *Wilderness and the American Mind,* Yale University Press, New Haven, 256 p.
Nicholson, M. 1970. *The Environmental Revolution,* McGraw-Hill, New York, 366 p.
Olson, S. F. 1970. Wilderness besieged, Audubon, July, p. 29–32.
Orme, A. R., and Bailey, R. G. 1970. The effect of vegetation conversion and flood discharge on stream channel geometry: the case of southern California watersheds: Assoc. Amer. Geog. Proc., v. 2. p. 101–106.
Osborn, F. 1948. *Our Plundered Planet:* Little, Brown, Boston, 217 p.
Poland, J. F., and Davis, G. H. 1969. Land subsidence due to withdrawal of fluids, in *Reviews in Engineering Geology,* Vol. II, Geol. Soc. Amer., p. 187–269.
Ramparts Editors. 1970. *Eco-Catastrophe,* Canfield Press, New York, 158 p.
Rienow, R., and Rienow, L. T. 1967. *Moment in the Sun,* Ballantine Books, New York, 305 p.
Roosevelt, N. 1970. *Conservation: Now or Never,* Dodd, Mead, New York, 238 p.
Ruhe, R. V. 1971. Stream regimen and man's manipulation, in *Environmental Geomorphology,* D. R. Coates, ed. (Publications in Geomorphology), State University of New York, Binghamton, p. 9–23.
Sargent, F. O. 1967. Scenery classification, Vermont Resources Research Center, Vermont Agric. Expt. Sta., Univ. Vermont Rept. 18.
Sato, A. 1964. Man and nature preservation of natural scenery in Japan, in *Shaping Tomorrow's Landscape,* Vol. 2, S. Crowe and Z. Miller, eds., Djambatan, Amsterdam, p. 50–54.
Searles, S. Z. 1969. Aesthetics in the law, New York State Bar Jour., v. 69, n. 3, p. 210–217.
Sears, P. L. 1947. *Deserts on the March,* Oklahoma University Press, Norman, 178 p.

Shafer, E. L., Jr., Hamilton, J. E., Jr., and Schmidt, E. A. 1969. Natural landscape preferences: a predictive model, Jour. Leisure Res., v. 1, p. 1–19.

Shaler, N. S. 1906. *Man and the Earth,* Fox, Duffield, New York, 233 p.

Sharpe, C. F. S. 1941. Geomorphic aspects of normal and accelerated erosion, in Symposium on dynamics of land-erosion, Amer. Geophys. Union Trans., Annual Meeting, p. 236–240.

Shoor, Y., Latte, H. H., and Mitchell, M. 1964. Israel's new landscape, in *Shaping Tomorrow's Landscape,* Vol. 2, S. Crowe and Z. Miller, eds., Djambatan, Amsterdam, p. 70–76.

Sil'vestrov, S. I. 1963. Erosion-control needs and the intensification of agriculture: Soviet Geog., v. 4, n. 2, p. 18–24.

Simms, D. H. 1970. *The Soil Conservation Service,* Praeger, New York, 238 p.

Sive, D. 1970. Some thoughts of an environmental lawyer in the wilderness of administrative law: Columbia Law Review, v. 70, n. 4, p. 612–651.

Smith, C. G. 1970. Water resources and irrigation development in the Middle East, Quart. Jour. Geographical Assoc., v. 55, p. 407–425.

Smith, H. M., and Stamey, W. L. 1965. Determining the range of tolerable erosion, Soil Science, v. 100, p. 414–424.

Sochava, V. B. 1971. Geographical ecology: Soviet Geog., v. 12, n. 5, p. 277–293.

Stallings, J. H. 1957. *Soil Conservation,* Prentice-Hall, Englewood Cliffs, N.J., 575 p.

Stamp, L. D. 1964. *Man and the Land,* Collins, London, 272 p.

Sternberg, H. O. 1956. Geography's contribution to the better use of resources, in *The Future of Arid Lands,* G. F. White, ed., Assoc. Advan. Sci. Pub. 43, p. 200–220.

Stoddart, D. R. 1968. The conservation of Aldabra: Geog. Jour., v. 134, p. 471–485.

Strahler, A. N. 1972. The environmental impact of ground water use on Cape Cod, Impact Study III, Assoc. for the Preservation of Cape Cod, Orleans, Mass., 68 p.

Swateck, P. 1970. *The User's Guide to the Protection of the Environment,* Ballantine Books, New York, 312 p.

Thomas, W. L., Jr., ed. 1956. *Man's Role in Changing the Face of the Earth,* University of Chicago Press, Chicago, 1193 p.

Udall, S. L. 1963. *The Quiet Crises,* Holt, Rinehart and Winston, New York, 209 p.

UNESCO. 1961. *Salinity Problems in the Arid Zones,* Proceedings of the Teheran Symposium, UNESCO, 395 p.

Union of South Africa. 1923. Final report of the Drought Investigation Commission: Cape Times Limited, Government Printers, Capetown, p. 14–15.

United Nations. 1948. Soil conservation: FAO United Nations, Agric. Studies 4, Washington, D.C., 189 p.

U.S. Congressional Record. 1913, 63rd Congr., 2nd Sess., 51 (Dec. 19), p. 1189.

U.S. Department of Agriculture. 1938. *Soils and Men,* Yearbook of Agriculture, U.S. Department of Agriculture, Washington, D.C., 1232 p.

U.S. Department of Agriculture. 1969. Summary of reservoir sediment deposition surveys made in the United States through 1965, U.S. Dept. Agric. Misc. Publ. 1143, 64 p.

U.S. Department of Interior. 1967. *Surface Mining and Our Environment,* Government Printing Office, Washington, D.C., 124 p.

U.S. Department of Interior. 1969. Report on the Florida jetport, U.S. Department of Interior, Washington, D.C.

U.S. Water Resources Council. 1968. *The Nation's Water Resources,* Government Printing Office, Washington, D.C.

Vendrov, S. L. 1964. Water management problems of Western Siberia: Soviet Geog., v. 5, n. 5, p. 13–24.

Young, A. 1968. Natural resource surveys for land development in the tropics: Quart. Jour. Geographical Assoc., v. 53, p. 229–248.

Author Citation Index

Subject Index

Milton Keynes UK
Ingram Content Group UK Ltd.
UKHW021908071024
449327UK00022B/1641